AF352767

Polymers Enhancing Bioavailability in Drug Delivery, 2nd Edition

Polymers Enhancing Bioavailability in Drug Delivery, 2nd Edition

Editor

Ana Isabel Fernandes

Basel • Beijing • Wuhan • Barcelona • Belgrade • Novi Sad • Cluj • Manchester

Editor
Ana Isabel Fernandes
Centro de Investigação
Interdisciplinar Egas Moniz
(CiiEM), Egas Moniz School
of Health & Science
Monte de Caparica
Portugal

Editorial Office
MDPI
St. Alban-Anlage 66
4052 Basel, Switzerland

This is a reprint of articles from the Special Issue published online in the open access journal *Pharmaceutics* (ISSN 1999-4923) (available at: www.mdpi.com/journal/pharmaceutics/special_issues/0JC3F78273).

For citation purposes, cite each article independently as indicated on the article page online and as indicated below:

Lastname, A.A.; Lastname, B.B. Article Title. *Journal Name* **Year**, *Volume Number*, Page Range.

ISBN 978-3-0365-9557-3 (Hbk)
ISBN 978-3-0365-9556-6 (PDF)
doi.org/10.3390/books978-3-0365-9556-6

Contents

About the Editor

Ana Isabel Fernandes

Ana I. Fernandes is an Associate Professor of Pharmaceutics and Head of the PharmSci Lab at the Egas Moniz School of Health and Science, Portugal. She holds a degree in Pharmaceutical Sciences (University of Lisbon, PT) and a Ph.D. in Drug Delivery (University of London, UK). She has been involved in the study of polymers and polymeric systems with biomedical applications, namely in the delivery of therapeutic proteins and conventional drugs. Her current research is related to pediatric formulations, drug solubility enhancement by co-amorphization, 3D printing of pharmaceuticals and nutraceuticals, and the use of lifestyle drugs. Over the years, she has been a principal investigator or collaborator in several externally funded projects (National Funding Agency—FCT), a scientific consultant, an editor of journals, and has published several international peer-reviewed papers.

 *pharmaceutics*

Editorial

Polymers Enhancing Bioavailability in Drug Delivery, 2nd Edition

Ana I. Fernandes

Egas Moniz Center for Interdisciplinary Research (CiiEM), Egas Moniz School of Health & Science, 2829-511 Caparica, Portugal; aifernandes@egasmoniz.edu.pt; Tel.: +351-212946823

Citation: Fernandes, A.I. Polymers Enhancing Bioavailability in Drug Delivery, 2nd Edition. *Pharmaceutics* **2023**, *15*, 2604. https://doi.org/10.3390/pharmaceutics15112604

Received: 3 October 2023
Accepted: 1 November 2023
Published: 8 November 2023

This Special Issue continues the previously published work [1] with the aim of updating the current status and presenting the emerging trends in polymeric drug delivery systems which are designed to improve drug bioavailability. As discussed there, many drugs are poorly bioavailable due to their low water solubility and/or low permeability across epithelia. Others are labile when administered orally due to acidic pHs or enzymatic degradation, precluding their use by the route most preferred by patients. Still, others, although administered through routes with reduced proteolytic activity, lack specificity and may require localization strategies at the site of action to maximize their therapeutic index. The use of polymers in drug delivery is well established [2–4] and essential to achieve improved drug pharmacokinetics and pharmacodynamics, as well as patient compliance [5].

A total of 12 original research papers and 1 literature review were published with open access; 21 papers were submitted for publication, representing an acceptance rate of 61.9%, ensuring the publication of high-quality work.

The current issue reflects the diversity of approaches from different research groups, ranging from drug encapsulation in nanostructures (e.g., nanoparticles, nanocrystals, nanosponges, micelles, dendrimers) to drug amorphization, co-amorphization or 3D-printed scaffolds for local drug delivery. Polymers are also used as hydrophilic coatings to provide biocompatibility or antimicrobial properties and are grafted/functionalized for target-specific delivery, as well as to provide stimuli-responsive controlled release of the drug. In fact, many of the reported strategies involve the synthesis of novel polymers/copolymers or the chemical conjugation of residues to tailor drug release or provide recognition (active targeting) or biomimetic or stealth (long blood circulation, passive targeting) properties, highlighting the importance of the chemical engineering of polymers [6] to improve performance.

The routes of administration of the systems described in this Special Issue are summarized in Figure 1. Oral and parenteral routes are the most common (nine publications); noteworthy is the representation (two papers) of medical devices such as catheters and scaffolds. Antimicrobials, antifungals, cytotoxics, anti-inflammatories, antidiabetics and immunosuppressants are among the investigated drugs.

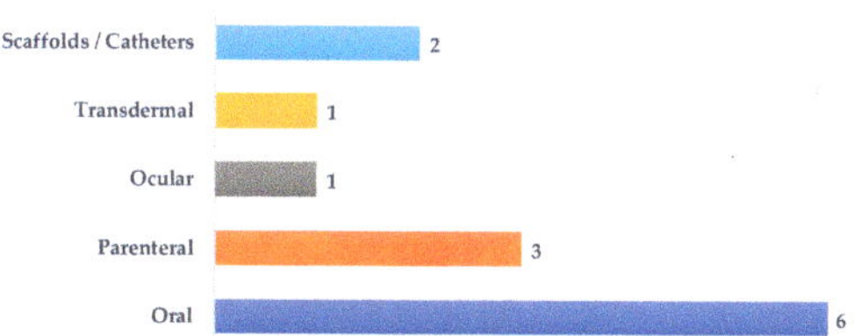

Figure 1. Number of papers published according to the route of administration or type of device.

An overview of the strategies described to treat and diagnose diseases or to provide advanced theranostic applications is given below.

The use of colloidal systems is explored in the following set of works. Itraconazole was encapsulated in core–shell-like polymeric nanoparticles functionalized with antibodies to provide both controlled release of the antifungal and targeting of macrophages. Drug-loaded poly-(lactic-co-glycolic acid) (PLGA, one of the most successful biodegradable polymers [7]) nanoparticles were prepared using a high-energy emulsification method and then covalently coupled to the antibodies using carbodiimide; their size was estimated as 226.66 ± 13.05 nm and their surface ζ-potential was estimated as -27.9 ± 0.26 mV, features that ensure a lack of particle aggregation and clearance by the mononuclear phagocytic system. The strategy provided improved drug stability, solubility and bioavailability, thus requiring lower doses for antifungal activity. Histoplasmosis was used as a model of intracellular infection, and antibody-directed nanoparticles showed significantly higher uptake in vitro by murine cells compared to bare nanoparticles, with no cytotoxic effect on macrophages. They also induced the elimination of *Histoplasma capsulatum*, an intracellular fungus, in co-cultures. In addition, the gene expression of anti-inflammatory and pro-inflammatory cytokines (IL-1, INF-γ, IL-6 and IL-10) on macrophages was modulated by the encapsulated drug, and the antibody coating provided an enhanced natural cellular response, synergistically preventing fungal growth at the intracellular level. Intracellular drug delivery is very promising since it reduces resistance to pathogens and prevents non-specific accumulation in other tissues, thereby reducing toxicity and side effects.

Superparamagnetic iron oxide nanoparticles (SPIONs), approved for diagnosis and treatment of malignant tumors, are biodegradable but require coating with biocompatible, non-cytotoxic materials to prevent embolism. In this work, a biocompatible copolyester—poly(globalide-co-ε-caprolactone) (PGlCL)—was synthesized and then modified with cysteine (Cys) via a thiol-ene reaction to obtain PGlCLCys. Due to the reduced crystallinity and increased hydrophilicity, the Cys-modified copolymer was used to coat the SPIONs. The free amine groups of Cys on the particles' surface were conjugated either to folic acid (targeting moiety to promote internalization, since receptors are overexpressed in many tumors) or methotrexate (targeting residue because it also specifically interacts with folate receptors and anti-cancer drug by blocking the use of folic acid by the body) with conjugation efficiencies of 62% and 60%, respectively. Protease incubation under mimicked lysosomal conditions (37 °C in phosphate buffer pH ~5.3, 72 h) resulted in the enzymatic release of 45% of the particle-conjugated drug. The breast carcinoma cell viability was reduced by 25% (MTT assay) after 72h of contact. The successful conjugation and triggered release of methotrexate in this model multifunctional platform pave the way for the development of more selective and less aggressive cancer treatments and/or diagnostics.

Neuroprotective particles targeted to the retina, developed by Colucci et al., have shown promise for the treatment of retinal degeneration. The retina is a highly metabolic tissue, very much affected by oxidative stress. The nerve growth factor (NGF) is a neurotrophin involved in the molecular response to oxidative damage. The neuroprotective and regenerative effects of the NGF are hampered by rapid degradation and clearance in vivo. Moreover, the anatomical and physiological barriers that protect the eye make the delivery of drugs to the posterior segment challenging, even using nanodevices which have been shown to increase the bioavailability of drugs. In this work, polyacrylamide nanoparticles (ANPs) were noncovalently functionalized with peanut agglutinin (PNA) for targeting and NGF for neuroprotection. The ANP:PNA:NGF nanoparticles were synthesized and characterized, and their bioactivity and protective effects were evaluated in vitro. The teleost zebrafish (*Danio rerio*) was used as an in vivo model organism to evaluate this new ocular therapeutic strategy. The nanoformulation (22.81 ± 3.17 nm) was targeted and demonstrated a prolonged residence time in the posterior segment of the eye; it was also shown to preserve human retinal cell viability under oxidative conditions and improve the visual function of zebrafish larvae, partially reducing oxidative-stress-triggered retinal cell apoptosis. The intravitreal administration of this nanosystem could overcome the need for multiple injections due to the unfavorable kinetics of the usual form of the drug, thereby reducing side effects.

Colon delivery of budesonide in nanoparticles was attempted to circumvent the limitations of oral administration [8]. Targeting topical anti-inflammatory drugs to the colon is effective in the treatment of inflammatory bowel diseases. Inflamed tissues accumulate nanoparticles due to the loosening of tight junctions. Chitosan (CS) is a biodegradable cationic polysaccharide with mucoadhesive properties to the negatively charged intestinal mucosa. As such, in this work, CS nanoparticles were loaded with budesonide by an ionic gelation technique, and characterized in terms of size, shape, ζ-potential, encapsulation efficiency and drug release. Nanoparticles were then pelletized by extrusion–spheronization and the pellets were coated with (1) two enteric polymers (Eudragit® L and S) to minimize early drug release in the upper GI tract and (2) time-dependent polymers (Eudragit® RS) for colon targeting. The size, morphology and mechanical properties of the pellets and in vitro drug release characteristics were evaluated. Wistar rats with induced colitis were used to evaluate the anti-inflammatory effect of the formulation and the relapse time after treatment discontinuation. Pelletized budesonide nanoparticles exhibited faster drug release in acidic pH; drug release in the GI tract was sustained but incomplete; and the anti-inflammatory effect lasted longer compared to conventional budesonide pellets.

Rutin is a plant flavonoid with a wide spectrum of clinical applications limited by a low solubility and bioavailability. To overcome these limitations, rutin nanocrystals (submicron colloidal dispersions of 100% drug) were prepared by the anti-solvent nanoprecipitation–ultrasonication method using a variety of stabilizers, including non-ionic surfactants and non-ionic polymers. The nanocrystals were characterized regarding their particle size, size distribution, morphology, ζ-potential, entrapment efficiency, colloidal stability, rutin photostability, dissolution rate and saturation solubility. After dispersion of the crystals in a hydroxypropyl methyl cellulose hydrogel base, the drug release kinetics and permeability through mouse skin were evaluated. The anti-inflammatory efficacy of rutin was investigated in a carrageenan-induced rat paw edema model. The size of the nanocrystals ranged from 270 to 500 nm with a polydispersity index of about 0.3–0.5. The size of the nanocrystals increased with storage time and the photostability of the rutin was improved approximately 2.3-fold. The use of hydroxypropyl beta-cyclodextrin (HP-β-CD) as a stabilizer resulted in the smallest nanocrystals size, the best colloidal stability, the highest drug entrapment efficiency and the highest drug photostability. Depending on the stabilizer, the aqueous solubility and dissolution rate of the drug were increased by 102- to 202-fold and 2.3- to 6.7-fold, respectively. Compared to the free drug hydrogel, a significantly higher percentage of drug was released from the HP-β-CD nanocrystal hydrogel and permeated through the mouse skin; the cumulative amount of drug penetration of the skin was 2.5 times higher. In vivo edema inhibition was also significantly higher than with the free rutin hydrogel and commercial diclofenac sodium gel. This study demonstrated the potential of nanocrystals to improve the solubility, dissolution rate and anti-inflammatory properties of drugs, and highlighted the importance of careful stabilizer selection.

Targeted β-cyclodextrin nanosponges containing fisetin have been developed for the effective management of breast cancer. Fisetin is a phytomedicine with a therapeutic potential that is significantly reduced by a low systemic bioavailability due to its highly lipophilic nature. In this work, lactoferrin-coated FS-loaded β-cyclodextrin nanosponges (LF-FS-NSs) were developed in an attempt to circumvent such a limitation. Nanosponges were produced through cross-linking of β-cyclodextrin by diphenyl carbonate and coated with lactoferrin for active targeting to transferrin receptors overexpressed in breast cancer cells. Fourier transform infrared (FTIR) and X-ray diffraction (XRD) confirmed the presence of nanosponges. Scanning electron microscopy (SEM) analysis revealed its mesoporous spherical structure with pore diameters of ~30nm, which was further confirmed by surface area determination. The selected LF-FS-NSs presented high loading efficiencies (96 ± 0.3%), good colloidal properties (size 52.7 ± 7.2 nm, dispersion index < 0.3, and ζ-potential 24 mV) and Fickian diffusion-controlled sustained drug release. The oral and intraperitoneal bioavailability of nanosponge-encapsulated fisetin was enhanced in rats (2.5- and 3.2-fold, respectively) compared to drug suspension rats. The antitumor efficacy of LF-FS-NSs evaluated in vitro (MDA-MB-231 cells) and in vivo (Ehrlich ascites mouse model) was significantly higher due to their superior activity and targetability compared to the free drug and uncoated formulation. The authors concluded that the formulation is promising for the effective management of breast cancer.

Novel hybrid block copolymers have been synthesized through ring-opening polymerization and used in the production of self-assembling micelles incorporating doxorubicin. The hybrid amphiphilic copolypeptides were synthesized from poly(ethylene oxide) (PEO), poly(l-histidine) (PHis) and poly(l-cysteine) (PCys). In aqueous media, they form micelles with an outer hydrophilic PEO corona and a redox- and pH-responsive hydrophobic layer, respectively, due to PHis and PCys. Crosslinking of PCys's thiol groups further stabilized the nanoparticles. Structural characterization using dynamic and static light scattering and transmission electron microscopy (TEM) revealed mainly a core–shell micellar structure. Structure and conformation studies of synthetic polypeptides were performed by circular dichroism; the neutral ζ-potential of particles indicated that PEO blocks were the outer, stealthy layer. Doxorubicin was encapsulated in the hydrophobic core and released in response to changes in pH and redox (using glutathione, which is elevated in tumors) mimicking the tumor microenvironment. The topology of PCys (either the middle block, the end block or randomly distributed along the PHis chain) significantly affected the structure of the micelles and the drug release profile. The antiproliferative activity of the doxorubicin-loaded nanocarriers, tested in three different breast cancer cell lines, was comparable to that of the free drug.

Insulin delivery to treat diabetes mellitus was attempted through pegylated ferrisilicate with large 3D pores. Due to the large surface area (804 m^2/g) and cubic pores (3.2 nm), the material was explored to improve drug encapsulation and loading efficiency. Ferrisilicate was coated with 400 D polyethylene glycol (PEG) to enhance drug stability and permeability across intestinal epithelia and provide pH sensitivity. The characterization methods included XRD, BET surface area analysis, FTIR and TEM analysis. After optimizing the insulin load, its release mechanism was studied using the dialysis membrane (MWCO = 14,000 Da) technique at pHs of 1.2, 6.8 and 7.4. The Korsmeyer–Peppas model was used to evaluate the kinetics and mechanism of drug release at different pH values. The in vitro cytotoxicity of the nanoformulations, evaluated in human foreskin fibroblast (HFF-1) cells using a 3-(4,5-dimethylthiazol-2-yl)-2,5-diphenyltetrazolium bromide (MTT) assay, was very low. On acute oral administration to diabetic Wistar rats, insulin/ferrisilicate/PEG (doses of 5 and 10 mg/kg body weight) demonstrated a hypoglycemic effect, significantly reducing blood glucose levels compared to the control. The insulin encapsulation/loading capacity (46%) and the smart, pH-sensitive kinetic release behavior observed demonstrate the potential of the system.

The use of nanosized Janus and dendrimer particles for the targeted delivery and enhanced bioavailability of pharmaceuticals has been extensively reviewed. The high surface-to-volume ratio and surface-charge-dependent behavior, as well as size- and shape-dependent tailorable properties, make nanoparticles particularly attractive for drug delivery. Janus nanoparticles are composed of two distinct components that differ in physical and chemical properties, representing a unique platform for the co-delivery of multiple drugs or the targeting of specific tissues for treatment and/or diagnosis of diseases. Their complex synthesis and residual solvent toxicity are significant limitations to their use. On the other hand, dendrimers are monodisperse, hyperbranched synthetic macromolecules with well-defined functional groups on the surface, and can be engineered for improved drug targeting (e.g., overexpressed receptors in tumors) and release. Scalability issues, a high non-specific toxicity and a low hydro-solubility are dendrimer-associated drawbacks. Despite the limitations, both nanocarriers are capable of improving drug solubility in water and drug stability, increasing the intracellular uptake and reducing the associated toxicity through control of the release rate. Hybrid systems incorporating Janus and dendrimer particles into composite materials show promise for improving the biocompatibility and drug delivery of pharmaceuticals, while overcoming the limitations of stand-alone nanoparticles. Janus–dendrimer particles require further investigation and fine-tuning to enter the clinical setting and improve therapeutic outcomes.

Urinary and central line catheters are prone to colonization by microorganisms in biological environments, which is one of the main causes of nosocomial infections in intensive care units. Duarte-Peña, Magaña and Bucio developed materials capable of simultaneously preventing bacterial adhesion and releasing an antibacterial drug, e.g., ciprofloxacin. These dual antimicrobial materials were produced by gamma-radiation-induced grafting of poly(vinyl chloride) catheters with 4-vinyl pyridine and later functionalization with 1,3-propane sultone. The surface properties of the materials were determined by thermogravimetric analysis, FTIR, swelling behavior, pH-responsiveness and contact angle measurements. The drug loading and release, inhibition of bacterial growth, bacterial and protein adhesion and cell growth stimulation were also evaluated. The hydrophilicity and pH sensitivity of the system were related to its ability to load and release ciprofloxacin. Reductions in bovine serum albumin and *E. coli* adhesion were enhanced by loading with ciprofloxacin. The modified materials were shown to be non-toxic, as the cell viability was not significantly affected. These hold promise for the manufacture of medical devices, such as catheters, for local prophylaxis and treatment of infections typically associated with their use, increasing their efficacy and reducing side effects.

Additive manufacturing offers the opportunity to design and develop innovative systems with complex geometries and programmed controlled release profiles that rely on the use of polymers [9]. In this next work, the antimicrobial efficacy of a novel 3D-printed scaffold containing vancomycin (Van) against Staphylococcus species was evaluated for use in drug delivery and tissue engineering. Local administration of the antibiotic has been shown to offer benefits over the traditional parenteral route in acute and chronic bone infections when ischemia is present. The medicated scaffold was based on polycaprolactone (PCL) and a chitosan (CS) hydrogel. The hydrophobicity of PCL (determined via contact angle measurements) was reduced by two cold plasma treatments to improve the adhesion of the CS hydrogel. The release of vancomycin was followed by high-performance liquid chromatography (HPLC). Surgical site and medical-device-associated infections are primarily caused by the opportunistic Gram-positive *Staphylococcus aureus* and *Staphylococcus epidermidis*—the microorganisms challenged in this study to evaluate the antibacterial efficacy. Finally, the biological response to the scaffolds was evaluated in terms of cytotoxicity (lactate dehydrogenase activity), proliferation and osteogenic differentiation (alkaline phosphatase activity; alizarin red staining) in a population of adult human bone-marrow-derived mesenchymal stem cells. The hybrid PCL/CS/Van scaffolds combined the biocompatibility, biodegradability and antibacterial properties of CS with the mechanical properties of PCL. Further in vivo preclinical studies using animal models are needed to confirm the usefulness of the systems in achieving a controlled and effective local release of vancomycin in bone infections.

Transformation of crystalline drugs into their more water-soluble amorphous counterparts may be achieved by producing amorphous solid dispersions (ASDs) with water-soluble polymeric carriers or co-amorphous dispersions (CADs) with low molecular weight entities, as presented in the next two papers. The advantages and disadvantages of each approach have recently been the subject of a review [10].

The molecular interactions established between delayed-release dugs and polymers used in the production of ASDs were investigated to simplify the development of such entities. At first, the compatibility of the drug and polymer in the solid state was investigated by molecular dynamics simulations, and used later on to identify ideal pairs. Afterwards, drug ASDs were produced by hot-melt extrusion (HME), an emergent processing technology for such purposes. In brief, a drug is dispersed in a polymeric matrix, heated at high temperatures (below the drug melting point) and mixed in an extruder to achieve homogeneity. The extrudates were milled and characterized by XRD, FTIR spectroscopy and dissolution studies. The potential of each drug–polymer pair considered was evaluated by determination of (a) the drug–polymer interaction energy (electrostatic, Lenard–Jones and total energies), (b) energy ratio (drug–polymer/drug–drug) and (c) drug–polymer hydrogen bonding. The best total energy values corresponded to naproxen–Eudragit® L100 (−143.38 kJ/mol), diclofenac sodium–hydroxypropyl methylcellulose phthalate (−348.04 kJ/mol), dimethyl fumarate–hydroxypropyl methylcellulose acetate succinate (−110.42 kJ/mol) and omeprazole–hydroxypropyl methylcellulose acetate succinate (−269.43 kJ/mol). The energy ratio trends of the pairs were consistent with these results and every drug–polymer pair established hydrogen bonds. Few drug–polymer pairs were extrudable by HME (up to 50% drug load); the extrudates did not release the drug at pH 1.2 (simulated gastric fluid) but released it at pH 6.8 (simulated intestinal fluid). The study not only demonstrates drug–polymer compatibility, but also that each drug requires a different enteric polymer. The atomic-level insights gathered are essential for the rational design and screening of ASD formulations.

Mohamed and co-workers developed CADs of tacrolimus (an immunosuppressive drug with low solubility) and sucrose acetate isobutyrate (SAIB) as a co-former. The in vitro and in vivo performance of the system was compared with that of a hydroxypropyl methylcellulose (HPMC)-based ASD. The solvent evaporation method was used to prepare both CADs and ASDs, which were characterized by FTIR, XRD and differential scanning calorimetry (DSC) and evaluated for dissolution, stability and oral pharmacokinetics in beagle dogs. Amorphization of tacrolimus was demonstrated by XRD and DSC, and more than 85% of the drug was dissolved within 90 min. Drug dissolution could be modulated by changing the drug/excipient ratio. After short-term stability testing (storage at 25 °C/60% RH and 40 °C/75% RH), no drug crystallization was observed in the thermograms and diffractograms; the dissolution profiles before and after storage were similar, again indicating the stability of the amorphous systems. The two formulations, the SAIB-based CAD and the HPMC-based ASD, were considered bioequivalent regarding the Cmax and AUC, meeting the established FDA confidence interval of 90–111.1% for narrow therapeutic index drugs. Their Cmax and AUC values were respectively 1.7–1.8 and 1.5–1.8 times higher than that of crystalline drug tablet formulations; thus, they are expected to be more bioavailable. Long-term stability evaluations and human clinical trials are required to demonstrate their superior pharmacokinetics and clinical efficacy.

The future of drug delivery lies in the use of increasingly complex, biodegradable, biocompatible and intelligent polymers [11]. Further basic and applied research is required to achieve greater control over the properties and performance of polymer-based drug delivery and targeting systems to fulfill their great potential in pharmaceutics.

Funding: A.I.F. was funded by Fundação para a Ciência e a Tecnologia, Lisbon, Portugal (Grants PTDC/CTM-BIO/3946/2014 and PTDC/CTM-CTM/30949/2017—Lisboa-010145-Feder-030949).

Institutional Review Board Statement: Not applicable.

Informed Consent Statement: Not applicable.

Data Availability Statement: Data are contained within the article.

Acknowledgments: The authors of this Special Issue who contributed their high-quality research and the excellent secretarial support in the MDPI office are gratefully acknowledged.

Conflicts of Interest: The author declares no conflict of interest.

List of Contributions

1. Mejía, S.P.; López, D.; Cano, L.E.; Naranjo, T.W.; Orozco, J. Antifungal Encapsulated into Ligand-Functionalized Nanoparticles with High Specificity for Macrophages. *Pharmaceutics* **2022**, *14*, 1932. https://doi.org/10.3390/pharmaceutics14091932.
2. Beltrame, J.M.; Ribeiro, B.B.P.; Guindani, C.; Candiotto, G.; Felipe, K.B.; Lucas, R.; Zottis, A.D.; Isoppo, E.; Sayer, C.; de Araújo, P.H.H. Coating of SPIONs with a Cysteine-Decorated

Copolyester: A Possible Novel Nanoplatform for Enzymatic Release. *Pharmaceutics* **2023**, *15*, 1000. https://doi.org/10.3390/pharmaceutics15031000.

3. Colucci, P.; Giannaccini, M.; Baggiani, M.; Kennedy, B.N.; Dente, L.; Raffa, V.; Gabellini, C. Neuroprotective Nanoparticles Targeting the Retina: A Polymeric Platform for Ocular Drug Delivery Applications. *Pharmaceutics* **2023**, *15*, 1096. https://doi.org/10.3390/pharmaceutics15041096.

4. Soltani, F.; Kamali, H.; Akhgari, A.; Ghasemzadeh Rahbardar, M.; Afrasiabi Garekani, H.; Nokhodchi, A.; Sadeghi, F. Preparation and Characterization of a Novel Multiparticulate Dosage Form Carrying Budesonide-Loaded Chitosan Nanoparticles to Enhance the Efficiency of Pellets in the Colon. *Pharmaceutics* **2022**, *15*, 69. https://doi.org/10.3390/pharmaceutics15010069.

5. Hassan, A.S.; Soliman, G.M. Rutin Nanocrystals with Enhanced Anti-Inflammatory Activity: Preparation and Ex Vivo/In Vivo Evaluation in an Inflammatory Rat Model. *Pharmaceutics* **2022**, *14*, 2727. https://doi.org/10.3390/pharmaceutics14122727.

6. Aboushanab, A.R.; El-Moslemany, R.M.; El-Kamel, A.H.; Mehanna, R.A.; Bakr, B.A.; Ashour, A.A. Targeted Fisetin-Encapsulated β-Cyclodextrin Nanosponges for Breast Cancer. *Pharmaceutics* **2023**, *15*, 1480. https://doi.org/10.3390/pharmaceutics15051480.

7. Stavroulaki, D.; Kyroglou, I.; Skourtis, D.; Athanasiou, V.; Thimi, P.; Sofianopoulou, S.; Kazaryan, D.; Fragouli, P.G.; Labrianidou, A.; Dimas, K.; et al. Influence of the Topology of Poly(L-Cysteine) on the Self-Assembly, Encapsulation and Release Profile of Doxorubicin on Dual-Responsive Hybrid Polypeptides. *Pharmaceutics* **2023**, *15*, 790. https://doi.org/10.3390/pharmaceutics15030790.

8. Jermy, B.R.; Salahuddin, M.; Tanimu, G.; Dafalla, H.; Almofty, S.; Ravinayagam, V. Design and Evaluation of Pegylated Large 3D Pore Ferrisilicate as a Potential Insulin Protein Therapy to Treat Diabetic Mellitus. *Pharmaceutics* **2023**, *15*, 593. https://doi.org/10.3390/pharmaceutics15020593.

9. Jeevanandam, J.; Tan, K.X.; Rodrigues, J.; Danquah, M.K. Target-Specific Delivery and Bioavailability of Pharmaceuticals via Janus and Dendrimer Particles. *Pharmaceutics* **2023**, *15*, 1614. https://doi.org/10.3390/pharmaceutics15061614.

10. Duarte-Peña, L.; Magaña, H.; Bucio, E. Catheters with Dual-Antimicrobial Properties by Gamma Radiation-Induced Grafting. *Pharmaceutics* **2023**, *15*, 960. https://doi.org/10.3390/pharmaceutics15030960.

11. López-González, I.; Hernández-Heredia, A.B.; Rodríguez-López, M.I.; Auñón-Calles, D.; Boudifa, M.; Gabaldón, J.A.; Meseguer-Olmo, L. Evaluation of the In Vitro Antimicrobial Efficacy against Staphylococcus aureus and epidermidis of a Novel 3D-Printed Degradable Drug Delivery System Based on Polycaprolactone/Chitosan/Vancomycin-Preclinical Study. *Pharmaceutics* **2023**, *15*, 1763. https://doi.org/10.3390/pharmaceutics15061763.

12. Gupta, K.M.; Chin, X.; Kanaujia, P. Molecular Interactions between APIs and Enteric Polymeric Excipients in Solid Dispersion: Insights from Molecular Simulations and Experiments. *Pharmaceutics* **2023**, *15*, 1164. https://doi.org/10.3390/pharmaceutics15041164.

13. Mohamed, E.M.; Dharani, S.; Nutan, M.T.H.; Cook, P.; Arunagiri, R.; Khan, M.A.; Rahman, Z. Application of Sucrose Acetate Isobutyrate in Development of Co-Amorphous Formulations of Tacrolimus for Bioavailability Enhancement. *Pharmaceutics* **2023**, *15*, 1442. https://doi.org/10.3390/pharmaceutics15051442.

References

1. Fernandes, A.I.; Jozala, A. Polymers Enhancing Bioavailability in Drug Delivery. *Pharmaceutics* **2022**, *14*, 2199. [CrossRef] [PubMed]

2. Sung, Y.K.; Kim, S.W. Recent advances in polymeric drug delivery systems. *Biomater. Res.* **2020**, *24*, 12. [CrossRef] [PubMed]

3. Bhatt, P.; Trehan, S.; Inamdar, N.; Mourya, V.K.; Misra, A. Polymers in drug delivery: An update. In *Applications of Polymers in Drug Delivery*; Misra, A., Shahiwala, A., Eds.; Elsevier: Amsterdam, The Netherlands, 2021; pp. 1–42, ISBN 9780128196595.

4. Campbell, S.; Smeets, N. Drug delivery: Polymers in the development of controlled release systems. In *Polymers and Polymeric Composites: A Reference Series*; Springer: Cham, Switzerland, 2019; pp. 1–29, ISBN 9783319765730.

5. Baryakova, T.H.; Pogostin, B.H.; Langer, R.; McHugh, K.J. Overcoming barriers to patient adherence: The case for developing innovative drug delivery systems. *Nat. Rev. Drug Discov.* **2023**, *22*, 387–409. [CrossRef] [PubMed]

6. Oh, J.K. Polymers in drug delivery: Chemistry and applications. *Mol. Pharm.* **2017**, *14*, 2459. [CrossRef] [PubMed]

7. Chavan, Y.R.; Tambe, S.M.; Jain, D.D.; Khairnar, S.V.; Amin, P.D. Redefining the importance of polylactide-co-glycolide acid (PLGA) in drug delivery. *Ann. Pharm. Fr.* **2022**, *80*, 603–616. [CrossRef] [PubMed]

8. Lou, J.; Duan, H.; Qin, Q.; Teng, Z.; Gan, F.; Zhou, X.; Zhou, X. Advances in oral drug delivery systems: Challenges and opportunities. *Pharmaceutics* **2023**, *15*, 484. [CrossRef] [PubMed]
9. Borandeh, S.; van Bochove, B.; Teotia, A.; Seppälä, J. Polymeric drug delivery systems by additive manufacturing. *Adv. Drug Deliv. Rev.* **2021**, *173*, 349–373. [CrossRef] [PubMed]
10. Vullendula, S.K.A.; Nair, A.R.; Yarlagadda, D.L.; Navya Sree, K.S.; Bhat, K.; Dengale, S.J. Polymeric solid dispersion Vs co-amorphous technology: A critical comparison. *J. Drug Deliv. Sci. Technol.* **2022**, *78*, 103980. [CrossRef]
11. Gao, J.; Karp, J.M.; Langer, R.; Joshi, N. The future of drug delivery. *Chem. Mater.* **2023**, *35*, 359–363. [CrossRef] [PubMed]

pharmaceutics

Article

Antifungal Encapsulated into Ligand-Functionalized Nanoparticles with High Specificity for Macrophages

Susana P. Mejía [1,2,†], Daniela López [2,†], Luz Elena Cano [2], Tonny W. Naranjo [2,3] and Jahir Orozco [1,*]

1   Max Planck Tandem Group in Nanobioengineering, Institute of Chemistry, Faculty of Natural and Exact Sciences, University of Antioquia, Complejo Ruta N, Calle 67 N° 52–20, Medellin 050010, Colombia
2   Experimental and Medical Micology Group, Corporación para Investigaciones Biológicas (CIB), UdeA, UPB, UdeS, Medellin 050010, Colombia
3   School of Health Sciences, Universidad Pontificia Bolivariana, Cl. 78b #72A-109, Medellin 050010, Colombia
*   Correspondence: grupotandem.nanobioe@udea.edu.co
†   These authors contributed equally to this work.

Abstract: Infectious diseases caused by intracellular microorganisms such as *Histoplasma capsulatum* represent a significant challenge worldwide. Drug encapsulation into functionalized nanoparticles (NPs) is a valuable alternative to improving drug solubility and bioavailability, preventing undesirable interactions and drug degradation, and reaching the specific therapeutic target with lower doses. This work reports on Itraconazole (ITZ) encapsulated into core-shell-like polymeric NPs and functionalized with anti-F4/80 antibodies for their targeted and controlled release into macrophages. Uptake assay on co-culture showed significant differences between the uptake of functionalized and bare NPs, higher with functionalized NPs. In vitro assays showed that F4/80-NPs with 0.007 µg/mL of encapsulated ITZ eliminated the *H. capsulatum* fungus in co-culture with macrophages effectively compared to the bare NPs, without any cytotoxic effect on macrophages after 24 h interaction. Furthermore, encapsulated ITZ modulated the gene expression of anti and pro-inflammatory cytokines (IL-1, INF-Y, IL-6 and IL-10) on macrophages. Additionally, the anti-F4/80 antibody-coating enhanced natural and adequate antifungal response in the cells, exerting a synergistic effect that prevented the growth of the fungus at the intracellular level. Functionalized NPs can potentially improve macrophage-targeted therapy, increasing NPs endocytosis and intracellular drug concentration.

Keywords: *Histoplasma capsulatum*; PLGA; Itraconazole; macrophage; functionalized nanoparticle; F4/80 receptor

Citation: Mejía, S.P.; López, D.; Cano, L.E.; Naranjo, T.W.; Orozco, J. Antifungal Encapsulated into Ligand-Functionalized Nanoparticles with High Specificity for Macrophages. *Pharmaceutics* **2022**, *14*, 1932. https://doi.org/10.3390/pharmaceutics14091932

Academic Editor: Maria Nowakowska

Received: 29 July 2022
Accepted: 2 September 2022
Published: 13 September 2022

Publisher's Note: MDPI stays neutral with regard to jurisdictional claims in published maps and institutional affiliations.

1. Introduction

Nanoencapsulation is making its way into the administration of drugs and subsequent site-specific release at the target cell or tissue through active functionalization of NPs with proteins, antibodies, carbohydrates, and peptides, among other ligands [1]. Ligand-functionalized NPs enhance uptake extent and can considerably increase the stability and systemic bioavailability of the drug since it protects its physicochemical characteristics, avoids possible degradation before reaching the therapeutic target, can achieve adequate doses at the intracellular level and, for example, facilitate the elimination of pathogens. The release of adequate drugs at the intracellular level is a strategy to avoid generating drug resistance to pathogens such as fungus and prevent non-specific accumulation in other tissues while reducing their toxicity and adverse side effects [2,3]. These adverse effects are due to the drug's action outside macrophages, the most important effector cells in host resistance, as in the case of histoplasmosis (HPM), used herein as a model of intracellular infection.

HPM is an endemic and systemic mycosis of primary pulmonary origin, affecting especially immunosuppressed and non-immunocompetent individuals [4]. HPM is caused by inhaling aerosols that contain the infecting particles (microconidia and small mycelial fragments) of the dimorphic fungus *Histoplasma capsulatum*. This mycosis has been reported in more than sixty countries on all continents, but it is endemic in the Americas [5]. The clinical presentation of HPM includes the acute, chronic pulmonary and progressive disseminated forms, the latter occurring especially in patients infected with HIV, considered of high prevalence with a 0.9% overall incidence of coinfection, reaching 27% in endemic areas and mortality rates up to 30% [6,7].

Although HPM can be treated with different antifungals, Itraconazole (ITZ) is considered the first choice to treat mild and moderate forms of HPM. Yet, at least twelve months of therapy is required for the acute and chronic forms and even longer times for the progressive disseminated form [8], causing adverse effects in patients. Hepatoxicity has been the most frequent failure, relevant for all antifungal agents, from mild anomalies in liver function to fatal fulminant liver failure [9,10]. Given orally, it causes nausea, vomiting, diarrhea, skin rashes, headache, etc. [11]. ITZ has other limitations such as high lipophilicity, low absorption capacity, low systemic bioavailability, high susceptibility of the active compound to gastrointestinal hydrolysis, and drug interactions, suggesting extreme caution when used as part of multiple drug therapy [12–15]. In addition, HPM treatments generally offer limited efficacy because many drugs degrade before reaching the target tissues or cells, hindering therapeutic levels [16–18]. As an alternative, Sporanox is the only commercial formulation of ITZ for oral administration, encapsulated with hydroxypropyl-β-cyclodextrin as an adjuvant that improves its solubility and biodistribution. However, it is contraindicated in patients with renal failure due to inefficient adjuvant removal [11].

It is then imperative to develop more efficient therapeutic strategies that shorten the treatment time and reduce the adverse side effects of drugs used to fight HPM. Different polymeric NPs have been reported in this context, searching for effective treatment with less toxicity and more patient-friendly regimes. These nanosystems have stability and high-load capacity and can transport one or more active principles (with similar or different physicochemical properties) or a combination of therapeutic and contrast image agents to be administered in the same formulation by various routes. Additionally, polymers can be chemically modified to achieve the optimal conditions required as active ingredient transport systems [19–21]. Among the most used, poly-(lactic-co-glycolic acid) (PLGA) is a highly biocompatible polymer approved by the Food and Drug Administration (FDA) that has a rate of adjustable biodegradation and modular mechanical properties [12].

Functionalization of the nanocarrier's surface has been widely explored for diagnosis, drug delivery, and other biomedical applications. Different ligands can be attached to the NPs surface, depending on the expected effect and whether intended for one or more functions [22–24]. For example, ligands containing bulky hydrophobic molecules can be attached to nanomaterial surfaces to prevent core agglomeration or be coated with water-soluble polymers, such as polyethylene glycol (PEG), to increase solubility required in physiological conditions, stealth properties, and biocompatibility. Ligands may also be attached to the surface of NPs to define their properties or act as "tags" for their recognition by target cells, either infected host cells or the microorganism directly [12].

Clinically important antifungals such as clotrimazole and econazole have been nanoencapsulated to improve their oral bioavailability [25]. ITZ encapsulated in PEG-functionalized poly-lactic acid (PLA) NPs has much better biocompatibility than commercial ITZ-cyclodextrin formulations, increasing drug solubility and stability in a wide range of concentrations and pHs [26]. Additionally, in vitro evaluations of the PLA-PEG-ITZ complex showed sustained drug release and growth inhibition of *H. capsulatum* [16]. Highly hydrophobic amphotericin B (AmtB), used to treat leishmaniasis and fungal infections, is hampered by its high toxicity. Encapsulation of AmptB into chitosan and PLGA-based NPs has reduced the drug's toxicity and increased the antiparasitic effect. Furthermore, AmtB-loaded NPs functionalized with mannan-type carbohydrates have shown to be an effective treatment against leishmaniasis. In addition to its antiparasitic effect, recognizing the carbohydrate by specific macrophage receptors causes an activation of the cell's defense mechanisms, increasing the treatment effectiveness [27,28]. PLGA NPs loaded with Fluconazole and coated with the cationic polymer polyethyleneimine (PEI) (FLZ-NP-PEI) also improved antifungal activity against four strains of clinically relevant *Candida* spp. [29,30]. Those are examples of antimicrobials encapsulated into functional (or not) polymeric nanocarriers reported in the literature but not in connection to PLGA-based NP-encapsulated ITZ against HPM.

We report a therapeutic solution to fight HPM intracellular infection. The therapeutic uses ITZ encapsulated in PLGA NPs, which functionalized with anti-F4/80 antibodies demonstrated for the first time increased antifungal effect on murine macrophages infected with *H. capsulatum* compared with bare NPs. Overall, results pave the way to design highly efficient nanocarriers for drug delivery against intracellular infections. Current studies are focused on directing the ITZ-loaded NPs to the lungs as the target organ by pegylation of NPs to improve their stealth properties, testing other macrophage-specific ligands and intranasal administration routes that can help increase the amount of NPs in the target organ. Additionally, the immunomodulation effect of functionalized nanoparticles was studied in vivo.

2. Materials and Methods

2.1. Reagents

Evonik (Essen, Germany) generously donated poly-lactic-co-glycolic acid (PLGA); LA: GA 75:25 (RG 752H) with 0.14–0.22 dl/g inherent viscosity (4 to 15 kDa). Sigma-Aldrich provided Nile Red (CAS 7385-67-3), Itraconazole (ITZ, CAS 84625-61-6), poloxamer 188 (CAS 9003-11-6, Kolliphor®), D-α-tocopherol polyethylene glycol 1000 succinate (vitamin E-TPGS, CAS 9002-96-4), voriconazole (CAS 137234-62-9), and phosphate-buffered saline solution (PBS D8573). MTT TOX-1 Kit, HMM Broth (F12 HAM nutrient medium, N6760), Janus Green (CAS 2869-83-2), sodium periodate (CAS 7790-28-5), Hoechst (CAS 23491-45-4), and TRIzol1 (Ref. 15596026) were purchased from Sigma-Aldrich (St. Louis, MO, USA). Fetal bovine serum (FBS) (Gibco, Ref. 16000044), Horse serum (HS) (Gibco, Ref. 16050122), Dulbecco's Modified Eagle Medium (DMEM, Ref. 10569010), and the antibiotic penicillin-streptomycin (Ref. 15140122), DNase I (Ref. EN0521), 3,3'-dihexyloxacarbocyanine iodide (DIOC6, CAS 53213-82-4), and RT-qPCR kit (Ref. A46109) were from Gibco (Gibco, Thermo Fisher Scientific, Inc., Waltham, MA, USA). Acetonitrile (CAS 75-05-8), ethyl acetate (CAS 141-78-6), ethanol (CAS 64-17-5), and dimethylsulfoxide (DMSO, CAS 67-68-5) were purchased from Merck. Citric acid (CAS 77-92-9) and sucrose (CAS 57-50-1) were purchased from VWR Chemicals. Potassium chloride (CAS 7447-40-7), sodium chloride (CAS 7647-14-5), and monobasic potassium phosphate (CAS 7778-77-0) were provided by JT Baker (JT Baker, Thermo Fisher Scientific, Inc., Waltham, MA, USA). Sodium citrate dihydrate (CAS 6132-04-3) and dibasic sodium phosphate (CAS 7558-79-4) were purchased from Panreac. Anti-F4/80 (ab100790) and Alexa flour 488 (ab150077) were obtained from Abcam (Abcam, Cambridge, UK). Brain Heart Infusion Agar (BHI) was obtained from Difco Laboratories (Ref. 211065, Thermo Fisher Scientific, Inc., Waltham, MA, USA). 0.1 M citrate buffer pH 4.5–5.5, 50 mM MES buffer pH 6.0–6.5, 50 mM HEPES buffer pH 7.0, and 1 M PBS buffer pH 6.5 were prepared by dissolving the reagents as received in Milli-Q water 18 MΩ·cm filtered through a 0.2 µm membrane.

2.2. Fungal Growth Conditions

Histoplasma capsulatum isolated Hc 1980 was obtained from the collection of the Corporation for Biological Research (CIB), Medellín, Colombia. Yeasts were grown on BHI agar, supplemented with 5% sheep red blood cells, 0.1% L-cystine and 1% glucose at 37 °C with 5% CO_2, and repot was performed every 5 days. Colonies were taken from the BHI culture and grown in a 50 mL conical tube with 10 mL of Ham's F-12 (Sigma-Aldrich, St. Louis, MO, USA). The base composition of the HMM medium was supplemented with 18.2 g/L glucose, 1 g/L glutamic acid, 6 g/L HEPES, and 8.4 mg/L cysteine at 37 °C after shaking at 150 rpm (Innova® 44, Thermo Fisher Scientific, Inc., Waltham, MA, USA), for 48 h to reach its exponential growth phase. The yeast suspension was passed 30 times through a 1 mL tuberculin syringe with a 27 G $\frac{1}{2}$ needle to separate the aggregated fungal cells and isolate the individual yeasts. The suspension was left upright for 1 min and then 5 mL was taken from the top. Cell viability was determined with Janus Green B vital stain and cell suspensions were adjusted to the required value of yeast/mL according to the hemocytometer count.

2.3. Cell Line and Culture Conditions

The mouse peritoneal macrophage cell line J774A.1, AMJC2-11b THP-1, CHO, and C2C12 were obtained from the American Type Culture Collection (ATCC, Manassas, VA, USA). The cells J774A.1, AMJC2-11b, CHO, and C2C12 were cultured in DMEM, and the THP-1 in RPMI, cultured medium containing glucose and L-glutamine, was supplemented with 10% heat-inactivated FBS or HS and 1% penicillin/streptomycin to avoid contamination with bacteria. The cultures were kept at 37 °C in 5% CO_2 to ensure a saturated humid atmosphere. The culture media were changed every 2–3 days and the cells were subcultured every time they reached 90% confluence.

2.4. Nanoparticle Assembly and Functionalization

NPs were prepared using a high-energy emulsification method using a PLGA polymer with a glycolic acid: lactic acid ratio 75:25 and a molecular weight of 7–14 kDa. Briefly, 75 mg of PLGA polymer and 6.6 mg of ITZ and 7.5 mg of tocopherol were dissolved in 5 mL of ethyl acetate (organic phase) and then poured 10 mL 1% Kolliphor® P188 in citrate buffer pH 5.

The NPs were functionalized by covalent coupling of the antibodies using the carbodiimide method. First, the carboxyl groups from the outermost NPs surface were activated by reaction with 4:1 EDC (400 mM):NHS (100 mM), followed by the anti-F4/80 (ab100790) polyclonal antibody with a 1:10 antibody: NPs ratio in PBS pH 6.5 for 2 h at 4 °C [31]. The functionalized NPs were characterized by changes in the size and ζ-potential obtained by DLS and ELS. The measures were taken in a Zetasizer-Pro (Malvern Instruments, Malvern, UK), at 25 °C 160 after adequate aqueous dilution by triplicate. The NPs were formed and functionalized based on our recent report [32].

2.5. Detection of Antibodies on NPs

A total of 20 µg/mL of goat antiRabbit IgG Alexa Fluor 488 (AF 488) antibodies were added to the solution of functionalized NPs, and then 1X sterile PBS was added; this solution was incubated at room temperature for 30 min with constant shaking. Once the incubation was completed, it was brought to a final volume of 200 µL to measure by fluorescence spectrophotometry. A calibration curve with different concentrations of AF488 was made between 25 and 0.195 µg/mL, where the fluorescence intensity values were plotted against concentration, and the values of the unknown antibody concentrations were calculated from the straight-line equation.

2.6. Determination of the Minimum Inhibitory Concentration (MIC)

To evaluate the antifungal activity of ITZ encapsulated into functionalized NPs and the different NPs precursors and free ITZ against Hc CIB1980, the microdilution method for yeast (M27-A3) described by the National Committee of Clinical Laboratory Standards (NCCLS) was used [33]. Once the fungus growth was determined by visible turbidity with the help of an inverted mirror, the viability of the fungus was evaluated using the MTT method. The yeast suspension was prepared in a histoplasma-macrophage medium (HMM) and a 96-well Round-(U)-bottom plate. A total of 1.5×10^5 yeast cells/well was added to 100 µL. The dilutions of each treatment were prepared in HMM medium with 0.5% DMSO adding 100 µL of the different treatments to each well at concentrations from 16 to 0.0009 µg/mL for a 200 µL/well final volume. However, empty NPs didn't encapsulate ITZ, so we calculated the amount of NPs that we added when encapsulated ITZ. Therefore, concerning each ITZ concentration, the amount of NPs added was 222.2 to 0.027 µg/mL.

The medium with 0.5% DMSO without the drug and untreated yeast was used as a control. The plates were incubated at 37 °C for 7 days with aeration on a mechanical shaker at 150 rpm. After the incubation time, the MIC was defined as the lowest concentration of treatment that can inhibit macroscopic fungal growth and low viability in MTT. Based on this result, the mean inhibitory concentration (IC50) (i.e., the concentration required for both free ITZ and encapsulated ITZ to inhibit Hc growth by 50) was calculated. The tests were in triplicate.

2.7. Evaluation of the Cytotoxicity of the Nanobioconjugate on Macrophages

A total of 3×10^4 macrophages J774A.1 or AMJC2-11b per well were adhered to a 96-well plate in 200 µL DMEM culture medium supplemented with 10% FBS for 17 h at 37 °C and 5% CO_2. THP-1 monocytes were differentiated to macrophages adding 1.2×10^5 cells per well and treated with 100 nM of PMA in 200 µL RPMI culture medium supplemented with 10% FBS and incubated 72 h at 37 °C, 5% CO_2. Then 100 µL of empty NPs, bare-NPs or F4/80-NPs were added to make different concentrations between 0.125 to 0.007 µg/mL. However, empty NPs didn't encapsulate ITZ, so we calculated the amount of NPs that would be added with encapsulated ITZ. Therefore, concerning each ITZ concentration, the amount of NPs added was 3.47 to 0.216 µg/mL. Each treatment was tested and incubated for 3, 5, 24, or 48 h at 37 °C, 5% CO_2. After the respective incubation time, 20 µL of MTT was added and incubated for 2 h at 37 °C in 5% CO_2. Then, 130 µL of solubilizer solution was added to each well to dissolve the formazan crystals, then incubated at room temperature in the dark overnight. The formazan crystal's optical density (OD) was measured spectrophotometrically at 595 nm in a BioTek ELx808 ELISA plate reader (BioTek, Winooski, VT, USA). Control cells (no treated) were maintained in DMEM-GlutaMax medium supplemented with 10% FBS.

2.8. Study of the Antifungal Activity of the Nanobioconjugate in an In Vitro Cell Model

The assay was performed in 96-well plates (Falcon®); 3×10^4 J774A.1 macrophages per well were adhered in 200 µL DMEM with 10% FBS for 17 h. Each macrophage monolayer was infected with 100 µL of yeast inoculum (1.5×10^5 yeast) prepared in an unsupplemented DMEM medium (1: 5 macrophage: yeast ratio) and incubated for 3 h at 37 °C with 5% CO_2. After, the non-phagocytosed yeasts were washed with PBS. Then, 0.031, 0.015, or 0.007 µg/mL free ITZ and encapsulated into NPs with and without functionalized anti-F4/80 antibodies were added and incubated for 6 h or 24 h. Also, empty NPs were added with the same conditions as a control [33]. However, we again calculated the amount of NPs we would add with encapsulated ITZ. Therefore, the amount of NPs added was 0.86, 0.43, and 0.216 µg/mL. After 6 h of incubation, the co-culture was washed twice with PBS to eliminate the non-endocytosed NPs and a new medium was added.

The plate was incubated again under the same conditions to complete 24 h. Then the supernatant was taken, diluted five times in PBS and seeded on supplemented BHI agar. On the other hand, 200 µL of cold, sterile water was added to each well to lyse the cells and recover the yeasts that the macrophages had phagocytosed. The lysate suspension was dissolved with PBS five times and plated on supplemented BHI agar. The BHI dishes with the lysates and supernatants of the co-cultures were incubated at 37 °C with 5% CO_2 for 7 days. After the incubation time, (CFU) was counted. The experiments were performed in two independent trials in triplicate.

2.9. In Vitro Assay of Specificity

A total of 3×10^4 J774A.1 cells were adhered onto a 96 well plate in DMEM with 10% FBS for 16 h. After, the cells were infected with 100 µL of *H. capsulatum* yeast inoculum (1.5×10^5 yeast) previously stained with calcofluor (100 µg/mL) and incubated for 3 h at 37 °C with 5% CO_2. Then, 100 µg/mL of functionalized (or bare) NPs with encapsulated Nile red were added and incubated at 37 °C for 6 h with 5% CO_2. Therefore, monolayers were washed three times with PBS to eliminate any endocytosed NPs. The cell cultures were characterized by fluorescence microscopy. Determination of Nile red inside the infected cells (area of Nile red) was measured indirectly to know the differences in ligands' specificity. To corroborate the NPs uptake, we used DIOC dye to stain cells' cytoplasm and localize the NPs and the fungus. The area of Nile red (%) was determined with ImageJ software version 64-bit Java 1.8.0 (National Institutes of Health, Bethesda, MD, USA). Fourteen images were captured with a 40X objective and analyzed randomly from different regions. Individual cells and Nile red areas were framed with the freehand selection to measure the inner region. These areas were taken as the total area, and the area of Nile red was calculated in Microsoft Excel 2016 (Redmond, WA, USA) package using the data.

Additionally, for verification of the intracellular localization of NPs, confocal laser scanning microscopy (cLSM) and flow cytometry were employed. The uptake of the NPs by the uninfected J774A.1 cells was quantified by flow cytometry after the treatment with 100 µg/mL for 3 h. The uptake was also visualized by confocal fluorescence microscopy after 3 h of the coincubation of the cells with a NPs suspension of 100 µg/mL. Details of these experiments are described in the Supporting Information. On the other hand, the specificity of NPs was evaluated in other types of cells, such as muscle cells and ovary cells (see supplementary material).

2.10. Gene Expression

Determination of cytokines in the co-culture of J774A.1 with *H. capsulatum* was evaluated after 6 h of incubation with 0.007 µg/mL ITZ encapsulated and functionalized (or not) with anti-F4/80 antibodies. Cells were removed from the culture plates and centrifuged at 3000 rpm at room T for 5 min. Total RNA was obtained using TRIzol1 and treating samples with DNase I. cDNA was synthesized using 500 ng of total RNA using the Maxima First Strand cDNA synthesis kit for RT-qPCR according to the manufacturer's instructions followed by real-time PCR using Maxima SYBR Green/Fluorescein qPCR Master Mix (2X). The CFX96 Real-Time PCR Detection System (Bio-Rad, Headquarters Hercules, CA, USA) was employed to measure gene expression levels. Analysis of the melting curve eliminated the possibility of non-specific amplification or primer-dimer formation. Validation of housekeeping genes for normalization of mRNA expression was performed before gene expression analysis. Validation of control genes for normalization of mRNA expression was performed before gene expression analysis. The expression of genes related to an inflammatory response IL-1 β, IL-6, TNF-α and IFN-γ and anti-inflammatory IL-10 was evaluated. Measurement of messenger RNA (mRNA) expression was obtained by relative quantification comparing the expression level of the target gene relative to GAPDH (glyceraldehyde-3-phosphate dehydrogenase, housekeeping gene). The expression level was measured in triplicate. Primers sequences used were GADPH Fw: 5′- CATGGCCTTCCGTGTTCCTA-3′ Rv: 5′- GCGGCACGTCAGATCCA-3′, IL-1β Fw: 5′-CTTCAAATCTCGCAGCAGCACATC-3′, Rv: 5′-TCCACGGGAAAGACACAGGTAGC-3′, TNF-α Fw: 5′-GACAAGGCTGCCCCGACTACG Rv: 5′-CTTGGGGCAGGGGCTCTTGAC-3′, IFNγ Fw: 5′- GACATGAAAATCCTGCAGAGCCAG -3′ Rv: 5′- TCGCCTTGCTGTTGCT-GAAGAAG -3′, IL-10 Fw: 5′-TGGGTTGCCAAGCCTTATCGG Rv: CTCACCCAGGGAATT-CAAATGCTC-3′, IL-6 Fw: 5′-CAACCACGGCCTTCCCTACTTC Rv:TCTCATTTCCACGAT-TTCCCAGAG-3′ [34].

2.11. Statistical Analysis

GraphPad Prism software (version 8, San Diego, CA, USA) was used for all statistical analyzes; the normal distribution was determined by the Shapiro-Wilk test. Medians and interquartile ranges were used to analyze cytokine expression levels. Analysis of variance between multiple experimental groups was performed using the ANOVA test, and differences between groups were analyzed using Tukey's test. p Values < 0.05 were considered significant. A p-value ≤ 0.01 was considered statistically significant.

3. Results

3.1. Functionalization of Nanocarriers

The antibodies were coupled to the NPs by the carbodiimide chemistry. The functionalized NPs were characterized by changes in the size and ζ-potential by dynamic light scattering (DLS) and electrophoretic light scattering (ELS), respectively. The anti-F4/80 antibody-functionalized NPs presented a larger size, 226.66 ± 13.05 nm, than bare NPs 160.3 ± 9.5 nm (Figure 1A). ζ-potential showed a significant change, i.e., reduced the negative charge from −39.53 ± 2.54 mV to −27.9 ± 0.26 mV after covalent coupling with the antibody (Figure 1B). The antibody's coupling was assessed by labeling it with the secondary antibody-Alexa fluor 488 and measured by fluorescence spectroscopy, obtaining 2.7% (0.059 µg antibody/mg of functionalized NPs); a similar result was reported previously [32].

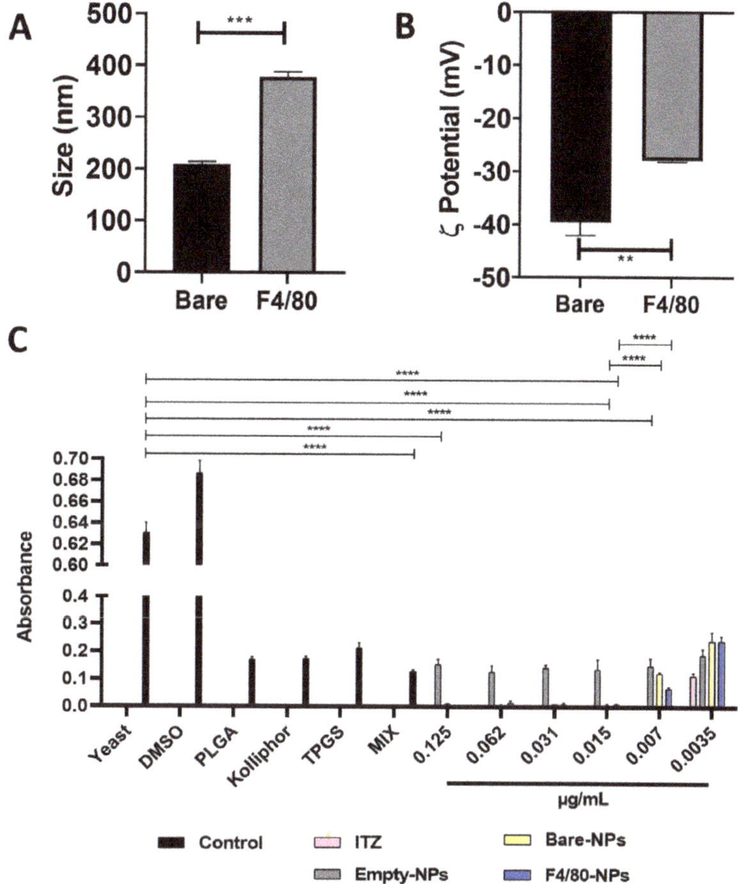

Figure 1. Comparison of size (**A**) and surface charge (**B**) of PLGA75: 25-TPGS-pH5-ITZ NPs functionalized or not (bare) with anti-F4/80 antibodies. (**C**) MIC strain *H. capsulatum* CIB1980 treated with free-ITZ, ITZ encapsulated into PLGA75:25 TPGS-pH5 NPs functionalized (or not) with anti-F4/80, empty PLGA75:25 TPGS-pH5 NPs and the corresponding amount of each NPs precursor (equivalent to the highest concentration of NPs that were used, i.e., 0.125 µg/mL). **, *** and **** indicate statistically significant differences with $p < 0.001$, $p < 0.0001$ and $p < 0.00001$, respectively.

3.2. Antifungal Activity

The minimum inhibitory concentration (MIC) of both free- and encapsulated-ITZ in the nanocarriers functionalized (or not) with anti-F4/80 antibodies was determined with a Colombian strain of *H. capsulatum* (CIB 1980). To determine the MIC, different concentrations ranging from 0.125 to 0.0018 µg/mL of free- and encapsulated-ITZ into functionalized (or not) PLGA75:25 TPGS-pH5 NPs were evaluated and empty NPs as a control. The macroscopic turbidity method and a 4,5-dimethylthiazol-2-yl) 2,5-diphenyltetrazolium bromide (MTT) assays (Figure 1C) showed a MIC down to 0.0035 µg/mL with free ITZ ($p < 0.0001$) and 0.007 µg/mL with both functionalized and ITZ-NPs ($p < 0.0001$), respectively. Empty-NPs in all concentrations and with all NPs precursors showed similar slightly inhibition-like fungus' growth ($p < 0.0001$). Likewise, compared to the yeast control, a significant concentration-independent decrease was observed in the groups treated with empty NPs ($p < 0.0001$) and their precursors, indicating that these molecules could be inhibiting the fungus growth favoring the effect of the drug.

3.3. Cytotoxicity

The cell survival plot for THP-1, J774A.1 and AMJC2-11b (Figure 2A–C) after 48 h of treatment with ITZ-NPs and empty NPs showed no reducing cell proliferation, with cellular viability higher than 80%. However, the viability after 5 h in AMJ2-C11 decreased depending on the treatment concentration with the bare NPs and free ITZ. Then, the cytotoxic effects of functionalized NPs in J774A.1 were evaluated, and a cell line was chosen to do the co-culture assay. Also, different times used to treat the co-culture (3 and 24 h) and higher, and lower NPs concentrations between the MIC concentration of functionalized NPs (0.062–0.007 µg/mL) were analyzed. Cell viability in all concentrations and with all types of NPs after 3 and 24 h of treatment was higher than 80% (Figure 3). However, the highest concentration of all treatments (0.031 µg/mL) significantly decreased cell viability compared to the lowest concentration (0.007 µg/mL) at both times evaluated (3 h and 24 h). Furthermore, when cells were treated for 24 h with the different formulations, a significant decrease in cell viability was observed compared to cells treated for 3 h, regardless of the concentration used. These results suggest that free ITZ and/or encapsulated into NPs and coated with F4/80 did not affect the viability of J774A.1 cells.

3.4. Antifungal Effect

The efficiency of ITZ encapsulated into PLGA75:25 TPGS-pH5 NPs to eliminate *H. capsulatum* in infected macrophages was evaluated by counting Colony Forming Units (CFU). For this purpose, co-culture of mouse macrophages J774A.1 (3×10^4 cell/well, 96-well microplates) with *H. capsulatum* yeasts (1.5×10^5 yeast/well, strain CIB 1980) were incubated under physiology conditions (37 °C with 5% of CO_2) for 3 h. Then, the co-cultures were washed with PBS to eliminate the yeasts that were not phagocytosed and treated with 0.031, 0.015, or 0.007 µg/mL of NPs with free- and encapsulated-ITZ and empty NPs, and incubated at 37 °C during 24 h (Figure 4A) or 6 h (Figure 4B). The co-culture with 6 h of treatment was washed to eliminate the not endocytosed NPs and incubated at 37 °C to complete 24 h. Each well was aspirated and washed with PBS when the incubation was completed. The stock solution of 1:10 dilutions of not phagocytosed (supernatant) and phagocytosed (lysed macrophages) yeasts were plated (10 or 20 µL, respectively) onto BHI supplemented agar, and the CFU values were counted after 7 days of culturing.

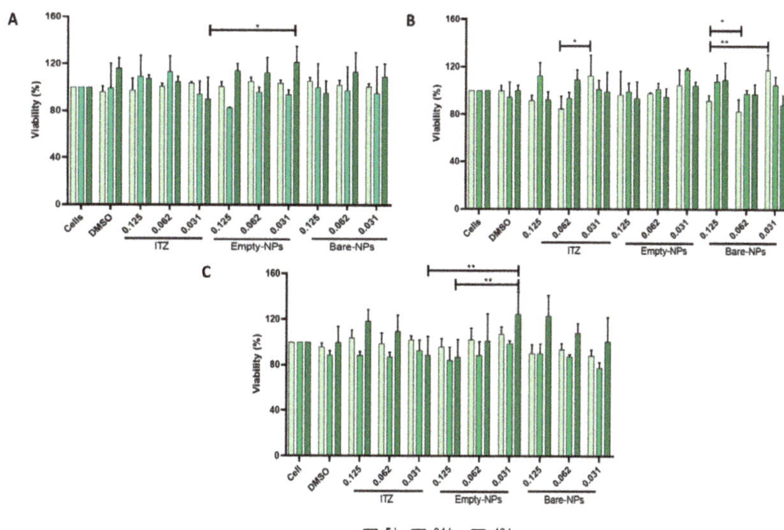

Figure 2. Cytotoxicity evaluation of ITZ encapsulated into PLGA75:25 TPGS-pH5 NPs by MTT with (**A**) J774A.1 (peritoneal mice macrophages), (**B**) AMJ2-C11 (alveolar mice macrophages), (**C**) THP-1 (human macrophages), with different concentrations and 0.125, 0.062, and 0.031 µg/mL ITZ for 5, 24, and 48 h, treatment with free ITZ and empty NPs and DMSO as controls. *, and ** indicate statistically significant differences with $p < 0.01$, and $p < 0.001$, respectively.

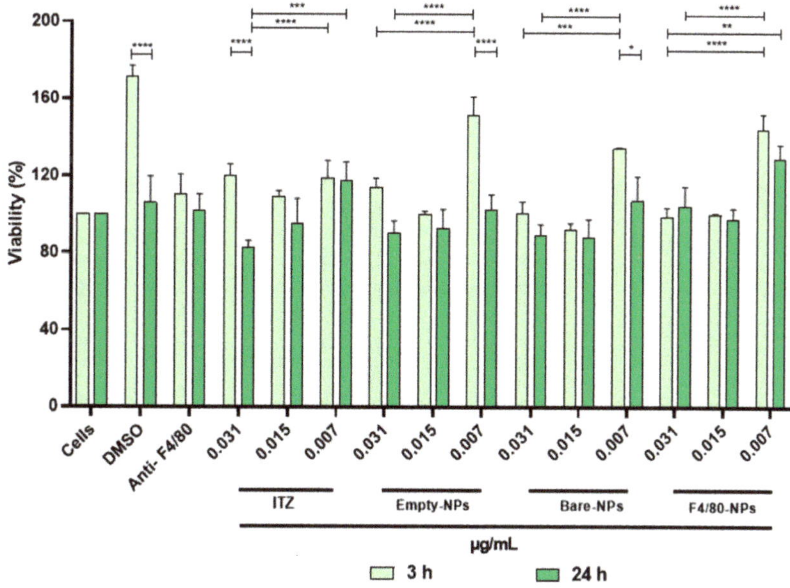

Figure 3. Cytotoxicity evaluation of ITZ encapsulated into functionalized PLGA75:25 TPGS-pH5 NPs with anti-F4/80 antibody on J774A.1 (peritoneal mice macrophages), with 0.031, 0.015, and 0.007 µg/mL of ITZ for 3 and 24 h, respectively by MTT and treatment with free ITZ and empty NPs or DMSO as controls. *, **, *** and **** indicate statistically significant differences with $p < 0.01$, $p < 0.001$, $p < 0.0001$, and $p < 0.0001$, respectively.

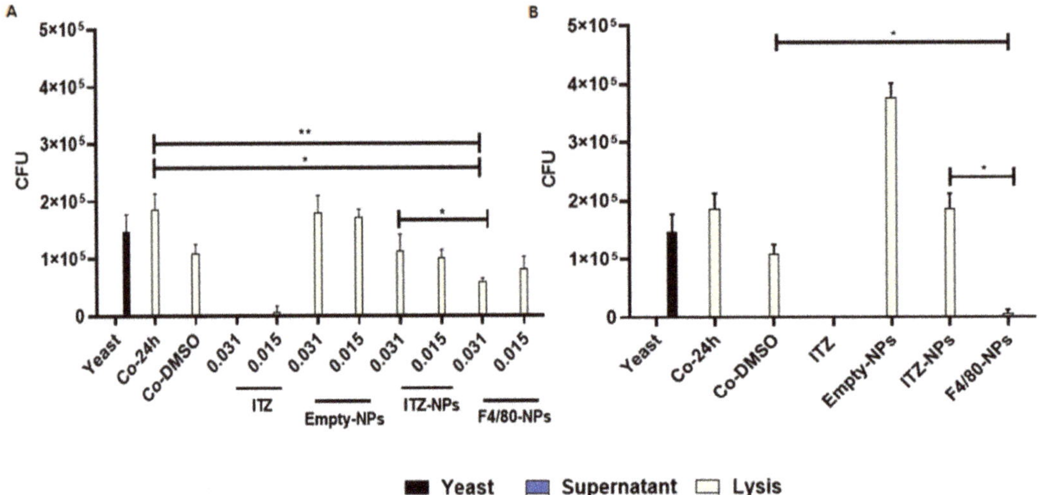

Figure 4. The efficiency of ITZ encapsulated into PLGA75:25-TPGS-pH5 NPs functionalized with the anti-F4/80 antibody to eliminate *H. capsulatum* yeast in co-culture (Co) with (**A**) J774A.1 by CFU counting of phagocytosed (lysed) or non-phagocytic yeasts (supernatant) after the treatment with 0.031 and 0.015 µg/mL of NPs with or without ITZ and free ITZ for 24 h (**B**) J774A.1 CFU counting of yeasts after the treatment with 0.007 µg/mL of NPs with or without ITZ and free ITZ for 6 h. *, and ** indicate statistically significant differences with $p < 0.01$, and $p < 0.001$, respectively.

Figure 4A shows that after 24 h with 0.031 and 0.015 µg/mL of ITZ encapsulated into NPs with F4/80, the CFU number from lysed macrophages significantly decreased compared to the control groups of untreated co-cultures and those treated with DMSO. Additionally, F4/80-NPs at a concentration of 0.031 µg/mL showed a significant reduction in the number of CFU compared to the values of bare NPs ($p < 0.01$). However, it is important to mention that the group treated with free ITZ significantly decreased CFU at all concentrations evaluated (0.031, 0.015, and 0.007 µg/mL) ($p < 0.0001$). On the other hand, in the lysates of the co-cultures washed after 6 h of treatment (Figure 4B), a significant reduction ($p < 0.0001$) was observed in the number of CFU obtained in the group treated with F4/80-NPs (0.007 µg/mL) in comparison with the control groups (co-cultures without treatment and treated with DMSO) and the group treated with bare-NPs. On the other hand, empty NPs- in co-culture did not show an antifungal effect (Figure 4A,B).

3.5. Specificity of Functionalized NPs for Macrophages

Nile red was used here as a model of a hydrophobic compound to evaluate the functionalized NPs specificity for J774A.1 macrophages infected with *H. capsulatum*. Fluorescence microscopy evaluated the intracellular co-localization of F4/80-coated-NPs and the fungus into the macrophages. The merge images in Figure 5A(down),C show that anti-F4/80-antibodies increased NPs endocytosis (Nile red-NP, blue fungus yeast and green cytoplasm cells), concerning bare-NPs that showed lower uptake by macrophages (Figure 5A(middle),B). Nile red fluorescence was estimated as described in the materials and methods section to confirm the uptake differences among the functionalized (or not) NPs. In this fashion, the macrophage's uptake extent was twice for those treated with F4/80-coated-NPs (2.14%) than those treated with bare-NPs (1.04%) (Figure 5C).

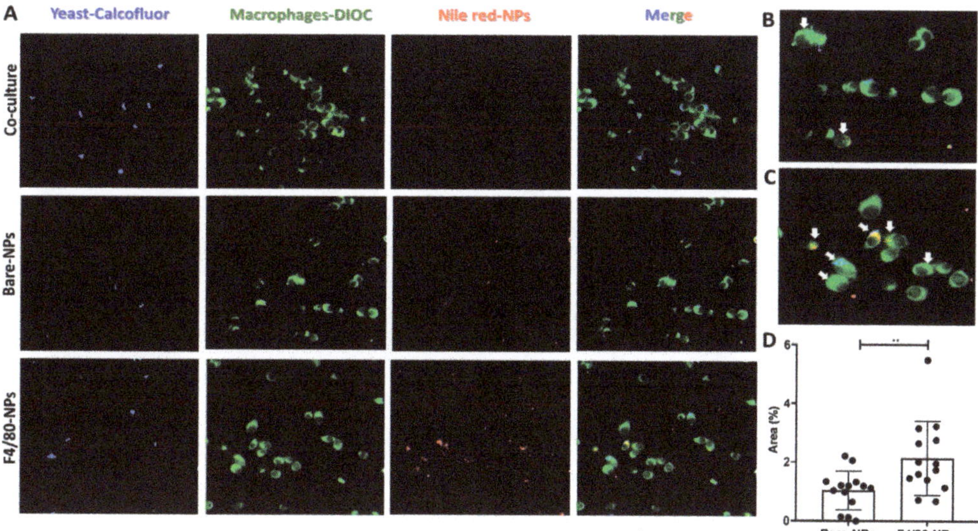

Figure 5. Endocytosis evaluation by fluorescence microscopy of anti-F4/80 antibody-functionalized PLGA75:25-TPGS-pH5-NPs encapsulating Nile red into J774A.1 macrophage *H. capsulatum* co-culture. (**A**) Images corresponding to the co-culture without NPs, bare-NPs, and F4/80-coated-NPs upon 6 h of incubation. Images from left to right were taken with DAPI filter (yeast-calcofluor stain), DIOC filter (macrophages-DIOC stain), TRITC filter (Nile red-encapsulated NPs), and merged (right), respectively. The scale is 20 μm. The zoomed-in (**B**) bare-NPs merged image and (**C**) F4/80-coated-NPs merged image. White arrows indicate macrophages infected with *H. capsulatum* with endocytosed NPs. (**D**) Endocyted NPs estimated by measuring Nile red by fluorescence intensity as described in the materials and methods section. ** indicates statistically significant differences with $p < 0.001$.

To confirm the uptake of nanoparticles into macrophages, we performed two additional confirmatory experiments with different methodologies, such as flow cytometry and confocal laser scanning microscopy (cLSM) (Figures S1 and S2). J774A.1 macrophages were treated with 100 μg/mL of NPs with encapsulated Nile red, functionalized (or bare) and incubated for 3 h. After 3 h, both flow cytometry and (cLSM) showed significant differences between the uptake of functionalized and bare NPs, higher with functionalized NPs (Figure S1, $p < 0.01$ and Figure S2, $p < 0.001$, respectively). Additionally, the cLSM images showed that the NPs were located in the cytoplasm and not in the membrane (the merged image didn't show yellow color) (Figure S2).

On the other hand, to corroborate the specificity of nanoconjugates C2Cl2 mice muscle cells and CHO hamster ovary cells were treated with F4/80-coated-NPs and bare-NPs for 3 h (Figures S3 and S4). Both cells did not show any NPs uptake.

3.6. Immunomodulation

Immunomodulation is understood as the modulation or modification of the immune response, not strictly of the immune system, referring to a living being. Therefore, different biomolecules can modulate the immune response [35]. The capacity of functionalized (or not) ITZ-encapsulated nanocarriers to modulate or modify the immune response in macrophages infected with *H. capsulatum* after 6 h of treatment at concentrations that presented an antifungal effect (0.007 μg/mL) was studied by analyzing the expression of the pro-inflammatory (IL-1β, IL-6, TNF-α, IFN-γ) and anti-inflammatory cytokines (IL-10) at the RNA level (Figure 6A–D). Results indicated that ITZ treatment significantly increased IL-1β expression ($p < 0.0001$) compared to co-cultures without treatment (Figure 6A).

However, IL-1 expression was reduced in the presence of ITZ nanoformulations, being more significant in the presence of F4/80-coated-NPs (Figure 6A). In the case of INF-γ (Figure 6B), empty-NPs significantly increased cytokine expression ($p < 0.0001$). In contrast, the levels of INF-γ significantly decreased with bare-NPs and F4/80-coated-NPs, but this decrease is more significant with the functionalized NPs treatment ($p < 0.001$). Additionally, it was observed that, compared to untreated co-cultures, treatment with free ITZ significantly increased IL-6 expression ($p < 0.01$) (Figure 5C). Contrary, treatments with ITZ nanoformulations (bare-NPs and F4/80-coated-NPs) significantly decreased the expression of IL-6 compared to free ITZ (Figure 6C). On the other hand, no statistically significant differences were observed between the different treatments and the expression of TNF-α.

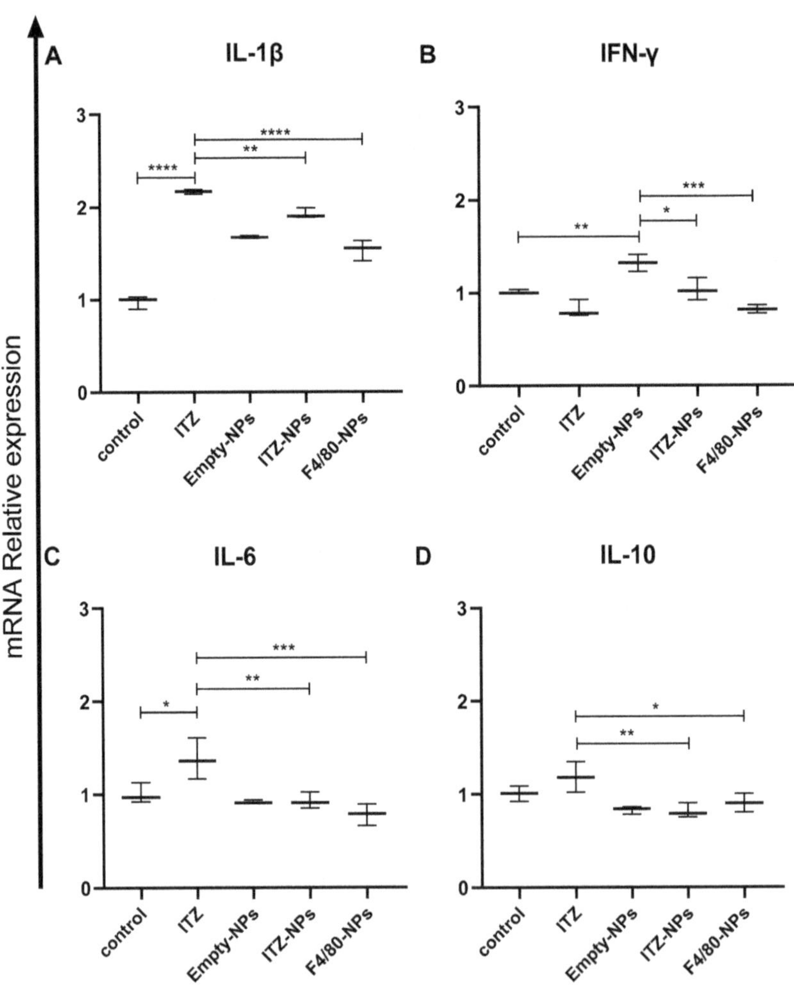

Figure 6. Immunomodulation effect of ITZ encapsulated into anti-F4/80-antibody-functionalized PLGA75:25-TPGS-pH5 NPs in co-culture of J774A.1 with *H. capsulatum* after 6 h of treatment. Expression of (**A**) IL-1β, (**B**) IFN-γ, (**C**) IL-6, and (**D**) IL-10. *, **, *** and **** indicate statistically significant differences with $p < 0.01$, $p < 0.001$, $p < 0.0001$, and $p < 0.0001$, respectively.

Furthermore, the co-cultures treated with the different nanoformulations (empty-, bare- and F4/80-coated NPs) significantly decreased the expression of IL-10 anti-inflammatory cytokine ($p < 0.001$) compared to the group without treatment.

4. Discussion

Intracellular microorganisms cause a wide range of infectious diseases with different degrees of clinical severity, representing significant morbidity and mortality worldwide [36]. Currently, infections caused by facultative intracellular microorganisms such as *H. capsulatum* are a great challenge because they have a more varied relationship with their hosts; the intracellular localization of the fungus makes it difficult to treat the infection, as macrophages can act as a barrier, preventing the antifungal drug from interacting with its target into the cell [37,38].

Fungal infections are often defined as challenging to treat due to the few options on the market and the multiple resistance that microorganisms are generating to existing treatments, the toxicity of antifungals, and their interaction with other drugs [39,40]. According to data reported by the FDA, since 1983, the generation of new drugs to fight infectious diseases, in general, has decreased, leaving us with fewer options. It is then imperative to develop a new efficient therapeutic strategy to improve the treatment time and reduce the adverse side effects and general issues to fight HPM. Encapsulation of microbial agents into NPs to selectively target pathogens has been shown to improve the efficacy and efficiency of therapeutic regimens. Thus, it is a strategy to overcome the challenges of conventional drug delivery systems that may actively drive drugs for site-specific and efficient drug uptaking, thereby readily reaching therapeutic intracellular levels [31,41–45]. Several NPs have been tested as potential drug delivery systems, including biodegradable polymeric NPs [1].

In this study, antifungal ITZ was encapsulated into PLGA NPs and further functionalized with anti-F4/80 antibodies, demonstrating an increased antifungal effect on murine macrophages infected with *H. capsulatum* compared with bare NPs. Increased size (Figure 1A) and decreased negative ζ-potential (Figure 1B) of the NPs suggest the antibody-NP surface binding without compromising its stability in maintaining a negative ζ-potential -29 mV (Figure 1B), avoiding NPs aggregation and remaining as a monodisperse colloidal system as reported [32]. Additionally, size remains very close to 200 nm (Figure 1A), which is essential to reduce the risk of rapid elimination by the endothelial reticulum system (RES), one of the main physiological barriers faced by NPs, affecting their biodistribution. Furthermore, several reports have shown that modifying the size and surface charge can regulate the interaction with physiological proteins (protein crown), modulating the type and quantity of proteins that bind to the surface; thus, recognition of NPs by specific receptors, facilitating (or not) the NPs uptake by the macrophages [12,46,47]. Therefore, the importance of functionalizing NPs with ligands for targeting cells and/or pegylation to neutralize the NPs surface charge and provide hydrophilic properties or steric hindrance stabilization, reduce the plasma proteins' adherence and unspecific phagocytosis, and increase the specificity of targeting ligands [1,12]. By evaluating MICs with free- and encapsulated ITZ with (or without) anti-F4/80 antibodies, it was possible to show that encapsulated ITZ into functionalized NPs preserved its antifungal activity against the fungus. Encapsulated ITZ required a similar concentration (0.015 µg/mL) concerning bare NPs but a higher concentration to free ITZ (0.007 µg/mL) (Figure 1C). The MIC difference between NPs and free ITZ is related to the type of kinetic release that the NPs presented [32], where after 7 h incubation, the NPs achieved a release of 40% of the encapsulated ITZ and the release continued sustainable until 72 h, achieving only 43% [32]. Therefore, to inhibit 100% of the fungus growth, more concentration of encapsulated ITZ would be necessary than the free one. Nevertheless, the diffusion of more ITZ from the hydrophobic NPs core to the hydrophilic phase is hindered because the concentration of ITZ in the assay with NPs may be achieving the maximum saturation point. Furthermore, the NPs components, considering the controls' results, didn't show any inhibition. On the contrary, the empty NPs and each of the components of the NPs (0.125 µg/mL) showed a slight inhibitory action against the fungal growth. The inhibition may be explained because a high concentration of NPs or their components binds easier to fungus molecules, including proteins, affecting their growth [32].

In vitro model, the PLGA75:25-TPGS-pH5 NPs loading with ITZ showed an initial burst release followed by a slower and more persistent release [32]. The degradation of PLGA-NPs in vitro is caused through a bulk erosion mechanism [48], which consists of three phases. In the first phase, the release is due to the cleavage of the ester bonds in the polymer, decreasing their molecular weight without loss of formulation mass. In the second phase, the polymer loses mass because of the microenvironment acidification, forming oligomers. Finally, the solubilization of the polymer is caused due to the fragmentation of oligomers in monomers [49]. However, in vivo model, the release of ITZ from the nanoparticles could be 100%. This fact is due to the degradation of the ester linkages of the polymer occurring acceleratedly because the tissue cells recognize the NPs as foreign agents, producing enzymes and free radicals that can promote the degradation of PLGA. On the other hand, PLGA oligomers in the blood are more soluble compared to the in vitro medium [50]. Additionally, the ITZ encapsulated in the in vivo model could produce a higher effect than free ITZ due to the nanosystem protecting the degradation of ITZ at low pH. Some reports showed the activity of ITZ reduced under pH 5, being the MIC higher than the MIC under pH 7 [51].

When evaluating cytotoxicity, bare- and empty-NPs did not affect the viability in human and mice macrophages at different concentrations (Figure 2A,B). The NPs material precursors are one of the main factors affecting NPs toxicity. The NPs evaluated in this study mainly consist of the PLGA polymer with low toxicity due to their high degree of biodegradability, biocompatibility, and stability under physiological conditions. PLGA is biodegraded to lactic and glycolic acid in the physiological environment by hydrolysis of its ester bonds, which can be further metabolized through the Krebs cycle in carbon dioxide and water. Therefore, different pharmaceutical products based on this polymer currently exist with a long history of clinical use [12,52–56]. Moreover, functionalized NPs did not show a cytotoxic effect, suggesting that the antibody didn't induce the wrong activation of cells. The results confirm the safe use of NPs in vitro and in vivo models (Figure 3).

The UFC assay showed the effectiveness of functionalized NPs in macrophages infected with *H. capsulatum* (Figure 4A,B). The functionalized NPs reduced the amount of UFC significantly concerning bare-NPs at a lower concentration of 0.007 µg/mL. This effect might be explained due to the significant increase of endocytosis in the macrophages of functionalized NPs, as we reported elsewhere [32] and also demonstrated by flow cytometry and the cLSM assay in this study (Figures S1 and S2). Furthermore, fluorescence microscopy images (Figure 5A,B) and the Nile red area quantification (Figure 5C) showed more macrophages with intracellular co-location of NPs and fungus. Therefore, we demonstrated that before 6 h of incubation, the NPs cell uptake was enough to get an antifungal effect compared with the bare-NPs even after 24 h (Figure 4B). Different types of macrophages, which are characterized by their heterogeneity, plasticity, and the type and amount of diverse receptors, are known to exist. In this sense, in mice, peritoneal macrophages present an intermediate expression of F4/80, low expression of mannose, and no expression of siglec-F receptor compared with lung macrophages [57].

It has been reported that the ability of NPs to modulate the immune response is a product of their physicochemical properties (size, charge, structure, composition, etc.) Other factors that also influence it are modifications in its surface and the therapeutic load [12,58,59]. The immune system cells recognize many of the components of NPs as foreign, triggering different immune responses through a complex process. Therefore, subsequent experiments aimed to determine analysis whether the administration of NPs-F4/80 alters the expression of genes related to the inflammatory and anti-inflammatory response in J774A.1 macrophages infected with *H. capsulatum* yeasts by PCR and explore whether this effect could be used as a mechanism that acts synergistically with the drug and enhances the effect of the treatment.

Macrophages are considered the most important effector cells in host resistance against histoplasmosis, participating in innate and adaptive immunity. Macrophages express a variety of pattern recognition receptors (PRRs) on their surface, including toll-like receptors (TLRs) and C-type lectin receptors (CRLs) that recognize pathogen-associated molecular patterns (PAMPs) of *H. capsulatum* such as Hsp60, Yps3, nucleic acids, glucans, mannan, among others, promoting its phagocytosis and the production of cytokines [60].

These cells express TLR2, the first documented TLR to bind to *H. capsulatum* and trigger an immune response. Its ligand is the surface protein Yps3. After its recognition, the adapter protein MyD88 is activated, a link between the extracellular receptor and the internal signaling pathway that leads to the transcription factor NF-κB entering the nucleus and activating several innate immune response genes such as TNF-α and IFN-γ [61]. They also express Dectin-1, a member of the C-type lectin receptor (CRL), which specifically binds to β-glucans, one of the main components of the *H. capsulatum* cell wall and the most inflammatory [62]. Upon recognition of β-glucans, Dectin-1 mediates Sky-dependent NFAT and NF-κB pathways and Sky-independent Raf-1 pathways to promote cytokine and chemokine production.

Our results show that free ITZ in co-cultures significantly increased the expression of IL-1β (Figure 6A). Other studies have reported that ITZ may directly modulate the immune response [63–65]. Similar results have been reported in other models; in an in vitro sepsis model, co-stimulation of ITZ and lipopolysaccharides (LPS) increased gene expression levels of IL-1β [65]. LPS are an essential component of the outer membrane of Gram-negative bacteria and act as a potent inducer of pro-inflammatory responses in monocytes and macrophages [66]. Similarly, this study's results indicated a co-stimulatory effect of *H. capsulatum* and ITZ (Figure 6A). It was also observed that when the co-cultures were treated with different nanoformulations, the expression of IL-1β was significantly reduced compared to free ITZ. It suggests that modulation of the expression of this cytokine may be more related to the nanocapsule than with direct activation by the anti-F4/80 antibody since empty and bare-NPs also showed a reduction of IL-1β expression (Figure 6A).

IFN-γ plays a central role in HPM control, being the crucial cytokine to initiate the effector phase of cell-mediated immunity by activating macrophages to enhance microbicidal activity [67]. The modulation of a pro-inflammatory immune response was shown to be beneficial in controlling intracellular pathogens [27]. Darwich L. et al. demonstrated in an in vitro model that macrophages derived from monocytes after treatment with IL-12 and IL-18 produced IFN-γ. Therefore, human macrophages and lymphoid cells contributed to the IFN-γ response, providing another link between innate and acquired immune responses [68]. In our model, we didn't measure the expression of IL-12 and IL-18. However, some reports demonstrated that the peritoneal macrophages express IL-18 precursor and usually activate during infectious diseases because most pathogens trigger the synthesis of mature IL-1β, which, in turn, activates caspase-1 to cleave pro-IL-18 to mature IL-18, which is then secreted [69,70]. Additionally, some reports showed macrophages infected with *H. capsulatum* produce IL-12 in low concentrations [71]. In agreement, it was observed that *H. capsulatum* in the macrophages induced the expression of IFN-γ (Figure 6B). Interestingly, treatment with bare-NPs and nanoencapsulated ITZ induced a higher cytokine expression in co-cultures, demonstrating the immunomodulatory role ITZ can exert. The immunomodulation effect in both treatments (free ITZ and bare-NPs) might be related to the modulation of IL-12 and IL-18, but more study is necessary to demonstrate it. However, functionalized NPs reduced the expression of this cytokine, indicating the possible role of the anti-F4/80 antibody as a binding molecule with immunomodulatory properties (Figure 6B). Consistent with our results, Warschkau et al., in an in vitro model of Listeriosis, showed that incubation with an antibody directed against the murine macrophage surface glycoprotein F4/80 reduced IFN-γ expression levels [72]. The F4/80 antigen is widely used as a specific marker for macrophages. Macrophages from the spleen (red pulp), lung, liver, peritoneal cavity, and nervous system express the F4/80 receptor. The interpreted amino acid sequence indicates a seven-transmembrane molecule with homologies to human EMR1 and CD97. As in vivo F4/80 is downregulated following infection with *Bacillus Calmette-Guerin*, its expression depends on the activation state of macrophages [73]. However, no specific ligands or biological functions of the F4/80 antigen have been described [72].

Additionally, the results with bare-NP can be related to the phenomenon where the endocytosis of macrophages of foreign objects can be modulated by characteristics such as size, shape, surface charge, and hydrophobicity/hydrophilicity because they regulate the interaction with physiological proteins and the protein corona formation [74–76]. In our case, it was the proteins from the Fetal Bovine serum (FBS). The type of proteins composing the protein corona may affect on endocytosis pathway [12,77]. After the NPs endocytosis, it depends on which way the macrophages are activated and produce different cytokines (pro-inflammatory or anti-inflammatory) [78,79]. Acharya D. et al. observed significant upregulation of pro-inflammatory mediators during phagocytosis. Using qPCR, they determined that complement receptor-mediated phagocytosis increased levels of TNF-α, IL-1β, IL-6, and MMP-9, compared to FcγR-mediated phagocytosis and control unstimulated cells [79].

The Th17 response is important in controlling many fungal infections. However, during PMH, Th17 is beneficial but not essential to controlling the fungus [44]. Free ITZ treatment significantly increased IL-6 expression (Figure 6C). Interestingly, treatment with the ITZ nanoformulations (bare-NPs and F4/80-coated-NPs) significantly reduced IL-6 expression compared to free ITZ (Figure 6C). The foregoing indicates that the encapsulation of ITZ into NPs allows a sustained and controlled drug release so that the immunomodulatory properties of free ITZ are potentially mitigated with nanoencapsulation.

TNF-α is a protective cytokine that exerts multiple effects during *H. capsulatum* infection, including activation of phagocytic cells, induction of apoptosis, and control of the CD4+ phenotype [67]. Blockade of TNF-α has been shown to reduce nitric oxide production by macrophages during primary infection. Neutralization of TNF-α suppresses the ability of murine T cells to mediate protection against *H. capsulatum* [80,81]. No statistically significant differences were observed between the different treatments and the expression of TNF-α. These results indicate that the treatments evaluated would not affect the immune response of the macrophage concerning the expression of this cytokine.

The expression of genes encoding anti-inflammatory molecules such as IL-10 was also evaluated. Although it is a crucial cytokine in limiting excessive immune activation, IL-10 exacerbates infection by preventing the clearance of *H. capsulatum*. Furthermore, it negatively affected the development of the protective Th1 response [82]. ITZ administrated in co-culture produced a significantly higher expression of IL-10 (Figure 6D), confirming that *H. capsulatum* and ITZ could exert a co-stimulatory effect. In addition, it was observed that the treatment of co-cultures with the different nanoformulations (empty-, bare- and F4/80-coated NPs) significantly decreased the expression of IL-10 compared to untreated co-cultures and co-cultures treated with ITZ. This indicates that the encapsulation of ITZ in NPs allows a sustained and controlled release of the drug, modulating its effect on inducing this cytokine in macrophages (Figure 6D). This result suggests that the NPs evaluated here can modulate the anti-inflammatory immune response during HPM, associated with a decreased inflammatory response and potential infection control.

Overall, this study opens the door to implementing new therapeutic strategies using antifungal encapsulated into functionalized NPs capable of directing the drug towards cells or organs infected by *H. capsulatum*, modulating the immune response, thus enhancing the effect of the therapy. Yet, it is necessary to continue improving the formulation and getting better knowledge about immunomodulation from this kind of therapy; it is crucial to know the scope and actual benefit of this treatment.

5. Conclusions

The encapsulation of antifungals into functionalized nanocarriers offers an up-and-coming treatment alternative to enhance the antimicrobial effect by site-specific targeting drugs in complex sites where pathogens are harbored, potentially reducing the dose and the therapy time. Our study demonstrated the improved antifungal effect by reducing the amount of ITZ to control *H. capsulatum* in co-culture with peritoneal macrophages. This effect can be related to the encapsulated ITZ NPs, which can generate a modulation in the immune response and antibody-coating enhanced natural and adequate antifungal response in the cells, exerting a synergistic effect that prevents the growth of the fungus at the intracellular level. Yet, it is crucial to study how nanosystems may influence the immune response more in-depth. However, it is necessary to develop new studies focused on further evaluating how this type of NPs interacts with macrophages and induces the expression of genes related to the different immunological profiles, allowing a deeper understanding of the type of immune response being generated. Similarly, it would be valuable to evaluate in vivo the antifungal and immunomodulatory effects of NPs using the experimental model of histoplasmosis in mice.

Supplementary Materials: The following supporting information can be downloaded at: https://www.mdpi.com/article/10.3390/pharmaceutics14091932/s1. Figure S1: Endocytosis evaluation of PLGA75:25-TPGS-pH5-NPs functionalized with (or without) anti-F4/80 antibody into J774A.1 macrophage by flow cytometry. Figure S2: Endocytosis evaluation of anti-F4/80 antibody-functionalized PLGA75:25-TPGS-pH5-NPs encapsulating Nile red into J774A.1 macrophage by confocal laser scanning microscopy. Figure S3: Endocytosis evaluation of anti-F4/80 antibody-functionalized PLGA75:25-TPGS-pH5-NPs encapsulating Nile red into C2C12 mice muscle cells by fluorescence microscopy. Figure S4: Endocytosis evaluation of anti-F4/80 antibody-functionalized PLGA75:25-TPGS-pH5-NPs encapsulating Nile red into CHO Hamster ovary cells by fluorescence microscopy.

Author Contributions: J.O., S.P.M., T.W.N. and L.E.C. participated in conceptualization. S.P.M. designed and performed most of the experiments, and D.L. contributed to the experimental design and the in vitro assay. S.P.M. made data curation and formal analysis and wrote the manuscript's first draft with D.L.'s contributions. J.O. supervised the work and edited the manuscript. J.O. and L.E.C. secured the funding. All authors have read and agreed to the published version of the manuscript.

Funding: This work was partially supported by Minciencias (Project 2213-777-57106), the University of Antioquia and the Max Planck Society through the Cooperation agreement 566-1, 2014. The authors thank Ruta N and EPM for hosting the tandem groups.

Institutional Review Board Statement: Not applicable.

Informed Consent Statement: Not applicable.

Data Availability Statement: All data relevant to the publication are included.

Acknowledgments: We thank Diana Zarate and Rogelio Hernandez Pando for their advice and discussions. We thank the young researcher Yeraldin López for all his technical support in the experimental part. We thank the Group of Volker Mailander of Max Planck Institute for Polymer Research, Mainz-Germany, for technical support. We also thank the Evonik Company for donating the polymers used in the study.

Conflicts of Interest: The authors declare no conflict of interest.

References

1. Binnemars-Postma, K.; Storm, G.; Prakash, J. Nanomedicine strategies to target tumor-associated macrophages. *Int. J. Mol. Sci.* **2017**, *18*, 979. [CrossRef] [PubMed]
2. Kapoor, D.N.; Bhatia, A.; Kaur, R.; Sharma, R.; Kaur, G.; Dhawan, S. PLGA: A unique polymer for drug delivery. *Ther. Deliv.* **2015**, *6*, 41–58. [CrossRef] [PubMed]
3. Dos Anjos Cassado, A. F4/80 as a Major Macrophage Marker: The Case of the Peritoneum and Spleen. *Results Probl. Cell Differ.* **2017**, *62*, 161–179. [CrossRef] [PubMed]
4. Adenis, A.A.; Aznar, C.; Couppié, P. Histoplasmosis in HIV-Infected Patients: A Review of New Developments and Remaining Gaps. *Curr. Trop. Med. Rep.* **2014**, *1*, 119–128. [CrossRef]

5. Ashraf, N.; Kubat, R.C.; Poplin, V.; Adenis, A.A.; Denning, D.W.; Wright, L.; McCotter, O.; Schwartz, I.S.; Jackson, B.R.; Chiller, T.; et al. Re-drawing the Maps for Endemic Mycoses. *Mycopathologia* **2020**, *185*, 843–865. [CrossRef]
6. Tobón, Á.M. Protocolo de estudio y manejo de histoplasmosis. *Infectio* **2012**, *16*, 126–128. [CrossRef]
7. Adenis, A.A.; Valdes, A.; Cropet, C.; McCotter, O.Z.; Derado, G.; Couppie, P.; Chiller, T.; Nacher, M. Burden of HIV-associated histoplasmosis compared with tuberculosis in Latin America: A modelling study. *Lancet. Infect. Dis.* **2018**, *18*, 1150–1159. [CrossRef]
8. Wheat, L.J.; Freifeld, A.G.; Kleiman, M.B.; Baddley, J.W.; McKinsey, D.S.; Loyd, J.E.; Kauffman, C.A. Clinical Practice Guidelines for the Management of Patients with Histoplasmosis: 2007 Update by the Infectious Diseases Society of America. *Clin. Infect. Dis.* **2007**, *45*, 807–825. [CrossRef]
9. Girmenia, C.; Iori, A.P. An update on the safety and interactions of antifungal drugs in stem cell transplant recipients. *Expert Opin. Drug Saf.* **2017**, *16*, 329–339. [CrossRef]
10. Kyriakidis, I.; Tragiannidis, A.; Munchen, S.; Groll, A.H. Clinical hepatotoxicity associated with antifungal agents. *Expert Opin. Drug Saf.* **2017**, *16*, 149–165. [CrossRef]
11. Zazo, H.; Colino, C.I.; Lanao, J.M. Current applications of nanoparticles in infectious diseases. *J. Control. Release* **2016**, *224*, 86–102. [CrossRef] [PubMed]
12. Sánchez, A.; Mejía, S.P.; Orozco, J. Recent Advances in Polymeric Nanoparticle-Encapsulated Drugs against Intracellular Infections. *Molecules* **2020**, *25*, 3760. [CrossRef] [PubMed]
13. Ashley, E.S.D.; Lewis, R.; Lewis, J.S.; Martin, C.; Andes, D. Pharmacology of Systemic Antifungal Agents. *Clin. Infect. Dis.* **2006**, *43*, S28–S39. [CrossRef]
14. National Center for Biotechnology Information. PubChem Database. Chloroquine, CID=2719. Available online: https://pubchem. ncbi.nlm.nih.gov/compound/Chloroquine (accessed on 5 May 2020).
15. Allen, D.; Wilson, D.; Drew, R.; Perfect, J. Azole antifungals: 35 years of invasive fungal infection management. *Expert Rev. Anti. Infect. Ther.* **2015**, *13*, 787–798. [CrossRef]
16. Ling, X.; Huang, Z.; Wang, J.; Xie, J.; Feng, M.; Chen, Y.; Abbas, F.; Tu, J.; Wu, J.; Sun, C. Development of an itraconazole encapsulated polymeric nanoparticle platform for effective antifungal therapy. *J. Mater. Chem. B* **2016**, *4*, 1787–1796. [CrossRef]
17. Tarudji, A.W.; Kievit, F.M. Chapter 3—Active targeting and transport. In *Nanoparticles for Biomedical Applications*; Micro and Nano Technologies; Chung, E.J., Leon, L., Rinaldi, C., Eds.; Elsevier: Amsterdam, The Netherlands, 2020; pp. 19–36, ISBN 978-0-12-816662-8.
18. Fernando, S.S.N.; Gunasekara, C.; Holton, J. Antimicrobial Nanoparticles: Applications and mechanisms of action. *Sri Lankan J. Infect. Dis.* **2018**, *8*, 2. [CrossRef]
19. Jayaraman, R. Antibiotic resistance: An overview of mechanisms and a paradigm shift. *Curr. Sci.* **2009**, *96*, 1475–1484.
20. Blecher, K.; Nasir, A.; Friedman, A. The growing role of nanotechnology in combating infectious disease. *Virulence* **2011**, *2*, 395–401. [CrossRef]
21. Ogier, J.; Arnauld, T.; Doris, E. Recent advances in the field of nanometric drug carriers. *Future Med. Chem.* **2009**, *1*, 693–711. [CrossRef]
22. Farokhzad, O.C.; Langer, R. Impact of Nanotechnology on Drug Delivery. *ACS Nano* **2009**, *3*, 16–20. [CrossRef]
23. Fernández, M.; Orozco, J. Advances in Functionalized Photosensitive Polymeric Nanocarriers. *Polymers* **2021**, *13*, 2464. [CrossRef] [PubMed]
24. Mena-Giraldo, P.; Orozco, J. Polymeric Micro/Nanocarriers and Motors for Cargo Transport and Phototriggered Delivery. *Polymers* **2021**, *13*, 3920. [CrossRef] [PubMed]
25. Pandey, R.; Ahmad, Z.; Sharma, S.; Khuller, G.K. Nano-encapsulation of azole antifungals: Potential applications to improve oral drug delivery. *Int. J. Pharm.* **2005**, *301*, 268–276. [CrossRef]
26. Yi, Y.; Yoon, H.J.; Kim, B.O.; Shim, M.; Kim, S.-O.; Hwang, S.-J.; Seo, M.H. A mixed polymeric micellar formulation of itraconazole: Characteristics, toxicity and pharmacokinetics. *J. Control. Release* **2007**, *117*, 59–67. [CrossRef]
27. Barros, D.; Costa Lima, S.A.; Cordeiro-da-Silva, A. Surface functionalization of polymeric nanospheres modulates macrophage activation: Relevance in leishmaniasis therapy. *Nanomedicine* **2015**, *10*, 387–403. [CrossRef] [PubMed]
28. Ribeiro, T.G.; Franca, J.R.; Fuscaldi, L.L.; Santos, M.L.; Duarte, M.C.; Lage, P.S.; Martins, V.T.; Costa, L.E.; Fernandes, S.O.A.; Cardoso, V.N.; et al. An optimized nanoparticle delivery system based on chitosan and chondroitin sulfate molecules reduces the toxicity of amphotericin B and is effective in treating tegumentary leishmaniasis. *Int. J. Nanomed.* **2014**, *9*, 5341–5353. [CrossRef]
29. Gómez-Sequeda, N.; Torres, R.; Ortiz, C. Synthesis, characterization, and in vitro activity against Candida spp. of fluconazole encapsulated on cationic and conventional nanoparticles of poly(lactic-co-glycolic acid). *Nanotechnol. Sci. Appl.* **2017**, *10*, 95–104. [CrossRef]
30. Ungaro, F.; d'Angelo, I.; Coletta, C.; d'Emmanuele di Villa Bianca, R.; Sorrentino, R.; Perfetto, B.; Tufano, M.A.; Miro, A.; La Rotonda, M.I.; Quaglia, F. Dry powders based on PLGA nanoparticles for pulmonary delivery of antibiotics: Modulation of encapsulation efficiency, release rate and lung deposition pattern by hydrophilic polymers. *J. Control. Release* **2012**, *157*, 149–159. [CrossRef]
31. Sun, Y.; Chen, D.; Pan, Y.; Qu, W.; Hao, H.; Wang, X.; Liu, Z.; Xie, S. Nanoparticles for antiparasitic drug delivery. *Drug Deliv.* **2019**, *26*, 1206–1221. [CrossRef]

32. Mejía, S.P.; Sánchez, A.; Vásquez, V.; Orozco, J. Functional Nanocarriers for Delivering Itraconazole Against Fungal Intracellular Infections. *Front. Pharmacol.* **2021**, *12*, 1520. [CrossRef]

33. Khan, M.A.; Jabeen, R.; Nasti, T.H.; Mohammad, O. Enhanced anticryptococcal activity of chloroquine in phosphatidylserine-containing liposomes in a murine model. *J. Antimicrob. Chemother.* **2005**, *55*, 223–228. [CrossRef]

34. Puerta, J.; Pino-Tamayo, P.; Arango, J.; Gonzalez, A. Depletion of Neutrophils Promotes the Resolution of Pulmonary Inflammation and Fibrosis in Mice Infected with Paracoccidioides brasiliensis. *PLoS ONE* **2016**, *11*, e0163985. [CrossRef]

35. Khatua, S.; Simal-Gandara, J.; Acharya, K. Understanding immune-modulatory efficacy in vitro. *Chem. Biol. Interact.* **2022**, *352*, 109776. [CrossRef] [PubMed]

36. Casadevall, A. Evolution of Intracellular Pathogens. *Annu. Rev. Microbiol.* **2008**, *62*, 19–33. [CrossRef]

37. Edwards, J.A.; Kemski, M.M.; Rappleye, C.A. Identification of an aminothiazole with antifungal activity against intracellular Histoplasma capsulatum. *Antimicrob. Agents Chemother.* **2013**, *57*, 4349–4359. [CrossRef] [PubMed]

38. Kischkel, B.; Rossi, S.A.; Santos, S.R.; Nosanchuk, J.D.; Travassos, L.R.; Taborda, C.P. Therapies and Vaccines Based on Nanoparticles for the Treatment of Systemic Fungal Infections. *Front. Cell. Infect. Microbiol.* **2020**, *10*, 463. [CrossRef]

39. Rautemaa-Richardson, R.; Richardson, M.D. Systemic fungal infections. *Medicine* **2017**, *45*, 757–762. [CrossRef]

40. Brunet, K.; Alanio, A.; Lortholary, O.; Rammaert, B. Reactivation of dormant/latent fungal infection. *J. Infect.* **2018**, *77*, 463–468. [CrossRef]

41. Costa-Gouveia, J.; Aínsa, J.A.; Brodin, P.; Lucía, A. How can nanoparticles contribute to antituberculosis therapy? *Drug Discov. Today* **2017**, *22*, 600–607. [CrossRef]

42. Singh, L.; Kruger, H.G.; Maguire, G.E.M.; Govender, T.; Parboosing, R. The role of nanotechnology in the treatment of viral infections. *Ther. Adv. Infect. Dis.* **2017**, *4*, 105–131. [CrossRef]

43. Gao, W.; Chen, Y.; Zhang, Y.; Zhang, Q.; Zhang, L. Nanoparticle-based local antimicrobial drug delivery. *Adv. Drug Deliv. Rev.* **2018**, *127*, 46–57. [CrossRef] [PubMed]

44. Batalha, I.L.; Bernut, A.; Schiebler, M.; Ouberai, M.M.; Passemar, C.; Klapholz, C.; Kinna, S.; Michel, S.; Sader, K.; Castro-Hartmann, P.; et al. Polymeric nanobiotics as a novel treatment for mycobacterial infections. *J. Control. Release* **2019**, *314*, 116–124. [CrossRef] [PubMed]

45. Reynolds, N.; Dearnley, M.; Hinton, T.M. Polymers in the Delivery of siRNA for the Treatment of Virus Infections. *Top. Curr. Chem.* **2017**, *375*, 38. [CrossRef] [PubMed]

46. Tonigold, M.; Simon, J.; Estupiñán, D.; Kokkinopoulou, M.; Reinholz, J.; Kintzel, U.; Kaltbeitzel, A.; Renz, P.; Domogalla, M.P.; Steinbrink, K.; et al. Pre-adsorption of antibodies enables targeting of nanocarriers despite a biomolecular corona. *Nat. Nanotechnol.* **2018**, *13*, 862–869. [CrossRef]

47. Duan, X.; Li, Y. Physicochemical characteristics of nanoparticles affect circulation, biodistribution, cellular internalization, and trafficking. *Small* **2013**, *9*, 1521–1532. [CrossRef]

48. Pitt, C.G.; Gratzl, M.M.; Kimmel, G.L.; Surles, J.; Schindler, A. Aliphatic polyesters II. The degradation of poly (DL-lactide), poly (epsilon-caprolactone), and their copolymers in vivo. *Biomaterials* **1981**, *2*, 215–220. [CrossRef]

49. Park, T.G. Degradation of poly(lactic-co-glycolic acid) microspheres: Effect of copolymer composition. *Biomaterials* **1995**, *16*, 1123–1130. [CrossRef]

50. Guo, L.-Y.; Yan, S.-Z.; Li, Q.; Xu, Q.; Lin, X.; Qi, S.-S.; YU, S.; Chen, S.-L. Poly(lactic-co-glycolic) acid nanoparticles improve oral bioavailability of hypocrellin A in rat. *RSC Adv.* **2017**, *7*, 42073–42082. [CrossRef]

51. Te Dorsthorst, D.T.A.; Mouton, J.W.; van den Beukel, C.J.P.; van der Lee, H.A.L.; Meis, J.F.G.M.; Verweij, P.E. Effect of pH on the in vitro activities of amphotericin B, itraconazole, and flucytosine against Aspergillus isolates. *Antimicrob. Agents Chemother.* **2004**, *48*, 3147–3150. [CrossRef]

52. Kamaly, N.; Yameen, B.; Wu, J.; Farokhzad, O.C. Degradable controlled-release polymers and polymeric nanoparticles: Mechanisms of controlling drug release. *Chem. Rev.* **2016**, *116*, 2602–2663. [CrossRef]

53. Peres, C.; Matos, A.I.; Conniot, J.; Sainz, V.; Zupančič, E.; Silva, J.M.; Graça, L.; Gaspar, R.S.; Préat, V.; Florindo, H.F. Poly (lactic acid)-based particulate systems are promising tools for immune modulation. *Acta Biomater.* **2017**, *48*, 41–57. [CrossRef] [PubMed]

54. Sharma, S.; Parmar, A.; Kori, S.; Sandhir, R. PLGA-based nanoparticles: A new paradigm in biomedical applications. *TrAC Trends Anal. Chem.* **2016**, *80*, 30–40. [CrossRef]

55. Bazile, D.V.; Ropert, C.; Huve, P.; Verrecchia, T.; Marlard, M.; Frydman, A.; Veillard, M.; Spenlehauer, G. Body distribution of fully biodegradable [14C]-poly(lactic acid) nanoparticles coated with albumin after parenteral administration to rats. *Biomaterials* **1992**, *13*, 1093–1102. [CrossRef]

56. Esmaeili, F.; Ghahremani, M.H.; Esmaeili, B.; Khoshayand, M.R.; Atyabi, F.; Dinarvand, R. PLGA nanoparticles of different surface properties: Preparation and evaluation of their body distribution. *Int. J. Pharm.* **2008**, *349*, 249–255. [CrossRef] [PubMed]

57. Gordon, S.; Plüddemann, A. Tissue macrophages: Heterogeneity and functions. *BMC Biol.* **2017**, *15*, 53. [CrossRef]

58. Elsabahy, M.; Wooley, K.L. Cytokines as biomarkers of nanoparticle immunotoxicity. *Chem. Soc. Rev.* **2013**, *42*, 5552–5576. [CrossRef]

59. Qasim, M.; Jang, Y.; Kang, S.; Moon, J.; Kim, W.; Park, H. Modulation of immune responses with nanoparticles and reduction of their immunotoxicity. *Biomater. Sci.* **2020**, *8*, 1490–1501. [CrossRef]

60. Ray, S.C.; Rappleye, C.A. Flying under the radar: Histoplasma capsulatum avoidance of innate immune recognition. *Semin. Cell Dev. Biol.* **2019**, *89*, 91–98. [CrossRef]

61. Aravalli, R.N.; Hu, S.; Woods, J.P.; Lokensgard, J.R. Histoplasma capsulatum yeast phase-specific protein Yps3p induces Toll-like receptor 2 signaling. *J. Neuroinflammation* **2008**, *5*, 30. [CrossRef]

62. Goyal, S.; Castrillón-Betancur, J.C.; Klaile, E.; Slevogt, H. The Interaction of Human Pathogenic Fungi with C-Type Lectin Receptors. *Front. Immunol.* **2018**, *9*, 1261. [CrossRef]

63. Puerta-Arias, J.D.; Pino-Tamayo, P.A.; Arango, J.C.; Salazar-Peláez, L.M.; González, A. Itraconazole in combination with neutrophil depletion reduces the expression of genes related to pulmonary fibrosis in an experimental model of paracoccidioidomycosis. *Med. Mycol.* **2018**, *56*, 579–590. [CrossRef] [PubMed]

64. Naranjo, T.; Lopera, D.; Zuluaga, A.; Cano, L. Immunomodulatory activity of itraconazole in lung. *Trop. J. Pharm. Res.* **2016**, *15*, 2603–2609. [CrossRef]

65. Muenster, S.; Bode, C.; Diedrich, B.; Jahnert, S.; Weisheit, C.; Steinhagen, F.; Frede, S.; Hoeft, A.; Meyer, R.; Boehm, O.; et al. Antifungal antibiotics modulate the pro-inflammatory cytokine production and phagocytic activity of human monocytes in an in vitro sepsis model. *Life Sci.* **2015**, *141*, 128–136. [CrossRef] [PubMed]

66. Zhongshuang, H.; Murakami, T.; Suzuki, K.; Tamura, H.; Kuwahara, K.; Iba, T.; Nagaoka, I. Antimicrobial Cathelicidin Peptide LL-37 Inhibits the LPS/ATP-Induced Pyroptosis of Macrophages by Dual Mechanism. *PLoS ONE* **2014**, *9*, e85765. [CrossRef]

67. Horwath, M.C.; Fecher, R.A.; Deepe, G.S.J. Histoplasma capsulatum, lung infection and immunity. *Future Microbiol.* **2015**, *10*, 967–975. [CrossRef]

68. Darwich, L.; Coma, G.; Peña, R.; Bellido, R.; Blanco, E.J.J.; Este, J.A.; Borras, F.E.; Clotet, B.; Ruiz, L.; Rosell, A.; et al. Secretion of interferon-gamma by human macrophages demonstrated at the single-cell level after costimulation with interleukin (IL)-12 plus IL-18. *Immunology* **2009**, *126*, 386–393. [CrossRef]

69. Dinarello, C.; Novick, D.; Kim, S.; Kaplanski, G. Interleukin-18 and IL-18 Binding Protein. *Front. Immunol.* **2013**, *4*, 289. [CrossRef]

70. Vecchié, A.; Bonaventura, A.; Toldo, S.; Dagna, L.; Dinarello, C.A.; Abbate, A. IL-18 and infections: Is there a role for targeted therapies? *J. Cell. Physiol.* **2021**, *236*, 1638–1657. [CrossRef]

71. Kroetz, D.N.; Deepe, G.S. The role of cytokines and chemokines in Histoplasma capsulatum infection. *Cytokine* **2012**, *58*, 112–117. [CrossRef]

72. Warschkau, H.; Kiderlen, A.F. A monoclonal antibody directed against the murine macrophage surface molecule F4/80 modulates natural immune response to Listeria monocytogenes. *J. Immunol.* **1999**, *163*, 3409–3416.

73. Ezekowitz, R.A.; Austyn, J.; Stahl, P.D.; Gordon, S. Surface properties of bacillus Calmette-Guérin-activated mouse macrophages. Reduced expression of mannose-specific endocytosis, Fc receptors, and antigen F4/80 accompanies induction of Ia. *J. Exp. Med.* **1981**, *154*, 60–76. [CrossRef]

74. Wolfram, J.; Yang, Y.; Shen, J.; Moten, A.; Chen, C.; Shen, H.; Ferrari, M.; Zhao, Y. The nano-plasma interface: Implications of the protein corona. *Colloids Surf. B. Biointerfaces* **2014**, *124*, 17–24. [CrossRef]

75. Yuan, Q.; Wang, Y.; Song, R.; Hou, X.; Yu, K.; Zheng, J.; Zhang, J.; Pu, X.; Han, J.; Zong, L. Study on Formulation, in vivo Exposure, and Passive Targeting of Intravenous Itraconazole Nanosuspensions. *Front. Pharmacol.* **2019**, *10*, 225. [CrossRef]

76. Lima, T.; Bernfur, K.; Vilanova, M.; Cedervall, T. Understanding the Lipid and Protein Corona Formation on Different Sized Polymeric Nanoparticles. *Sci. Rep.* **2020**, *10*, 1129. [CrossRef]

77. Rennick, J.J.; Johnston, A.P.R.; Parton, R.G. Key principles and methods for studying the endocytosis of biological and nanoparticle therapeutics. *Nat. Nanotechnol.* **2021**, *16*, 266–276. [CrossRef]

78. Kapetanovic, R.; Nahori, M.-A.; Balloy, V.; Fitting, C.; Philpott, D.; Cavaillon, J.-M.; Adib-Conquy, M. Contribution of Phagocytosis and Intracellular Sensing for Cytokine Production by Staphylococcus aureus-Activated Macrophages. *Infect. Immun.* **2007**, *75*, 830–837. [CrossRef]

79. Acharya, D.; Li, X.R.; Heineman, R.E.-S.; Harrison, R.E. Complement Receptor-Mediated Phagocytosis Induces Proinflammatory Cytokine Production in Murine Macrophages. *Front. Immunol.* **2020**, *10*, 3049. [CrossRef]

80. Allendoerfer, R.; Deepe, G.S.J. Blockade of endogenous TNF-alpha exacerbates primary and secondary pulmonary histoplasmosis by differential mechanisms. *J. Immunol.* **1998**, *160*, 6072–6082.

81. Deepe, G.S.J.; Gibbons, R.S. T cells require tumor necrosis factor-alpha to provide protective immunity in mice infected with Histoplasma capsulatum. *J. Infect. Dis.* **2006**, *193*, 322–330. [CrossRef]

82. Deepe, G.S.J.; Gibbons, R.S. Protective and memory immunity to Histoplasma capsulatum in the absence of IL-10. *J. Immunol.* **2003**, *171*, 5353–5362. [CrossRef]

 *pharmaceutics*

Article

Coating of SPIONs with a Cysteine-Decorated Copolyester: A Possible Novel Nanoplatform for Enzymatic Release

Jeovandro Maria Beltrame [1,*], Brena Beatriz Pereira Ribeiro [1], Camila Guindani [2], Graziâni Candiotto [3], Karina Bettega Felipe [4], Rodrigo Lucas [5], Alexandre D'Agostini Zottis [6], Eduardo Isoppo [7], Claudia Sayer [1,*] and Pedro Henrique Hermes de Araújo [1]

1. Department of Chemical Engineering and Food Engineering, Federal University of Santa Catarina (EQA/UFSC), Florianópolis 88040-900, SC, Brazil
2. Chemical Engineering Program, COPPE, Federal University of Rio de Janeiro (PEQ/COPPE/UFRJ), Rio de Janeiro 21941-972, RJ, Brazil
3. Institute of Physics, Federal University of Rio de Janeiro, Rio de Janeiro 21941-909, RJ, Brazil
4. Department of Clinic Analysis, Federal University of Paraná DAC/UFPR, Curitiba 80.210-170, PR, Brazil
5. Foundation Osvaldo Cruz (FIOCRUZ), Curitiba 81350-010, PR, Brazil
6. Academic Department of Health and Services, NANOTEC Group, Federal Institute of Santa Catarina, Florianópolis 88020-300, SC, Brazil
7. Central Laboratory of Electron Microscopy, Federal University of Santa Catarina, LCME/UFSC, Florianópolis 88040-900, SC, Brazil
* Correspondence: jbeltrame20@gmail.com (J.M.B.); claudia.sayer@ufsc.br (C.S.)

Citation: Beltrame, J.M.; Ribeiro, B.B.P.; Guindani, C.; Candiotto, G.; Felipe, K.B.; Lucas, R.; Zottis, A.D.; Isoppo, E.; Sayer, C.; de Araújo, P.H.H. Coating of SPIONs with a Cysteine-Decorated Copolyester: A Possible Novel Nanoplatform for Enzymatic Release. *Pharmaceutics* **2023**, *15*, 1000. https://doi.org/10.3390/pharmaceutics15031000

Academic Editors: Ana Isabel Fernandes and Ortensia Ilaria Parisi

Received: 1 February 2023
Revised: 25 February 2023
Accepted: 17 March 2023
Published: 20 March 2023

Abstract: Superparamagnetic iron oxide nanoparticles (SPIONs) have their use approved for the diagnosis/treatment of malignant tumors and can be metabolized by the organism. To prevent embolism caused by these nanoparticles, they need to be coated with biocompatible and non-cytotoxic materials. Here, we synthesized an unsaturated and biocompatible copolyester, poly (globalide-*co*-ε-caprolactone) (PGlCL), and modified it with the amino acid cysteine (Cys) via a thiol-ene reaction (PGlCLCys). The Cys-modified copolymer presented reduced crystallinity and increased hydrophilicity in comparison to PGlCL, thus being used for the coating of SPIONS (SPION@PGlCLCys). Additionally, cysteine pendant groups at the particle's surface allowed the direct conjugation of (bio)molecules that establish specific interactions with tumor cells (MDA-MB 231). The conjugation of either folic acid (FA) or the anti-cancer drug methotrexate (MTX) was carried out directly on the amine groups of cysteine molecules present in the SPION@PGlCLCys surface (SPION@PGlCLCys_FA and SPION@PGlCLCys_MTX) by carbodiimide-mediated coupling, leading to the formation of amide bonds, with conjugation efficiencies of 62% for FA and 60% for MTX. Then, the release of MTX from the nanoparticle surface was evaluated using a protease at 37 °C in phosphate buffer pH~5.3. It was found that 45% of MTX conjugated to the SPIONs were released after 72 h. Cell viability was measured by MTT assay, and after 72 h, 25% reduction in cell viability of tumor cells was observed. Thus, after a successful conjugation and subsequent triggered release of MTX, we understand that SPION@PGlCLCys has a strong potential to be treated as a model nanoplatform for the development of treatments and diagnosis techniques (or theranostic applications) that can be less aggressive to patients.

Keywords: copolyester; SPION; cysteine; bioconjugation; and enzymatic release

1. Introduction

Traditional cancer treatments typically involve the use of non-selective agents, that besides action on destroying the tumor cells, also damage healthy tissues and organs, causing side effects that many times can be life threatening [1]. The use of nanotechnology for cancer treatment allows the engineering of nanostructures for specific targeting, together with the controlled release, being able to reach places of very difficult access inside the human body [2]. Thus, it has been the focus of many research groups around the globe. In this context, superparamagnetic iron oxide nanoparticles (SPIONs) have their use approved for the diagnosis/treatment of malignant tumors and can be metabolized by the organism [3].

To prevent embolism caused by SPIONs, these need to be coated with biocompatible and non-cytotoxic materials [4]. Synthetic polymers, such as non-biodegradable poly (methyl methacrylate), have been used for this purpose. However, polyesters derived from lactones and macrolactones appear as promising alternatives due to their biocompatibility, and biodegradability. Poly (ε-caprolactone) (PCL) is a great example of a biodegradable and biocompatible polyester, which is already applied in commercial biomedical devices [5]. On the other hand, unsaturated macrolactones provide (co)polymers that can be functionalized by thiol-ene "click reactions" with different (bio)molecules, such as amino acids and their derivatives, drugs, proteins, antibodies, etc. [6]. This kind of modification is a topic of great interest in biomedical applications, being useful as tools in targeting breast cancer in both diagnosis [7] and treatment [8,9]. Globalide is an unsaturated macrolactone that is gaining attention as a monomer to obtain unsaturated polyesters of great interest in the medical field [10]. Recent advances in polymerization and modification methods involving unsaturated polyesters have aroused increasing interest, as they allow obtaining polymers with modified characteristics and properties, such as lower crystallinity, and higher hydrophilicity, in addition to specific interactions with specific cells, for example [11–14].

Cysteine (Cys) is an amino acid that was reported to interact on the surface of the SPIONs through carboxylic acid, thiol, and amine groups [15]. Therefore, the modification of unsaturated polyesters with Cys is a promising strategy, since the cysteine-modified polyester can be applied in the coating of SPIONs, given the good compatibility between Cys and the SPIONs. Furthermore, the presence of Cys on the surface of the SPIONs provide amine groups available for further modification, which would give a specific function to the nanoparticle. Additionally, cysteine-modified polyesters tend to present low crystallinity and high hydrophilicity, which would increase its degradability, as previously reported in polymer modification studies using N-acetylcysteine [16].

There are many examples of (bio) molecules that could be conjugated to the amine groups present in Cys, giving a specific function to the SPIONs. Folic acid (FA) is a naturally occurring molecule and an essential nutrient for humans [17]. FA can be used as a targeting agent to cancer cells, since several tumor cells overexpress FA receptors. Thus, the use of FA for surface functionalization promotes cellular internalization of SPIONs [18]. In this way, the mechanism of endocytosis mediated by the folate receptor and the subcellular traffic of nanoparticles [19] could be achieved. Methotrexate (MTX) is another example. MTX is an anti-cancer drug that also has specific interaction for folate receptors in tumor cells, and acts blocking the body's use of FA, which is responsible for the rapid division of cells, thus controlling the growth of the tumor [20].

In this context, the present work (Scheme 1) is focused on the preparation of SPIONs coated by an unsaturated copolymer poly(globalide-*co*-caprolactone) (PGlCL) modified with cysteine (SPION@PGlCLCys). Cysteine, besides providing the proper protection of the SPION core, offers important amine groups as anchoring points for further conjugation with FA or with MTX. FA and MTX were used (separately) for the production of conjugates (SPION@PGlCLCys_FA and SPION@PGlCLCys_MTX) focused on applications for cancer treatment/diagnosis. Both molecules were conjugated by amidation reaction between their carboxylic acid and the amine group of cysteine. Release assays for MTX were performed in the presence of a protease, since proteases, that can cleave amide bonds, are present in the lysosome of the tumor cells [21,22]. Cell viability assays (MTT assay) were also carried out in order to evaluate the performance of the conjugate SPION@PGlCLCys_MTX in acting against tumor cells (MDA-MB 231). To the best of our knowledge, this is the first report on the preparation of a nanoplatform made of SPION@PGlCLCys, which has a multifunctional purpose and has a strong potential to be applied in targeted drug delivery, hyperthermia, diagnosis of cancer, or even in theranostics.

Modification of PGlCL with amino acid cysteine

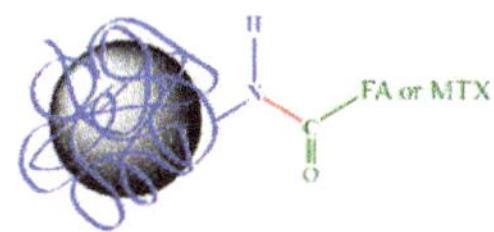

Poly(globalide-*co*-ε-caprolactone) (PGlCL)

Poly(globalide-*co*-ε-caprolactone)-Cys (PGlCLCys)

Fe(II)/Fe(III)
90°C, N$_2$

Conjugation with FA or MTX

Preparation of SPIONs coated with PGlCLCys by co-precipitation method

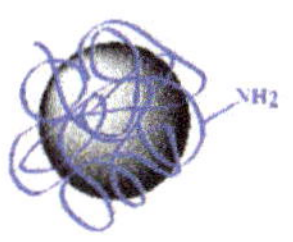

EDC/NHS
pH 7.4

SPION@PGlCLCys_FA or SPION@PGlCLCys_MTX

SPION@PGlCLCys

Scheme 1. Synthesis of PGlCLCys, preparation of SPIONs, and conjugation with either folic acid (FA) or methotrexate (MTX).

2. Materials and Methods

2.1. Materials

Acetone P.A 99.8%, chloroform P.A. 99.8%, dichloromethane P.A. 99.8% (DCM), dimethylformamide P.A 99.5%, dimethyl sulfoxide P.A 99.7%, ethanol P.A. 99.8% (EtOH), tetrahydrofuran P.A. 99.8% (THF), and toluene P.A. 99.0% were purchased from Merck (Rio de Janeiro, Brazil). The photoinitiator 2,2-dimethoxy-2-phenylacetophenone (DMPA) CAS:24650-42-8 was kindly donated by IGM resins (Valinhos, Brazil). Iron (III) chloride hexahydrate (FeCl$_3$·6H$_2$O), iron (II) chloride tetrahydrate (FeCl·4H$_2$O), and ammonium hydroxide (NH$_4$OH) were purchased from Vetec (Duque de caxias, Brazil). Folic acid (FA) 98%, 1-ethyl-3-(3-dimethyl aminopropyl) carbodiimide hydrochloride (EDC), N-Hydroxysuccinimide (NHS), methotrexate hydrate (MTX), fluorescein isothiocyanate (FITC), and enzyme Bromelain were purchased from Sigma-Aldrich (Cotia, Brazil). Cysteine hydrochloride 99.8% (Cys) was purchased from Gemini (Anápolis, Brazil). Novozym 435 (commercial lipase B from Candida Antarctica immobilized on cross-linked polyacrylate beads) was kindly donated by Novozymes, Brazil, A/S. Globalide (Gl) was a kind gift from Symrise (Cotia, Brazil), while ε-caprolactone (CL) was purchased from Sigma-Aldrich (Cotia, Brazil). Globalide and ε-caprolactone were dried under vacuum for 24 h and kept in a desiccator over silica and 4 Å molecular sieves. Water was purified by a Milli-Q water purification system.

2.2. Experimental Procedure

2.2.1. Enzymatic Synthesis of Poly(Globalide-*co*-ε-caprolactone)

The synthesis of PGlCL was carried out according to the method used by Guindani et al. [23], with modifications. The polymerization was performed in a 50/50 mass ratio (Gl/CL). Toluene was used as solvent (toluene:monomers = 1:2, w/w), and the system was maintained at a temperature of 65 °C for 2 h. The enzyme content was fixed at 5 wt.% relative to the total mass of monomer. After polymerization, the immobilized enzyme (N-435) was filtered off and the copolymer was precipitated in a mixture of cold EtOH: acetone (70:30 v:v), being dried under vacuum up to constant weight.

2.2.2. Modification of PGlCL with Cysteine via Thiol-Ene Reaction

The post-polymerization modification of PGlCL was carried out by photopolymerization using thiol-ene reactions using a DMPA photoinitiator, directly on PGlCL unsaturations [24]. Cysteine was chosen as a functionalizing molecule because it contains a thiol group and because its presence as a pendent group on PGlCL chains confers desirable hydrophilic characteristic and, furthermore, enables the covalent conjugation of high interest biological interest molecules to the nanoparticle surface. For the modification procedure, the copolymer PGlCL (0.300 g) and Cys (0.224 g) were placed in a flask with the photoinitiator DMPA (0.016 g), using a mixture of chloroform (4 mL) and DMF (2 mL) as a solvent, under nitrogen atmosphere. The reaction was carried out in a UV chamber for 4 h, under continuous magnetic stirring. The amount of Cys used was established to be twice the minimum required to react with all double bonds.

2.2.3. Synthesis of Superparamagnetic Iron Oxide Nanoparticles (SPIONs) and Coating with PGlCLCys

SPIONs were synthesized by the Fe_3O_4 co-precipitation method [25] and PGlCLCys was used as the coating agent. First, solutions of $FeCl_2 \cdot 4H_2O$ (2.0 mmol) and $FeCl_3 \cdot 6.H_2O$ (4.0 mmol) salts were prepared in the proportion of 1:2 (mol/mol) dissolved in 1 mL of 1M HCl solution under an inert atmosphere, to avoid oxidation of Fe (II). Then, 100 mL of deionized water was heated up to 80 °C, and the solution containing Fe^{3+} and Fe^{2+} was added under nitrogen flow. When the solution reached 90 °C, 40 mL of NH_4OH (25% v:v) was added, reaching pH~10. Next, the modified copolymer (PGlCLCys) was added (Fe^{3+} PGlCLCys = 8:1 mol/mol), and the system was kept under constant stirring at 90 °C during 1 h. The appearance of a dark brown/black color in the solution is indicative of the formation of coated iron oxide, forming SPION@PGlCLCys. After that time, SPIONs were separated with a magnet and washed with distilled water. Afterwards, the sample was frozen and lyophilized for the subsequent analyses.

2.2.4. Conjugation of SPION@PGlCLCys with Folic Acid

Folic acid was conjugated to SPION@PGlCLCys by the carbodiimide approach [26]. First, FA was activated with NHS for further conjugation with the SPION@PGlCLCys. FA (2 mmol), NHS (2 mmol), and EDC·HCl (2.2 mmol) were dissolved in 100 mL DMSO. The mixture was purged with nitrogen, and the reaction was carried out overnight, under constant stirring, at room temperature, and protected from light. DMSO and unreacted FA were removed by dialysis (Spectra-Por 100-500 Da, Biontech CE Tubing) in a buffer solution (PBS, pH 8.0) for four days, replacing the buffer solution daily. After dialysis, the FA-NHS solution was lyophilized. For the conjugation of FA-NHS to the SPION@PGlCLCys surface, the amount of cysteine in the copolymer was calculated based on the consumption of the double bonds of the copolymer, determined by ^1H NMR analysis. The proportion between carboxylic acid groups (from FA-NHS) and amine groups (from PGlCLCys) was varied: it was tested a stoichiometric ratio of NH_2:COOH = 1:1; and an excess of carboxylic groups (FA-NHS), where with a ratio of NH_2:COOH = 1:2. In sequence, 5 mg of SPION@PGlCLCys was dispersed in phosphate buffer (pH~8.0) with the help of a sonicator. After the complete dispersion of the SPION@PGlCLCys, the amounts of FA-NHS 0.160 mg (NH_2:COOH = 1:1) and 0.260 mg (NH_2:COOH = 1:2) were added to falcon tubes containing SPION@PGlCLCys. The system was purged with nitrogen and left in a Klein-type shaker for 24 h. Afterwards, magnetic separation was performed and samples were submitted to 3 washing cycles, collecting the supernatant for FA quantification, and calculation of the conjugation efficiency. The amount of FA conjugated to the SPION@PGlCLCys was calculated based on a FA calibration curve (λ = 283 nm) using UV–vis equipment (Perkin Elmer, Shelton, CT, USA).

2.2.5. Conjugation of SPION@PGlCLCys with Methotrexate

For the conjugation of the drug MTX to the surface of SPION@PGlCLCys, the same approach described in Section 2.2.4 for the conjugation of SPIONs with FA was applied.

The amount of MTX conjugated to SPION@PGlCLCys was calculated based on an MTX calibration curve (λ = 303 nm) using UV–vis equipment.

2.3. Physicochemical Characterizations

Proton nuclear magnetic resonance (^{1}H NMR) analyses were performed on Bruker AC-200F NMR equipment, operating at 200 MHz. Chemical shifts are reported in ppm relative to tetramethylsilane (TMS) 0.01% (vol%) (δ = 0.00). All samples were dissolved in CDCl$_3$ (δ = 7.27 for 1H NMR). Poly(globalide-*co*-ε-caprolactone) ^{1}H NMR (CDCl$_3$ 200 MHz): δ(ppm) 5.49–5.32 (m, CH=CH); 4.10–4.04 (m, CH$_2$O(C=O)). Poly(globalide-*co*-ε-caprolactone) -Cys RMN ^{1}H (CDCl$_3$ 200 MHz): δ(ppm) 5.49–5.32 (m, CH=CH); 4.10–4.04 (m, CH$_2$O(C=O)); 2.90–2.70 (m, S–CH$_2$). Molecular weight distributions were determined by Gel Permeation Chromatography (GPC). Therefore, 0.02 g of the copolymer was dissolved in 4 ml of tetrahydrofuran (THF) and the solution obtained was filtered (pore: 0.45 µm, diameter: 33 mm) before the analysis. The analysis was performed using a high-performance liquid chromatography equipment (HPLC, model LC 20-A, Shimadzu, Kyoto, Japan) and Shim Pack columns of the GPC800 series (GPC 801, GPC 804, and GPC 807), also from Shimadzu. As eluent, THF was used at a volumetric flow rate of 1 mL min^{-1} at 40 °C. Calibration was achieved using polystyrene standards. Polystyrene standards with molecular weights ranging from 580 to 9.225 × 10^6 g·mol^{-1} were used to construct the standard curve. Fourier transform infrared spectroscopy (FTIR) was used to identify the chemical structure of the cysteine-modified copolymer and SPION@PGlCLCys. The analysis was performed on a Prestige 21 spectrophotometer (Shimadzu IR, Kyoto, Japan) using the KBr tableting technique for obtaining a transparent tablet. The spectra were recorded in a wavenumber range of 400 to 4000 cm^{-1}. Differential scanning calorimetry (DSC) was used to measure melting temperatures and melting enthalpy, used to determine the degree of crystallinity of the polymeric materials using a Jade DSC (Perkin-Elmer, Shelton, CT, USA). For the analysis, approximately 5 mg of the dried polymer was analyzed under nitrogen atmosphere (20 mL·min^{-1}), with temperatures ranging from 0 to 150 °C and a heating rate of 10 °C min^{-1}. The thermal history was removed before the analyses (for pure polymer samples) at a heating rate of 20 °C min^{-1} and a cooling rate of 10 °C min^{-1}. The second heating runs were used to obtain the thermal properties of the polymer. For the contact angle assay, PGlCL and PGlCL-Cys films were produced, by depositing copolymer solutions (100 mg:mL—Chloroform:DMF 9:1) on microscope slides. The contact angle between the polymeric films and the water drops was measured in a goniometer Ramé-Hart 250 (Ramé-Hart Instrument Co., Succasunna, NJ, USA). For this assay, samples were analyzed in triplicate at room temperature, using 10 µL drop volume. Transmission electron microscopy and selected area electron diffraction (TEM/SAED) images were obtained on a JEM-1011 TEM (JEOL, Peabody, MA, USA) microscope at an acceleration voltage of 100 kV. Powder X-ray diffraction (XRD) analyses were performed on a Rigaku MiniFlex 600 (Rikagu, Tokyo, Japan) diffractometer with graphite monochromatized CuKα (λ = 1.5418 A°), with maximum voltage and current at 40 kV and 40 mA, respectively, with a 2θ°/min scan rate in the range of 20 to 80° with 0.05° steps. Magnetic properties of SPION@ PGlCLCys were measured in a vibrating sample magnetometer (VSM) EV9 Model (MicroSense, Lowell, MA, USA). For the VSM analysis, the samples were dried, pressed, and held in a quartz cylinder holder. Thermogravimetric analysis (TGA) was performed under a nitrogen atmosphere at a heating rate of 10 °C min^{-1} from room temperature to 900 °C, in a STA 449-F3 Jupiter (2012) (Netzsch, Hanau, HE, Germany) equipment. The particle size distribution was measured by dynamic light scattering (DLS) and zeta potential using a Zetasizer 3000 HSA (Malvern Instruments, Malvern, Worcs, United Kingdom).

2.4. Cell Culture

All biological assays used normal L929 and breast carcinoma-derived MDA-MB 231 cells lines. L929 and MDA-MB 231 were cultivated in RPMI 1640 medium (GIBCO, Baltimore, MD, USA) supplemented with fetal bovine serum (10%), penicillin (100 U/mL)

(GIBCO, Baltimore, MD, USA), streptomycin (100 µg/mL) (GIBCO, Baltimore, MD, USA), and glutamine 2 mM. All cells were maintained under controlled conditions (37 °C in a 5% CO_2 atmosphere with 95% air humidity) for 24 and 72 h. Results are expressed as an average of three independent experiments.

2.4.1. In Vitro Cell Viability of the SPION@PGlCLCys, SPION@PGlCLCys_FA, and SPION@PGlCLCys_MTX

The cell viabilities of L929 and MDA-MB 231 exposed to SPION@PGlCLCys and SPION@PGlCLCys_FA were measured using the MTT [27] assay. For this, 7.5×10^3 cells/well (24 h assay) and 5.0×10^3 cells/well (72 h assay) were seeded in 96-well plates. Following the same procedure, the cell viability of MDA-MB 231 was also evaluated when exposed to SPION@PGlCLCys_MTX and free MTX, for 72 h. At confluence, cells were exposed to different free MTX (8.52×10^{-6}, 8.52×10^{-5}, 8.52×10^{-4}, 8.52×10^{-3}, 8.52×10^{-2}, 8.52×10^{-1}, and 8.52 µg·mL^{-1}) and nanoparticle (0.0001, 0.001, 0.01, 0.1, 1, 10, and 100 µg·mL^{-1}) concentrations and incubated at 37 °C and pH 7.4 for 24 h and 72 h for each cell. After incubation, cells were washed twice with PBS and incubated for 2 h with MTT (0.5 mg·mL^{-1}). Subsequently, the formed formazan crystals were dissolved by the addition of DMSO (100 µL/well), and colored solutions were read at 570 nm. For all tested conditions, the respective blanks were analyzed. The experiments were performed independently for each cell line and in triplicate, and the results are presented as cell viability.

2.4.2. Enzymatic Release of Folic Acid and Methotrexate

The release of the FA conjugated to the SPION@PGlCLCys was studied under the lysosomal condition (pH = 5.3), taking 400 µg of SPION@PGlCLCys_FA and 400 µg of Bromelain protease in 2 mL of PBS (0.1 M solution, pH = 5.3) at 37 °C. The FA release was evaluated for 12 h, 24 h, 48 h, and 72 h [28]. Magnetic separation was performed and the samples were submitted to 3 washing cycles, collecting the supernatant for FA quantification [28]. The amount of released FA was calculated based on a FA standard curve built using UV–vis (283 nm) equipment. The same approach was applied for the release of MTX and the amount of MTX released was calculated based on an MTX standard curve constructed using a UV–vis instrument (303 nm).

2.5. Computational Section

The theoretical values of the logarithm of partition coefficient (log P) obtained in the present work were based on the Density Functional Theory (DFT) [29,30]. The simulated molecules were generated and analyzed by the software Avogadro [31] version (1.2.0).

First, the molecular geometry of PGlCL, PGlCLCys, FA, and PGlCLCys_FA were optimized to the ground state geometry of these molecules in different media such as gas phase, n-octanol and water. The optimized structures were confirmed as real minima by vibration analysis (no imaginary frequency was detected) [31]. Through the thermodynamic properties obtained by the DFT calculations, it is possible to obtain the Gibbs free energy in different media and the partition coefficient.

The partition coefficient (P) can be defined as the ratio between the concentration of a solute in two phases of a mixture that contains two immiscible solvents at equilibrium [32]. The logarithm of partition coefficients for n-octanol/water mixtures (log $P^{O/W}$) can be obtained according to [33]:

$$\log P^{O/W} = \frac{\Delta G^{W}_{solv} - \Delta G^{O}_{solv}}{2.303RT} \tag{1}$$

where $\Delta Gsolv = G^X - G$ is the Gibbs free energy of solvation, G and G^X are respectively the Gibbs free energy in gas phase and in the solvent. The superscripts (X = W and X = O) represent respectively the water and n-octanol solvents, while R is the ideal gas constant (8.314 J·K^{-1} mol^{-1}).

The results were obtained at the reaction conditions (25 °C and 1.0 atm), using the Becke-three-parameter Lee–Yang–Parr (B3LYP) model as hybrid functional, along with a split-valence double-zeta polarized basis set, based on Gaussian type orbitals (6-31G**) [34]. The Gibbs free energy of solvation in water and n-octanol were computed using the solvation model based on electronic density (SMD) [35]. All DFT calculations were performed using the ORCA 5.0.2 package [36].

3. Results and Discussion

Scheme 1 shows the steps involved in the development of the nanoplatform. In a first stage, the copolymer PGlCL was synthesized via enzymatic ring-opening polymerization (e-ROP). Subsequently, the modification of the copolymer with the amino acid cysteine was carried out through photoinitiated thiol-ene reactions. The modified copolymer PGlCLCys was then used for the coating of the SPIONS, and the amine groups from cysteine in the SPION@PGlCLCys allow the conjugation of either FA or MTX via carbodiimide chemistry.

3.1. Synthesis, Modification, and Characterization PGlCLCys

The synthesis of PGlCL (50:50) was carried out via e-ROP using toluene as solvent at a fixed mass ratio of 2:1 (monomer:solvent). The number (Mn) and weight (Mw) average molecular weights of PGlCL determined by gel permeation chromatography (GPC) were 24.500 g·mol^{-1} and 57.000 g·mol^{-1}, respectively. In sequence, the copolyester was modified with cysteine via thiol-ene reaction by photoinitiation with DMPA (PGlCLCys).

The modification of the polymer with cysteine was verified by FT-IR (Figure 1A). For PGlCL a band was observed in 1635 cm^{-1} referring to the elongation of the alkene group (–C=C–) present in the PGlCL. The same band could not be observed in the PGlCLCys spectrum. Additionally, the appearance of bands at 674 cm^{-1} (–C–S–C–) and at 1580 cm^{-1} (–C–NH$_2$) confirms the incorporation of cysteine to the copolymer. The nonappearance of bands at 2550 cm^{-1} (–C–SH) indicates the absence of free cysteine in the modified polymer.

The degree of modification of PGlCL with cysteine was determined by ^{1}H NMR analysis (Figure 1B) based on the peak area of the double bond (–C=C–, at 5.40 ppm) present in globalide repeat units in the copolymer (results in Table S1, see Supplementary Materials (SM)). For integration, the peak of ester methylene at 4.06 ppm was used as a reference, since the amount of ester methylene remains constant. An 18% consumption of double bonds was observed after the reaction with cysteine, indicating the coupling of cysteine to 18% of the globalide repeating units of PGlCL. The modification was also evidenced by the emergence of a peak at 2.90–2.70 ppm, that can be assigned to the –S–CH$_2$– group. It is important to emphasize that due to the hydrophobic character of PGlCL and the hydrophilic character of cysteine hydrochloride, solubilizing both reactants simultaneously is a great challenge, and it hinders the modification reaction. However, in spite of the relatively low degree of modification, the success of the modification reaction is evidenced by the changes in physicochemical properties of the copolymer, reported below.

The degree of crystallinity was determined by the relationship between the ΔH_m measured by DSC of each sample and the theoretical value of a 100% crystalline PCL sample (more details in Table S2 in SM). The melting temperature (T_m) for PGlCL was 38 °C and after modification with cysteine, T_m was reduced to 32 °C. At the same time, the crystallinity decreased from 44% to 30% after the reaction with the amino acid (Figure 1C). The modification of the polyester with cysteine also affected its hydrophobic character, reducing the value of the contact angle with water from 77° to 56° (Figure 1C). The decrease in the melting point and in the degree of crystallinity are related to the presence of cysteine as a pendant group in the polymer chains. Cysteine acts as a "spacer" molecule, which disturbs the crystalline arrangement, consequently decreasing the degree of crystallinity. The decrease in the melting temperature, in turn, is due to the increase in free volume caused by the presence of cysteine, which reduces the energetic level necessary to overcome the secondary intermolecular forces between the chains of the crystalline phase [37]. The increase in the hydrophilicity of the polymer, on the other hand, is endorsed by the prediction of the partition coefficient (P, also represented in the logarithm form, log P) through DFT calculations (for details, see Table S3 on SM). The partition coefficient is defined as the ratio between the concentration of a solute in two phases of a mixture of two immiscible solvents at equilibrium (in this case, n-octanol and water, log $P^{O/W}$) [32], being a form to evaluate the hydrophobicity or hydrophilicity of a solute. The polymer PGlCL presented log $P^{O/W}$ = 5.53, while after the modification with cysteine, this value decreased to log $P^{O/W}$ = 3.54, indicating an increase in hydrophilicity, which is mostly affected by the polar character of the carboxyl group [37,38].

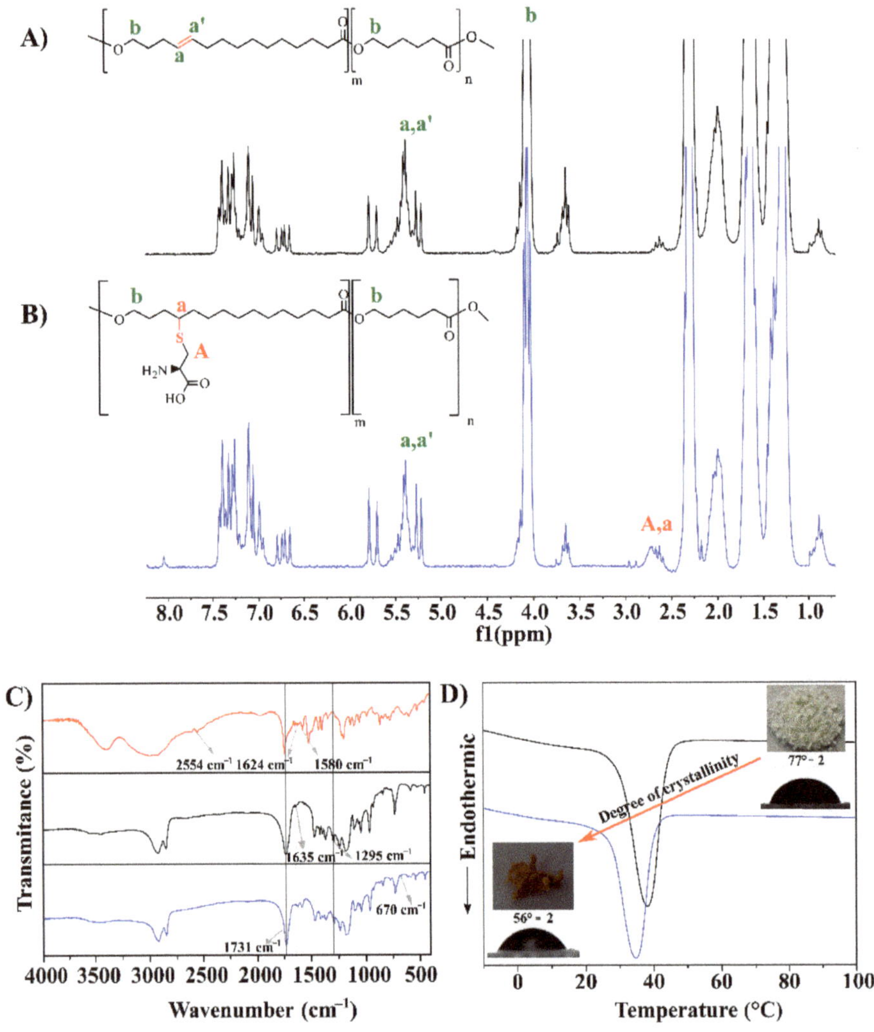

Figure 1. (**A**) ^{1}H NMR spectra of PGlCL with peak assignments (a, a′, b). (**B**) ^{1}H NMR spectra of PGlCLCys (50/50 Gl/CL weight ratio) with peak assignments (a, a′, b, a, A). (**C**) FTIR spectra of cysteine (red) and respective copolymers PGlCL (black) and PGlCLCys (blue). (**D**) Second run DSC heating curves of PGlCL (black) and PGlCLCys (blue) with different respective contact angle.

3.2. Preparation of SPIONs Coated with PGlCLcys by Co-Precipitation Method

The preparation of SPIONs coated with PGlCLcys (SPION@PGlCLCys) was performed according to the co-precipitation approach with modifications [25,39]. The SPION@PGlCLCys exhibited an average particle diameter of 7.2 ± 2.2 nm (Figure 2A,D), determined by transmission electron microscopy (TEM-100 kV), and a monomodal distribution, as well as high crystallinity (Figure 2B) determined by selected area electron diffraction (SAED) technique. The average crystalline domains and core size were around 7.6 nm as estimated from the results of X-ray powder diffraction (XRPD) analysis (Figure 2C). The lattice parameter obtained after Rietvield refinement (a = 8.36154 Å) applied to the XRPD data (0.252 nm to (311)) is consistent with the value of 8.46 Å based on the (220) interplanar spacing of 0.299 nm obtained using SAED (Figures S3 and S4, see SM). Thus, the results agree with those of other studies [40] related to the spinel structure of Fe_3O_4 and standard measurements (a = 8.3940 Å) (for more details, see Table S4, in SM).

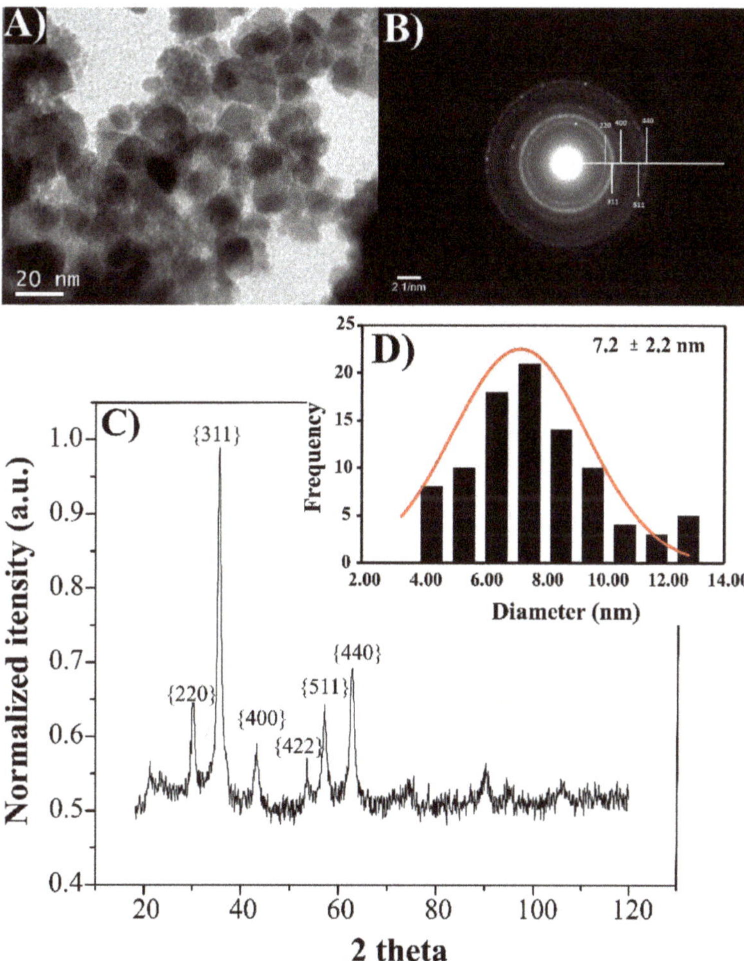

Figure 2. SPION@PGlCLCys (**A**) Low magnification TEM image; (**B**) SAED; (**C**) XRPD diffractogram with diffraction peaks indexed to the spinel iron oxide phase. (**D**) Particle size distribution of the magnetic nanoparticles obtained from the TEM images.

The superparamagnetic behavior of SPION@PGlCLCys observed by VSM analysis corroborates the core size below 15 nm [41] and the successful coating of the SPIONs, presenting a saturation magnetization of 60 emu·g^{-1}, which is consistent with saturation magnetization values obtained for magnetite in literature [42,43] (for details, see Figure S7, in SM). The coating of the SPIONs by PGlCLCys was also evidenced by the FTIR spectra (Figure S5, in SM), indicated by vibration frequencies at 1736 cm^{-1} (elongation –C=O), associated with a carboxyl group and elongation of the amine group at 1643 cm^{-1} (–NH$_2$).

TGA results (Figure S6) gives important information on the SPION@PGlCLCys composition. The thermogram shows a mass loss of 1.7%, related to water evaporation, followed by a mass loss of 13.3%, assigned to PGlCLCys decomposition, and finally the 85% that remained corresponds to the iron oxide presence. The proportion of Fe3O4:PGlCL used in the preparation of the SPION@PGlCLCys was 8:1 (w/w), which means theoretically there are 11.1% PGlCLCys and 88.9% Fe$_3$O$_4$ in SPION@PGlCLCys. Therefore, one can say that the experimental results is in accordance with the theoretical composition values.

SPION@PGlCLCys were resuspended in buffer solution (pH = 8.0, close to physiological conditions) and Dynamic Light Scattering (DLS) measurements were performed (see Figure S8, in SM). The intensity average particle diameter measured was 145 nm with a PDI of 0.2. This result shows that after being dispersed, SPION@PGlCLCys formed some aggregates, but the unimodal size distribution and PDI value of 0.2 indicate that such aggregates are stable. A possible cause for this aggregation is the small amount of PGlCLCys used to coat the SPIONs, which provides a very thin layer of polymer coating the magnetite core. On one hand, this has a very positive impact on the saturation magnetization of SPION@PGlCLCys (60 emu·g^{-1}). On the other hand, such a thin layer does not avoid a strong interaction between the nuclei and provides some aggregation of SPIONs after the lyophilization process. However, it is also important to emphasize that hydrodynamic diameter of 145 nm with a PDI of 0.2, which is an acceptable size for drug delivery purposes. The SPION@PGlCLCys also showed a zeta potential value of −35.4 mV (pH = 8.0, close to physiological conditions), suggesting moderate colloidal stability [44]. Such negative value is much likely related to the presence of cysteine in the polymer, suggesting that in addition to developing the role of protecting the SPIONs core, PGlCLCys also helps in some extent to maintain some colloidal stability in the system. The isoelectric point of the amino acid cysteine is widely reported in literature as being 5.07 [45]. That is, in pH values above 5.07, cysteine structure is deprotonated, assuming an overall negative charge.

3.3. Conjugation of FA or MTX on the SPION@PGLCLCys, Enzymatic Release, and Cell Viability

Since SPION@PGlCLCys is stable in aqueous medium, its surface was covalently conjugated with FA by carbodiimide approach, aiming to target the folate receptors in breast cancer. The amount of FA covalently bonded to the SPION@PGlCLCys was determined by UV–vis spectrometry, by using a FA calibration curve, dissolved in pH = 8.0 buffer at 286 nm (further details see Figure S10A, in SM). Considering NH$_2$:COOH ratios of 1:1 (stoichiometric proportion) and 1:2 (excess of FA), the conjugation efficiency of FA on the SPIONs surface was 62% and 183.5%, respectively, relative to the available NH$_2$ sites for conjugation. A conjugation efficiency higher than 100% is explained by the adsorption of the FA in excess on the surface of the SPION@PGlCLCys. For the release assays and cell viability assays, we chose to proceed with the samples synthesized using a NH$_2$:COOH ratio of 1:1, containing approximately 8.27 µg FA/mg iron oxide.

DFT calculations were also performed in order to provide theoretical information on the hydrophilicity of PGlCLCys_FA, which presented an estimated log $P^{O/W}$ value of 0.81, a much lower value than PGlCL (log $P^{O/W}$ = 5.53) and PGlCLCys (log $P^{O/W}$ = 3.54), indicating that the presence of FA makes SPION@PGlCLCys_FA quite hydrophilic. Such hydrophilic characteristic must be highlighted, since the hydrophilization of nanoparticles is commonly reported as a strategy to increase their circulation time in the blood stream [46].

The cytotoxicity of SPION@PGlCLCys and SPION@PGlCLCys_FA was evaluated by the MTT assay using L929 and MDA-MB 231 cell lines (for details, see Figures S11 and S12 in SM). Cell viability reached 100% for all tested samples, in the concentration range from 1.10^{-4} ppm to 100 ppm, for 24 h and 72 h, and for both tested cell lines. In sequence, the release profile of conjugated FA on the surface of SPION@PGlCLCys_FA was evaluated. This way, bromelain was used as a protease to simulate a possible release triggered via lysosomal protease at 37 °C and pH~5.3, similar to cellular pH, to verify the cleavage of the amide bond formed in the conjugation of FA to the surface of SPION@PGlCLCys (more details, see Figure S13, in SM). Results showed that protease cleaves approximately ~28% of the amide bonds within 24 h, and ~35% in 72 h. This is the main route of drug release, since for the assay at the same pH but without enzyme, there is no release.

Finally, the conjugation of the drug MTX on SPION@PGlCLCys was performed, and the conjugation efficiency was determined by the UV–vis method (calibration curves in Figure S10, SM, dissolved in pH = 8.0 buffer, at 303 nm). The results obtained for MTX conjugation efficiency are 60.3% (1:1 stoichiometric proportion) and 132.4% (1:2 excess of MTX), relative to the available NH_2 sites for conjugation. A conjugation efficiency higher than 100% is explained by the adsorption of the MTX in excess on the surface of the SPION@PGlCLCys Again, for the release assays and cell viability assays, we chose to proceed with the samples synthesized using a NH_2:COOH ratio of 1:1, which provided the highest conjugation efficiency, containing approximately 3.20 μg MTX/mg iron oxide.

Similar to FA, the enzymatic release assay for SPION@PGlCLCys_MTX was carried out at lysosomal pH. The results (Figure 3A) showed that protease cleaves approximately ~38% of the amide bonds within 24 h, and ~45% in 72 h. This can be considered a very positive result, since other works in literature that investigated MTX encapsulation in polyesters nanoparticles (e.g., polycaprolactone) often report a release of around 40% in 72 h, or even less depending on how hydrophobic the polymer matrix is [47,48]. Therefore, it is important to emphasize that in our conjugate (SPION@PGlCLCys_MTX), MTX is not only acting as a drug itself, but also as a molecule that has specific interactions for folate receptors, promoting specific targeting in tumor cells, forming the basis of our proposal of a multifunctional nanoplatform for cancer treatment.

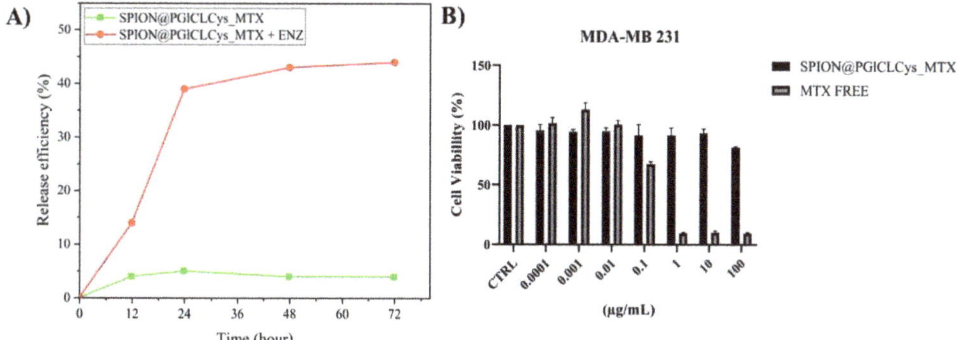

Figure 3. SPION@PGlCLCys_MTX: (**A**) drug delivery assay at lysosomal pH (pH 5.3) with or without protease (ENZ); (**B**) breast carcinoma-derived MDA-MB 231 cells viability after 72 h at different MTX concentrations.

Regarding the MTT cell viability assay, for 72 h incubation, a strong reduction in breast carcinoma-derived MDA-MB 231 cell viability was observed in free MTX for concentrations above 0.1 ug·mL^{-1}. For the conjugates SPION@PGlCLCys_MTX, this decrease was smoother reaching a reduction of approximately 20% at the highest evaluated concentration (Figure 3B).

The superior behavior of free MTX in comparison to SPION@PGlCLCys_MTX in the cell viability assays was expected, since free MTX is immediately available to interact with the cells, while MTX attached to SPION@PGlCLCys releases about 45% of MTX in 72 h.

4. Conclusions

In this work, we have successfully prepared SPIONs coated with a polyester modified with the amino acid cysteine (SPION@PGlCLCys). The cysteine molecules present in the SPIONs' coating can be used as anchoring points for the conjugation of a wide variety of molecules (e.g., peptides, antibodies, drugs, etc.). The main goal here was to prepare SPION@PGlCLCys and test their potential as a multifunctional nanoplatform for future developments to the traditional cancer treatments. In order to test such potential, we conjugated the SPION@PGlCLCys with two different molecules, separately, in order to understand their behavior when conjugated to SPION@PGlCLCys. FA was chosen due to its characteristic of stablishing specific interactions to folate receptors (overexpressed in tumor cells), as well as methotrexate, an anti-cancer drug that also has specific interaction for folate receptors and acts controlling the growth of the tumor. Both FA and MTX presented satisfactory conjugation efficiencies (above 65%). Release assays were performed in the presence of a protease, and MTX presented 45% release after 72 h, which is a very positive result in comparison to other works that studied the encapsulation and release of MTX from polymeric nanoparticles. MTT assay revealed that after 72 h, the conjugates are capable of reducing the tumor cell viability in about 20%, which is also noted as a very positive result, considering the small amount of MTX conjugated to SPION@PGlCLCys. The results obtained in this work can be seen as the first step for the development of a promising nanoplatform that can be easily modified and improved for future applications in less aggressive cancer treatments, allying targeting of tumor cells, controlled drug delivery, hyperthermia, and eventually diagnosis (theranostics).

Supplementary Materials: The following Supplementary Materials can be downloaded at: https://www.mdpi.com/article/10.3390/pharmaceutics15031000/s1. Figure S1. Representative scheme of the reaction steps to obtain PGlCL and its modification with cysteine followed by coating of SPIONs. A) Enzymatic Ring-Opening Polymerization reaction (e-ROP). B) Modification reaction with the amino acid cysteine (Cys) via thiol-ene reaction to obtain PGlCLCys. C) Synthesis and stabilization of SPIONs with PGlCLCys. D) Conjugation of SPIONs with FA. E) Conjugation of SPIONs with MTX. Figure S2: TEM dark field image showing individual crystalline particles (DF-TEM) of SPION@PGlCLCys. Figure S3: Selected area electron diffraction (SAED) image showing indexed diffraction rings corresponding to magnetite crystallographic planes of SPION@PGlCLCys. Figure S4: SAED image with Indexed diffraction pattern of magnetite and simulated diffraction ring pattern matching the results for the SPION@PGlCLCys sample. Figure S5: FT-IR spectra for SPIONs, and modified copolymer (PGlCLCys) and after coating of magnetic nanoparticles (SPION@PGlCLCys). Figure S6: Thermogravimetric analysis of SPION@PGlCLCys. Figure S7: VSM analysis of SPION@PGlCLCys. Figure S8: DLS analysis of SPION@PGlCLCys. Figure S9: Zeta potential surface analysis of SPION@PGlCLCys ($\zeta = -35.4$ mV) dispersed in buffer solution (pH = 8.0). Figure S10: UV-vis calibration curve for: A) folic acid (FA), and B) methotrexate (MTX) in buffer solution (pH = 8.0). Figure S11: MTT assay of SPION@PGlCLCys and SPION@PGlCLCys_FA showing cells viability as a function of nanoparticles concentration (0.0001 to 100 μg·mL−1) for 24 h. All SPIONs tested at different concentrations did not exert difference (ANOVA) in relations to the control group n = 3. Figure S12: MTT assay of SPION@PGlCLCys and SPION@PGlCLCys_FA showing cells viability as a function of nanoparticles concentration (0.0001 to 100 μg·mL−1) for 72 h. All SPIONs tested at different concentrations did not exert difference (ANOVA) in relations to the control group n = 3. Figure S13: FA assay at lysosomal pH (pH 5.3). Table S1: Thiol-ene reaction conversion calculated based on the consumption of the double bonds present in PGlCL chains, determined by 1H NMR. Table S2: Thermal properties of the polymers, determined by DSC. Table S3: Gibbs free energy calculated at 1 atm and 25 °C using DFT/B3LYP/6-31G** with water and n-octanol solvents in SMD model. Table S4: Interplanar distance.

Author Contributions: J.M.B.: investigation, methodology, writing—original draft, writing—review and editing. B.B.P.R.: formal analysis, investigation, methodology, writing—original draft, writing—review and editing. C.G.: conceptualization, writing—original draft, writing—review and editing, project administration, funding acquisition, supervision. G.C.: formal analysis, investigation, methodology, writing—original draft, writing—review and editing. K.B.F.: formal analysis, investigation, methodology, writing—original draft, writing—review and editing. R.L.: formal analysis, investigation, methodology, writing—original draft, writing—review and editing. A.D.Z.: formal analysis, investigation, methodology, writing—original draft, writing—review and editing. E.I.: formal analysis, investigation, methodology, writing—original draft, writing—review and editing. C.S.: conceptualization, project administration, funding acquisition, supervision. P.H.H.d.A.: conceptualization, project administration, funding acquisition, supervision. All authors have read and agreed to the published version of the manuscript.

Funding: This research was funded by Conselho Nacional de Desenvolvimento Científico e Tecnológico CNPq (project number 406078/2018-1), CAPES PRINT Program (project number 88887.310560/2018-00). C.G. and G.C. thank FAPERJ (Fundação Carlos Chagas Filho de Amparo à Pesquisa do Estado do Rio de Janeiro), process number E-26/201.911/2020, E-26/201.912/2020, E-26/200.627/2022 and E-26/210.391/2022. A.D.A.Z. is grateful to FAPESC (Fundação de Amparo à Pesquisa e Inovação do Estado de Santa Catarina), project number IFS2020201000004.

Institutional Review Board Statement: Not applicable.

Informed Consent Statement: Not applicable.

Data Availability Statement: The data presented in this study are available on request from the corresponding author.

Acknowledgments: The authors thank the Central Laboratory of Electron Microscopy (LCME) at UFSC (TEM analysis) as well as the Laboratory for Characterization Magnetic Materials (LabCAM) for the VSM measurements. In addition, special thanks are due to the Laboratory of Nanotechnology (LINDEN-UFSC) at UFSC. The authors gratefully acknowledge the computational support of Núcleo Avançado de Computação de Alto Desempenho (NACAD/COPPE/UFRJ), Sistema Nacional de Processamento de Alto Desempenho (SINAPAD), Centro Nacional de Processamento de Alto Desempenho em São Paulo (CENAPAD-SP) and the Startup SMMOL (Rio de Janeiro, RJ, Brazil) for the support. We also would like to thank the Program for Technological Development in Tools for Health-PDTIS-FIOCRUZ, the researcher Marco Augusto Stimamiglio, and the laboratory Stem Cells Basic Biology Laboratory, Carlos Chagas Institute—FIOCRUZ/PR for their collaboration and availability of infrastructure.

Conflicts of Interest: The authors declare no conflict of interest.

References

1. Senapati, S.; Mahanta, A.K.; Kumar, S.; Maiti, P. Controlled drug delivery vehicles for cancer treatment and their performance. *Signal Transduct. Target. Ther.* **2018**, *3*, 1–19. [CrossRef] [PubMed]
2. Grodzinski, P.; Kircher, M.; Goldberg, M.; Gabizon, A. Integrating Nanotechnology into Cancer Care. *ACS Nano* **2019**, *13*, 7370–7376. [CrossRef] [PubMed]
3. Palanisamy, S.; Wang, Y.M. Superparamagnetic iron oxide nanoparticulate system: Synthesis, targeting, drug delivery and therapy in cancer. *Dalt. Trans.* **2019**, *48*, 9490–9515. [CrossRef] [PubMed]
4. Shen, S.; Kong, F.; Guo, X.; Wu, L.; Shen, H.; Xie, M.; Wang, X.; Jin, Y.; Ge, Y. CMCTS stabilized Fe3O4 particles with extremely low toxicity as highly efficient near-infrared photothermal agents for in vivo tumor ablation. *Nanoscale* **2013**, *5*, 8056–8066. [CrossRef] [PubMed]
5. Albertsson, A.C.; Varma, I.K. Recent developments in ring opening polymerization of lactones for biomedical applications. *Biomacromolecules* **2003**, *4*, 1466–1486. [CrossRef]
6. Ates, Z.; Thornton, P.D.; Heise, A. Side-chain functionalisation of unsaturated polyesters from ring-opening polymerisation of macrolactones by thiol-ene click chemistry. *Polym. Chem.* **2011**, *2*, 309–312. [CrossRef]
7. Dadfar, S.M.; Roemhild, K.; Drude, N.I.; von Stillfried, S.; Knüchel, R.; Kiessling, F.; Lammers, T. Iron oxide nanoparticles: Diagnostic, therapeutic and theranostic applications. *Adv. Drug Deliv. Rev.* **2019**, *138*, 302–325. [CrossRef]
8. Qu, J.; Tian, Z.; Wang, Q.; Peng, S.; Luo, J.B.; Zhou, Q.H.; Lin, J. Surface design and preparation of multi-functional magnetic nanoparticles for cancer cell targeting, therapy, and imaging. *RSC Adv.* **2018**, *8*, 35437–35447. [CrossRef]
9. Yao, X.; Mu, J.; Zeng, L.; Lin, J.; Nie, Z.; Jiang, X.; Huang, P. Stimuli-responsive cyclodextrin-based nanoplatforms for cancer treatment and theranostics. *Mater. Horiz.* **2019**, *6*, 846–870. [CrossRef]

10. Van Der Meulen, I.; De Geus, M.; Antheunis, H.; Deumens, R.; Joosten, E.A.J.; Koning, C.E.; Heise, A. Polymers from functional macrolactones as potential biomaterials: Enzymatic ring opening polymerization, biodegradation, and biocompatibility. *Biomacromolecules* **2008**, *9*, 3404–3410. [CrossRef]

11. Ates, Z.; Heise, A. Functional films from unsaturated poly(macrolactones) by thiol–ene cross-linking and functionalisation. *Polym. Chem.* **2014**, *5*, 2936. [CrossRef]

12. de Oliveira, F.C.S.; Amaral, R.J.F.C.D.; Santos, L.E.C.D.; Cummins, C.; Morris, M.M.; Kearney, C.J.; Heise, A.J. Versatility of unsaturated polyesters from electrospun macrolactones: RGD immobilization to increase cell attachment. *Biomed. Mater. Res. Part A* **2021**, 1–9. [CrossRef] [PubMed]

13. Beltrame, J.M.; Guindani, C.; Novy, M.G.; Felipe, K.B.; Sayer, C.; Pedrosa, R.C.; De Araújo, P.H.H. Covalently Bonded N-Acetylcysteine-polyester Loaded in PCL Scaffolds for Enhanced Interactions with Fibroblasts. *ACS Appl. Bio Mater.* **2021**, *4*, 1552–1562. [CrossRef] [PubMed]

14. Guindani, C.; Frey, M.-L.; Simon, J.; Koynov, K.; Schultze, J.; Ferreira, S.R.S.; Araújo, P.H.H.; Oliveira, D.; Wurm, F.R.; Mailänder, V.; et al. Covalently Binding of Bovine Serum Albumin to Unsaturated Poly(Globalide-Co-ε-Caprolactone) Nanoparticles by Thiol-Ene Reactions. *Macromol. Biosci.* **2019**, 1900145. [CrossRef]

15. Lee, N.; Schuck, P.J.; Nico, P.S.; Gilbert, B.J. Surface enhanced Raman spectroscopy of organic molecules on magnetite (Fe_3O_4) nanoparticles. *Phys. Chem. Lett.* **2015**, *6*, 970–974. [CrossRef]

16. Guindani, C.; Candiotto, G.; Araújo, P.H.H.; Ferreira, S.R.S.; de Oliveira, D.; Wurm, F.R.; Landfester, K. Controlling the biodegradation rates of poly(globalide-*co*-ε-caprolactone) copolymers by post polymerization modification. *Polym. Degrad. Stab.* **2020**, *179*, 109287. [CrossRef]

17. Pommerville, J.C. *Alcamo's Fundamentals of Microbiology: Body Systems*; Jones & Bartlett Publishers: Burlington, MA, USA, 2009.

18. Shetty, V.; Jakhade, A.; Shinde, K.; Chikate, R.; Kaul-Ghanekar, R. Folate mediated targeted delivery of cinnamaldehyde loaded and FITC functionalized magnetic nanoparticles in breast cancer: In vitro, in vivo and pharmacokinetic studies. *New J. Chem.* **2021**, *45*, 1500–1515. [CrossRef]

19. Allard-Vannier, E.; Hervé-Aubert, K.; Kaaki, K.; Blondy, T.; Shebanova, A.; Shaitan, K.V.; Ignatova, A.A.; Saboungi, M.L.; Feofanov, A.V.; Chourpa, I. Folic acid-capped PEGylated magnetic nanoparticles enter cancer cells mostly via clathrin-dependent endocytosis. *Biochim. Biophys. Acta Gen. Subj.* **2017**, *1861*, 1578–1586. [CrossRef]

20. Gonen, N.; Assaraf, Y.G. Antifolates in cancer therapy: Structure, activity and mechanisms of drug resistance. *Drug Resist. Updat.* **2012**, *15*, 183–210. [CrossRef]

21. Turk, V.; Stoka, V.; Vasiljeva, O.; Renko, M.; Sun, T.; Turk, B.; Turk, D. Cysteine cathepsins: From structure, function and regulation to new frontiers. *Biochim. Biophys. Acta Proteins Proteomics* **2012**, *1824*, 68–88. [CrossRef] [PubMed]

22. Clark, A.C. Caspase Allostery and Conformational Selection. *Chem. Rev.* **2016**, *116*, 6666–6706. [CrossRef] [PubMed]

23. Guindani, C.; Dozoretz, P.; Veneral, J.G.; Silva, D.M.; Araújo, P.H.H.; Ferreira, S.R.S.; Oliveira, D.J. Enzymatic ring opening copolymerization of globalide and ε-caprolactone under supercritical conditions. *Supercrit. Fluids* **2017**, *128*, 404–411. [CrossRef]

24. Polloni, A.E.; Chiaradia, V.; Amaral, R.J.F.C.D.; Kearney, C.; Gorey, B.; De Oliveira, D.; De Oliveira, J.V.; De Araújo, P.H.H.; Sayer, C.; Heise, A. Polyesters with main and side chain phosphoesters as structural motives for biocompatible electrospun fibres. *Polym. Chem.* **2020**, *11*, 2157–2165. [CrossRef]

25. Zottis, A.D.A.; Beltrame, J.M.; Lara, L.R.S.; Costa, T.G.; Feldhaus, M.J.; Pedrosa, R.C.; Ourique, F.; de Campos, C.E.M.; de A Isoppo, E.; da Silva Miranda, F.; et al. Pheomelanin-coated iron oxide magnetic nanoparticles: A promising candidate for negative T 2 contrast enhancement in magnetic resonance imaging. *Chem. Commun.* **2015**, *51*, 11194–11197. [CrossRef]

26. Guo, X.; Shi, C.; Wang, J.; Di, S.; Zhou, S. PH-triggered intracellular release from actively targeting polymer micelles. *Biomaterials* **2013**, *34*, 4544–4554. [CrossRef] [PubMed]

27. Mosmann, T. Rapid colorimetric assay for cellular growth and survival: Application to proliferation and cytotoxicity assays. *J. Immunol. Methods* **1983**, *65*, 55–63. [CrossRef]

28. Gupta, J.; Bhargava, P.; Bahadur, D.J. Methotrexate conjugated magnetic nanoparticle for targeted drug delivery and thermal therapy. *Appl. Phys.* **2014**, *115*, 2012–2015. [CrossRef]

29. Chu, C.H.; Leung, C. The convolution equation of Choquet and Deny on [IN]-groups. *Integr. Equations Oper. Theory* **2001**, *40*, 391–402. [CrossRef]

30. Jung, J.Y.; Park, J.H.; Jeong, Y.J.; Yang, K.H.; Choi, N.K.; Kim, S.H.; Kim, W.J. Involvement of Bcl-2 family and caspases cascade in sodium fluoride-induced apoptosis of human gingival fibroblasts. *Korean J. Physiol. Pharmacol.* **2006**, *10*, 289–295.

31. Hanwell, M.D.; Curtis, D.E.; Lonie, D.C.; Vandermeersch, T.; Zurek, E.; Hutchison, G.R. Avogadro: An advanced semantic chemical editor, visualization, and analysis platform. *J. Cheminform.* **2012**, *4*, 17. [CrossRef]

32. Garrido, N.M.; Jorge, M.; Queimada, A.J.; Macedo, E.A.; Economou, I.G. Using molecular simulation to predict solute solvation and partition coefficients in solvents of different polarity. *Phys. Chem. Chem. Phys.* **2011**, *13*, 9155. [CrossRef]

33. Nedyalkova, M.A.; Madurga, S.; Tobiszewski, M.; Simeonov, V.J. Calculating the Partition Coefficients of Organic Solvents in Octanol/Water and Octanol/Air. *Chem. Inf. Model.* **2019**, *59*, 2257–2263. [CrossRef] [PubMed]

34. Candiotto, G.; Giro, R.; Horta, B.A.C.; Rosselli, F.P.; De Cicco, M.; Achete, C.A.; Cremona, M.; Capaz, R.B. Emission redshift in DCM2-doped Alq3 caused by nonlinear Stark shifts and Förster-mediated exciton diffusion. *Phys. Rev. B* **2020**, *102*, 1–7. [CrossRef]

35. Bauer, K.N.; Liu, L.; Wagner, M.; Andrienko, D.; Wurm, F.R. Mechanistic study on the hydrolytic degradation of polyphosphates. *Eur. Polym. J.* **2018**, *108*, 286–294. [CrossRef]
36. Neese, F. The ORCA program system. *Wiley Interdiscip. Rev. Comput. Mol. Sci.* **2012**, *2*, 73–78. [CrossRef]
37. Guindani, C.; Dozoretz, P.; Araújo, P.H.H.; Ferreira, S.R.S.; de Oliveira, D. N-acetylcysteine side-chain functionalization of poly(globalide-*co*-ε-caprolactone) through thiol-ene reaction. *Mater. Sci. Eng. C* **2019**, *94*, 477–483. [CrossRef]
38. Pachence, J.M.; Bohrer, M.P.; Kohn, J. *Principles of Tissue Engineering*, 3rd ed.; Lanza, R., Langer, R., Vacanti, J., Eds.; Elsevier: Amsterdam, The Netherlands, 2007; pp. 323–339.
39. Bee, A.; Massart, R.; Neveu, S.J. Synthesis of very fine maghemite particles. *Magn. Magn. Mater.* **1995**, *149*, 6–9. [CrossRef]
40. Habibi, N. Preparation of biocompatible magnetite-carboxymethyl cellulose nanocomposite: Characterization of nanocomposite by FTIR, XRD, FESEM and TEM. *Spectrochim. Acta Part A Mol. Biomol. Spectrosc.* **2014**, *131*, 55–58. [CrossRef]
41. Lee, N.; Hyeon, T. Designed synthesis of uniformly sized iron oxide nanoparticles for efficient magnetic resonance imaging contrast agents. *Chem. Soc. Rev.* **2012**, *41*, 2575–2589. [CrossRef]
42. Hadadian, Y.; Masoomi, H.; Dinari, A.; Ryu, C.; Hwang, S.; Kim, S.; Cho, B.K.; Lee, J.Y.; Yoon, J. From Low to High Saturation Magnetization in Magnetite Nanoparticles: The Crucial Role of the Molar Ratios between the Chemicals. *ACS Omega* **2022**, *7*, 15996–16012. [CrossRef]
43. Celis, J.A.; Mejía, O.F.O.; Cabral-Prieto, A.; García-Sosa, I.; Derat-Escudero, R.; Saitovitch, E.M.B.; Camarena, M.A. Alzamora Camarena, M. Synthesis and characterization of nanometric magnetite coated by oleic acid and the surfactant CTAB: Surfactant coated nanometric magnetite/maghemite. *Hyperfine Interact.* **2017**, *238*, 43. [CrossRef]
44. Kumar, A.; Dixit, C.K. Methods for characterization of nanoparticles. In *Advances in Nanomedicine for the Delivery of Therapeutic Nucleic Acids*; Elsevier: Amsterdam, The Netherlands, 2017; pp. 43–58.
45. Gunguli, N.C. Paper disk electrophoresis in the determination of isoelectric points of amino acids. *Fresenius' Zeitschrift Anal. Chem.* **1956**, *154*, 161–167. [CrossRef]
46. Yoo, J.-W.; Chambers, E.; Mitragotri, S. Factors that Control the Circulation Time of Nanoparticles in Blood: Challenges, Solutions and Future Prospects. *Curr. Pharm. Des.* **2010**, *16*, 2298–2307. [CrossRef] [PubMed]
47. Brandt, J.V.; Piazza, R.D.; Santos, C.C.D.; Vega-Chacón, J.; Amantéa, B.E.; Pinto, G.C.; Jafelicci, M.; Marques, R.F.C. Synthesis of core@shell nanoparticles functionalized with folic acid-modified PCL-*co*-PEGMA copolymer for methotrexate delivery. *Nano-Struct. Nano-Objects* **2021**, *25*, 100675. [CrossRef]
48. Piazza, R.D.; Brandt, J.V.; Gobo, G.G.; Tedesco, A.C.; Primo, F.L.; Marques, R.F.C.; Junior, M.J. mPEG-*co*-PCL nanoparticles: The influence of hydrophobic segment on methotrexate drug delivery. *Colloids Surf. A Physicochem. Eng. Asp.* **2018**, *555*, 142–149. [CrossRef]

pharmaceutics

Article

Neuroprotective Nanoparticles Targeting the Retina: A Polymeric Platform for Ocular Drug Delivery Applications

Patrizia Colucci [1,2,3,4,*], Martina Giannaccini [1], Matteo Baggiani [1], Breandán N. Kennedy [3,4], Luciana Dente [1], Vittoria Raffa [1] and Chiara Gabellini [1,*]

1 Department of Biology, University of Pisa, 56127 Pisa, Italy
2 Department of Biotechnology, Chemistry and Pharmacy, University of Siena, 53100 Siena, Italy
3 UCD Conway Institute, University College Dublin, D04 V1W8 Dublin, Ireland
4 UCD School of Biomolecular and Biomedical Science, University College Dublin, D04 V1W8 Dublin, Ireland
* Correspondence: patrizia.colucci@student.unisi.it or patrizia.colucci1@ucd.ie (P.C.); chiara.gabellini@unipi.it (C.G.)

Abstract: Neuroprotective drug delivery to the posterior segment of the eye represents a major challenge to counteract vision loss. This work focuses on the development of a polymer-based nanocarrier, specifically designed for targeting the posterior eye. Polyacrylamide nanoparticles (ANPs) were synthesised and characterised, and their high binding efficiency was exploited to gain both ocular targeting and neuroprotective capabilities, through conjugation with peanut agglutinin (ANP:PNA) and neurotrophin nerve growth factor (ANP:PNA:NGF). The neuroprotective activity of ANP:PNA:NGF was assessed in an oxidative stress-induced retinal degeneration model using the teleost zebrafish. Upon nanoformulation, NGF improved the visual function of zebrafish larvae after the intravitreal injection of hydrogen peroxide, accompanied by a reduction in the number of apoptotic cells in the retina. Additionally, ANP:PNA:NGF counteracted the impairment of visual behaviour in zebrafish larvae exposed to cigarette smoke extract (CSE). Collectively, these data suggest that our polymeric drug delivery system represents a promising strategy for implementing targeted treatment against retinal degeneration.

Keywords: polymeric nanoparticles; drug delivery and targeting; ocular posterior segment; oxidative stress; retinal degeneration; nerve growth factor; peanut agglutinin; zebrafish

Citation: Colucci, P.; Giannaccini, M.; Baggiani, M.; Kennedy, B.N.; Dente, L.; Raffa, V.; Gabellini, C. Neuroprotective Nanoparticles Targeting the Retina: A Polymeric Platform for Ocular Drug Delivery Applications. *Pharmaceutics* **2023**, *15*, 1096. https://doi.org/10.3390/pharmaceutics15041096

Academic Editor: Ana Isabel Fernandes

Received: 14 February 2023
Revised: 24 March 2023
Accepted: 28 March 2023
Published: 29 March 2023

1. Introduction

Diseases affecting the posterior segment of the eye, including glaucoma, age-related macular degeneration (AMD) and retinopathies, represent a debilitating and increasingly frequent worldwide health problem, leading to partial or complete blindness [1]. Oxidative stress plays a key role in the pathogenesis of ocular degenerative diseases. In fact, the retina is extensively exposed to light and is a highly metabolic tissue requiring enormous oxygen consumption. Consequently, this tissue is strongly affected by oxidative stress [2]. Moreover, in vitro studies showed that exposure to cigarette smoke extract (CSE), a mixture of genotoxic and carcinogenic compounds, induced oxidative stress in human retinal cells and injured primary retinal ganglion cells in rats [3,4]. Systemic treatment with CSE alters the retinal structures as well as the visual behaviour of zebrafish larvae [5], thus furthering the contribution of smoking in the pathogenesis of degenerative diseases.

Neurotrophins are extensively involved in the molecular response against oxidative damage [6–8]. Nerve growth factor (NGF), the most well-characterised member of the family, as well as its receptors are widely expressed in the visual system, thus suggesting a crucial role in promoting retinal development and cell survival [9,10]. Several experimental studies demonstrated that NGF exerts neuroprotective and regenerative effects on multiple retinal cell populations [11–14] through the modulation of several molecular pathways [15,16]. However, the clinical applicability of neurotrophins such as NGF is limited by rapid degradation and clearance in vivo [17]. Moreover, the delivery of drugs to the posterior segment of the eye is a challenging goal due to the anatomical and physiological barriers that protect this organ [18]. Nanoscale materials represent a minimally invasive method in ophthalmology for overcoming such limitations, offering a wide range of physicochemical characteristics to increase the half-life and solubility of drugs, thus improving their pharmacokinetic and pharmacodynamic profiles [19,20]. One of the most appealing characteristics of a nanodevice for the ocular posterior segment is the capability to accumulate in a specific retinal layer and to be retained for a sustained drug release over time, hence reducing the risks associated with multiple injections and side effects caused by a bolus dosage of drugs. Indeed, the number of patents for nanotechnology-based formulations used for ocular drug delivery is increasing and includes a wide range of materials with different properties [21]. However, the development of patient-friendly drug delivery systems for controlled release of therapeutics at the posterior segment of the eye is still a major challenge [22].

The teleost zebrafish (*Danio rerio*) represents an appealing model organism to study human degenerative diseases in a complex biological system and to validate new therapeutic strategies for the posterior segment of the eye. Interestingly, the zebrafish eye shares equivalent anatomical and cell-layered retinal structures with humans, and visual function is well established after only 3 days post-fertilisation (dpf), thus allowing short-term studies [23–25]. Giannaccini and coauthors demonstrated that magnetic nanoparticles (MNPs) are able to localise at the ocular posterior segment after intravitreal injection in *Danio rerio* (zebrafish) larvae [26–28]. In a retinal degeneration model previously developed by our group, protection against oxidative stress by NGF as well as by brain-derived neurotrophic factor (BDNF) is notably improved when they are bound to MNPs compared to the free neurotrophins, which have short half-lives and weak efficacy after in vivo administration. Thus, MNPs not only are good carriers to reach the retinal layers, but they also protect molecules from fast degradation [28].

Although MNPs are biocompatible and already approved for use in humans, they pose some concerns because of their inorganic nature [29,30]. Organic-based polymeric nanoparticles are the most used nanosystems to overcome these limitations. They can be developed from natural and synthetic biocompatible polymers and exploited in ocular drug and gene delivery [31–33]. Among them, polymeric acrylic nanoparticles possess exclusive properties that make them suitable for drug delivery applications [34,35]. The high stability and biocompatibility of polyacrylic derivatives have been widely studied, and the presence of numerous functional groups perfectly addresses the need to load drugs and enable sustained release, further enhanced by their unique properties of being responsive to external and internal stimuli [36,37].

Here, we describe a polymeric nanocarrier based on acrylic monomers, specifically designed for targeting the ocular posterior segment thanks to functionalisation with peanut agglutinin (PNA). In particular, polyacrylamide nanoparticles (ANPs) were synthesised and characterised, and their high binding efficiency was exploited for implementing a polymer-based, ocular-targeting, neuroprotective drug delivery system. The biological activity of the ANP:PNA-conjugated NGF as well as its neuroprotective efficacy have been validated in vitro. To deeply investigate the protective effect of our nanoformulation as a therapeutical strategy for the ocular posterior segment, we performed in vivo studies in zebrafish. Using the previously established zebrafish model of visual impairment induced by oxidative stress [28], we demonstrated by both behavioural and functional assessments in zebrafish larvae that only the ANP:PNA-conjugated NGF, but not the free neurotrophin, is able to protect the retina from the oxidative damage induced by the intravitreal injection of hydrogen peroxide. We additionally assessed our NGF-coated polymeric nanocarrier as a neuroprotective treatment against the visual impairment of zebrafish larvae caused by the pharmacological exposure to CSE, using a model recently described [5]. Collectively, these data suggest that our polymeric drug delivery system represents a promising strategy for implementing treatment against retinal degeneration.

2. Materials and Methods

2.1. Chemicals

N-Isopropylacrylamide (NIPAM; #731129), acrylamide (AAm; #A9099), allylamine hydrochloride (AH; #735132), *N,N′*-Methylenebisacrylamide (BIS; #M1533), sodium dodecyl sulfate (SDS; #436143), ammonium persulfate (APS; #A3678), N,N,N′,N′-Tetramethyl ethylenediamine (TEMED; #T9281), FITC-conjugated PNA from *Arachis hypogaea* (PNA; #L7381), poly-L-lysine (#P4707), DMEM F-12 Ham (#D6421), hydrogen peroxide (H_2O_2; #H1009), thiazolyl blue tetrazolium bromide (#M5655) and dimethyl sulfoxide (DMSO; #276855) were purchased from Merck (Darmstadt, Germany). Fluorescent dyes Alexa Fluor 488 (#A20000), Alexa Fluor 594 (#A20004), Hoechst 33342 (#H3570) and secondary antibody (#A21244) were purchased from Invitrogen (Carlsbad, CA, USA). Native mouse NGF 2.5 S protein (NGF; #N-100) was acquired from Alomone labs (Jerusalem, Israel). Dulbecco's Modified Eagle's Medium (DMEM; #21969), horse serum (HS; #16050), fetal bovine serum (FBS; #10270), penicillin/streptomycin (Pen Strep; #15140), GlutaMAX™ (#35050) and L-Glutamine (#25030) were purchased from Gibco (Thermo Fisher Scientific, Waltham, MA, USA). OCT embedding compound for cryostat (#05-9801) was acquired from Bio-Optica (Milano, Italy). Cigarette smoke extract (CSE) was purchased from Murty Pharmaceuticals (Lexington, KY, USA). Cleaved caspase-3 primary antibody (#9661) was acquired from Cell Signaling Technology (Danvers, MA, USA).

2.2. Nanoparticle Synthesis

2.2.1. Synthesis of Fluorescent Polyacrylamide Nanoparticles (ANPs)

Fluorescent polyacrylamide nanoparticles were synthesised by optimising the radical polymerisation protocol developed by Rahimi and colleagues [38]. A total of 222 mg of NIPAM, 28.6 mg of AAm, 76 mg of AH and 262 µL of BIS were dissolved in 10 mL of deionised water previously purged with argon at room temperature and under stirring. Next, 11.5 µL of 10% SDS solution was added, and the solution was purged with argon for 30 min. Then, 1 mg of Alexa Fluor 488 or Alexa Fluor 594 dyes was dissolved to stop the argon flow, and 15.6 mg of APS and 20 µL of TEMED were added. Finally, deionised water was added to the final volume of 20 mL, and the reaction was carried out at room temperature for 3 h under continuous stirring in darkness. Dialysis was performed by transferring the sample into a 10 kDa cut-off membrane kept in deionised water under stirring and replaced with new water 4 times per hour. Finally, the sample was concentrated in a 30 kDa Vivaspin® tube (Sartorius, Goettingen, Germany) by centrifuging at $3000\times g$ until 1 mL of nanoparticle suspension was obtained, and then kept at 4 °C.

2.2.2. Functionalisation of Nanoparticles with Peanut Agglutinin (ANP:PNA) and Nerve Growth Factor (ANP:PNA:NGF)

ANPs coated with FITC-conjugated PNA (ANP:PNA) were obtained through a non-covalent functionalisation of the particles by incubating a 1:1 ratio of ANPs and PNA at 4 °C for 1.5 h under stirring. The sample was concentrated using a 100 kDa Vivaspin® tube (Sartorius), washed with deionised water and kept at −20 °C in a 20% glycerol solution, while the supernatant derived from the washing steps was stored for further analysis. The functionalisation with NGF and PNA was carried out by incubating a 1:1:1 ratio of ANPs, PNA and NGF (ANP:PNA:NGF), respectively, following the same protocol as above.

2.3. Nanoparticle Characterisation

The concentration of PNA conjugated to the ANPs was estimated by BCA assay (Pierce™ BCA Protein Assay Kit; #23227, Thermo Scientific, Waltham, MA, USA), according to the manufacturer's protocol. An ELISA assay (Mouse beta-NGF ELISA Kit; #RAB1119, Merck, Darmstadt, Germany) was performed to evaluate the amount of NGF bound to the ANP:PNA. Both quantifications were performed by using the supernatant derived from the washing steps of the functionalisation reactions. The concentration of PNA and NGF attached to the ANPs' surface was extrapolated through subtraction by measuring the supernatants' absorbance at 562 and 450 nm, respectively. The protein loading efficiency of the ANPs was calculated by using the following formula: (mass of protein in nanoformulation/total mass of protein) $\times$ 100.

The size distribution and polydispersity index of naked, single and double-functionalised ANPs were assessed by dynamic light scattering (DLS) using a Zetasizer NS (Malvern Panalytical, Malvern, UK).

2.4. In Vitro Studies

2.4.1. Assessment of ANP:PNA:NGF Bioactivity

Rat pheochromocytoma PC12 cells were obtained from the American Type Culture Collection (ATCC). Cells were cultured in DMEM supplemented with 10% HS, 5% FBS, 100 IU/mL penicillin, 100 µg/mL Pen Strep and 1% GlutaMAX™ in T25 flasks coated with 10 µg/mL poly-L-lysine. Cells were incubated in a saturated humidity atmosphere at 37 °C and 5% CO_2. For the differentiation, PC12 cells were seeded at low density (2.5×10^4 cells per cm^2) and maintained in DMEM with 1% FBS and 100 ng/mL NGF or ANP:PNA:NGF for 4 days. Cells untreated or incubated with ANP:PNA were used as controls. Then, cells were washed with DPBS and fixed in 2% paraformaldehyde (PFA) at room temperature for 10 min. Images were acquired at $10\times$ magnification with a Nikon Eclipse TE2000-U microscope, and the length of 180 neurites (randomly selected) was evaluated by using the plugin NeuronJ available on ImageJ software (version 1.53t). Independent experiments were performed, $n = 3$.

2.4.2. Assessment of ANP:PNA:NGF Protective Effect

Human retinal pigment epithelial ARPE-19 cells were obtained from the American Type Culture Collection (ATCC). Cells were maintained in DMEM F-12 Ham supplemented with 10% FBS, 2% Pen Strep and 2 mM L-Glutamine in a humidified atmosphere at 37 °C and 5% CO_2. Cells were seeded at density of 10×10^3 cells per well in a 96-well plate. The next day, cells were co-incubated with 100 ng/mL of free or ANP:PNA-conjugated NGF and 300 µM of H_2O_2. Untreated cells were used as negative control, while the positive control group was only treated with 300 µM of H_2O_2. After 24 h, the treatment solution was removed, and an MTT assay was performed. Briefly, 100 µL of 0.5 mg/mL thiazolyl blue tetrazolium bromide solution was added, and cells were incubated for 2.5 h. Then, 100 µL of DMSO was added and incubated for a few minutes on a plate shaker to dissolve the water-insoluble formazan salt. Quantification was carried out using a CLARIOstar microplate reader, and the absorbance was measured at 595 nm. Independent experiments were performed, $n = 3$.

2.5. In Vivo Studies Using Zebrafish

2.5.1. Animal Handling

Animals were handled in compliance with protocols approved by the Italian Ministry of Public Health, the local Ethical Committee of the University of Pisa and the University College Dublin Animal Research Ethics Committee (AREC), in conformity with EU legislation (Directive 2010/63/EU). Zebrafish embryos were obtained by natural mating of wild-type fish and maintained in the incubator at 28 °C in E3 media according to the ZFIN procedures. Before any injection at 3, 4, and 5 days post-fertilisation (dpf), larvae were anesthetised in 0.02% tricaine and embedded in 0.3% agarose.

2.5.2. Nanoparticle Localisation Studies

A total of 2 nL of green fluorescent Alexa Fluor 488-loaded ANPs or DPBS, as control, were intravitreally (IVT) microinjected into the left eye of 3 dpf larvae. Larvae were firstly observed under a stereomicroscope using both bright field and FITC channel to evidence the Alexa Fluor 488 fluorescence in vivo. Then, they were fixed after 4, 8 and 24 h post-injection (hpi) in a 4% PFA solution at room temperature for 10 min, embedded in OCT and sliced with a cryostat, obtaining 10 µm cross-sections. The sections were stained with

Hoechst 33342 (1:1000) to distinguish the retinal layers and imaged with a Nikon Eclipse-Ti microscope at 20× magnification. The presence of green fluorescent nanoparticles at the ganglion cell layer (GCL), inner nuclear layer (INL), photoreceptors layer (PR) and retinal pigment epithelium (RPE) was qualitatively evaluated. The retinal distribution of ANPs was assessed on 15 larvae per group, $n = 3$.

To evaluate the effect of PNA conjugation on nanoparticle localisation, 2 nL of naked Alexa Fluor 594-positive ANPs or FITC-labelled ANP:PNA or DPBS, as control, were IVT microinjected into the left eye of larvae at 3 dpf. The injected eyes were in vivo imaged at 4 and 24 hpi with a Nikon SMZ18 stereomicroscope, using bright field, TRITC and FITC channels to check the fluorescence of Alexa Fluor 594 and the PNA: FITC-conjugated, respectively. Independent experiments were carried out on 15 larvae per group, $n = 3$.

2.5.3. Neuroprotection Studies

A volume of 2 nL, corresponding to 1 ng of free NGF or ANP:PNA:NGF or DPBS, was IVT microinjected into the left eye of larvae at 4 dpf. At the later stage of 5 dpf, larvae were injected in the same eye with 2 nL of 1 M H_2O_2 to induce oxidative damage as previously described by Giannaccini and coauthors [28]. In addition, 2 nL of DPBS was injected as negative control.

Similarly, 1 ng of free NGF or ANP:PNA:NGF or DPBS was IVT microinjected into one eye of zebrafish larvae at 4 dpf. After 16 hpi, larvae were immersed in E3 embryo medium containing 20 µg/mL cigarette smoke extract (CSE) or 0.05% DMSO as control for 24 h, following the protocol described by Gómez Sánchez and colleagues [5].

2.5.4. Optokinetic Response (OKR) Assay

Visual function was assessed 8 h after the injection of hydrogen peroxide via optokinetic response (OKR) assay, as previously described [39]. Briefly, the larvae were embedded in 6% methylcellulose in a 3.5 mm Petri dish and then were placed in the centre of a drum with black and white stripes rotating at 11 rpm, and the ocular movements (saccades) of the left eye per minute were quantified. Each experiment was performed on at least 9 larvae per group, $n = 4$.

Optokinetic response was assessed 24 h after the exposure to CSE, according to the protocol previously published by Gómez Sánchez and collaborators [40]. Each experiment was performed on at least 8 larvae per group, $n = 3$.

2.5.5. Whole-Mount Immunostaining

At 8 h after the induction of oxidative stress, larvae were fixed in 4% PFA overnight at 4 °C and maintained in bleach solution (0.18 M KOH and 3% H_2O_2 in ddH_2O) at room temperature for 40 min to remove the pigmentation of the RPE. The whole-mount immunostaining was carried out following an optimised protocol for zebrafish retinal samples from Inoue and Wittbrodt [41]. To enhance tissue permeabilisation, larvae were first heated in a 150 mM Tris-HCl pH = 9 solution at 70 °C for 15 min and then incubated with acetone at −20 °C for 20 min. Then, they were blocked (10% goat serum, 0.8% TritonX, 1% BSA and 1% DMSO in 1X PBS + 0.1% Tween) at 4 °C for 3 h and incubated with cleaved caspase-3 primary antibody (1:250) diluted in 1% goat serum, 0.8% Triton X-100 and 1% BSA in 1X PBS + 0.1% Tween at 4 °C for three days. Samples were sequentially washed, then incubated with the secondary antibody (1:500) and Hoechst 33342 (1:100) in the dark for two and half days. After washing, larvae were mounted on glass slides using Aqua/PolyMount and a #1.5 glass coverslip. All images were acquired by a Nikon A1 confocal microscope at 40× magnification with 2.5 µm interval Z-stacks. Quantification of active caspase 3 (aCasp3)-positive cells was performed using ImageJ software (version 1.53t) (Figure S1). Each experiment was performed on at least 6 larvae per group, $n = 3$.

2.6. Statistical Analysis

All data were reported as the mean ± standard error of the mean and analysed on GraphPad Prism 7. Neurites' length was analysed by t-test for unpaired data followed by Kolmogorov–Smirnov's test. Cell viability was analysed by one-way ANOVA followed by Tukey's multiple comparisons test. OKR data were analysed by Kruskal–Wallis's test, followed by Dunn's test, one-way ANOVA and finally by Tukey's test. Apoptosis data were analysed by one-way ANOVA followed by Dunnett's test. Normal distribution of differences was tested by the D'Agostino and Pearson test, and significance levels were $p < 0.05$ (*), $p < 0.01$ (**), $p < 0.001$ (***) and $p < 0.0001$ (****).

3. Results

3.1. Nanoparticle Characterisation

To exploit polyacrylamide nanoparticles as nanocarriers for ocular drug delivery application, peanut agglutinin (PNA) and nerve growth factor (NGF) were conjugated to the nanocarrier through noncovalent functionalisation. PNA, a lectin from *Arachis hypogea*, is widely known to selectively recognise and bind cone photoreceptors [42], while NGF is a neurotrophin with established activity in counteracting retinal degeneration [11–13]. To estimate the loading of proteins onto the nanoparticles, BCA and ELISA assays were carried out for quantifying conjugated PNA and NGF, respectively. The PNA and NGF binding efficiency was 62% and 91%, respectively, as the loaded content of proteins was 0.62 µg of PNA and 0.91 µg of NGF per µL of particles suspension.

Size distribution and polydispersity index of the functionalised nanoparticles were also characterised by DLS measurements (Figure 1).

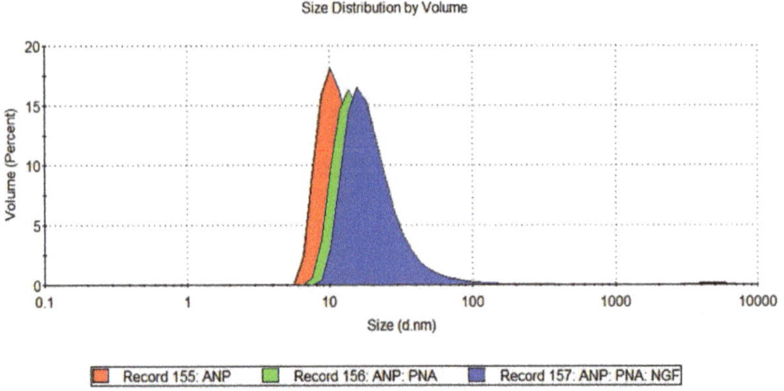

Figure 1. Nanoparticle characterisation through dynamic light scattering measurements. The hydrodynamic diameter increase of single- (ANP:PNA, green curve) and double- (ANP:PNA:NGF, blue curve) functionalised polyacrylamide nanoparticles resulted in a shift of size distribution compared to naked ANPs (red curve).

ANP:PNA:NGF showed an increase in size distribution (22.81 ± 3.17 nm) compared to the PNA-conjugated ANPs (19.97 ± 2.41 nm) and naked ANPs (13.89 ± 1.74 nm), as well as an improvement in the polydispersity index that was 0.18 ± 0.03 for ANP:PNA:NGF, 0.47 ± 0.09 for ANP:PNA and 0.44 ± 0.03 for the naked ANPs (Table 1).

Table 1. Nanoparticle characterisation. Hydrodynamic diameter and polydispersity index of naked and functionalised polyacrylamide nanoparticles. Mean ± SEM, *n* = 3.

Nanoparticles	Functionalisation	Hydrodynamic Diameter	Polydispersity Index
ANPs	Alexa Fluor 594	13.89 ± 1.74 nm	0.44 ± 0.03
ANP:PNA	Alexa Fluor 594; PNA	19.97 ± 2.41 nm	0.47 ± 0.09
ANP:PNA:NGF	Alexa Fluor 594; PNA; NGF	22.81 ± 3.17 nm	0.18 ± 0.03

3.2. Peanut Agglutinin Is Essential for Targeting the Ocular Posterior Segment In Vivo

To study the localisation of the polymeric unbound nanocarrier and investigate its retinal distribution over time, 3 dpf larvae were IVT injected with ANPs containing the green fluorescent dye Alexa Fluor 488, then fixed at 4, 8 and 24 hpi, cryosectioned and imaged in DAPI and FITC channels (Figure 2a). By analysing cross-sections at different time points, we observed a different profile of particle distribution through the retinal layers over time. At 4 hpi, green fluorescent spots, corresponding to Alexa Fluor 488-positive ANPs, were found to localise at both the GCL and INL. At 8 hpi, they were also observed at the PR and the RPE layers, where the Alexa Fluor 488 signal was mainly present at 24 hpi and no particles were found at the GCL, thus suggesting their spontaneous migration over time (Figure 2b). Interestingly, no ANPs were found in the right eye used as uninjected control (Figure S2).

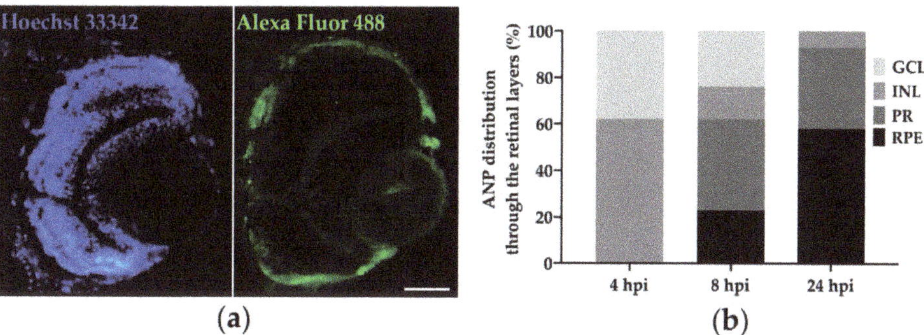

Figure 2. Nanoparticle distribution through retinal layers after intravitreal injection in zebrafish larvae. (**a**) Representative images of a retinal cross-section of a zebrafish larva IVT injected with Alexa Fluor 488-positive polyacrylamide nanoparticles, analysed at 4 h post-injection (hpi). Hoechst staining allows identification of the retinal layers. In FITC channel, the intense green spots clearly highlight the presence of Alexa Fluor 488-positive nanoparticles in the injected eye. (**b**) Quantitative evaluation of the distribution of Alexa Fluor 488-positive polyacrylamide nanoparticles in the retinal layers at different time points. *n* = 15 injected eyes from 3 independent experiments. GCL, ganglion cell layer; INL, inner nuclear layer; PR, photoreceptors; RPE, retinal pigment epithelium. Scale bar: 50 μm.

Nevertheless, preliminary in vivo whole-mount imaging suggested the presence of green fluorescent ANPs not only in the injected eye but also at the pronephros, where they already accumulated by 4 hpi, indicating that the nanocarrier was not stably retained in the ocular environment. To confirm that the PNA conjugation enhances the ANPs capability to target the ocular posterior segment, nanoparticles containing a red fluorescent Alexa Fluor 594 dye were functionalised with an FITC-conjugated PNA (ANP:PNA) in order to identify the distribution profile of the carrier (red fluorescent ANPs) and the cargo (green fluorescent PNA). Naked and PNA-conjugated ANPs were IVT injected in 3 dpf larvae, and the fluorescence was checked after 4 and 24 hpi (Figure 3). After 4 hpi, a strong red signal was observed in the left eye of both injected groups. However, only the larvae injected with the naked ANPs showed the presence of nanoparticles at the level of the pronephros at 24 hpi, confirming the evidence previously described and suggesting their instability in the ocular environment. On the contrary, the injected PNA-conjugated ANPs were retained in the eye until 24 hpi, as observed in 100% of injected larvae. Indeed, the stable localisation of ANP:PNA was indicated by the presence of green fluorescence of the FITC-conjugated PNA in the left eye at all the time points and its absence in any other tissue of the larvae, thus suggesting a crucial improvement of the polymeric nanoformulation as an ocular drug delivery nanocarrier. No fluorescence was found in control larvae injected with saline solution, as expected (Figure 3).

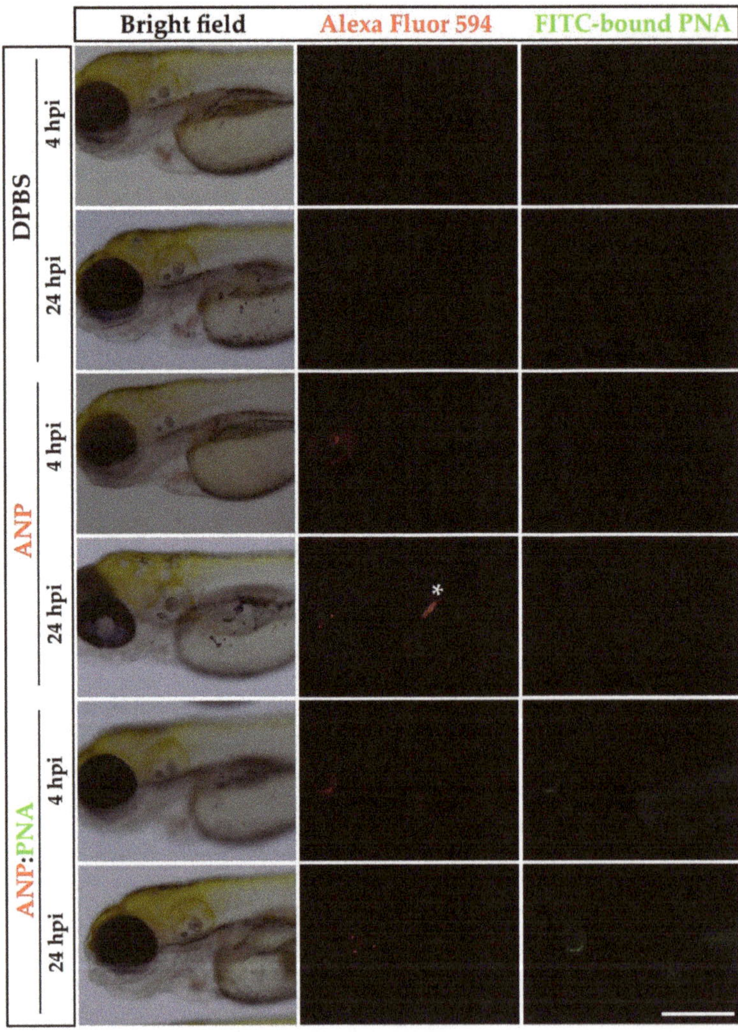

Figure 3. Prolonged localisation of nanoparticles at the site of injection. Representative images of 3 days post-fertilisation (dpf) larvae IVT injected with naked Alexa Fluor 594-positive nanoparticles (ANPs) or FITC-labelled PNA-conjugated polyacrylamide nanoparticles (ANP:PNA), analysed at 4 and 24 h post-injection (hpi). As control, zebrafish larvae were IVT injected with DPBS. White asterisk indicates the pronephros. n = 45 injected eyes from 3 independent experiments. Scale bar: 300 μm.

3.3. Nanoformulated NGF Preserves Human Retinal Cells Viability in Oxidative Conditions

The biological activity of the ANP:PNA-conjugated NGF was assessed on pheochromocytoma PC12 cells, since they differentiate in a neuron-like phenotype in response to the NGF treatment [43]. After 4 days of treatment, PC12 cells displayed the presence of long neurites in both groups treated with free or conjugated NGF, thus confirming the NGF bioactivity through its ability of triggering neurite formation (Figure 4a). As expected, untreated and ANP:PNA-treated control cells maintained a round-shape morphology. Additionally, neurite length was comparable between the groups exposed to free NGF (127.5 $\pm$ 4.17 μm) and the nanoformulation (129 $\pm$ 3.74 μm), as shown in Figure 4b, suggesting that the functionalisation process did not affect the biological activity of the NGF.

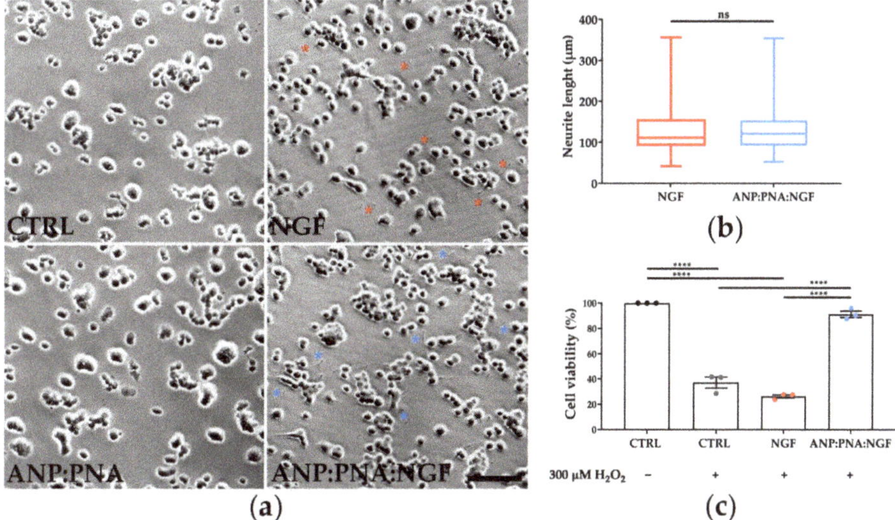

Figure 4. NGF exerts its biological activity upon nanoformulation. (**a**) Representative images of PC12 cells incubated for 4 days with 100 ng/mL of free NGF or with the same concentration of NGF bound to PNA-conjugated polyacrylamide nanoparticles (ANP:PNA:NGF). As control, PC12 cells were incubated with PNA-conjugated nanoparticles (ANP:PNA) or DPBS. Red and blue asterisks indicate the presence of neurites in cells exposed to NGF and ANP:PNA:NGF, respectively. Scale bar: 300 μm. (**b**) Quantitative evaluation of neurite length in PC12 cells incubated with NGF or ANP:PNA:NGF. n = 180 neurites from 3 biological replicates, t-test for unpaired data followed by Kolmogorov–Smirnov test. (**c**) Protective effect of NGF-conjugated polyacrylamide nanoparticles against oxidative stress. ARPE-19 cells were cotreated with 100 ng/mL of free NGF (NGF) or nanoformulated NGF (ANP:PNA:NGF) and 300 μM of H_2O_2 for 24 h. MTT assay was performed to evaluate cell viability. Absorbance values of all the groups were normalised to those of the untreated control sample (CTRL), which was set at 100%. n = 3 biological replicates, one-way ANOVA applying Tukey's multiple comparisons test. (**b**,**c**) ns, $p > 0.05$ and ****, $p < 0.0001$.

To validate the neuroprotective effect of our NGF-based nanoformulation, human retinal pigment epithelium ARPE-19 cells were incubated with hydrogen peroxide, to mimic an oxidative condition as occurs in retinal degeneration (Figure 4c). Cell viability was assessed by measuring the metabolic activity by performing an MTT assay after 24 h of co-incubation of cells with hydrogen peroxide and NGF-conjugated nanoparticles or free NGF. As expected, data showed a cytotoxic effect in the H_2O_2-treated group compared to the untreated control (****, $p < 0.0001$) as oxidative damage arose. While free neurotrophin was not able to counteract H_2O_2-induced cell mortality, cellular metabolic activity was preserved by nanoformulated NGF (ANP:PNA:NGF) compared to the H_2O_2-treated group (****, $p < 0.0001$), suggesting a pronounced protective effect only with the NGF nanoformulation.

3.4. Nanoformulated NGF Improves Visual Function in Zebrafish Larvae and Partially Protects Retinal Cells from Oxidative Stress-Triggered Apoptosis

To further confirm the neuroprotective activity of the ANP:PNA-conjugated NGF, in vivo experiments resembling a condition of retinal degeneration were performed in zebrafish larvae, by inducing a retinal oxidative damage previously developed by our research group [28]. Free NGF or nanoformulated neurotrophin (ANP:PNA:NGF) was IVT injected as a preventive treatment against an oxidative injury obtained through the intravitreal injection of hydrogen peroxide. To evaluate the effect of the nanoformulated neurotrophin in preventing vision impairment, visual function of injected larvae was evaluated by the optokinetic response (OKR) assay (Figure 5a).

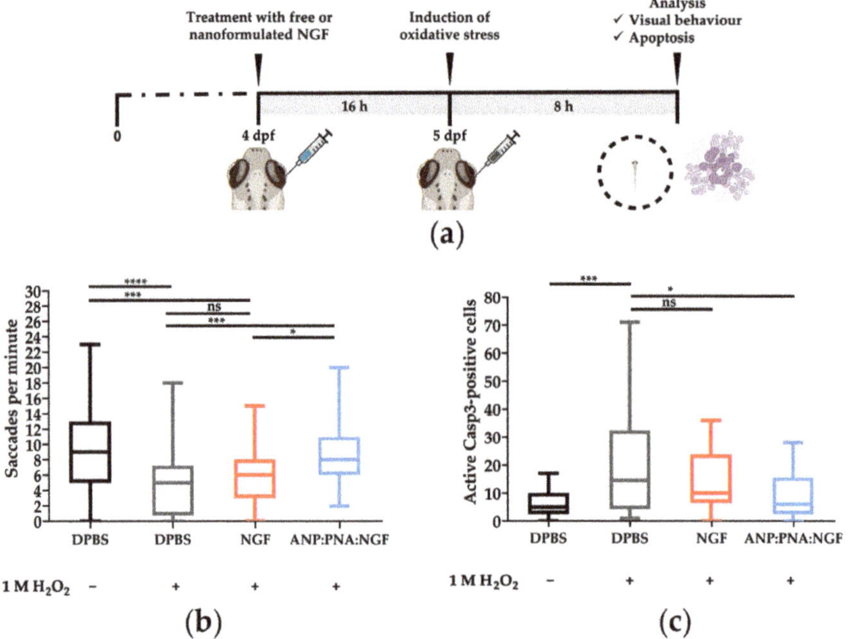

Figure 5. Nanoformulated NGF improves visual function impaired by oxidative stress in zebrafish larvae. (**a**) Experimental timeline relative to the intravitreal injection of free NGF or NGF bound to PNA-conjugated polyacrylamide nanoparticles (ANP:PNA:NGF) in zebrafish larvae at 4 days post-fertilisation (dpf), followed by H_2O_2 exposure to induce oxidative stress in the retina. As control, zebrafish larvae were IVT injected with DPBS. (**b**) Evaluation of visual function by OKR assay in zebrafish larvae exposed to the treatments shown in panel (**a**). $n \geq 39$ larvae for each group, Kruskal–Wallis followed by Dunn's multiple comparisons test. (**c**) Quantitative evaluation of apoptosis through the detection of active caspase 3 (Casp3)-positive cells by immunofluorescence in zebrafish larvae exposed to the treatments shown in panel (**a**). $n \geq 20$ larvae for each group, one-way ANOVA using Dunnett's multiple comparisons test. (**b**,**c**) ns, $p > 0.05$; *, $p < 0.01$; ***, $p < 0.001$ and ****, $p < 0.0001$.

Data showed a statistically significant reduction (****, $p < 0.0001$) in the ocular movements (saccades) recorded in the larvae injected with H_2O_2 (4.65 ± 0.51 saccades per minute) compared to the negative control group (9.42 ± 0.63 saccades per minute), thus confirming a visual impairment of larvae exposed to the oxidative stress. Injection of free neurotrophin was not able to prevent the visual dysfunction induced by hydrogen peroxide (5.81 ± 0.49 saccades per minute; ns, $p > 0.05$). Interestingly, the number of saccades recorded in ANP:PNA:NGF-injected larvae (8.59 ± 0.69 saccades per minute) was significantly higher compared to both the group treated with only hydrogen peroxide (***, $p = 0.0001$) and the one pretreated with free NGF (*, $p = 0.0179$), thus suggesting a neuroprotective effect of the NGF only when nanoformulated with the ANP:PNA nanocarrier. Additionally, no differences were observed between the larvae injected with ANP:PNA:NGF and the negative control group (ns, $p > 0.05$). In contrast, the number of saccades displayed by the larvae pretreated with free NGF was significantly lower than those of the negative control group (***, $p = 0.0003$) (Figure 5b).

The preventive treatment with neurotrophin showed promising results also against visual impairment induced by the exposure of zebrafish larvae to cigarette smoke extract (CSE). The number of saccades recorded in the negative control (28.44 ± 1.13 saccades per minute) was significantly higher (****, $p < 0.0001$) compared to the CSE-treated positive control group (13.68 ± 0.98 saccades per minute), in line with reported data [5]. Preventive injections of free NGF (21.96 ± 1.09 saccades per minute) or its nanoformulation ANP:PNA:NGF (20.19 ± 1.33 saccades per minute) significantly improved visual function impaired by exposure to the CSE (****, $p < 0.0001$ and ***, $p = 0.0004$, respectively) (Figure S3). This rescue was only partial since there was still a significant difference compared to the vehicle control group (**, $p = 0.0017$ and ****, $p < 0.0001$, respectively).

To study the mechanism underlying the recovery of compromised visual function induced by oxygen peroxide exposure, the neuroprotective effect of the NGF-conjugated nanoparticles was investigated at a functional level through the detection of active caspase 3 (aCasp3)-positive cells as a marker of apoptosis, as a consequence of an oxidative stress condition [44,45]. In agreement with the OKR data, the number of apoptotic cells was significantly higher (***, $p = 0.0003$) in the H_2O_2-treated group (20.8 ± 4.4 aCasp3-positive cells) compared to the negative control (6.2 ± 0.9 aCasp3-positive cells), hence corroborating cytotoxic damage induced by the injection of hydrogen peroxide. Interestingly, the reduction of apoptotic cells in the retina was only observed in the H_2O_2-injected larvae pretreated with ANP:PNA:NGF (9.7 ± 2.02 aCasp3-positive cells; *, $p = 0.0113$) while no neuroprotection was observed in the larvae pretreated with the free neurotrophin (13.8 ± 2.32 aCasp3-positive cells), since the number of apoptotic cells was not statistically different (ns, $p > 0.05$) to the group exposed to hydrogen peroxide (Figure 5c). These data are in line with the visual impairment revealed by the optokinetic response assay and show a promising effect of our neuroprotective NGF-based nanoformulation against the retinal degeneration induced by oxidative stress.

4. Discussion

This study aimed to develop a neuroprotective, polymeric-based drug delivery system as a preventive strategy against retinal degeneration or dysfunction. Due to the high loading efficiency, biocompatibility and biodegradability, we synthesised and optimised polyacrylamide-based nanoparticles for implementing an ocular drug delivery system for the posterior segment of the eye. To our knowledge, this is the first polymer-based nanoformulation implemented for delivering to the ocular posterior segment by both exploiting peanut agglutinin as a targeting molecule and the neurotrophin NGF as a neuroprotective compound.

The advancement of nanotechnology-based strategies in ophthalmology is one of the most popular approaches for achieving successful therapeutic options against vision loss, thanks to their significant abilities in improving drug delivery [20]. In particular, polymers are extensively employed as materials for drug delivery systems because of their noticeable biodegradability, stimuli-responsiveness, mucosal adhesive properties and biocompatibility [46,47]. Indeed, the number of polymer-based FDA-approved nanotechnological products for ocular treatment is increasing, but topical delivery still represents the preferred route of administration [47]. Unfortunately, the static and dynamic barriers that protect the eye pose many challenges, and only a small fraction of administered drugs (less than 5% of the instilled dose) reaches the retinal layers [48]. Consequently, topical instillation is not the optimal route for treating the posterior segment. Moreover, the majority of available ocular nanoformulations in the pharmaceutical market focus on the treatment of the anterior segment of the eye (dry eye disease, inflammation, acute keratitis, bacterial conjunctivitis and uveitis), thus not addressing retina degeneration in the posterior eye [21]. Since oxidative stress is the leading cause of cellular death and visual impairment, and it represents a common pathological mechanism in ocular degenerative diseases, great attention is addressed to developing neuroprotective systems for preventing retinal degeneration [1].

Here, we developed a multifunctional polymer-based ocular drug delivery system, and we assessed its neuroprotective effect in vivo, using a zebrafish model of retinal degeneration induced by oxidative stress, as previously developed by our group [28]. The synthesis of polyacrylamide nanoparticles embedding a fluorescent dye led to obtaining a polymeric nanosystem that was easily traceable in transparent zebrafish larvae, allowing in vivo imaging to study the localisation of the particles after intravitreal injections in larvae at 3 days post-fertilisation. Our results indicated that our polymeric nanoparticles spontaneously migrate through the retinal layers over time, as suggested by their presence mainly at the level of the ganglion cell layer after 4 h post-injection and their autonomous distribution towards photoreceptor and retinal pigment epithelium layers after 24 hpi. Importantly, no particles were found to diffuse in the contralateral eye. Unfortunately, in vivo imaging highlighted the presence of nanoparticles at the level of the pronephros, the ancestral kidney of vertebrate embryos, where they started accumulating from 4 hpi. Due to their instability in the eye, we initiated an optimisation of the nanocarrier by exploiting the easy functionalisation of the polyacrylamide nanoparticles, thanks to the presence of several functional groups on their surface [36,38]. Therefore, we selected peanut agglutinin (PNA) to take advantage of its specific and well-characterised high affinity for the galactose-galactosamine disaccharide residues of cone photoreceptors, as demonstrated in humans as well as in fish [42]. In fact, this is the first study where the PNA was nanoformulated for both stabilising a polymeric nanocarrier in the ocular environment and conferring an active targeting capacity, thus improving its applicability as an ocular drug delivery system. After the characterisation of the red fluorescent ANP:PNA, revealing both an increase in the hydrodynamic diameter compared to the naked formulation and good conjugation properties, in vivo studies showed a promising modification in the biodistribution of our nanocarrier. The presence of the fluorescent dye Alexa Fluor 594 contained in the polyacrylamide matrix and of an FITC fluorophore conjugated to the PNA easily allowed us to simultaneously follow the destiny of our nanocarrier and its cargo, respectively. After the intravitreal injection in 3 dpf larvae, the PNA-conjugated polymeric nanoparticles stably localised in the injected eye over time, without spreading in the contralateral eye or in any other larval tissues. On the contrary, naked ANPs were found in the pronephros at 24 hpi, thus suggesting their migration from the eye. The strong improvement in the localisation profile of the PNA-conjugated ANPs strongly suggests the ability of the lectin to prolong the residence time of the polymeric nanocarrier and its stability. Although the red signal from Alexa Fluor 594-positive nanoparticles and the green fluorescence associated with the FITC-conjugated PNA perfectly colocalise in the eye of larvae after 4 h post-injection, the two fluorescent signals showed an incomplete overlapping at 24 hpi. This can be explained by partial detachment of PNA from the ANPs occurring at 24 hpi, which can be ascribed to the fact that the functionalisation reaction involved noncovalent interactions and the ANP:PNA are exposed to physiological processes at the humour vitreous of the zebrafish eye.

Taking advantage of the evidence collected, we decided to perform a double conjugation onto the polyacrylamide nanoparticles' surface using the NGF as a neuroprotective molecule against retinal degeneration. The role of this neurotrophin in the differentiation and maintenance of neurons is widely acknowledged. Indeed, several studies demonstrated that the NGF exerts prosurvival and regenerative effects in preclinical and clinical models of glaucoma [13,49,50], retinitis pigmentosa [51,52] and AMD [53] through the activation of different molecular pathways [14–16]. Although there is a robust body of evidence reporting the beneficial effect of the neurotrophin against retinal degeneration, NGF-based therapies are at the very early stages of market approval and clinical applicability due to several drawbacks that still require a solution, such as poor solubility, low delivery efficiency, short half-life and off-target effect [17]. To overwhelm these limitations, our group previously implemented an inorganic nanoformulation carrying the NGF [28]. Our previous results demonstrated increased stability and protection from degradation, resulting in increased prevention of retinal ganglion cell loss, upon conjugation of NGF with magnetic nanoparticles. Although these NGF-functionalised magnetic nanoparticles emerged as a promising ocular-tailored therapeutical nanotool, we decided to move towards an organic drug delivery system to overcome some disadvantages associated with the lack of evidence concerning biodegradability, toxicity and clearance mechanisms, which still pose some issues to the use of iron-based nanomaterials for the treatment of retinal disorders/degeneration.

Hence, our PNA-targeted polyacrylamide-based nanocarrier has been further functionalised with NGF via a noncovalent reaction, revealing a shift in the hydrodynamic diameter, an improved polydispersity index and a high conjugation efficiency. The bioactivity assessment of the nanoformulated NGF demonstrated its unaltered capacity, in triggering the differentiation of PC12 cells in a neuron-like phenotype compared to the free factor, thus suggesting that the conjugated NGF was biologically active. Further, to test the neuroprotective effect of our nanocarrier, we induced a condition of oxidative stress in human RPE cells by co-incubating hydrogen peroxide with free or ANP:PNA-conjugated NGF. In sharp contrast to evidence reported in the literature [54], we found that only the nanoformulated NGF, but not the free form, was able to exert a protective effect against the hydrogen peroxide-elicited oxidative damage, as indicated by the higher percentage of metabolically active cells in the ANP:PNA:NGF cotreated group.

To corroborate these encouraging results regarding the biological activity of the nanoformulation, we decided to validate the neuroprotective ability of NGF-conjugated ANPs in vivo using zebrafish larvae. Zebrafish represent a popular and well-established model for studying retinal degeneration or dysfunction due to the high genome homology and the same retinal stratification of different cell types shared with humans [23–25]. Moreover, since its development is extremely rapid, the visual function starts only 3 days post-fertilisation, thus allowing visual behavioural assessments and short-term studies in the early stages of development. Indeed, zebrafish constitute a valuable model organism for biomedical research and are increasingly being employed to conduct nanotoxicological studies and preclinical validation of innovative drug delivery nanoformulations [55,56]. Consequently, the ethical impact is significantly reduced as zebrafish are considered a non-sentient organism until 5 dpf.

The neuroprotective activity of our polymeric NGF-conjugated nanoformulation was tested using an ocular model of oxidative stress in zebrafish, previously developed by our group [28]. As expected, the visual behavioural assessment carried out through the OKR assay revealed impaired vision in larvae IVT injected with hydrogen peroxide compared to the saline-injected group. Most importantly, only the larvae pretreated with nanoformulated NGF showed a significantly higher number of saccades, compared to both the positive control and the free NGF-pretreated group. Conversely, the preventive intraocular injection of the free NGF was not able to improve the visual function impaired by the induction of oxidative injury. These results are in line with our previous findings that demonstrated an improved neuroprotection capability of the neurotrophin loaded in a drug delivery system, thus guiding us to speculate on increased stability and bioavailability of the nanoformulated NGF compared to the free form. In line with the data collected from the behavioural analysis, we also found a strong increase in the number of apoptotic cells in the retina of larvae IVT injected with hydrogen peroxide compared to the negative control group. These results are in line with evidence reported in the literature, according to which the apoptotic event and, in particular, the activation of specific caspase is strictly related to retinal degeneration as a consequent result of oxidative stress [44,45]. Most encouragingly, the preventive treatment with the nanoformulated NGF, but not the injection of the free NGF, resulted in a statistically significant reduction of active caspase 3-positive cells in the retina of injected larvae compared to the positive control, thus suggesting the ability of our nanoformulated NGF to prevent oxidative stress-triggered apoptosis.

Moreover, using a zebrafish model of CSE-induced visual impairment, we demonstrated that our NGF-based nanoformulation was able to improve the optokinetic response of zebrafish larvae compromised by acute systemic treatment with CSE. Thus, taking advantage of a zebrafish model of pharmacological modulation of visual behaviour, we were able to further assess the neuroprotective efficacy of our nanocarrier, paving the way for the development of neurotrophin-based therapeutical strategies against CSE-induced retinal degeneration. Although no difference emerged between the use of free NGF and its polymeric nanoformulation, this was not unexpected due to the acute experimental time frame. Notably, this is the first study demonstrating neurotrophin NGF rescue of visual function in zebrafish larvae negatively affected by systemic exposure to CSE. Additional studies will be performed to clarify the beneficial role of our polymeric NGF-based nanoformulation compared to the free neurotrophin in providing neuroprotection against a chronic exposure to CSE.

5. Conclusions

In this study, we developed a novel multifunctional polymeric nanoformulation, which represents a promising ocular drug delivery alternative for preventing retinal degeneration. Particularly, this is the first study demonstrating that the lectin PNA not only improves the targeting to the posterior segment of the eye but also prolongs the residence time of our NGF-conjugated nanocarriers, thus resulting in a neuroprotective response against oxidative stress, as observed both in vitro and in vivo in zebrafish. The intravitreal administration route of a nano-based system could overcome the requirement of multiple injections due to the unfavourable kinetics of the traditional form of drugs, thereby reducing side effects. However, further experiments are needed to investigate how long our nanocarrier can protect NGF and ensure its sustained neuroprotective release, as well as to perform a deeper toxicological assessment. Moreover, we demonstrated for the first time that nanoformulated NGF improves visual function of zebrafish larvae exposed to cigarette smoke extract, thus offering a potential treatment to be further validated and exploited against retinal dysfunction. Future perspective aims to extend the study to rodent models to acquire new evidence about our nanoformulation's safety and efficacy profile.

Supplementary Materials: The following supporting information can be downloaded at https: //www.mdpi.com/article/10.3390/pharmaceutics15041096/s1, Figure S1: the intravitreal injection of hydrogen peroxide induces apoptosis in the zebrafish retina, Figure S2: nanoparticles do not diffuse in the contralateral eye after intravitreal injections, Figure S3: intravitreal injection of ANP:PNA-conjugated NGF attenuates cigarette smoke extract (CSE)-induced visual impairment in zebrafish larvae.

Author Contributions: Conceptualisation, P.C., M.G., V.R. and C.G.; methodology, M.G., P.C. and V.R.; validation, P.C., M.B. and M.G.; formal analysis, P.C.; investigation, P.C.; data curation, P.C.; writing—original draft preparation, P.C. and C.G.; writing—review and editing, P.C., M.G., M.B., B.N.K., L.D., V.R. and C.G.; supervision, L.D., V.R., B.N.K. and C.G.; funding acquisition, L.D., V.R., B.N.K. and C.G. All authors have read and agreed to the published version of the manuscript.

Funding: This research was funded by the University Research Projects programme of the University of Pisa, grant number PRA2020 and Local Funds from the University of Pisa. This work is part of the PhD project of Patrizia Colucci funded by the Regione Toscana Pegaso Scholarship. The research in UCD was funded by the European Union's Horizon 2020 research and innovation programme under the MSCA ORBITALITN (grant agreement—No 813440) and the MSCA-RISE CRYSTAL^3 (grant agreement—No 101007931).

Institutional Review Board Statement: The animal study protocol was approved by the Italian Ministry of Public Health, the local Ethical Committee of the University of Pisa (authorisation n. 99/2012-A, 19/04/2012) and the University College Dublin Animal Research Ethics Committee (AREC), in conformity with the Directive 2010/63/EU.

Informed Consent Statement: Not applicable.

Data Availability Statement: The data presented in this study are available on request from the corresponding authors.

Conflicts of Interest: The authors declare no conflict of interest. The funders had no role in the design of the study; in the collection, analyses, or interpretation of data; in the writing of the manuscript; or in the decision to publish the results.

References

1. Dammak, A.; Huete-Toral, F.; Carpena-Torres, C.; Martin-Gil, A.; Pastrana, C.; Carracedo, G. From Oxidative Stress to Inflammation in the Posterior Ocular Diseases: Diagnosis and Treatment. *Pharmaceutics* **2021**, *13*, 1376. [CrossRef] [PubMed]
2. Ahmad, A.; Ahsan, H. Biomarkers of inflammation and oxidative stress in ophthalmic disorders. *J. Immunoass. Immunochem.* **2020**, *41*, 257–271. [CrossRef] [PubMed]
3. Bertram, K.M.; Baglole, C.J.; Phipps, R.P.; Libby, R.T. Molecular regulation of cigarette smoke induced-oxidative stress in human retinal pigment epithelial cells: Implications for age-related macular degeneration. *Am. J. Physiol. Cell Physiol.* **2009**, *297*, C1200–C1210. [CrossRef]
4. Lee, K.; Hong, S.; Seong, G.J.; Kim, C.Y. Cigarette Smoke Extract Causes Injury in Primary Retinal Ganglion Cells via Apoptosis and Autophagy. *Curr. Eye Res.* **2016**, *41*, 1367–1372. [CrossRef] [PubMed]
5. Sánchez, A.G.; Colucci, P.; Moran, A.; López, A.M.; Colligris, B.; Álvarez, Y.; Kennedy, B.N. Systemic treatment with cigarette smoke extract affects zebrafish visual behaviour, intraocular vasculature morphology and outer segment phagocytosis [version 1; peer review: Awaiting peer review]. *Open Res. Europe* **2023**, *3*, 48. [CrossRef]
6. Tang, L.-L.; Wang, R.; Tang, X.-C. Huperzine A protects SHSY5Y neuroblastoma cells against oxidative stress damage via nerve growth factor production. *Eur. J. Pharmacol.* **2005**, *519*, 9–15. [CrossRef]
7. Dong, A.; Shen, J.; Krause, M.; Hackett, S.F.; Campochiaro, P.A. Increased expression of glial cell line-derived neurotrophic factor protects against oxidative damage-induced retinal degeneration. *J. Neurochem.* **2007**, *103*, 1041–1052. [CrossRef]
8. Yoo, J.-M.; Lee, B.D.; Sok, D.-E.; Ma, J.Y.; Kim, M.R. Neuroprotective action of N-acetyl serotonin in oxidative stress-induced apoptosis through the activation of both TrkB/CREB/BDNF pathway and Akt/Nrf2/Antioxidant enzyme in neuronal cells. *Redox Biol.* **2017**, *11*, 592–599. [CrossRef]
9. Götz, R.; Schartl, M. The conservation of neurotrophic factors during vertebrate evolution. *Comp. Biochem. Physiol. Part C Pharmacol. Toxicol. Endocrinol.* **1994**, *108*, 1–10. [CrossRef]
10. Garcia, T.B.; Hollborn, M.; Bringmann, A. Expression and signaling of NGF in the healthy and injured retina. *Cytokine Growth Factor Rev.* **2017**, *34*, 43–57. [CrossRef]
11. Lambiase, A.; Aloe, L.; Centofanti, M.; Parisi, V.; Bao, S.N.; Mantelli, F.; Colafrancesco, V.; Manni, G.L.; Bucci, M.G.; Bonini, S.; et al. Experimental and clinical evidence of neuroprotection by nerve growth factor eye drops: Implications for glaucoma. *Proc. Natl. Acad. Sci. USA* **2009**, *106*, 13469–13474. [CrossRef]
12. Sacchetti, M.; Mantelli, F.; Rocco, M.L.; Micera, A.; Brandolini, L.; Focareta, L.; Pisano, C.; Aloe, L.; Lambiase, A. Recombinant Human Nerve Growth Factor Treatment Promotes Photoreceptor Survival in the Retinas of Rats with Retinitis Pigmentosa. *Curr. Eye Res.* **2017**, *42*, 1064–1068. [CrossRef]
13. Guo, L.; Davis, B.M.; Ravindran, N.; Galvao, J.; Kapoor, N.; Haamedi, N.; Shamsher, E.; Luong, V.; Fico, E.; Cordeiro, M.F. Topical recombinant human Nerve growth factor (rh-NGF) is neuroprotective to retinal ganglion cells by targeting secondary degeneration. *Sci. Rep.* **2020**, *10*, 3375. [CrossRef]
14. Li, B.; Ning, B.; Yang, F.; Guo, C. Nerve Growth Factor Promotes Retinal Neurovascular Unit Repair: A Review. *Curr. Eye Res.* **2022**, *47*, 1095–1105. [CrossRef]
15. Salinas, M.; Diaz, R.; Abraham, N.G.; de Galarreta, C.M.R.; Cuadrado, A. Nerve Growth Factor Protects against 6-Hydroxydopamine-induced Oxidative Stress by Increasing Expression of Heme Oxygenase-1 in a Phosphatidylinositol 3-Kinase-dependent Manner. *J. Biol. Chem.* **2003**, *278*, 13898–13904. [CrossRef]
16. Elsherbiny, N.M.; Abdel-Mottaleb, Y.; Elkazaz, A.Y.; Atef, H.; Lashine, R.M.; Youssef, A.M.; Ezzat, W.; El-Ghaiesh, S.; Elshaer, R.E.; El-Shafey, M.; et al. Carbamazepine Alleviates Retinal and Optic Nerve Neural Degeneration in Diabetic Mice via Nerve Growth Factor-Induced PI3K/Akt/mTOR Activation. *Front. Neurosci.* **2019**, *13*, 1089. [CrossRef]
17. Alastra, G.; Aloe, L.; Baldassarro, V.A.; Calzà, L.; Cescatti, M.; Duskey, J.T.; Focarete, M.L.; Giacomini, D.; Giardino, L.; Giraldi, V.; et al. Nerve Growth Factor Biodelivery: A Limiting Step in Moving Toward Extensive Clinical Application? *Front. Neurosci.* **2021**, *15*, 695592. [CrossRef]
18. Yavuz, B.; Kompella, U.B. Ocular Drug Delivery. In *Handbook of Experimental Pharmacology*; Springer: Berlin/Heidelberg, Germany, 2017; Volume 242, pp. 57–93. [CrossRef]

19. Meng, T.; Kulkarni, V.; Simmers, R.; Brar, V.; Xu, Q. Therapeutic implications of nanomedicine for ocular drug delivery. *Drug Discov. Today* **2019**, *24*, 1524–1538. [CrossRef]

20. Qamar, Z.; Qizilbash, F.F.; Iqubal, A.; Ali, A.; Narang, J.K.; Ali, J.; Baboota, S. Nano-Based Drug Delivery System: Recent Strategies for the Treatment of Ocular Disease and Future Perspective. *Recent Pat. Drug Deliv. Formul.* **2019**, *13*, 246–254. [CrossRef]

21. Kagkelaris, K.; Panayiotakopoulos, G.; Georgakopoulos, C.D. Nanotechnology-based formulations to amplify intraocular bioavailability. *Ther. Adv. Ophthalmol.* **2022**, *14*, 25158414221112356. [CrossRef]

22. Tawfik, M.; Chen, F.; Goldberg, J.L.; Sabel, B.A. Nanomedicine and drug delivery to the retina: Current status and implications for gene therapy. *Naunyn Schmiedebergs Arch. Pharmacol.* **2022**, *395*, 1477–1507. [CrossRef] [PubMed]

23. Bibliowicz, J.; Tittle, R.K.; Gross, J.M. Toward a better understanding of human eye disease insights from the zebrafish, Danio rerio. *Prog. Mol. Biol. Transl. Sci.* **2011**, *100*, 287–330. [CrossRef] [PubMed]

24. Richardson, R.; Tracey-White, D.; Webster, A.; Moosajee, M. The zebrafish eye—A paradigm for investigating human ocular genetics. *Eye* **2017**, *31*, 68–86. [CrossRef] [PubMed]

25. Hong, Y.; Luo, Y. Zebrafish Model in Ophthalmology to Study Disease Mechanism and Drug Discovery. *Pharmaceuticals* **2021**, *14*, 716. [CrossRef] [PubMed]

26. Giannaccini, M.; Giannini, M.; Calatayud, M.P.; Goya, G.F.; Cuschieri, A.; Dente, L.; Raffa, V. Magnetic Nanoparticles as Intraocular Drug Delivery System to Target Retinal Pigmented Epithelium (RPE). *Int. J. Mol. Sci.* **2014**, *15*, 1590–1605. [CrossRef]

27. Giannaccini, M.; Pedicini, L.; De Matienzo, G.; Chiellini, F.; Dente, L.; Raffa, V. Magnetic nanoparticles: A strategy to target the choroidal layer in the posterior segment of the eye. *Sci. Rep.* **2017**, *7*, 43092. [CrossRef]

28. Giannaccini, M.; Usai, A.; Chiellini, F.; Guadagni, V.; Andreazzoli, M.; Ori, M.; Pasqualetti, M.; Dente, L.; Raffa, V. Neurotrophin-conjugated nanoparticles prevent retina damage induced by oxidative stress. *Cell. Mol. Life Sci.* **2018**, *75*, 1255–1267. [CrossRef]

29. He, X.; Hahn, P.; Iacovelli, J.; Wong, R.; King, C.; Bhisitkul, R.; Massaro-Giordano, M.; Dunaief, J.L. Iron homeostasis and toxicity in retinal degeneration. *Prog. Retin. Eye Res.* **2007**, *26*, 649–673. [CrossRef]

30. Schneider-Futschik, E.; Reyes-Ortega, F. Advantages and Disadvantages of Using Magnetic Nanoparticles for the Treatment of Complicated Ocular Disorders. *Pharmaceutics* **2021**, *13*, 1157. [CrossRef]

31. Tsai, C.-H.; Wang, P.-Y.; Lin, I.-C.; Huang, H.; Liu, G.-S.; Tseng, C.-L. Ocular Drug Delivery: Role of Degradable Polymeric Nanocarriers for Ophthalmic Application. *Int. J. Mol. Sci.* **2018**, *19*, 2830. [CrossRef]

32. Chien, Y.; Hsiao, Y.-J.; Chou, S.-J.; Lin, T.-Y.; Yarmishyn, A.A.; Lai, W.-Y.; Lee, M.-S.; Lin, Y.-Y.; Lin, T.-W.; Hwang, D.-K.; et al. Nanoparticles-mediated CRISPR-Cas9 gene therapy in inherited retinal diseases: Applications, challenges, and emerging opportunities. *J. Nanobiotechnology* **2022**, *20*, 511. [CrossRef]

33. Luo, L.-J.; Nguyen, D.D.; Lai, J.-Y. Dually functional hollow ceria nanoparticle platform for intraocular drug delivery: A push beyond the limits of static and dynamic ocular barriers toward glaucoma therapy. *Biomaterials* **2020**, *243*, 119961. [CrossRef]

34. Arkaban, H.; Barani, M.; Akbarizadeh, M.R.; Chauhan, N.P.S.; Jadoun, S.; Soltani, M.D.; Zarrintaj, P. Polyacrylic Acid Nanoplatforms: Antimicrobial, Tissue Engineering, and Cancer Theranostic Applications. *Polymers* **2022**, *14*, 1259. [CrossRef]

35. Khiev, D.; Mohamed, Z.; Vichare, R.; Paulson, R.; Bhatia, S.; Mohapatra, S.; Lobo, G.; Valapala, M.; Kerur, N.; Passaglia, C.; et al. Emerging Nano-Formulations and Nanomedicines Applications for Ocular Drug Delivery. *Nanomaterials* **2021**, *11*, 173. [CrossRef]

36. Li, K.; Zang, X.; Cheng, M.; Chen, X. Stimuli-responsive nanoparticles based on poly acrylic derivatives for tumor therapy. *Int. J. Pharm.* **2021**, *601*, 120506. [CrossRef]

37. Chakraborty, M.; Banerjee, D.; Mukherjee, S.; Karati, D. Exploring the advancement of polymer-based nano-formulations for ocular drug delivery systems: An explicative review. *Polym. Bull.* **2022**. [CrossRef]

38. Rahimi, M.; Kilaru, S.; Sleiman, G.E.H.; Saleh, A.; Rudkevich, D.; Nguyen, K. Synthesis and Characterization of Thermo-Sensitive Nanoparticles for Drug Delivery Applications. *J. Biomed. Nanotechnol.* **2008**, *4*, 482–490. [CrossRef]

39. E Brockerhoff, S. Measuring the optokinetic response of zebrafish larvae. *Nat. Protoc.* **2006**, *1*, 2448–2451. [CrossRef]

40. Gómez Sánchez, A.; Álvarez, Y.; Colligris, B.; Kennedy, B.N. Affordable and Effective Optokinetic Response Methods to Assess Visual Acuity and Contrast Sensitivity in Larval to Juvenile Zebrafish. Available online: https://open-research-europe.ec.europa.eu/articles/1-92 (accessed on 6 January 2022).

41. Inoue, D.; Wittbrodt, J. One for All—A Highly Efficient and Versatile Method for Fluorescent Immunostaining in Fish Embryos. *PLoS ONE* **2011**, *6*, e19713. [CrossRef]

42. Blanks, J.C.; Johnson, L.V. Specific binding of peanut lectin to a class of retinal photoreceptor cells. A species comparison. *Investig. Opthalmol. Vis. Sci.* **1984**, *25*, 546–557.

43. Greene, L.A.; Tischler, A. Establishment of a noradrenergic clonal line of rat adrenal pheochromocytoma cells which respond to nerve growth factor. *Proc. Natl. Acad. Sci. USA* **1976**, *73*, 2424–2428. [CrossRef] [PubMed]

44. Zhou, J.; Chen, F.; Yan, A.; Xia, X. Madecassoside protects retinal pigment epithelial cells against hydrogen peroxide-induced oxidative stress and apoptosis through the activation of Nrf2/HO-1 pathway. *Biosci. Rep.* **2020**, *40*, BSR20194347. [CrossRef] [PubMed]

45. Zhang, J.; Zhou, H.; Chen, J.; Lv, X.; Liu, H. Aloperine protects human retinal pigment epithelial cells against hydrogen peroxide–induced oxidative stress and apoptosis through activation of Nrf2/HO-1 pathway. *J. Recept. Signal Transduct.* **2022**, *42*, 88–94. [CrossRef]

46. Berillo, D.; Zharkinbekov, Z.; Kim, Y.; Raziyeva, K.; Temirkhanova, K.; Saparov, A. Stimuli-Responsive Polymers for Transdermal, Transmucosal and Ocular Drug Delivery. *Pharmaceutics* **2021**, *13*, 2050. [CrossRef] [PubMed]

47. Chiaretti, A.; Eftimiadi, G.; Soligo, M.; Manni, L.; Di Giuda, D.; Calcagni, M.L. Topical delivery of nerve growth factor for treatment of ocular and brain disorders. *Neural Regen. Res.* **2021**, *16*, 1740–1750. [CrossRef]

48. Awwad, S.; Ahmed, A.H.A.M.; Sharma, G.; Heng, J.S.; Khaw, P.T.; Brocchini, S.; Lockwood, A. Principles of pharmacology in the eye. *Br. J. Pharmacol.* **2017**, *174*, 4205–4223. [CrossRef]

49. Mesentier-Louro, L.A.; Rosso, P.; Carito, V.; Mendez-Otero, R.; Santiago, M.F.; Rama, P.; Lambiase, A.; Tirassa, P. Nerve Growth Factor Role on Retinal Ganglion Cell Survival and Axon Regrowth: Effects of Ocular Administration in Experimental Model of Optic Nerve Injury. *Mol. Neurobiol.* **2019**, *56*, 1056–1069. [CrossRef]

50. Beykin, G.; Stell, L.; Halim, M.S.; Nuñez, M.; Popova, L.; Nguyen, B.T.; Groth, S.L.; Dennis, A.; Li, Z.; Atkins, M.; et al. Phase 1b Randomized Controlled Study of Short Course Topical Recombinant Human Nerve Growth Factor (rhNGF) for Neuroenhancement in Glaucoma: Safety, Tolerability, and Efficacy Measure Outcomes. *Am. J. Ophthalmol.* **2022**, *234*, 223–234. [CrossRef]

51. Falsini, B.; Iarossi, G.; Chiaretti, A.; Ruggiero, A.; Manni, L.; Galli-Resta, L.; Corbo, G.; Abed, E. NGF eye-drops topical administration in patients with retinitis pigmentosa, a pilot study. *J. Transl. Med.* **2016**, *14*, 8. [CrossRef]

52. Rocco, M.L.; Calzà, L.; Aloe, L. NGF and Retinitis Pigmentosa: Structural and Molecular Studies. *Adv. Exp. Med. Biol.* **2021**, *133*, 255–263. [CrossRef]

53. Telegina, D.V.; Kolosova, N.G.; Kozhevnikova, O.S. Immunohistochemical localization of NGF, BDNF, and their receptors in a normal and AMD-like rat retina. *BMC Med. Genom.* **2019**, *12*, 48. [CrossRef]

54. Cao, G.-F.; Liu, Y.; Yang, W.; Wan, J.; Yao, J.; Wan, Y.; Jiang, Q. Rapamycin sensitive mTOR activation mediates nerve growth factor (NGF) induced cell migration and pro-survival effects against hydrogen peroxide in retinal pigment epithelial cells. *Biochem. Biophys. Res. Commun.* **2011**, *414*, 499–505. [CrossRef]

55. Haque, E.; Ward, A.C. Zebrafish as a Model to Evaluate Nanoparticle Toxicity. *Nanomaterials* **2018**, *8*, 561. [CrossRef]

56. Pensado-López, A.; Fernández-Rey, J.; Reimunde, P.; Crecente-Campo, J.; Sánchez, L.; Andón, F.T. Zebrafish Models for the Safety and Therapeutic Testing of Nanoparticles with a Focus on Macrophages. *Nanomaterials* **2021**, *11*, 1784. [CrossRef]

pharmaceutics

Article

Preparation and Characterization of a Novel Multiparticulate Dosage Form Carrying Budesonide-Loaded Chitosan Nanoparticles to Enhance the Efficiency of Pellets in the Colon

Fatemeh Soltani [1], Hossein Kamali [2], Abbas Akhgari [1,2], Mahboobeh Ghasemzadeh Rahbardar [3], Hadi Afrasiabi Garekani [1,3], Ali Nokhodchi [4,5,*] and Fatemeh Sadeghi [1,2,*]

[1] Department of Pharmaceutics, School of Pharmacy, Mashhad University of Medical Sciences, Mashhad 9177899191, Iran
[2] Targeted Drug Delivery Research Center, Pharmaceutical Technology Institute, Mashhad University of Medical Sciences, Mashhad 9177899191, Iran
[3] Pharmaceutical Research Center, Pharmaceutical Technology Institute, Mashhad University of Medical Sciences, Mashhad 9177899191, Iran
[4] Lupin Pharmaceutical Research Center, Coral Springs, FL 33065, USA
[5] School of Life Sciences, University of Sussex, Brighton BN1 9RH, UK
* Correspondence: AliNokhodchi@lupin or a.nokhodchi@sussex.ac.uk (A.N.); sadeghif@mums.ac.ir (F.S.)

Citation: Soltani, F.; Kamali, H.; Akhgari, A.; Ghasemzadeh Rahbardar, M.; Afrasiabi Garekani, H.; Nokhodchi, A.; Sadeghi, F. Preparation and Characterization of a Novel Multiparticulate Dosage Form Carrying Budesonide-Loaded Chitosan Nanoparticles to Enhance the Efficiency of Pellets in the Colon. *Pharmaceutics* 2023, 15, 69. https://doi.org/10.3390/pharmaceutics 15010069

Academic Editor: Ana Isabel Fernandes

Received: 10 November 2022
Revised: 14 December 2022
Accepted: 20 December 2022
Published: 26 December 2022

Abstract: An attempt was made to conquer the limitation of orally administered nanoparticles for the delivery of budesonide to the colon. The ionic gelation technique was used to load budesonide on chitosan nanoparticles. The nanoparticles were investigated in terms of size, zeta potential, encapsulation efficiency, shape and drug release. Then, nanoparticles were pelletized using the extrusion–spheronization method and were investigated for their size, mechanical properties, and drug release. Pellets were subsequently coated with a polymeric solution composed of two enteric (eudragit L and S) and time-dependent polymers (eudragit RS) for colon-specific delivery. All formulations were examined for their anti-inflammatory effect in rats with induced colitis and the relapse of the colitis after discontinuation of treatment was also followed. The size of nanoparticles ranged between 288 ± 7.5 and 566 ± 7.7 nm and zeta potential verified their positive charged surface. The drug release from nanoparticles showed an initial burst release followed by a continuous release. Pelletized nanoparticles showed proper mechanical properties and faster drug release in acidic pH compared with alkaline pH. It was interesting to note that pelletized budesonide nanoparticles released the drug throughout the GIT in a sustained fashion, and had long-lasting anti-inflammatory effects while rapid relapse was observed for those treated with conventional budesonide pellets. It seems that there is a synergistic effect of nanoformulation of budesonide and the encapsulation of pelletized nanoparticles in a proper coating system for colon delivery that could result in a significant and long-lasting anti-inflammatory effect.

Keywords: budesonide; nanoparticles; chitosan; colon delivery; eudragit; pellets

1. Introduction

Targeting drug delivery is advantageous for the treatment of inflammatory bowel disease (IBD), including Crohn's disease and ulcerative colitis (UC) [1–3]. An appropriate drug delivery system could allow delivery of the sufficient amounts of drugs with appropriate release rates to the site of its action [4]. Budesonide (BU) is a topical anti-inflammatory steroid, which has proven to be effective in the treatment of IBD [5]. Therefore, targeting this drug at the site of inflammation is an interesting perspective for clinical applications.

Nanoparticles have shown an encouraging and promising outlook for drug or gene delivery in IBD treatment [6–8]. The design of drug-loaded nanoparticles for delivery to the colon via oral administration has been used as a strategy to further amplify drug uptake into the inflamed tissue of the colon [1,9]. Nanoparticles have a great chance to reach and accumulate in inflamed tissues due to the loosening of tight junctional complexes [10]. A study on the rat intestinal loop model demonstrated that nanoparticles in the size range of 100 nm allow better penetration in the submucosal layers compared with those in the size range of 500 nm which mainly showed localized targeting to the epithelial lining [11]. Additionally, it has been reported that nano-sized particles showed favored uptake by immune cells whose numbers are high in inflamed tissue [12].

In some studies, triggering drug release into the colon was reported by encapsulation of prednisolone in nanocarriers [13]. Additionally, 5-aminosalicylic acid nanoparticles demonstrated a significant improvement in targeting inflamed tissue in a mouse with UC [14]. Many studies demonstrated that budesonide-loaded nanoparticles have shown promising results for targeted delivery to intestinal mucosa with inflammation [15–17].

Limitations such as burst drug release, premature nanoparticle uptake or lack of pH sensitivity have been noticed as obstacles to the efficient transport of drugs to the colon upon oral administration of nanoparticles [4]. For example, the orally administered tacrolimus-nanoparticles showed minor therapeutic effects due to the slow rate of drug release, degradation in an enzymatic environment of the upper parts of the GI tract, or uptake into the systemic circulation followed by hepatic metabolism [1,12].

Based on this consideration, additional strategies such as surface charge-dependent nanoparticles [18], PEGylation-dependent nanoparticles [19], pH-dependent nanoparticles [1,20], hydrogel-based targeting [21], ligand-receptor-mediated targeting [22] and reactive oxygen species [23] are being explored to enhance drug delivery to the areas of inflammation and achieve maximal retention time in diseased tissues for orally administered nanoparticles. Amongst the methods mentioned above, the design of a pH-sensitive nano-delivery system is one of the simplest [24] and most commonly used approaches for the selective delivery of nanoparticles to the site of inflammation [6,25,26]. Many studies demonstrated that pH-dependent BU nanoparticles could alleviate the colitis better than plain nanoparticles [27–29].

Despite the promising results observed for pH-dependent nanoparticles in targeting the colon, some concerns including the inter- and intra-individual discrepancies in pH in GIT and the disease-related variations in luminal pH could be obstacles to the successful performance of these systems. Encapsulation of nanoparticles in dual pH and time-dependent polymers has been applied as a strategy to decrease initial drug release and activity in the upper parts of GIT in comparison with plain nanoparticles [29]. While such systems demonstrated better therapeutic effects than single pH or time-dependent systems, premature drug absorption in the upper region of GI due to their small size should also be considered [30]. It should also be mentioned that the systems described to date such as entrapped nanoparticles in enteric microparticles require two steps of emulsification and solvent evaporation techniques for preparation [31,32]. On the other hand, the use of organic solvents for solubilizing pH-sensitive polymers could probably increase the leaching of the drug and result in a decrease in the drug content of the nanoparticles [33], and a lack of reproducibility [34]. To overcome these limitations researchers embedded BU lipid-based nanoparticles in enteric-coated pellets to reach the exact site of action [30].

Since lipid-based nanoparticles might be unstable in the GI tract [35], and also due to the superiority of polymeric to lipid-based formulations of BU nanoparticles in the treatment of IBD [36], polymer-based nanocarriers for specific delivery of BU to the colon would be desirable.

Chitosan (CS) as a cationic natural polysaccharide with desirable biodegradability, biocompatibility and mucoadhesive properties [16] has been extensively used in colon targeting delivery systems [37–39]. Chitosan-based nanoparticles could adhere to the surface of intestinal mucosa in inflamed tissues via the interaction between the positively charged nanoparticles and the negatively charged intestinal mucosa [40] and for drugs such as BU with local action this perspective might be beneficial.

In this study, to overcome the limitations of orally administered nanoparticles for delivery of BU to the colon, an attempt was made to design a coated multiparticulate dosage form (pellets) carrying budesonide-loaded chitosan nanoparticles. The coating was composed of combined pH and time-dependent eudragits to minimize the early drug release and absorption in the upper sections of the GI tract and maximize delivery of the drug-loaded nanoparticles to the colon.

2. Materials and Methods

2.1. Materials

Budesonide (BU) was acquired from Jaber Ebne Hayyan Pharmaceutical Company (Tehran, Iran). Lactose monohydrate and Avicel pH 102 were obtained from Merck, Frankfurt, Germany. Polyvinylpyrrolidone (PVP K30) (Rahavard Tamin, Saveh, Iran), eudragit S PO, (Evonik Industries AG, Hanau, Germany), eudragit L PO, (Evonik Industries AG, Hanau, Germany), eudragit RS PO, (Evonik Industries AG, Hanau, Germany), talc, (Merck, Frankfurt, Germany), triethyl citrate (TEC) (Merck, Frankfurt, Germany), chitosan (medium-molecular-weight, 75–85% deacetylated) (Sigma-Aldrich, St. Louis, USA), three polyphosphate (Merk, Frankfurt, Germany), acetic acid (Dr. Mojallaly, Tehran, Iran), ethanol (Dr. Mojallaly, Tehran, Iran), isopropyl alcohol (2-propanol) (Dr. Mojallaly, Tehran, Iran) and sodium lauryl sulfate (SLS) (Scharlau, Barcelona, Spain) were utilized. All other reagents and solvents were of analytical grades.

2.2. Preparation of Nanoparticles

Budesonide-loaded chitosan nanoparticles (BCN) were made based on the ionotropic gelation technique [41]. First, 100 mg of chitosan was dispersed and dissolved in 50 mL of deionized water containing 1% v/v acetic acid (pH 4.7) to obtain the homogeneous solution (2 mg/mL). Second, ethanol was employed to dissolve BU (the minimum volume of ethanol was used). The BU ethanolic solution was added to chitosan solution at a different drug: chitosan ratios (1:10, 2:10, 3:10 and 4:10 w/w), followed by magnetic stirring at 700 rpm for 30 min (Heidolph Instruments, MR Hei-Tec, Schwabach, Germany). This solution was heated to 60 °C for 10 min. An aqueous solution of three polyphosphate (TPP) (1 mg/mL at 8 °C), was added dropwise to the warm chitosan solution under continuous stirring (700 rpm) at room temperature. The weight ratio of chitosan to TPP in all formulations was kept at 3:1 based on previous studies [42,43]. After the addition of TPP, stirring was continued for 1 more hour at room temperature. The resulting suspension was centrifuged at 21,000 rpm at 4 °C for 30 min (Sigma, 3–30 K, Schnelldorf, Germany) to separate nanoparticles. The nanoparticles were washed with deionized water and then freeze-dried for 48 h (Heto, Dw 3, Allerd, Denmark) using sucrose 1.5% w/v as lyoprotectant. The freeze-dried nanoparticles were maintained in a closed container for further use. The same procedure was followed for the preparation of drug-free chitosan nanoparticles (CN) for comparison purposes.

2.3. Physicochemical Characterization of Nanoparticles

2.3.1. Particle Size Distribution Study

The particle size distribution and zeta potential of nanoparticles were assessed by dynamic light scattering (DLS) (Zetasizer 5000, Malvern, UK). The obtained nanoparticles were diluted 10 times by adding 900 µL of distilled water to 100 µL of the solution containing nanoparticles before measurement. All measurements were carried out in triplicate at room temperature.

2.3.2. Encapsulation Efficiency and Yield

The amount of BU entrapped within the nanoparticles was measured directly by dispersing 20 mg of nanoparticles in a solvent mixture containing 5 mL of acetic acid 1% v/v (pH 4.7) and 5 mL of ethanol. This suspension was shaken at 50 rpm by a mechanical shaker (TUV-NORD, Tehran, Iran) for 72 h at room temperature, and then filtered via a disposable syringe filter (pore size 0.22 μm). The concentration of BU in the filtrate was examined by a UV spectrophotometer at 246 nm (Shimadzu, UV/1204, Tokyo, Japan). Specimens were prepared in triplicate, and encapsulation efficiency (%) (EE) was analyzed by Equation (1) below:

$$\text{Encapsulation efficiency (\%)} \frac{Amount\ of\ budesonide\ in\ NPs}{Amount\ of\ budesonide\ initially\ added} \times 100 \qquad (1)$$

Additionally, the production yield (%) for the selected formulation was calculated using the following equation:

$$\text{Yield (\%)} = \frac{Weight\ of\ nanoparticles}{Initial\ weight\ of\ polymer\ and\ drug} \times 100 \qquad (2)$$

2.3.3. Transmission Electron Microscopy (TEM)

In order to observe the morphology of the selected nanoparticles (nanoparticles with 1:10 drug:chitosan) transmission electron microscopy (TEM) (Jeol JEM-1400, JEOL Ltd.; Tokyo, Japan) was employed. A drop of the aqueous dispersion of the washed nanoparticles was deposited in a mesh copper grid and air-dried at room temperature. Then, the grid was subjected to a 60 kV acceleration voltage.

2.3.4. Scanning Electron Microscopy (SEM)

To evaluate the external morphology of the selected nanoparticles after freeze-drying scanning electron microscope (Leo, VP1450, Neu-Isenburg, Germany) was employed. Freeze-dried nanoparticles were spread on a black steel grid and then coated with a thin film of gold using a gold sputter (Polaron, SC7620 sputter coater, Laughton, England) for 180 s under argon atmosphere. Voltages of 5 kV were selected for accelerating the electrons from electron gun onto the specimen.

2.3.5. Differential Scanning Calorimetry (DSC)

To evaluate the thermal property of nanoparticles, a differential scanning calorimeter calibrated with indium standard was used (Mettler Toledo, DSC 822e, Greifensee, Switzerland). The samples (3–5 mg) were hermetically sealed in DSC pans and were scanned in the temperature range of 25–300 °C at a rate of 10 °C/min under a nitrogen flow of 80 mL/min.

2.3.6. X-ray Powder Diffraction (XRPD)

XRPD study of nanoparticles with 1:10 drug:chitosan was carried out by employing an X-ray powder diffractometer (GNR, Explorer, Milan, Italy). The instrument was operated at 40 kV and 30 mA in the range (2θ) of 5 to 55° using a step size of 0.01° 2θ and step-time of 3 s, with a non-stop mode.

2.3.7. FTIR

Selected nanoparticles (1:10 drug:chitosan) were analyzed with an FTIR spectrometer (Thermo Nicolet, AVATAR 370, Walthman, USA) from 400 to 4000 cm^{-1} at room temperature by the KBr disc method.

2.3.8. In Vitro Drug Release Studies

The release pattern of BU (powder), as well as the selected BCN (1:10 drug:chitosan) corresponding to 9 mg of BU ($n = 6$) were studied in 250 mL of simulated gastric fluid (SGF) pH 1.2 and simulated colonic fluid (SCF) pH 6.8 comprising 0.25% w/v SLS. USP dissolution apparatus I (Pharmatest, PTWS 3E, Hainburg, Germany) was employed to study the release pattern of BU from the selected formulations. The dissolution was operated under the rotation speed of 75 rpm, at 37 ± 0.5 °C. In the case of BU powder and freeze-dried BCN, an accurately weighed sample was gently dispersed in the medium. At pre-defined time intervals, samples (5 mL) were withdrawn from the release medium and filtered through a disposable syringe filter (pore size 0.22 µm). Subsequently, the amount of BU in 1 mL of filtrate sample was quantified using HPLC (Shimadzu, Kyoto, Japan) equipped with a Teknokroma column (BRISA LC2 C18 250 mm × 4.6 mm, 5 µm). The mobile phase consisted of an acetate buffer (pH 3.9) and acetonitrile mixture (35:65) flowing at a rate of 1.5 mL/min and BU was measured at 240 nm.

2.4. Pelletization of Budesonide Nanoparticles

The extrusion–spheronization technique was used to load selected BCN (the formulation with the smallest size and highest encapsulation efficiency) into the pellet formulation. For the preparation of pellets containing budesonide-loaded nanoparticles (BCNP) with 1% *w/w* of the drug, a powder blend containing 25% *w/w* freeze-dried selected BCN, 2% *w/w* PVP K30, 56% *w/w* Lactose mono hydrate and 17% *w/w* Avicel® PH 102, were mixed by a kitchen mixer (FUMA, Fu-1877 Hand Mixer, Tokyo, Japan) for twenty minutes. The mixture was wetted by adding distilled water. The wet mass was passed through an axial screw extruder (Dorsa Tech, EX-01, Tehran, Iran). The extrusion was performed under the rotation speed of 100 rpm using flat sieves of 1 mm aperture size. Then, the extrudates were spheronized (Dorsa Tech, EX-01, Tehran, Iran) for 5 min using a cross-hatched friction plate rotated at 1200 rpm. The obtained pellets were dried in an oven (24 h at 40 °C) and followed by sifting through 1180 and 850 µm. The sifted pellets between 1180 and 850 µm were collected.

Conventional BU pellets (CP) were also manufactured according to the method established in our previous study [44]. In brief, all powdered components, including 1.5% *w/w* BU, 5% *w/w* PVP K30, 68.5% *w/w* lactose mono hydrate and 25% *w/w* Avicel® PH 102 were mixed for 20 min and then turned to cohesive mass using distilled water. The wet mass was extruded and spheronized according to the methods described for BCNP.

2.5. Evaluation of Pellets

2.5.1. In Vitro Drug Release Studies

The release profiles of uncoated BCNP corresponding to 9 mg of BU (n = 6) were studied as described in Section 2.3.8.

2.5.2. Pellet Morphology Studies

The morphology of BCNP and CP was analyzed by measuring the sphericity and aspect ratio of pellets. Pictures were taken by a stereomicroscope (Kyowa, Tokyo, Japan) equipped with a computer system and video camera (Sony, Tokyo, Japan) and then analyzed by an image analyzing software (ImageJ 1/50 for windows).

2.5.3. Particle Size Analysis of Pellets

The particle size of BCNP and CP was analyzed using the sieve method. A sample of 25 g of pellets was shaken on top of the series of standard sieves (150, 180, 250, 425, 850, 1000, and 1180 µm) using a sieve shaker (Azmun test, 50410, Tehran, Iran) for 10 min. The mass of pellets that remained on each sieve was determined and used to calculate the geometric mean particle size (dg) and geometric standard deviation (σg) from the plot of the cumulative percentage of undersize on the probability scale versus the log of particle diameter.

2.5.4. Pellet Mechanical Properties

The mechanical characteristics of BCNP and CP were examined by testing twenty pellets in the size range of 850–1180 µm using a Material Testing Machine (Hounsfield, H50KS, London, England). Force–displacement graphs were prepared by a computer program connected to the device (Hounsfield, QMAT, London, England). The 1 kN load cell was set up at a moving speed of 1 mm/min to determine the pellet crushing strengths and elastic modulus.

2.6. Coating of Pellets

CP and BCNP were coated using a Wurster column fluid bed coater (Haltingen-Binzen, UNI-Glatt, Binzen, Germany). A solution of 48% w/w eudragit S, 12% w/w eudragit L and 40% w/w eudragit RS was prepared in a mixture of isopropanol and distilled water (9:1) under agitation. Triethyl citrate (TEC) was included as a plasticizer (10% w/w based on the weight of polymers) and stirred for 1 h. Talc was then added as an antiadhesion agent (5% w/w based on polymer weight). The coating solution was utilized onto 50 g pellets where the inlet air and outlet air temperatures were adjusted at 43–40 °C and 39–36 °C, respectively. In this experiment, the atomization pressure and spray rate were set up at 2 bar and 0.5 g/min, respectively. The coating process was continued till the coating mass reached 10% (w/w).

2.7. In Vitro Drug Release Studies of Coated Pellets

The evaluation of the drug release pattern from coated pellets was performed according to the continuous mode of the dissolution test at 37 $\pm$ 0.5 °C (USP dissolution apparatus I, Pharmatest, PTWS 3E, Hainburg, Germany). The amounts of weighed pellets ($n = 6$) equal to 9 mg of BU were tested in 250 mL of a medium comprising 0.25% w/v SLS at a rotation speed of 75 rpm. Dissolution tests were performed at various time intervals of 2 h (pH 1.2), 1 h (pH 6.5), 2 h (6.8), 1 h (pH 7.2) and 10 h (pH 6.8). The dissolution was also performed in the medium containing 4% w/v rat cecal content with pH 6.8 to simulate GIT media and the transit time of the pellets from various sections (i.e., stomach, duodenum, jejunum, ileum, and colon). The media supplemented by cecum content was continuously bubbled with CO2. After each incubation time for each medium, the baskets with their pellets were moved rapidly from one medium to another. At defined time intervals, samples (5 mL) were withdrawn manually and analyzed by HPLC according to the method explained in Section 2.5.1. In the case of media supplemented by cecum content, 1 mL of the medium was removed and centrifuged at 15,000 rpm at 4 °C for 30 min. Then, the supernatant was taken and diluted with acetonitrile at a ratio of 1:3 and the mixture was filtered using a disposable syringe filter (pore size 0.22 μm) before the sample is injected to the HPLC column [45].

2.8. Morphology of Coated Pellets

The surface morphology as well as cross section of the coated BCNP were characterized using scanning electron microscope (Leo, VP1450, Neu-Isenburg, Germany). The pellets were fixed on an aluminum stub and sputter-coated (Polaron, SC7620 sputter coater, Laughton, England) with a thin layer of gold for 180 s using a sputter coating machine under argon atmosphere, and then analyzed using SEM. Voltages of 5 kV were selected for accelerating the electrons from electron gun onto the specimen. The cross-sectional samples were prepared using a razor. These samples were then covered with a thin layer of gold.

2.9. Animal Treatment

To investigate the efficacy of different formulations of BU in the treatment of colonic damage, the rat model of ulcerative colitis was employed. Wistar male rats aged 10–12 weeks and weighing 250–300 g were used in this part of the study, under institutional guidelines granted by the institutional ethics committee of Mashhad University of Medical Sciences (IR.MUMS.REC.1399.011).

All animals employed in the study were kept in a clean room, with air-conditioned (22 $\pm$ 3 °C), controlled humidity (40 $\pm$ 5%) and light–dark cycles of 12 h; 24 h before the colitis induction, all the rats were weighed, examined to be healthy and kept fasted except for water. For induction of colitis, first of all, an intraperitoneal injection of ketamine (50 mg/kg) and xylazine (5 mg/kg) solution was used to anaesthetize rats [46,47]. Thereafter, acetic acid (2 mL 3% v/v in normal saline) was administered rectally 8 cm deep using a 2 mm diameter cannula [48]. The healthy control group received 2 mL of normal saline solution intrarectally. Rats were kept in an upside-down position for 1 min and then they were taken back to their cages. All animals were left for 3 days without treatment to let the colitis to be developed while they were free for consumption of water and food [49]. The colitis-induced rats were then divided randomly into 7 groups, with 6 animals per group. One group of rats were used as the control group (untreated group), whilst others were treated with conventional coated pellets (coated CP), pellets containing BU loaded nanoparticles (BCNP), coated BCNP, BU nanoparticle (BCN), free drug chitosan nanoparticles (CN) and free BU powder (BU) three days after induction of colitis. Each treated group received orally an equal dose of BU (0.15 mg/day) for 7 consecutive days [50]. In the case of BCN, CN and BU, the dose was suspended in 1 mL of purified water before administration. The healthy and the untreated control groups received 1 mL of normal saline. The weight and stool consistency of the rats were evaluated during this period. The rats were killed by CO_2 asphyxiation either 24 h or 6 days after the last dose administration. Then, the colon, from the cecum to the anus was cut for further studies.

2.10. Evaluation of Colitis Treatment

2.10.1. Colitis Activity Index

The degree of inflammation was checked daily by determining the colitis activity index (CAI) which consists of three clinical parameters including weight loss, stool consistency, and anal bleeding [51]. Briefly, the score scale from 0 to 4 indicates the degree of inflammation (from healthy to maximal inflammation). Scores of 0, 1, 2, 3 and 4 correspond to no weight loss, 1–5% weight loss, 5–10% weight loss, 10–20% weight loss and more than 20% weight loss, respectively. Similarly, for consistency of stool, a well-formed pellet gets 0, pasty and semi-formed stools with no sign of stickiness to the animal's anus gets 2, and liquid stools with stickiness to the anus gets 4. In addition, bleeding scores were 0 for no blood, 2 for positive findings, and 4 for gross anal bleeding. The average of these points was reported as CAI.

2.10.2. Colon/Body Weight Ratio

To determine the colon/body weight ratio, rats were weighed before scarification. Colon tissue samples were resected, cut longitudinally, and washed with an iced phosphate-buffered solution (pH 7.4). The weight of washed resected colon was also determined and the ratio of the wet colon weight to the body weight of each rat was defined as an index of colonic inflammation [31].

2.10.3. Weight/Length Ratio of Colon

The length and wet weight of the washed colon tissue samples was assessed, and the weight/length ratio was determined as a sensitive and reliable indicator of the severity of the colonic inflammation [52].

2.10.4. Glutathione Content of the Colon Tissue

The amount of reduced glutathione in the colon tissue was defined spectrophotometrically based on the appearance of yellow color after the addition of DTNB (5,5'-dithiobis-(2-nitrobenzoic acid)) to compounds bearing sulfhydryl groups [53]. In brief, 0.5 mL of tissue homogenate 10% w/v in phosphate-buffered saline (0.1 M, pH 7.4) was blended with 0.5 mL of 10% w/v trichloro acetic acid (TCA) and vortexed. The blend was then centrifuged at $2500 \times g$ for 10 min at 4 °C and 0.5 mL of supernatants was withdrawn and mixed with reaction mixtures consisting of 2.5 mL phosphate buffer (pH 8) and 0.5 mL DTNB. The absorbance of the mixture was determined at 412 nm spectrophotometrically (Jenway 6105 UV/Vis, Staffordshire, UK) and GSH contents (nmol/g tissue) were assayed against GSH standard curve prepared in the blank medium [54].

2.10.5. Malondialdehyde Content of the Colon Tissue

The malondialdehyde (MDA) concentration of the colon tissue which is related to the severity of colitis was measured [55]. Briefly, colon specimens were frozen in liquid

N$_2$ rapidly after scarification and stored for subsequent evaluation. To determine MDA concentration (nmol/g tissue), 3 mL phosphoric acid (1% w/v) and 1 mL thiobarbituric acid (TBA) (0.6% w/v) were introduced to colon tissue homogenate (10% w/v) in KCl (1.15% w/v). The mixtures were heated in a water bath for 45 min. Then, 4 mL of n-butanol was added to the cooled mixture and vortexed for 1 min. The mixture was then centrifuged at $3000\times g$ for 10 min. Then, absorbance of the supernatant at 532 nm was determined by a spectrophotometer (Jenway 6105 UV/Vis, Staffordshire, UK) and the MDA concentration was calculated based on the constructed standard curve obtained in the blank medium [56].

2.10.6. Histological Assessment of Colitis Severity

A segment of the washed colon tissue specimens was fixed in a 10% formalin solution and inserted in paraffin. Cutting sections obtained by microtome were stained with hematoxylin and eosin (H&E) and examined by optical microscope to assess the severity of colitis. The colitis severity was scored from 0 to 4 based on the microscopic observation of cross-sections of the colon. The severity of colitis was scored 0 when no or minor inflammation was observed. When there was a trace region of focal inflammatory cell infiltration, the severity of colitis was scored 1. When a major part of the colon tissue was exposed to inflammation and sign of smooth muscle thickening was observed, the severity of colitis was scored 2. If ulcerated regions and inflammatory cell infiltration were formed in the tissue sections following the inflammation, the severity of colitis was assigned a score of 3. Finally, the highest colitis severity score (i.e., 4) was assigned when the entire tissue damage appeared as necrosis and gangrene [45].

2.10.7. Blood Glucose Level

The blood glucose level in rats was measured from the 12th day until the 17th day of the study by sampling the tail vein blood with an 18-gauge needle in the strip of a blood glucose meter (Accu-chek® Active, Roche, Mannheim, Germany). This experiment could provide some comparison amongst specific colon delivery potential of different formulations investigated [57,58], and also information regarding the potency of nanoparticles for accumulation in inflamed regions. Data on blood glucose levels are stated in mg/dL and compared with untreated animals [48].

2.11. Statistical Analysis

To analyze the obtained data statistically, Graph Pad Prism software (Graph Pad Prism, version 7, San Diego, CA, USA) was employed. One and two-way analyses of variance (ANOVA) followed by Tukey–Kramer test was employed to compare the differences between means ($p < 0.05$ is an indication of a significant difference).

3. Results and Discussion

3.1. Preparation and Characterization of Nanoparticles

The anti-inflammatory effect of chitosan against IBD and UC has been reported in previous studies [59] where high-molecular-weight chitosan had better therapeutic effects in UC compared with-low molecular-weight chitosan [60]. On the other hand, an increase in the molecular weight of chitosan could result in a decline in the drug release which could be due to the greater viscosity of the gel layer created around the drug particles upon contact with the dissolution medium [61,62]. Since BU is a highly hydrophobic drug with low water solubility [63], its slow release could provide an insufficient concentration of the drug at the site of action. Furthermore, it has been reported that in the ionic gelation procedure for producing chitosan nanoparticles, medium-molecular-weight chitosan has shown more encapsulation efficacy compared with high-molecular-weight chitosan [64]. Therefore, medium-molecular-weight chitosan had been used in this study in an attempt to prepare BU nanoparticles by the ionic gelation technique [41,42]. This method has great potential for industrial application due to the lack of need for use of organic solvents, sonication or high temperatures and also the relative easiness of scaling up [65]. The production of nanoparticles with a lower polymer content could lead to a more cost-effective formulation; therefore, in the current study, the effect of the drug-to-chitosan ratio in the preparation of nanoparticles was investigated to find out if it is possible to achieve desirable properties for nanoparticles at low concentrations of chitosan in formulations.

The size characteristics of nanoparticles have been found to affect drug accumulation in the inflamed colon [66]. Despite the fact that larger size nanoparticles (500 nm) are mainly localized in the epithelial lining [11,67], it has been demonstrated that excellent mucoadhesive properties belong to the nanoparticles with a particle size range between 200 and 300 nm [68]. In this study, the particle size of the different formulations ranged between 288 ± 7.5 and 566 ± 7.7 nm with small PDI values indicating the production of monodispersed particles. Table 1 shows that with an enhancement in the ratio of drug: chitosan (from 1:10 to 4:10), larger nanoparticles with lower encapsulation efficacy (37% to 2%, respectively) were obtained. The nanoparticles are formed through inter and intramolecular interactions between the negative groups of TPP and the positive charged amino groups of chitosan [69]. It has been found that chitosan molecules have a spread conformation in solution due to the electrostatic repulsion forces between free amino groups along the molecular chain [70]. These free amino groups are also responsible for the electrostatic interactions between the nanoparticles, helping to reduce particle sizes [70]. Upon the addition of the BU to the chitosan solution under constant stirring, oxygen atoms of BU interact non-covalently with the free amino groups of chitosan chains [41], leading to a decrease in the number of free amino groups. Since the concentration of free amino groups can have an impact on the creation of nanoparticles [62] and the encapsulation efficiency depends on the number of available sites present in the formulating material [70], the higher concentration of BU resulted in the formation of larger nanoparticles as well as lower encapsulation efficacy. Similar results have been reported in other studies [41,70,71].

Table 1. Size, PDI, zeta potential and EE for different nanoparticles.

Formulation	Size (nm)	PDI	Zeta (mV)	EE (%)
BCN-1	288 ± 7.5	0.33 ± 0.02	+26.6 ± 1.1	37.2 ± 3.1
BCN-2	323 ± 5.1	0.43 ± 0.05	+27.9 ± 0.9	29.1 ± 2.4
BCN-3	416 ± 6.8	0.44 ± 0.03	+30.6 ± 1.4	10.8 ± 1.1
BCN-4	566 ± 7.7	0.45 ± 0.01	+31.0 ± 1.2	2.2 ± 0.9

All prepared formulations showed positive zeta-potential ranges from +26.6 ± 1.1 to +31.0 ± 1.2 mV. Since BU is a slightly anionic molecule [72], positive zeta potentials are due to the positively charged amino groups of chitosan molecules present at the surface of particles. It can be postulated that nanoparticles with zeta potential below −25 mV and above +25 mV can be considered stable at physiological pH [73]; therefore, all nanoparticles produced in the current study can be expected to be stable under physiological conditions.

According to data shown in Table 1, the formulation with a drug:chitosan ratio of 1:10 (BCN-1) provided the smallest particle size and highest encapsulation efficiency (37.2%). Thus, this formulation was chosen, as a representative candidate, for additional studies and was assigned as BCN in the next paragraphs. The yield of the production process of this formulation was 51.0 ± 1.2% which was considered satisfying, taking into account the initial low amounts of the processed material.

The morphology of the selected nanoparticle (BCN), as well as freeze-dried BCN, was observed by TEM and SEM, respectively. As it can be seen in Figure 1A,B, the nanoparticles were spherical with almost smooth surfaces. The microscopic images of BCN nanoparticles demonstrated a smaller average diameter in comparison with the DLS results. Such this difference which was also observed in other studies [16,74,75], could be attributed to the fact that in DLS the hydrodynamic diameter of the nanoparticles which is a combination of the diameter of nanoparticles plus the solvent layer attached to them [75] are measured, while in the microscopic image the hydration layer is not present [76]. The absence of major changes in nanoparticle diameter before and after freeze drying (TEM images in comparisons with SEM images) could be related to the proper concentration and good functioning of the lyoprotectant used [77].

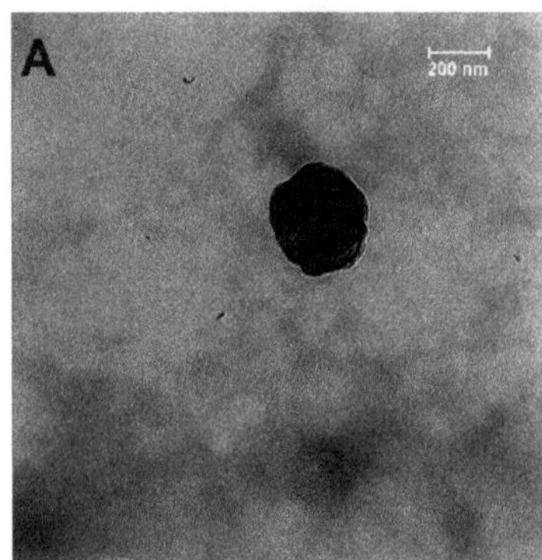
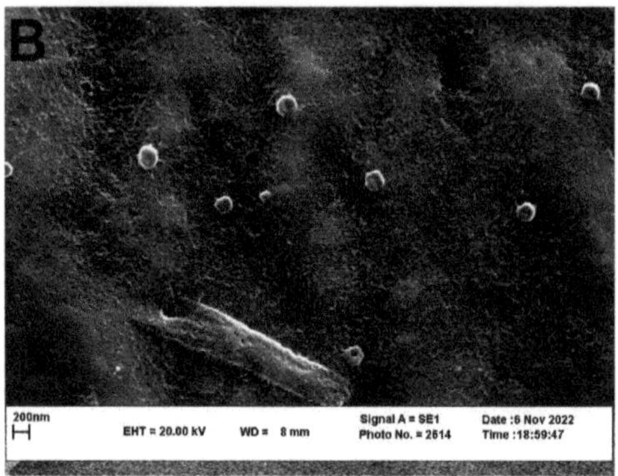

Figure 1. (**A**) TEM image of BCN; and (**B**) SEM images of the freeze-dried BCN.

Figure 2A shows thermograms of the drug powder as well as the chitosan, sucrose, BCN and physical mixture of the excipients used in the preparation of BCN matrix. Budesonide and sucrose are crystalline compounds. Budesonide, as stated by the literature, demonstrated a sharp endothermic melting peak at around 261 °C [41], and sucrose showed an endothermic melting peak at around 190 °C [78]. In contrast, chitosan showed a broad endothermic peak at 100 °C (corresponding to the loss of water) and an exothermic peak at around 310 °C (related to the decomposition of amine units) [79]. The peak related to the melting of sucrose as well as the decomposition peak of the chitosan could be observed in the thermogram of the physical mixture, while the chitosan decomposition peak shifted towards lower temperatures in the BCN thermogram. This shift could be ascribed to the rearrangement of an intermolecular and intramolecular network of the chitosan, due to the crosslinking with the TPP ions as reported elsewhere [41] and also probable interaction between chitosan and the drug. The characteristic peaks of BU disappeared in both BCN and physical mixture thermograms probably either due to the very low content of the drug or presenting in the molecularly dispersed state [44,57].

Figure 2B shows the XRPD spectra of the drug as well as the chitosan, sucrose and BCN. Chitosan is a semi-crystalline polymeric material with two characteristic peaks at 2θ of 11 and 21 [80]. The characteristic peaks of BU appeared at 2θ of 5.99, 11.95, 14.42, 15.36 and 15.96 which confirmed the crystalline nature of the drug have also been reported elsewhere [28]. The XRPD spectra of BCN exhibited the characteristic peak of BU with reduced intensity confirming the reduced crystallinity of the drug in nanoparticles.

FTIR analysis was performed to characterize the interactions between chitosan and TPP as well as possible interaction between chitosan and BU in nanoparticles (Figure 3). The budesonide spectrum presented peaks at 3495 and 2957 cm^{-1} that correspond to the OH groups and CH2 groups, respectively. Additionally, at 1723 and 1667 cm^{-1} the characteristic stretching peaks of C=O groups as well as stretching peaks of the double bond C=C at 1625 cm^{-1} could be seen [81]. Chitosan showed absorption bands for O-H and N-H2 stretching at 3378 cm^{-1}, C-H stretching at 2871 cm^{-1}, amides and primary amines at 1663 and 1601 cm^{-1}, respectively, C-N bond at 1319 cm^{-1} and ultimately C-O bonds at 1076 and 1029 [41]. The characteristic absorbance peaks of the TPP, related to the phosphate groups presented at 898 cm^{-1} [82], were shifted to lower values (892 cm^{-1}) in BCN spectra. This shift could result from interactions between amino groups of chitosan and the phosphate groups of TPP [82]. Almost all characteristic peaks of the BU shifted to the lower values in the FTIR spectrum of the BCN which could account for the interaction between BU and the amino or hydroxyl groups of chitosan. Furthermore, in the BCN spectrum, a new peak at 1565.4 cm^{-1} could be seen that was neither observed in BU nor in chitosan spectrum. This new peak could be related to the N-H2 stretching bond and confirms the presence of hydrogen bonds between BU and chitosan in BCN and could be considered as evidence of the loading of BU into the chitosan nanoparticles.

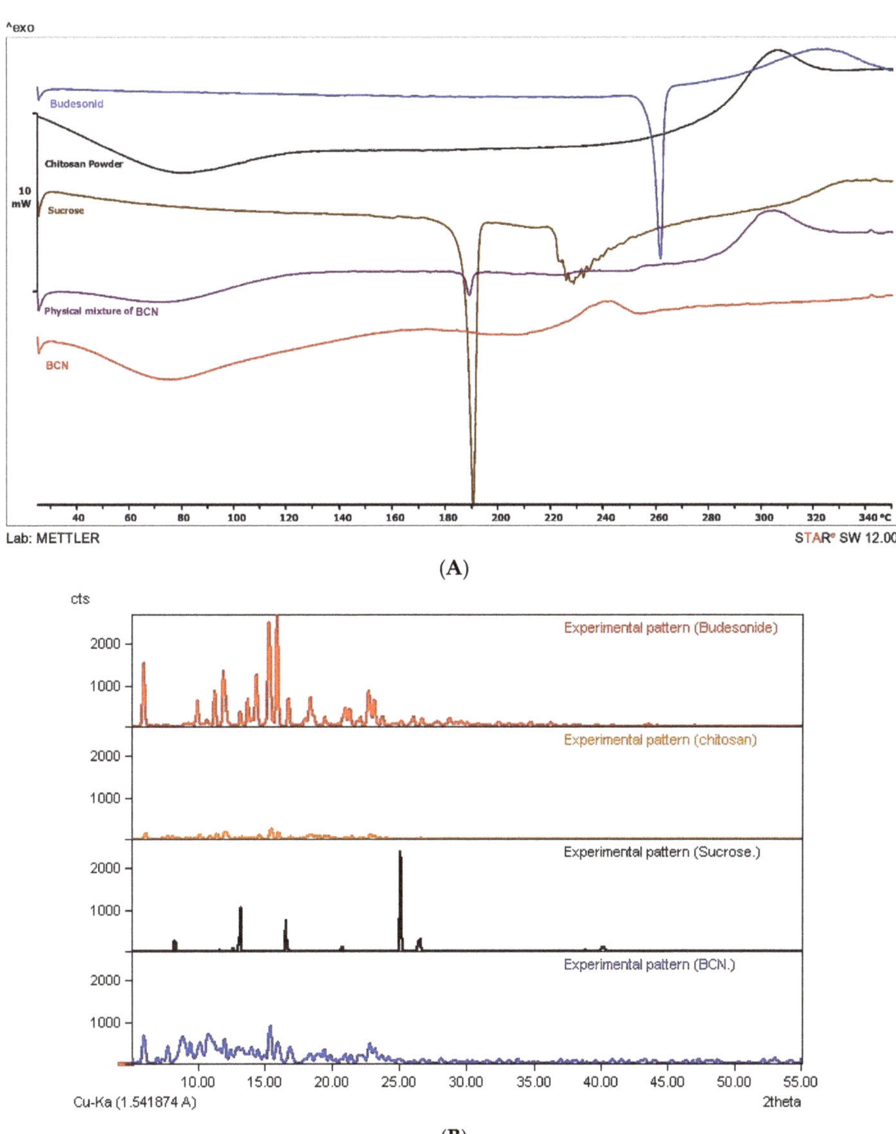

Figure 2. (**A**) DSC thermograms and (**B**) XRPD spectra of budesonide, chitosan, sucrose and BCN.

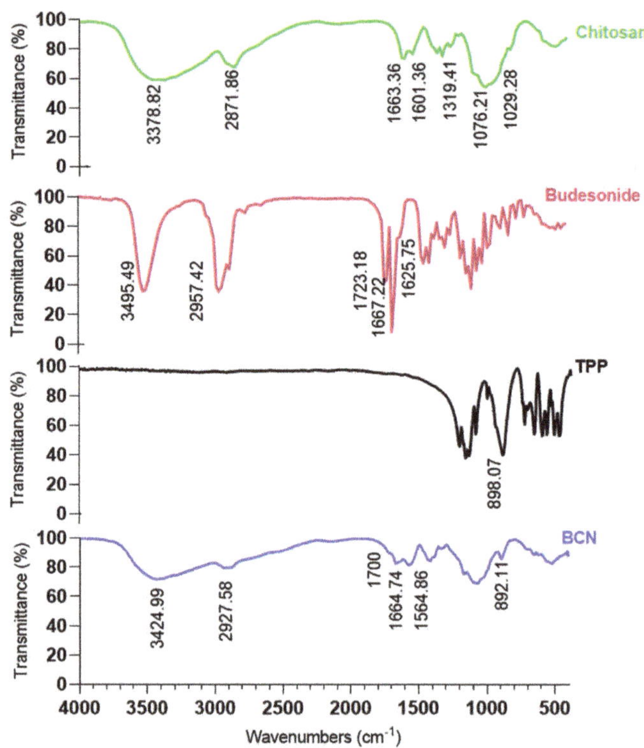

Figure 3. FTIR spectra of chitosan, budesonide, TPP and BCN.

3.2. Preparation and Characterization of Pellets

The freeze-dried BCN was loaded into the pellets in order to improve the local stability within the gastrointestinal tract and prevent premature uptake of nanoparticles. The extrusion–spheronization technique is a promising technology for converting nanoparticles into solid dosage forms [83]. Conventional pellets of BU were also made for comparison purposes. The formulation of conventional pellets was developed in our previous study [44].

Pellets containing nanoparticles (BCNP) were nearly spherical and had acceptable roundness with the mean sphericity and aspect ratios of 0.98 ± 0.06 and 1.14 ± 0.07, respectively [84]. The geometric mean diameter and standard deviation of BCNP were 1082.1 ± 1.3 µm which indicates the suitability of these formulations for consideration as multiparticulate dosage forms [85]. Additionally, the results of crushing strength (6.3 ± 0.9 N) and elastic modulus (1447.0 ± 170.2 MPa) of BCNP showed that these pellets had suitable mechanical properties for coating steps [86].

The pellets were subsequently coated with combined pH and time-dependent eudragits to prepare a colonic drug delivery system with a sustained release of the drug. Since the pH variations in the intestine could affect the efficacy of single pH-dependent drug delivery systems [29], such a combined system was selected for delivering the drug or nanoparticles to the desired site. The adopted coating process operative conditions, coating level and coating material compositions that were used in this study were previously set up by our team.

3.3. Release Studies

3.3.1. Release of Budesonide from BCN and Uncoated BCNP

Figure 4 shows the release profiles of BU powder, budesonide-loaded nanoparticles (BCN) and uncoated BCNP in simulated gastric fluid (SGF, pH 1.2) and simulated colonic fluid (SCF, pH 6.8). It could be observed that the loading of BU in chitosan nanoparticles resulted in a decline in the drug release rate. Similar findings were observed in other studies when drugs were encapsulated in polymeric matrices [62,68]. According to the results, the release profiles of BCN were pH-dependent which is not expected as BU solubility is not pH-dependent [58] and also it dissolves completely within the first 2 h at SGF and SCF. The unexpected pH-dependency of BCN formulations could be described by the properties of chitosan which is soluble in an acidic medium and conversely, but it has poor solubility in neutral pH [87], resulting in a faster release at pH 1.2 compared with pH 6.8. Release profiles of BCN in both pH levels presented a burst release of about 45% and 30% in pH 1.2 and 6.8, respectively during the first hour, followed by a sustained release behavior for up to 12 h. Initial burst release might be due to the dissolution of the drug which is located at or close to the nanoparticle surface which is more accessible to the release medium, and the hydrophilic nature of chitosan [88]. It can be seen that after 12 h, only 74% and 67% of the encapsulated drug were released from BCN in pH 1.2 and pH 6.8, respectively. The incomplete drug release from nanoparticles could be explained by the drug release mechanism from the chitosan matrix. Upon the exposure of chitosan to the aqueous media, a gel layer is formed around the drug particles [62]. Slow drug diffusion through this high viscose gel layer and chitosan matrix makes the dissolution profile sustained and incomplete. A slower release profile from pelletized nanoparticles compared with the BCN demonstrated that nanoparticles were not adversely affected by mechanical stresses during the extrusion and spheronization process.

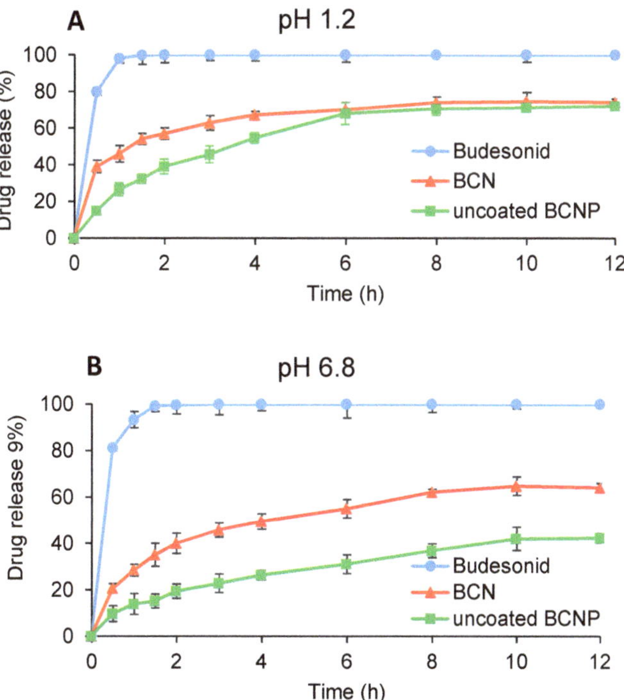

Figure 4. The drug release profile of budesonide, BCN and uncoated BCNP in (**A**) pH 1.2 and (**B**) pH 6.8.

3.3.2. Release of Budesonide from Coated BCNP

As it can be seen in Figure 4, uncoated BCNP released more than 50% of the loaded drug during the first 2 h in acidic pH. Thus, to avoid early drug release and deliver most of the drug into the colon, pellets were coated with a combined pH and time-dependent coating system. In this case, continuous dissolution testing was performed in a pH progression medium for 16 h for coated CP and BCNP. The coated pellets exhibited a pH-dependent release profile (Figure 5A) as expected. No drug release was observed when coated conventional pellets passed through media with pH 1.2, 6.5, and 6.8 for 2, 1, and 2 h residence times, respectively. It was observed that while coated CP released almost 90% of the drug in the medium simulating the pH of the colonic environment, in the case of pellets containing nanoparticles such a coating system was less effective for delivering most of the drug to the colon. Coated BCNP released almost 30% and 35% of the drug in simulating pH of gastric and small intestine fluid, respectively; thus, only 35% of the drug was released in the final part of the continuous dissolution test. The lower efficiency of this coating system in the colon delivery of pellets containing nanoparticles can be attributed to the swelling properties of chitosan [89] and the composition of the coating. Although the coating composition is insoluble in the acidic medium it is permeable to it due to the presence of eudragit RS. As the acidic medium penetrates the pellets, chitosan would be swelled and could damage the functional coating. The swelling ratio of chitosan at low pH values is higher than neutral pH due to repulsion between similarly charged groups [90]. It seems that by preventing the entry of the acidic medium into the pellet, it is possible to prevent the destruction of the coating so a larger amount of drug could be delivered to the colon. To prove this hypothesis, the same continuous dissolution test except for the initial 2 h in pH 1.2 was performed with and without 4% w/v rat cecal content in the last medium with pH 6.8. As illustrated in Figure 5B, when the formulation was directly placed in the medium with pH 6.5 (bypassing the acidic environment) only 30% of the drug was released before reaching the colon-simulated media. Although it seems that the drug release from coated BCNP in such conditions might be incomplete (60% during 14 h), the presence of rat cecal content in the SCF medium increased the drug dissolution (80% during 14 h). This confirms the effect of the bacteria in the rat cecal as well as the pH sensitivity of the system to the drug release. This suggests that the degradation of chitosan in the pellets' matrix might happen upon the action of the bacterial enzymes in the SCF [91], resulting in an increase in the release rate. The overall results indicated that in the case of coated BCNP, where the target of drug delivery is the colon, drug release in the stomach would be considered as a drawback [92], so more strategies such as using enteric-coated capsule [93] filled with the coated BCNP should be considered for oral delivery of this type of formulation where the target site is the colon area. However, when drug delivery is aimed at the entire GIT (Crohn's disease) coated BCNP could provide a continuous release throughout the GIT.

3.4. Morphological Characteristics of Coated BCNP

Figure 6 illustrated the SEM images of the coated BCNP. These pellets were spherical and had smooth surfaces (Figure 6A). The cross-sectional image showed a uniform coat with approximately 30 μm thickness (Figure 6B).

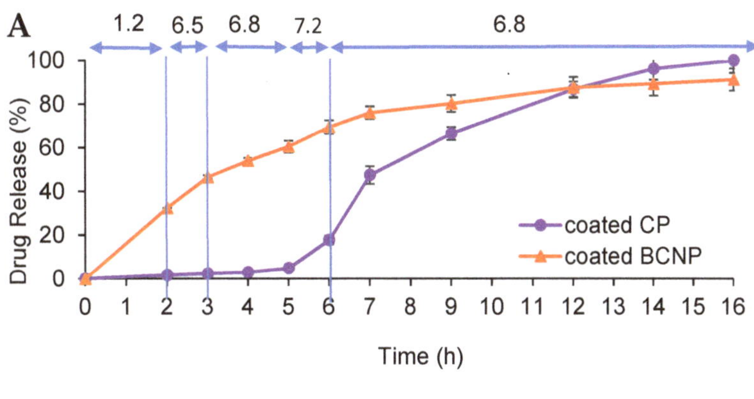

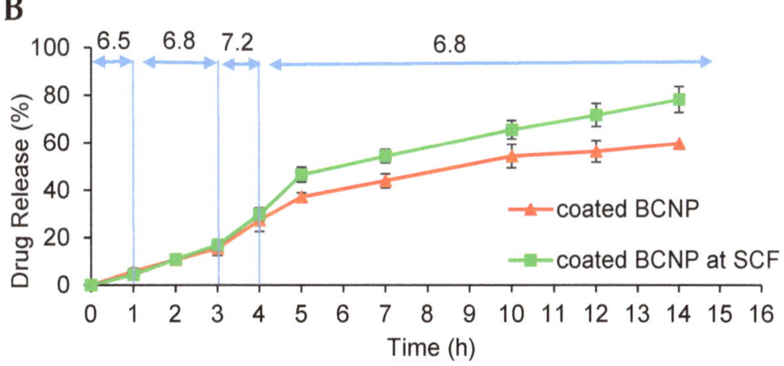

Figure 5. The drug release profile of coated pellets (**A**) in the continuous mode of dissolution test and (**B**) in the continuous mode of dissolution test except for the initial 2 h in pH 1.2.

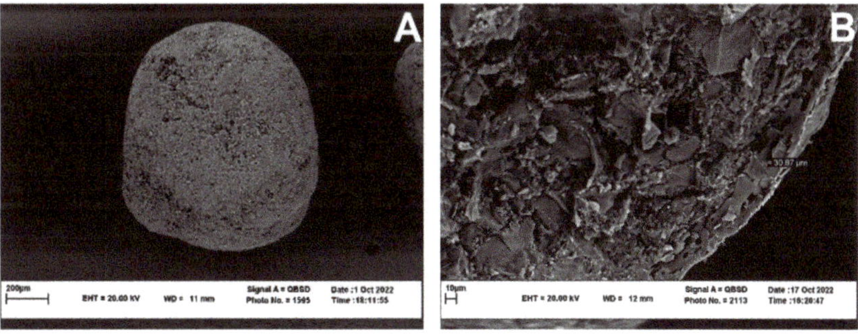

Figure 6. (**A**) Scanning electron microscopy of the coated BCNP and (**B**) the cross-sectional image of the coated BCNP.

3.5. In Vivo Therapeutic Efficacy in Rats

The therapeutic effects of the selected formulations were characterized by determining some indexes of colonic tissue edema (colitis activity index (CAI), colon/body weight, weight/length ratio of colon, the level of MDA and amount of GSH in colon tissue) [94], as well as histological studies.

The CAI is a clinical tool for the evaluation of the severity of inflammation in living animals [14]. After inducing experimental colitis, CAI increased rapidly (Figure 7); all animals showed loss of weight and diarrhea over the next 3 days indicating that all animals presented with colitis. All treatment groups showed a consistent decrease in CAI throughout the entire treatment period (days 5 to 11). On the day of sacrifice, all treated rats showed statistically lower CAI compared with the untreated control group ($p < 0.05$) in response to the decrease in intestinal inflammation [95]. While the group treated with free BU showed little improvement in the CAI, the coated BCNP-treated group demonstrated the most prominent decline in this index (two-fold reductions compared with the untreated group). For those rats which were not sacrificed on day 12, during the next 5 days of monitoring with no treatment (days 13 to 17), the increase in the inflammation severity was observed for both the free BU and coated CP treated groups, so that the CAI value significantly increased from day 12 to day 17, while the difference between CAI in this period was not significant for BCN, BCNP and CN treated groups. In the case of coated BCNP-treated rats, even though the administration of the drug was discontinued, surprisingly the healing process continued (Figure 7). A similar finding has been reported in another study in which the nanoparticle-treated rats continued to heal after stopping the treatment [96].

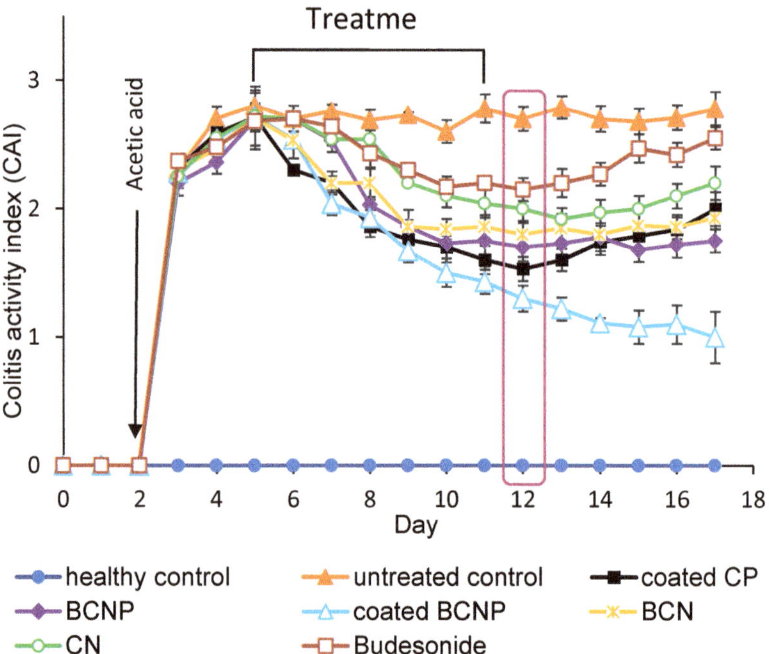

Figure 7. Colitis activity index for different rat groups.

As can be seen in Figure 8A, healthy rats had a colon length of 19 cm, while the colon length significantly decreased in untreated rats with values of ~11 cm. Additionally, untreated colitis rats showed signs of localized inflammation and ulceration and thickening of the colon wall. Although the conditions of the rats did not return to normal during the experiment period, it seems that the inflammatory signs were decreased in all of the treated groups when compared with the untreated group (Figure 8A). Based on the visual comparison, the most increase in colon length was observed in coated BCNP-treated group (almost 9 cm), and the colon length of this group was much close to the healthy group.

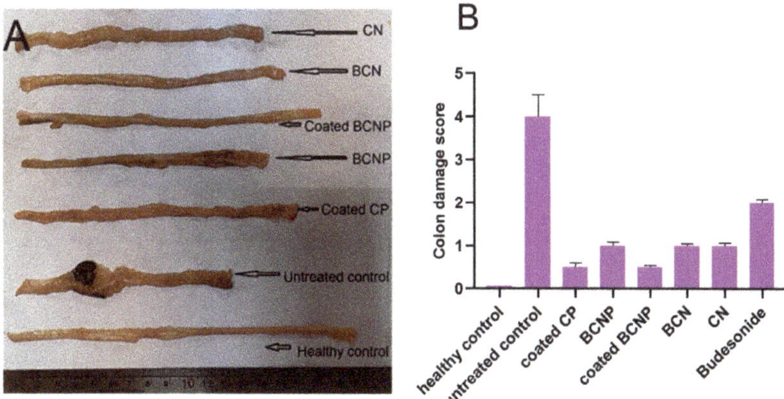

Figure 8. (**A**) Photographs for macroscopically examination and (**B**) colon damage score of the colon specimens resected from rats 24 h after the last drug administration.

As shown in Figure 9A,B, the colon/body weight ratio and weight/length ratio of the colon declined significantly in all of the treated groups when compared with the untreated control group ($p < 0.05$). These two indexes usually increase in response to the severity of inflammation [96]; therefore, the obtained results could indicate that all of the prescribed formulations had some anti-inflammatory effects. Regarding these two inflammatory indexes data, again, the most effective formulation for the treatment of inflammation was coated BCNP whilst free BU was less effective than others. Similar to the results obtained for CAI, BU and coated CP-treated groups showed a statistical difference when these two items (the colon/body weight ratio and weight/length ratio) were compared 6 days after the last drug administration ($p < 0.05$) while no significant differences for these items were observed in groups treated with nanoparticle-based formulations.

It is well-known that the amount of GSH is an indication of the anti-oxidative potential of the cells upon treatment and MDA is a marker indicating tissue injury [56]. A significant increase in the amount of GSH (Figure 9C), as well as a significant decrease in the level of MDA (Figure 9D), was observed in all treated groups compared with the untreated group. These results demonstrated that inflammation significantly decreased upon treatment of rats.

Figure 10 shows the histopathological characteristics of the colon's tissues and Figure 8B presents a colitis damage score. Light microscopic observation of the colon tissue of healthy rats demonstrated normal colon histology with no signs of unusual tissue architecture or abnormal tissue morphology (Figure 10A) which was scored as zero for colitis. By contrast, in untreated colitis rats the utmost tissue damage, and goblet cell depletion associated with disruption and necrotic mucosal structure was seen (Figure 10B) which corresponds to the highest score (four) of colitis. Although the histological sections of the colonic tissues in all treatment groups showed some signs of improvement compared with the untreated group (Figure 10C–H), in the colon tissue of free budesonide-treated groups signs of smooth muscle thickening were seen (Figure 10H) which was scored as two for colitis. A much closer condition to the normal colon histology was seen in rats treated with coated BCNP and coated CP (Figure 10C,E). Observation of colon tissue of other treated groups (BCNP, BCN and CN) demonstrated normal colon histology; however, trace regions of focal inflammatory cell infiltration were seen (Figure 10D,F,G) which was scored one for colitis.

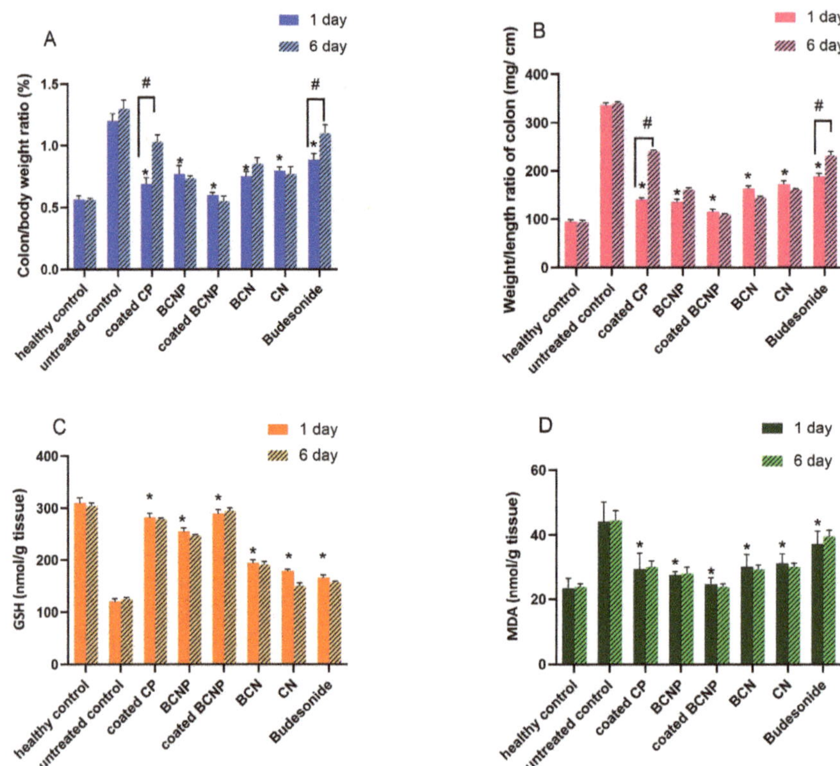

Figure 9. (**A**) Colon/body weight ratio; (**B**) Weight/length ratio of colon; (**C**) GSH and (**D**) level of MDA; determined 1 day or 6 days after final drug administration. * $p < 0.05$ comparison of treatment groups with untreated group after "1 day". # $p < 0.05$ comparison of each group "1 day" and "6 day" after treatment.

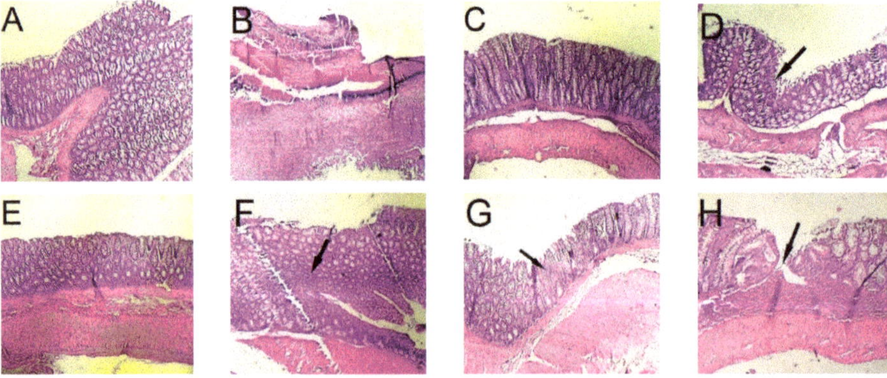

Figure 10. Histopathological characteristics of the colon tissues of the rats subjected to different treatments 24 h after the last drug administration. (**A**) Rat with normal colonic tissues; (**B**) untreated rat; (**C**) rat treated with coated CP; (**D**) rat treated with BCNP; (**E**) rat treated with coated BCNP; (**F**) rat treated with BCN; (**G**) rat treated with CN and (**H**) rat treated with Budesonide. (arrows point to area of inflammation).

Overall, results from the in vivo studies demonstrated that all prescribed formulations significantly are effective in colitis treatment. A decrease in inflammation in CN-treated rats could be related to the anti-inflammatory effects of chitosan in IBD [59].

Formulation containing nanoparticles was found to be more effective than free BU formulation. This could be due to the absorption of a higher amount of free drug in the upper sites of GIT which resulted in inadequate amounts of drug delivered to the inflamed colon [97]. The high efficacy of nanoparticles in the treatment of colitis indicates the presence of higher amounts of the drug at the site of inflammation, confirming the greater accumulation of small-sized particles at the site of inflammation [13,98] and the ease of their adherence to the thick mucus layer [99]. In addition, polymeric nanoparticles would protect drug release in the upper site of GIT and allow for a prolonged residence at the surface of the release site [68].

In this study, nanoparticles were turned into pellet dosage form to increase the clinical effects by reducing the probability of premature uptake in the small intestine. Although histological characteristics of BCN and BCNP-treated rats demonstrated similar therapeutic effects, according to the other in vivo results the superiority of BCNP is obvious. Degradation of free nanoparticles in oral delivery and during their passage in upper parts of the GIT could result in a lower drug amount delivered to the colon [6] while incorporating the nanoparticles in the matrix of pellets could improve local stability within the gastrointestinal tract [100].

During the treatment period, coated BCNP formulation showed the most anti-inflammatory effects; however, regarding the amount of GSH, level of MDA and histological assessment of colon tissues, significant differences with the coated CP were not clearly visible. The superiority of coated BCNP formulation in in vivo studies in rats could be related to the shorter and faster movement of delivery systems in the intestine in rats compared with humans [101], which could take more drugs to the site of action.

The higher efficacy of coated systems (coated BCNP and coated CP) compared with others is due to their coating system which allows the maximum delivery of the drug to the target site without early release in the stomach or upper sections of the gastrointestinal tract after taking the formulation orally.

Although it seems that with a proper coating system colon targeting of BU pellets could effectively be achieved, the advantages of using pelletized BU nanoparticles would be bolded after stopping drug administration. After treatment discontinuation, the group receiving coated CP showed a fast relapse, whereas this was not the case in rats treated with coated BCNP. Similar findings previously had been observed for PLGA nanoparticles [96,98]. These differences may suggest a higher drug accumulation in the inflamed tissue after treatment with nanoparticles and remaining drug carriers in the inflamed regions for up to several days. This claim could be confirmed by an examination of blood glucose levels.

3.6. Blood Glucose Level

Glucocorticoids are known to increase blood glucose levels due to their potential for enhancing intestinal sugar uptake [58]. Thus, for determination of the potency of formulations which their aim is local drug delivery to the intestinal mucosa and accumulation in it, the level of blood glucose of the animals could be considered as an indicator. As it can be seen in Figure 11, the blood glucose level of all rats treated with a formulation containing the drug increased progressively compared with untreated rats. These results indicate that a certain amount of drug was available in the inflamed region on the test day. However, glucose levels were significantly higher in the group of rats treated with nanoparticles containing BCN and BCNP compared with the groups treated with free drugs. The increased glucose level could suggest more extent of targeted drug delivery to the site of inflammation by using nanotechnology. Similar results had been seen previously in other studies [48,57]. The blood glucose level was significantly higher in the rats treated with coated CP and coated BCNP compared with the other treated groups (110 and 109 mg/dl, respectively), which is related to the higher amounts of drug that were delivered to the colon by avoiding release in the upper parts of GIT or premature uptake of the drug due to the coating system. However, it was observed that animals treated with coated CP revealed a remarkable decrease in blood glucose concentration within 24 h after the first examination, while in animals treated with coated BCNP, this reduction become significant after four days. These results also confirm higher drug accumulation after treatment with nanoparticles in comparison with other groups which may last for up to several days and are in agreement with other in vivo results.

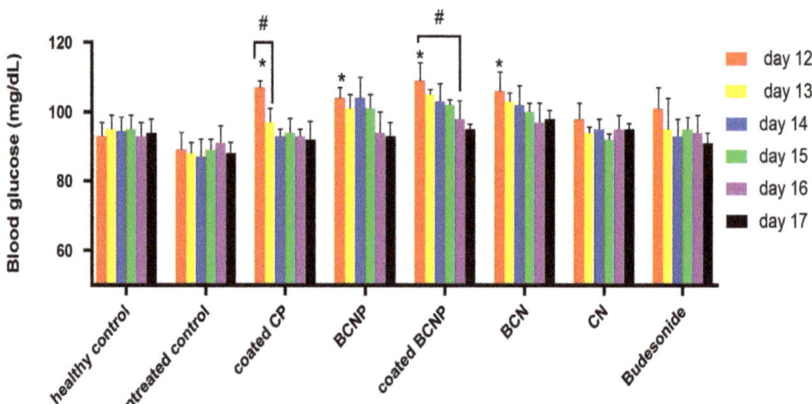

Figure 11. Blood glucose level (mg/dL) in animals on 12th day to 17th day of study. * $p < 0.05$ comparison of treatment groups with untreated group. # $p < 0.05$ comparison of each group on different days.

4. Conclusions

To overcome the limitations of oral delivery of budesonide nanoparticles in colon delivery, a novel formulation based on functional coating of pelletized nanoparticles was proposed. Although in vitro dissolution test results did not indicate the superiority of this formulation over conventional coated pellets (due to the effect of acidic conditions on chitosan), in vivo results demonstrated the supplementary effect of nanoformulation of budesonide and the encapsulation in a properly coated pellet system for colon delivery that resulted in a synergistic and long-lasting anti-inflammatory effect in comparison with coated pellets containing free drug.

Author Contributions: Conceptualization, F.S. (Fatemeh Sadeghi), H.A.G., A.A. and A.N.; methodology, F.S. (Fatemeh Sadeghi), H.A.G., M.G.R., H.K. and A.A.; software, A.A. and H.K.; formal analysis, F.S. (Fatemeh Soltani), F.S. (Fatemeh Sadeghi), H.A.G. and A.A.; resources, F.S. (Fatemeh Sadeghi); data curation, F.S. (Fatemeh Soltani), F.S. (Fatemeh Sadeghi) and A.A.; writing original draft preparation, F.S. (Fatemeh Soltani); writing—review and editing, F.S. (Fatemeh Sadeghi) and A.N.; supervision, F.S. (Fatemeh Sadeghi), H.A.G., A.A. and AN. All authors have read and agreed to the published version of the manuscript.

Funding: This research was funded by the Vice Chancellor for Research and Technology of Mashhad University of Medical Sciences, grant number 981658.

Institutional Review Board Statement: The animal study protocol was approved by the Ethics Committee of Mashhad University of Medical Sciences (IR.MUMS.REC.1399.011) for studies involving animals.

Informed Consent Statement: Not applicable.

Data Availability Statement: Not applicable.

Conflicts of Interest: The authors declare no conflict of interest. The funders had no role in the design of the study; in the collection, analyses, or interpretation of data; in the writing of the manuscript; or in the decision to publish the results.

References

1. Hua, S.; Marks, E.; Schneider, J.J.; Keely, S. Advances in oral nano-delivery systems for colon targeted drug delivery in inflammatory bowel disease: Selective targeting to diseased versus healthy tissue. *Nanomed. Nanotechnol. Biol. Med.* **2015**, *11*, 1117–1132. [CrossRef] [PubMed]
2. Yan, Y.; Ren, F.; Wang, P.; Sun, Y.; Xing, J. Synthesis and evaluation of a prodrug of 5-aminosalicylic acid for the treatment of ulcerative colitis. *Iran. J. Basic Med. Sci.* **2019**, *22*, 1452–1461. [CrossRef] [PubMed]
3. Chen, M.; Yu, Y.; Yang, S.; Yang, D. Pretreatment with licochalcone a enhances therapeutic activity of rat bone marrow mesenchymal stem cells in animal models of colitis. *Iran. J. Basic Med. Sci.* **2021**, *24*, 1050–1057. [CrossRef]
4. Ali, H.; Weigmann, B.; Neurath, M.; Collnot, E.; Windbergs, M.; Lehr, C.-M. Budesonide loaded nanoparticles with pH-sensitive coating for improved mucosal targeting in mouse models of inflammatory bowel diseases. *J. Control. Release* **2014**, *183*, 167–177. [CrossRef] [PubMed]
5. Lahiff, C.; Kane, S.; Moss, A.C. Drug development in inflammatory bowel disease: The role of the FDA. *Inflamm. Bowel Dis.* **2011**, *17*, 2585–2593. [CrossRef] [PubMed]
6. Meißner, Y.; Pellequer, Y.; Lamprecht, A. Nanoparticles in inflammatory bowel disease: Particle targeting versus pH-sensitive delivery. *Int. J. Pharm.* **2006**, *316*, 138–143. [CrossRef]
7. Xiao, B.; Laroui, H.; Ayyadurai, S.; Viennois, E.; Charania, M.A.; Zhang, Y.; Merlin, D. Mannosylated bioreducible nanoparticle-mediated macrophage-specific TNF-α RNA interference for IBD therapy. *Biomaterials* **2013**, *34*, 7471–7482. [CrossRef]
8. Ding, S.; Zhang, N.; Lyu, Z.; Zhu, W.; Chang, Y.-C.; Hu, X.; Du, D.; Lin, Y. Protein-based nanomaterials and nanosystems for biomedical applications: A review. *Mater. Today* **2021**, *43*, 166–184. [CrossRef]
9. Li, D.-F.; Yang, M.-F.; Xu, H.-M.; Zhu, M.-Z.; Zhang, Y.; Tian, C.-M.; Nie, Y.-Q.; Wang, J.-Y.; Liang, Y.-J.; Yao, J.; et al. Nanoparticles for oral delivery: Targeted therapy for inflammatory bowel disease. *J. Mater. Chem. B* **2022**, *10*, 5853–5872. [CrossRef]
10. Press, A.G.; A Hauptmann, I.; Hauptmann, L.; Fuchs, B.; Ewe, K.; Fuchs, M.; Ramadori, G.; Ewe, K. Gastrointestinal pH profiles in patients with inflammatory bowel disease. *Aliment. Pharmacol. Ther.* **1998**, *12*, 673–678. [CrossRef]
11. Desai, M.P.; Labhasetwar, V.; Amidon, G.L.; Levy, R.J. Gastrointestinal Uptake of Biodegradable Microparticles: Effect of Particle Size. *Pharm. Res.* **1996**, *13*, 1838–1845. [CrossRef] [PubMed]
12. Lamprecht, A.; Yamamoto, H.; Takeuchi, H.; Kawashima, Y. Nanoparticles Enhance Therapeutic Efficiency by Selectively Increased Local Drug Dose in Experimental Colitis in Rats. *J. Pharmacol. Exp. Ther.* **2005**, *315*, 196–202. [CrossRef] [PubMed]
13. Brusini, R.; Varna, M.; Couvreur, P. Advanced nanomedicines for the treatment of inflammatory diseases. *Adv. Drug Deliv. Rev.* **2020**, *157*, 161–178. [CrossRef] [PubMed]
14. Tang, H.; Xiang, D.; Wang, F.; Mao, J.; Tan, X.; Wang, Y. 5-ASA-loaded SiO_2 nanoparticles-a novel drug delivery system targeting therapy on ulcerative colitis in mice. *Mol. Med. Rep.* **2017**, *15*, 1117–1122. [CrossRef] [PubMed]
15. Ali, H.; Weigmann, B.; Collnot, E.-M.; Khan, S.A.; Windbergs, M.; Lehr, C.-M. Budesonide Loaded PLGA Nanoparticles for Targeting the Inflamed Intestinal Mucosa—Pharmaceutical Characterization and Fluorescence Imaging. *Pharm. Res.* **2015**, *33*, 1085–1092. [CrossRef]
16. Vafaei, S.Y.; Esmaeili, M.; Amini, M.; Atyabi, F.; Ostad, S.N.; Dinarvand, R. Self assembled hyaluronic acid nanoparticles as a potential carrier for targeting the inflamed intestinal mucosa. *Carbohydr. Polym.* **2016**, *144*, 371–381. [CrossRef]
17. Beloqui, A.; Coco, R.; Alhouayek, M.; Solinís, M.; Rodríguez-Gascón, A.; Muccioli, G.G.; Préat, V. Budesonide-loaded nanostructured lipid carriers reduce inflammation in murine DSS-induced colitis. *Int. J. Pharm.* **2013**, *454*, 775–783. [CrossRef]
18. Han, H.-K.; Shin, H.-J.; Ha, D.H. Improved oral bioavailability of alendronate via the mucoadhesive liposomal delivery system. *Eur. J. Pharm. Sci.* **2012**, *46*, 500–507. [CrossRef]
19. Cu, Y.; Saltzman, W.M. Controlled Surface Modification with Poly(ethylene)glycol Enhances Diffusion of PLGA Nanoparticles in Human Cervical Mucus. *Mol. Pharm.* **2009**, *6*, 173–181. [CrossRef]
20. Barea, M.J.; Jenkins, M.J.; Lee, Y.S.; Johnson, P.; Bridson, R.H. Encapsulation of Liposomes within pH Responsive Microspheres for Oral Colonic Drug Delivery. *Int. J. Biomater.* **2012**, *2012*, 1–8. [CrossRef]
21. Laroui, H.; Dalmasso, G.; Nguyen, H.T.T.; Yan, Y.; Sitaraman, S.V.; Merlin, D. Drug-Loaded Nanoparticles Targeted to the Colon With Polysaccharide Hydrogel Reduce Colitis in a Mouse Model. *Gastroenterology* **2010**, *138*, 843–853. [CrossRef] [PubMed]
22. Zhang, M.; Merlin, D. Nanoparticle-Based Oral Drug Delivery Systems Targeting the Colon for Treatment of Ulcerative Colitis. *Inflamm. Bowel Dis.* **2018**, *24*, 1401–1415. [CrossRef] [PubMed]
23. Wilson, D.S.; Dalmasso, G.; Wang, L.; Sitaraman, S.V.; Merlin, D.; Murthy, N. Orally delivered thioketal nanoparticles loaded with TNF-α–siRNA target inflammation and inhibit gene expression in the intestines. *Nat. Mater.* **2010**, *9*, 923–928. [CrossRef] [PubMed]
24. Ashford, M.; Fell, J.T.; Attwood, D.; Sharma, H.; Woodhead, P.J. An in vivo investigation into the suitability of pH dependent polymers for colonic targeting. *Int. J. Pharm.* **1993**, *95*, 193–199. [CrossRef]
25. Wang, N.; Shao, L.; Lu, W.; Fang, W.; Zhang, Q.; Sun, L.; Gao, S.; Zhu, Q.; Chen, S.; Hu, R. 5-aminosalicylic acid pH sensitive core-shell nanoparticles targeting ulcerative colitis. *J. Drug Deliv. Sci. Technol.* **2022**, *74*, 103578. [CrossRef]

26. Cai, X.; Wang, X.; He, M.; Wang, Y.; Lan, M.; Zhao, Y.; Gao, F. Colon-targeted delivery of tacrolimus using pH-responsive polymeric nanoparticles for murine colitis therapy. *Int. J. Pharm.* **2021**, *606*, 120836. [CrossRef]
27. Zhou, H.; Qian, H. Preparation and characterization of pH-sensitive nanoparticles of budesonide for the treatment of ulcerative colitis. *Drug Des. Dev. Ther.* **2018**, *12*, 2601–2609. [CrossRef]
28. Turanlı, Y.; Acartürk, F. Fabrication and characterization of budesonide loaded colon-specific nanofiber drug delivery systems using anionic and cationic polymethacrylate polymers. *J. Drug Deliv. Sci. Technol.* **2021**, *63*, 102511. [CrossRef]
29. Naeem, M.; Choi, M.; Cao, J.; Lee, Y.; Ikram, M.; Yoon, S.; Lee, J.; Moon, H.R.; Kim, M.; Jung, Y.; et al. Colon-targeted delivery of budesonide using dual pH- and time-dependent polymeric nanoparticles for colitis therapy. *Drug Des. Dev. Ther.* **2015**, *9*, 3789–3799. [CrossRef]
30. Sinhmar, G.K.; Shah, N.N.; Chokshi, N.V.; Khatri, H.N.; Patel, M.M. Process, optimization, and characterization of budesonide-loaded nanostructured lipid carriers for the treatment of inflammatory bowel disease. *Drug Dev. Ind. Pharm.* **2018**, *44*, 1078–1089. [CrossRef]
31. Lamprecht, A.; Yamamoto, H.; Takeuchi, H.; Kawashima, Y. A pH-sensitive microsphere system for the colon delivery of tacrolimus containing nanoparticles. *J. Control. Release* **2005**, *104*, 337–346. [CrossRef] [PubMed]
32. Rodriguez, M.; Antúnez, J.A.; Taboada, C.; Seijo, B.; Torres, D. Colon-specific delivery of budesonide from microencapsulated cellulosic cores: Evaluation of the efficacy against colonic inflammation in rats. *J. Pharm. Pharmacol.* **2001**, *53*, 1207–1215. [CrossRef] [PubMed]
33. Krishnamachari, Y.; Madan, P.; Lin, S. Development of pH- and time-dependent oral microparticles to optimize budesonide delivery to ileum and colon. *Int. J. Pharm.* **2007**, *338*, 238–247. [CrossRef] [PubMed]
34. Patel, M.M.; Amin, A.F. Process, optimization and characterization of mebeverine hydrochloride loaded guar gum microspheres for irritable bowel syndrome. *Carbohydr. Polym.* **2011**, *86*, 536–545. [CrossRef]
35. Barea, M.; Jenkins, M.; Gaber, M.; Bridson, R. Evaluation of liposomes coated with a pH responsive polymer. *Int. J. Pharm.* **2010**, *402*, 89–94. [CrossRef] [PubMed]
36. Leonard, F.; Ali, H.; Collnot, E.-M.; Crielaard, B.J.; Lammers, T.; Storm, G. Screening of budesonide nanoformulations for treatment of inflammatory bowel disease in an inflamed 3D cell-culture model. *Altex—Altern. Anim. Exp.* **2012**, *29*, 275–285. [CrossRef]
37. Smitha, K.; Anitha, A.; Furuike, T.; Tamura, H.; Nair, S.V.; Jayakumar, R. In vitro evaluation of paclitaxel loaded amorphous chitin nanoparticles for colon cancer drug delivery. *Colloids Surf. B* **2013**, *104*, 245–253. [CrossRef]
38. Chen, M.-C.; Mi, F.-L.; Liao, Z.-X.; Hsiao, C.-W.; Sonaje, K.; Chung, M.-F.; Hsu, L.-W.; Sung, H.-W. Recent advances in chitosan-based nanoparticles for oral delivery of macromolecules. *Adv. Drug Deliv. Rev.* **2013**, *65*, 865–879. [CrossRef]
39. Ahmed, O.; Abdel-Halim, M.; Farid, A.; Elamir, A. Taurine loaded chitosan-pectin nanoparticle shows curative effect against acetic acid-induced colitis in rats. *Chem. Interact.* **2022**, *351*, 109715. [CrossRef]
40. Coco, R.; Plapied, L.; Pourcelle, V.; Jérôme, C.; Brayden, D.J.; Schneider, Y.-J.; Préat, V. Drug delivery to inflamed colon by nanoparticles: Comparison of different strategies. *Int. J. Pharm.* **2013**, *440*, 3–12. [CrossRef]
41. Michailidou, G.; Ainali, N.M.; Xanthopoulou, E.; Nanaki, S.; Kostoglou, M.; Koukaras, E.N.; Bikiaris, D.N. Effect of Poly(vinyl alcohol) on Nanoencapsulation of Budesonide in Chitosan Nanoparticles via Ionic Gelation and Its Improved Bioavailability. *Polymers* **2020**, *12*, 1101. [CrossRef] [PubMed]
42. de Pinho Neves, A.L.; Milioli, C.C.; Müller, L.; Riella, H.G.; Kuhnen, N.C.; Stulzer, H.K. Factorial design as tool in chitosan nanoparticles development by ionic gelation technique. *Colloids Surf. A Physicochem. Eng. Asp.* **2014**, *445*, 34–39. [CrossRef]
43. Koukaras, E.; Papadimitriou, S.A.; Bikiaris, D.N.; Froudakis, G.E. Insight on the Formation of Chitosan Nanoparticles through Ionotropic Gelation with Tripolyphosphate. *Mol. Pharm.* **2012**, *9*, 2856–2862. [CrossRef] [PubMed]
44. Soltani, F.; Kamali, H.; Akhgari, A.; Garekani, H.A.; Nokhodchi, A.; Sadeghi, F. Different trends for preparation of budesonide pellets with enhanced dissolution rate. *Adv. Powder Technol.* **2022**, *33*, 103684. [CrossRef]
45. Sardou, H.S.; Akhgari, A.; Mohammadpour, A.H.; Namdar, A.B.; Kamali, H.; Jafarian, A.H.; Garekani, H.A.; Sadeghi, F. Optimization study of combined enteric and time-dependent polymethacrylates as a coating for colon targeted delivery of 5-ASA pellets in rats with ulcerative colitis. *Eur. J. Pharm. Sci.* **2022**, *168*, 106072. [CrossRef]
46. de Assis, P.O.A.; Guerra, G.C.B.; de Souza Araújo, D.F.; de Araújo Júnior, R.F.; Machado, T.A.D.G.; de Araújo, A.A.; de Lima, T.A.S.; Garcia, H.E.M.; de Fátima Leal Interaminense de Andrade, L.; de Cássia Ramos do Egypto Queiroga, R. Intestinal anti-inflammatory activity of goat milk and goat yoghurt in the acetic acid model of rat colitis. *Int. Dairy J.* **2016**, *56*, 45–54. [CrossRef]
47. Guerra, G.C.; Araújo, A.; Lira, G.A.; Melo, M.N.; Souto, K.K.; Fernandes, D.; Silva, A.L.; Araújo, R.F., Jr. Telmisartan decreases inflammation by modulating TNF-α, IL-10, and RANK/RANKL in a rat model of ulcerative colitis. *Pharmacol. Rep.* **2015**, *67*, 520–526. [CrossRef]
48. Qelliny, M.R.; Aly, U.F.; Elgarhy, O.H.; Khaled, K.A. Budesonide-Loaded Eudragit S 100 Nanocapsules for the Treatment of Acetic Acid-Induced Colitis in Animal Model. *AAPS PharmSciTech* **2019**, *20*, 237. [CrossRef]
49. Sardou, H.S.; Akhgari, A.; Mohammadpour, A.H.; Kamali, H.; Jafarian, A.H.; Garekani, H.A.; Sadeghi, F. Application of inulin/Eudragit RS in 5-ASA pellet coating with tuned, sustained-release feature in an animal model of ulcerative colitis. *Int. J. Pharm.* **2021**, *597*, 120347. [CrossRef]

50. Ferri, D.; Costero, A.M.; Gaviña, P.; Parra, M.; Merino, V.; Teruel, A.H.; Sancenón, F.; Martínez-Máñez, R. Efficacy of budesonide-loaded mesoporous silica microparticles capped with a bulky azo derivative in rats with TNBS-induced colitis. *Int. J. Pharm.* **2019**, *561*, 93–101. [CrossRef]

51. Dai, C.; Zheng, C.-Q.; Meng, F.-J.; Zhou, Z.; Sang, L.-X.; Jiang, M. VSL#3 probiotics exerts the anti-inflammatory activity via PI3k/Akt and NF-κB pathway in rat model of DSS-induced colitis. *Mol. Cell. Biochem.* **2012**, *374*, 1–11. [CrossRef] [PubMed]

52. Alhouayek, M.; Lambert, D.M.; Delzenne, N.M.; Cani, P.D.; Muccioli, G.G. Increasing endogenous 2-arachidonoylglycerol levels counteracts colitis and related systemic inflammation. *FASEB J.* **2011**, *25*, 2711–2721. [CrossRef] [PubMed]

53. Bayat, A.; Dorkoosh, F.A.; Dehpour, A.R.; Moezi, L.; Larijani, B.; Junginger, H.E.; Rafiee-Tehrani, M. Nanoparticles of quaternized chitosan derivatives as a carrier for colon delivery of insulin: Ex vivo and in vivo studies. *Int. J. Pharm.* **2008**, *356*, 259–266. [CrossRef] [PubMed]

54. Gareb, B.; Dijkstra, G.; Kosterink, J.G.; Frijlink, H.W. Development of novel zero-order release budesonide tablets for the treatment of ileo-colonic inflammatory bowel disease and comparison with formulations currently used in clinical practice. *Int. J. Pharm.* **2018**, *554*, 366–375. [CrossRef] [PubMed]

55. Uchiyama, M.; Mihara, M. Determination of malonaldehyde precursor in tissues by thiobarbituric acid test. *Anal. Biochem.* **1978**, *86*, 271–278. [CrossRef]

56. Mehri, S.; Meshki, M.A.; Hosseinzadeh, H. Linalool as a neuroprotective agent against acrylamide-induced neurotoxicity in Wistar rats. *Drug Chem. Toxicol.* **2015**, *38*, 162–166. [CrossRef]

57. Yoo, J.-W.; Naeem, M.; Cao, J.; Choi, M.; Kim, W.; Moon, H.R.; Lee, B.L.; Kim, M.-S.; Jung, Y. Enhanced therapeutic efficacy of budesonide in experimental colitis with enzyme/pH dual-sensitive polymeric nanoparticles. *Int. J. Nanomed.* **2015**, *10*, 4565–4580. [CrossRef]

58. Thiesen, A.; E Wild, G.; A Tappenden, K.; Drozdowski, L.; Keelan, M.; A Thomson, B.K.; I McBurney, M.; Clandinin, M.T.; Thomson, A.B.R. The locally acting glucocorticosteroid budesonide enhances intestinal sugar uptake following intestinal resection in rats. *Gut* **2003**, *52*, 252–259. [CrossRef]

59. Ribeiro, J.C.V.; Forte, T.C.M.; Tavares, S.J.S.; Andrade, F.K.; Vieira, R.S.; Lima, V. The effects of the molecular weight of chitosan on the tissue inflammatory response. *J. Biomed. Mater. Res. Part A* **2021**, *109*, 2556–2569. [CrossRef]

60. Niu, W.; Dong, Y.; Fu, Z.; Lv, J.; Wang, L.; Zhang, Z.; Huo, J.; Ju, J. Effects of molecular weight of chitosan on anti-inflammatory activity and modulation of intestinal microflora in an ulcerative colitis model. *Int. J. Biol. Macromol.* **2021**, *193*, 1927–1936. [CrossRef]

61. Xu, Y.; Du, Y. Effect of molecular structure of chitosan on protein delivery properties of chitosan nanoparticles. *Int. J. Pharm.* **2002**, *250*, 215–226. [CrossRef] [PubMed]

62. Yang, H.-C.; Hon, M.-H. The effect of the molecular weight of chitosan nanoparticles and its application on drug delivery. *Microchem. J.* **2009**, *92*, 87–91. [CrossRef]

63. Bhatt, H.; Naik, B.; Dharamsi, A. Solubility Enhancement of Budesonide and Statistical Optimization of Coating Variables for Targeted Drug Delivery. *J. Pharm.* **2014**, *2014*, 1–13. [CrossRef] [PubMed]

64. Kouchak, M.; Avadi, M.; Abbaspour, M.; Jahangiri, A.; Boldaji, S.K. Effect of different molecular weights of chitosan on preparation and characterization of insulin loaded nanoparticles by ion gelation method. *Int. J. Drug Dev. Res.* **2012**, *4*, 0975–9344.

65. Desai, K.G. Chitosan Nanoparticles Prepared by Ionotropic Gelation: An Overview of Recent Advances. *Crit. Rev. Ther. Drug Carr. Syst.* **2016**, *33*, 107–158. [CrossRef]

66. Schmidt, C.; Lautenschlaeger, C.; Collnot, E.-M.; Schumann, M.; Bojarski, C.; Schulzke, J.-D.; Lehr, C.-M.; Stallmach, A. Nano- and microscaled particles for drug targeting to inflamed intestinal mucosa—A first in vivo study in human patients. *J. Control. Release* **2013**, *165*, 139–145. [CrossRef]

67. Dube, A.; Nicolazzo, J.A.; Larson, I. Chitosan nanoparticles enhance the intestinal absorption of the green tea catechins (+)-catechin and (−)-epigallocatechin gallate. *Eur. J. Pharm. Sci.* **2010**, *41*, 219–225. [CrossRef]

68. Elzatahry, A.A.; Eldin, M.M. Preparation and characterization of metronidazole-loaded chitosan nanoparticles for drug delivery application. *Polym. Adv. Technol.* **2008**, *19*, 1787–1791. [CrossRef]

69. Fan, W.; Yan, W.; Xu, Z.; Ni, H. Formation mechanism of monodisperse, low molecular weight chitosan nanoparticles by ionic gelation technique. *Colloids Surf. B Biointerfaces* **2012**, *90*, 21–27. [CrossRef]

70. Gomathi, T.; Sudha, P.; Florence, J.A.K.; Venkatesan, J.; Anil, S. Fabrication of letrozole formulation using chitosan nanoparticles through ionic gelation method. *Int. J. Biol. Macromol.* **2017**, *104*, 1820–1832. [CrossRef]

71. Joseph, J.J.; Sangeetha, D.; Gomathi, T. Sunitinib loaded chitosan nanoparticles formulation and its evaluation. *Int. J. Biol. Macromol.* **2016**, *82*, 952–958. [CrossRef] [PubMed]

72. Bouwman, A.M.; Heijstra, M.P.; Schaefer, N.C.; Duiverman, E.J.; Lesouëf, P.N.; Devadason, S.G. Improved Efficiency of Budesonide Nebulization Using Surface-Active Agents. *Drug Deliv.* **2006**, *13*, 357–363. [CrossRef] [PubMed]

73. Abouelhag, H.; Sivakumar, S.; Bagul, U.; Eltyep, E.M.; Safhi, M. Preparation and physical characterization of cisplatin chitosan nanoparticles by zeta nano sizer "prime step for formulation and development". *Int. J. Pharm. Sci. Res.* **2017**, *8*, 4245–4249. [CrossRef]

74. Souza, T.G.F.; Ciminelli, V.S.T.; Mohallem, N.D.S. A comparison of TEM and DLS methods to characterize size distribution of ceramic nanoparticles. *J. Phys. Conf. Ser.* **2016**, *733*, 012039. [CrossRef]

75. Tuoriniemi, J.; Johnsson, A.-C.J.H.; Perez Holmberg, J.; Gustafsson, S.; Gallego-Urrea, J.A.; Olsson, E.; Pettersson, J.B.C.; Hassellöv, M. Intermethod comparison of the particle size distributions of colloidal silica nanoparticles. *Sci. Technol. Adv. Mater.* **2014**, *15*, 35009. [CrossRef] [PubMed]

76. Vafaei, S.Y.; Dinarvand, R.; Esmaeili, M.; Mahjub, R.; Toliyat, T. Controlled-release drug delivery system based on fluocinolone acetonide–cyclodextrin inclusion complex incorporated in multivesicular liposomes. *Pharm. Dev. Technol.* **2015**, *20*, 775–781. [CrossRef] [PubMed]

77. Colby, A.H.; Liu, R.; Doyle, R.P.; Merting, A.; Zhang, H.; Savage, N.; Chu, N.-Q.; Hollister, B.A.; McCulloch, W.; Burdette, J.E.; et al. Pilot-scale production of expansile nanoparticles: Practical methods for clinical scale-up. *J. Control. Release* **2021**, *337*, 144–154. [CrossRef]

78. Kumar, A.; Sawant, K. Encapsulation of exemestane in polycaprolactone nanoparticles: Optimization, characterization, and release kinetics. *Cancer Nanotechnol.* **2013**, *4*, 57–71. [CrossRef]

79. Ferrero, F.; Periolatto, M. Antimicrobial finish of textiles by chitosan UV-curing. *J. Nanosci. Nanotechnol.* **2012**, *12*, 4803–4810. [CrossRef]

80. Ntohogian, S.; Gavriliadou, V.; Christodoulou, E.; Nanaki, S.; Lykidou, S.; Naidis, P.; Mischopoulou, L.; Barmpalexis, P.; Nikolaidis, N.; Bikiaris, D.N. Chitosan Nanoparticles with Encapsulated Natural and UF-Purified Annatto and Saffron for the Preparation of UV Protective Cosmetic Emulsions. *Molecules* **2018**, *23*, 2107. [CrossRef]

81. Bruni, G.; Maggi, L.; Tammaro, L.; Canobbio, A.; Di Lorenzo, R.; D'Aniello, S.; Domenighini, C.; Berbenni, V.; Milanese, C.; Marini, A. Fabrication, Physico-Chemical, and Pharmaceutical Characterization of Budesonide-Loaded Electrospun Fibers for Drug Targeting to the Colon. *J. Pharm. Sci.* **2015**, *104*, 3798–3803. [CrossRef] [PubMed]

82. Papadimitriou, S.; Bikiaris, D.; Avgoustakis, K.; Karavas, E.; Georgarakis, M. Chitosan nanoparticles loaded with dorzolamide and pramipexole. *Carbohydr. Polym.* **2008**, *73*, 44–54. [CrossRef]

83. Aghrbi, I.; Fülöp, V.; Jakab, G.; Kállai-Szabó, N.; Balogh, E.; Antal, I. Nanosuspension with improved saturated solubility and dissolution rate of cilostazol and effect of solidification on stability. *J. Drug Deliv. Sci. Technol.* **2021**, *61*, 102165. [CrossRef]

84. Chopra, R.; Podczeck, F.; Newton, J.; Alderborn, G. The influence of pellet shape and film coating on the filling of pellets into hard shell capsules. *Eur. J. Pharm. Biopharm.* **2002**, *53*, 327–333. [CrossRef] [PubMed]

85. Davis, S.S.; Hardy, J.G.; Fara, J.W. Transit of pharmaceutical dosage forms through the small intestine. *Gut* **1986**, *27*, 886–892. [CrossRef] [PubMed]

86. Kanwar, N.; Kumar, R.; Sinha, V. Preparation and Evaluation of Multi-Particulate System (Pellets) of Prasugrel Hydrochloride. *Open Pharm. Sci. J.* **2015**, *2*, 74–80. [CrossRef]

87. Motwani, S.K.; Chopra, S.; Talegaonkar, S.; Kohli, K.; Ahmad, F.; Khar, R.K. Chitosan–sodium alginate nanoparticles as submicroscopic reservoirs for ocular delivery: Formulation, optimisation and in vitro characterisation. *Eur. J. Pharm. Biopharm.* **2008**, *68*, 513–525. [CrossRef]

88. Egusquiaguirre, S.P.; Manguán-García, C.; Pintado-Berninches, L.; Iarriccio, L.; Carbajo, D.; Albericio, F.; Royo, M.; Pedraz, J.L.; Hernández, R.M.; Perona, R.; et al. Development of surface modified biodegradable polymeric nanoparticles to deliver GSE24.2 peptide to cells: A promising approach for the treatment of defective telomerase disorders. *Eur. J. Pharm. Biopharm.* **2015**, *91*, 91–102. [CrossRef]

89. Huanbutta, K.; Cheewatanakornkool, K.; Terada, K.; Nunthanid, J.; Sriamornsak, P. Impact of salt form and molecular weight of chitosan on swelling and drug release from chitosan matrix tablets. *Carbohydr. Polym.* **2013**, *97*, 26–33. [CrossRef]

90. Karaaslan, M.A.; Tshabalala, M.A.; Buschle-Diller, G. Wood hemicellulose/chitosan-based semi-interpenetrating network hydrogels: Mechanical, swelling and controlled drug release properties. *BioResources* **2010**, *5*, 1036–1054.

91. Zhang, H.; Neau, S.H. In vitro degradation of chitosan by bacterial enzymes from rat cecal and colonic contents. *Biomaterials* **2002**, *23*, 2761–2766. [CrossRef] [PubMed]

92. Kobayashi, T.; Siegmund, B.; Le Berre, C.; Wei, S.C.; Ferrante, M.; Shen, B.; Bernstein, C.N.; Danese, S.; Peyrin-Biroulet, L.; Hibi, T. Ulcerative colitis. *Nat. Rev. Dis. Prim.* **2020**, *6*, 1–20. [CrossRef]

93. Sonaje, K.; Chen, Y.-J.; Chen, H.-L.; Wey, S.-P.; Juang, J.-H.; Nguyen, H.-N.; Hsu, C.-W.; Lin, K.-J.; Sung, H.-W. Enteric-coated capsules filled with freeze-dried chitosan/poly(γ-glutamic acid) nanoparticles for oral insulin delivery. *Biomaterials* **2010**, *31*, 3384–3394. [CrossRef] [PubMed]

94. Mura, C.; Nácher, A.; Merino, V.; Merino-Sanjuan, M.; Carda, C.; Ruiz, A.; Manconi, M.; Loy, G.; Fadda, A.; Diez-Sales, O. N-Succinyl-chitosan systems for 5-aminosalicylic acid colon delivery: In vivo study with TNBS-induced colitis model in rats. *Int. J. Pharm.* **2011**, *416*, 145–154. [CrossRef]

95. Walsh, A.J.; Bryant, R.V.; Travis, S.P. Current best practice for disease activity assessment in IBD. *Nat. Rev. Gastroenterol. Hepatol.* **2016**, *13*, 567–579. [CrossRef]

96. Makhlof, A.; Tozuka, Y.; Takeuchi, H. pH-Sensitive nanospheres for colon-specific drug delivery in experimentally induced colitis rat model. *Eur. J. Pharm. Biopharm.* **2009**, *72*, 1–8. [CrossRef]

97. Varshosaz, J.; Emami, J.; Fassihi, A.; Tavakoli, N.; Minaiyan, M.; Ahmadi, F.; Mahzouni, P.; Dorkoosh, F. Effectiveness of budesonide-succinate-dextran conjugate as a novel prodrug of budesonide against acetic acid-induced colitis in rats. *Int. J. Color. Dis.* **2010**, *25*, 1159–1165. [CrossRef]

98. Lamprecht, A.; Ubrich, N.; Yamamoto, H.; Schäfer, U.; Takeuchi, H.; Maincent, P.; Kawashima, Y.; Lehr, C.-M. Biodegradable nanoparticles for targeted drug delivery in treatment of inflammatory bowel disease. *J. Pharmacol. Exp. Ther.* **2001**, *299*, 775–781.

99. Lamprecht, A.; Schäfer, U.; Lehr, C.-M.M. Size-Dependent bioadhesion of micro- and nanoparticulate carriers to the inflamed colonic mucosa. *Pharm. Res.* **2001**, *18*, 788–793. [CrossRef]
100. Schmidt, C.; Bodmeier, R. Incorporation of polymeric nanoparticles into solid dosage forms. *J. Control. Release* **1999**, *57*, 115–125. [CrossRef]
101. McConnell, E.L.; Basit, A.W.; Murdan, S. Measurements of rat and mouse gastrointestinal pH, fluid and lymphoid tissue, and implications for in-vivo experiments. *J. Pharm. Pharmacol.* **2008**, *60*, 63–70. [CrossRef] [PubMed]

 *pharmaceutics*

Article

Rutin Nanocrystals with Enhanced Anti-Inflammatory Activity: Preparation and Ex Vivo/In Vivo Evaluation in an Inflammatory Rat Model

Abeer S. Hassan [1] and Ghareb M. Soliman [2,3,*]

[1] Department of Pharmaceutics, Faculty of Pharmacy, South Valley University, Qena 83523, Egypt
[2] Department of Pharmaceutics, Faculty of Pharmacy, University of Tabuk, Tabuk 71491, Saudi Arabia
[3] Department of Pharmaceutics, Faculty of Pharmacy, Assiut University, Assiut 71526, Egypt
* Correspondence: ghareb.soliman@aun.edu.eg or gh.soliman@ut.edu.sa; Tel.: +20-101-342-7311; Fax: +20-882-080-774

Citation: Hassan, A.S.; Soliman, G.M. Rutin Nanocrystals with Enhanced Anti-Inflammatory Activity: Preparation and Ex Vivo/In Vivo Evaluation in an Inflammatory Rat Model. *Pharmaceutics* **2022**, *14*, 2727. https://doi.org/10.3390/pharmaceutics14122727

Academic Editor: Ana Isabel Fernandes

Received: 3 November 2022
Accepted: 1 December 2022
Published: 6 December 2022

Publisher's Note: MDPI stays neutral with regard to jurisdictional claims in published maps and institutional affiliations.

Abstract: Rutin is a polyphenolic flavonoid with an interestingly wide therapeutic spectrum. However, its clinical benefits are limited by its poor aqueous solubility and low bioavailability. In an attempt to overcome these limitations, rutin nanocrystals were prepared using various stabilizers including nonionic surfactants and nonionic polymers. The nanocrystals were evaluated for particle size, zeta potential, drug entrapment efficiency, morphology, colloidal stability, rutin photostability, dissolution rate, and saturation solubility. The selected nanocrystal formulation was dispersed in a hydrogel base and the drug release kinetics and permeability through mouse skin were characterized. Rutin's anti-inflammatory efficacy was studied in a carrageenan-induced rat paw edema model. The nanocrystals had a size in the range of around 270–500 nm and a polydispersity index of around 0.3–0.5. Nanocrystals stabilized by hydroxypropyl beta-cyclodextrin (HP-β-CD) had the smallest particle size, highest drug entrapment efficiency, best colloidal stability, and highest drug photostability. Nanocrystals had around a 102- to 202-fold and 2.3- to 6.7-fold increase in the drug aqueous solubility and dissolution rate, respectively, depending on the type of stabilizer. HP-β-CD nanocrystals hydrogel had a significantly higher percent of drug released and permeated through the mouse skin compared with the free drug hydrogel. The cumulative drug amount permeated through the skin was 2.5-fold higher than that of the free drug hydrogel. In vivo studies showed that HP-β-CD-stabilized rutin nanocrystals hydrogel had significantly higher edema inhibition compared with the free drug hydrogel and commercial diclofenac sodium gel. These results highlight the potential of HP-β-CD-stabilized nanocrystals as a promising approach to enhance drug solubility, dissolution rate, and anti-inflammatory properties.

Keywords: rutin; nanocrystals; anti-inflammatory; hydroxypropyl beta-cyclodextrin; nanoparticles

1. Introduction

Rutin, also known as vitamin P, is a polyphenolic flavonoid found in plants such as buckwheat, green tea, citrus fruits (e.g., orange, grapefruit, lemon, lime), and apples [1,2]. Chemically, rutin is known as (2-(3,4-dihydroxyphenyl)-4,5-dihydroxy-3-[3,4,5-trihydroxy-6-[(3,4,5-trihydroxy-6-methyl-oxan-2-yl)oxymethyl]oxan-2-yl]oxy-chromen-7-one) or quercetin-3-rutinoside. Rutin has several attractive features as a drug such as its natural source, safety, cost-effectiveness, and wide spectrum of pharmacological actions [3–5]. Moreover, several reports have shown rutin to have antioxidant, antidiabetic, anticancer, anti-inflammatory, antibacterial, anti-arthritic, and neuroprotection activities [5–8]. In addition, rutin was also shown to have antihypertensive, cardioprotective, antispasmodic, anti-thrombotic, and anti-hyperlipidemia actions [9–12]. Despite these interesting pharmacological properties, rutin suffers from some drawbacks such as poor aqueous solubility and stability which result in poor oral bioavailability [13,14]. This poses some challenges for the formulation of oral dosage forms of rutin whether for its application as a nutritional supplement or a therapeutic agent [2]. To overcome these shortcomings and improve rutin's therapeutic efficacy, several formulation strategies and drug delivery systems have been exploited. For instance, dried rutin nanocrystals have been prepared and incorporated into tablets [2]. The nanocrystal tablets achieved complete rutin dissolution in 30 min compared to only 71% and 55% dissolution from the microcrystal tablets and the commercialized tablets, respectively. Rutin-loaded silver nanoparticles were fabricated and their anti-thrombotic activity was evaluated [12]. Rutin-silver nanoparticles had prolonged activated partial thromboplastin time and prothrombin time. Quinoa and maize starch nanoparticles were also used to encapsulate rutin and improve its bioavailability [15]. In a simulated in vitro digestion test, the nanoparticles were able to increase rutin's bioavailability and improve its antioxidant activity. Rutin was also loaded into liquid crystalline nanoparticles (LCNs) and its anticancer activity against non-small cell lung cancer was tested in an A549 human lung epithelial carcinoma cell line [16]. Rutin-LCNs showed promising anti-proliferative and anti-migratory activities. In addition, they induced apoptosis in the A549 cells and inhibited colony formation.

Nanocrystals (NCs) are sub-micron colloidal dispersions composed of 100% drug material. They were first introduced in the 1990s as a means of improving dissolution rate, aqueous solubility, and thus the bioavailability of poorly water-soluble drugs [17]. NCs have a particle size in the nanometer range and are typically produced by the milling of bulk drug material [18]. They are stabilized by surfactants, polymers, or both [19]. The enhanced dissolution rate and solubility are believed to be due to the decrease in particle size, which in turn increases the surface area available for dissolution according to the Noyes–Whitney equation [20]. The enormous increase in surface area and saturation solubility result in improved drug oral bioavailability and permeability through biological membranes making them an attractive drug delivery approach for poorly water soluble drugs [18,19,21]. These interesting features culminated in the approval of Rapamune® (sirolimus, Pfizer), Megace ES® (megestrol acetate, Elan/Par Pharm), Emend® (aprepitant, MSD), Tricor® (fenofibrate, AbbVie), and Invega Sustenna® (paliperidone palmitate, Elan/Johnson and Johnson) for oral administration [18,22]. Nanocrystals have also found interesting applications for dermal and transdermal drug delivery since they were reported to increase drug penetration through the skin [18,23]. This is usually achieved by virtue of nanocrystals' ability to create a higher concentration gradient across the skin due to increased drug saturation solubility leading to increased passive diffusion [24]. They also increase drug delivery through the hair follicles due to particle size reduction, as well as adhesion to the skin [23]. However, no nanocrystal-based topical products have been approved for clinical use so far thus showing that further research is still needed in this area.

The literature shows a very limited number of reports on using nanocrystals to improve the dermal and transdermal delivery of rutin. Thus, Pelikh et al. prepared rutin and hesperetin nanocrystals with a size in the range of 160–700 nm and studied the effect of particle size and vehicle type on the skin penetration of the drugs [18]. The results showed that oleogels and creams were better than hydrogels in improving rutin nanocrystals skin penetration. In addition, smaller particles achieved better skin penetration. Furthermore, Pyo et al. showed that particle size also influenced the antioxidant efficacy of rutin nanocrystals where the nanocrystals with a mean dimeter of 300 nm had the highest antioxidant capacity compared to the drug microparticles (33 μm) and commercial cosmetic drug products [25]. In another study, rutin nanocrystals were suspended in Carbopol gel and tested as an anti-photoaging agent [26]. The nanocrystal gel achieved 3-fold higher drug permeability through mice skin compared with the coarse drug gel and prevented UV irradiation-induced photoaging and tissue damage. For cosmetic applications, rutin nanocrystal formulation is commercially available under the tradename Juvedical® (Juvena) [27].

To date no reports, however, have studied the effects of nanocrystal formulation on rutin's anti-inflammatory properties following topical application. Therefore, the aim of this study was to prepare rutin nanocrystals using different stabilizers such as Pluronic F-17, Tween 80, hydroxypropyl beta-cyclodextrin, and PEG 6000 and test their ability to enhance rutin's permeability through mice skin. In addition, the in vivo anti-inflammatory properties of the selected rutin nanocrystal formulations were examined using a carrageenan-induced rat paw edema model.

2. Materials and Methods

Rutin (RT) (purity > 95%) was purchased from Oxford Lab Fine Chem LLP, Vasai, India. Hydroxypropyl beta-cyclodextrin (HP-β-CD), and Pluronic F-127 was obtained from Sigma-Aldrich (St. Louis, MO, USA). Potassium dihydrogen phosphate, PEG 200, PEG 6000, Tween 80, hydroxypropyl methyl cellulose (HPMC), disodium hydrogen phosphate, and sodium hydroxide were obtained from United Company for Chem. and Med. Prep., Cairo, Egypt.

2.1. Preparation of Rutin Nanocrystals (RT-NCs)

RT-NCs were prepared by modifying the anti-solvent nanoprecipitation–ultrasonication method reported in the literature [28]. In brief, RT was dissolved in ethanol to prepare the organic phase. The anti-solvent phase was prepared by dissolving the stabilizers (Pluronic F-17, Tween 80, HP-β-CD, or PEG 6000) at a concentration of 0.2%, *w/v* in distilled water with 1 mL of PEG 200 as a co-stabilizer (Table 1). The organic phase was added drop-wise by a syringe into the specified volume of anti-solvent phase and the dispersion was stirred on a magnetic stirrer at a speed of 3000 RPM for 2 h at room temperature to remove the organic solvent. The obtained suspension was subjected to ultrasonication using a probe sonicator (Cole-Parmer, Vernon Hills, IL, USA) at 5 s pauses, 5 s ON at an amplitude of 45% to form nanosized particles [29]. A drug suspension without a stabilizer was prepared for comparison studies.

Table 1. Composition of RT-NCs prepared using different stabilizers at a weight ratio of 2:1, *w/w* (drug: stabilizer).

Ingredients	RT-NC1	RT-NC2	RT-NC3	RT-NC4
Rutin (mg)	30	30	30	30
Pluronic F-127 (mg)	60	-	-	-
HP-β-CD (mg)	-	60	-	-
Tween 80 (mg)	-	-	60	
PEG 6000 (mg)	-	-	-	60
PEG 200 (mL)	1	1	1	1
Ethanol (mL)	5	5	5	5
Water (mL)	30	30	30	30

2.2. Particle Size, Polydispersity Index, and Zeta Potential Measurements

The mean particle size, polydispersity index (PDI), and zeta potential were determined by dynamic light scattering using a Malvern Zetasizer Nanoseries ZS® instrument (Malvern Instruments, Malvern, UK) equipped with a backscattered light detector operating at 173°. The measurements were performed in triplicate at room temperature.

2.3. Determination of Percent Drug Entrapment Efficiency (%EE)

The entrapment efficiency of RT nanocrystals was evaluated indirectly by estimating the unentrapped RT. Briefly, the unentrapped drug was separated from nanocrystals by centrifugation at 15,000 RPM for 30 min at $4 \pm 0.5\,°C$ using cooling ultracentrifuge. The concentration of the drug in the supernatant was measured spectrophotometrically at λ_{max} of 359 nm (LISCO GmbH, Bargteheide, Germany) [30] and using a calibration curve (y = 0.0151x + 0.1129, where y is the absorbance and x is rutin concentration, $R^2 = 0.9992$). The drug entrapment efficiency (%EE) was determined by applying the following equation:

$$EE(\%) = \frac{(\text{Total drug} - \text{Drug in supernatant})}{\text{Total drug}} \times 100 \tag{1}$$

2.4. Stability Studies for RT-NCs

2.4.1. Assessment of RT-NCs Physical Stability

RT-loaded NC dispersions prepared with different stabilizers were stored in a dark place at an ambient temperature for up to three weeks. The physical stability of RT-NC formulations was evaluated by visual appearance and settlement volume ratio (F). The settlement volume ratio is the ratio of volume or height before and after sedimentation for a given period [31,32]. It was calculated using the following equation:

$$F = \frac{V}{V_0} = \frac{H}{H_0} \tag{2}$$

where H_0 is the height of suspension before sedimentation and H is the height of sediment surface after sedimentation. V and V_0 are the suspension volumes after and before sedimentation, respectively.

2.4.2. Storage Stability

An aqueous dispersion of the selected formulation (RT-NC2) was transferred into sealed brown glass bottles and stored at two different storage conditions: the room temperature and refrigerated conditions of (4 °C) for 60 days. At various time intervals, aliquots were withdrawn and analyzed for their particle size, polydispersity index, and percent drug entrapment efficiency.

2.4.3. Storage Chemical Photostability

Free RT and RT-NC2 aqueous dispersions were transferred into transparent glass vials and sealed by rubber stoppers. The vials were exposed to sunlight for 1 month at room temperature. At various time intervals, aliquots were withdrawn and methanol was added to dissolve RT followed by filtration to remove undissolved materials [33]. The RT concentration in the filtrate was measured spectrophotometrically at 359 nm.

The percent of RT remaining after different light exposure times was calculated using the following equation:

$$\text{RT remaining } (\%) = \frac{C_t}{C_0} \times 100 \tag{3}$$

where C_0 and C_t are the concentrations of RT at zero time and various time intervals, respectively.

2.5. Lyophilization of RT-NCs

RT-NCs were lyophilized to convert them to a dry form. The formulations were transferred to glass flasks then frozen over night at -80 ± 1 °C and lyophilized over a period of 48 h using a FreeZone freeze drier (Labconco Inc., Kansas City, MO, USA). The dried nanocrystals were stored in a desiccator until further investigations.

2.6. Characterization of Freeze-Dried Powder of RT-NCs

2.6.1. Fourier Transform Infrared Spectroscopy (FT-IR) Studies

The FT-IR spectra of RT alone, HP-β-CD alone, their physical mixture (1:1, w/w), and selected RT–nanocrystal formulation (RT-NC2) were recorded using a Shimadzu IR-470 spectrophotometer (Shimadzu, Seisakusho Ltd., Kyoto, Japan) at a wavenumber range of 4000–400 cm^{-1}. The potassium bromide (KBr) disc method was used. The samples were ground, mixed thoroughly with KBr, and compressed into discs using an IR compression machine.

2.6.2. Saturation Solubility

An excess amount of sample (equivalent to 5 mg) was placed in a screw-capped glass vial containing 500 mL of phosphate buffer pH 6.5 and shaken in a thermostatically controlled shaking water bath (DAIHAN Scientific Co., Seoul, Republic of Korea) at 50 RPM and 37 °C for 48 h until equilibrium was attained [34]. The suspensions were filtered using a membrane disc filter (0.45 μm) and the drug concentration in the filtrate was determined spectrophotometrically at λ_{max} of 359 nm.

2.6.3. In Vitro Drug Dissolution Studies

Dissolution studies were performed in pH 6.5 phosphate buffer containing 0.25% v/v ethanol using an USP XXIV type II dissolution apparatus. Free RT and freeze-dried RT-NCs were dispersed into 500 mL of the dissolution medium and stirred at 50 RPM at 37 ± 0.5 °C. At various time intervals (0, 5, 15, 30, 60, 90, and 120 min), an aliquot (5 mL) was withdrawn and replaced immediately with the same volume of fresh dissolution medium. The drug concentration in the withdrawn samples was measured spectrophotometrically at 359 nm. The dissolution experiments were conducted in triplicate.

2.6.4. Scanning Electron Microscopy (SEM) Studies

The surface morphology of RT-NCs was examined using scanning electron microscopy (SEM) (Jeol, JSM-5200, Tokyo, Japan). A sample of selected formulation (RT-NC2) was prepared by applying a droplet of RT-NCs onto an aluminum specimen stub, dried overnight, and sputter-coated with gold prior to imaging. An acceleration voltage of 15 kV was utilized.

2.7. Preparation and Characterization of Free RT and RT-NCs Hydrogels

An aqueous warm dispersion of a weighed amount of hydroxypropyl methyl cellulose (HPMC 15000) (5%, w/w) was developed with continuous stirring until plain gel was formed. This concentration of HPMC was selected based on previous work to produce hydrogels with desirable viscosity and homogeneity. The dispersion was sonicated for 15 min to remove air bubbles. RT dispersion in distilled water was added slowly to 10 mL of aqueous HPMC dispersion while stirring took place until a homogenous RT hydrogel was formed. The calculated amount of freeze-dried selected RT-NC2 formulation was incorporated into HPMC plain gel 5%, w/v by magnetic stirring and the final weight of the gel was adjusted to 10 g with distilled water. The RT concentration in the free RT and RT-NC2 hydrogels was 0.5%, w/w. The prepared free RT and RT-NC2 hydrogels were left in the fridge for further studies. The viscosity of the hydrogel was measured by a Brookfield Digital Viscometer (Model DV-II Brookfield Engineering Laboratories, Inc., Stoughton, MA, USA). The pH of the free RT and RT-NC2 hydrogels was measured using a pH meter (3500 pH meter, Jenway, UK). The RT content of the hydrogels was measured by dissolving 0.5 g in methanol and the drug concentration was measured spectrophotometrically at λ_{max} of 359 nm.

2.8. In Vitro Drug Release Studies from Hydrogels

The release of RT from free RT and RT-NC2 hydrogels was characterized using the dialysis method through a semi-permeable cellophane membrane (molecular weight cutoff 12,000–14,000, Sigma Aldrich, St. Louis, MO, USA) as mentioned previously with slight modification [35]. Briefly, the tested formulation (1 g of hydrogel equivalent to 5 mg RT) was placed over a previously soaked cellophane membrane fitted at the bottom of a glass tube open at both sides. The glass tube was immersed in a beaker containing 100 mL of phosphate buffer pH 6.5 with 0.25%, v/v ethanol. The beakers were placed in a thermostatically controlled shaking water bath, (DAIHAN Scientific Co., Seoul, South

Korea) operating at 50 RPM and $37 \pm 0.5\ ^\circ$C. Aliquots of 5 mL were withdrawn at intervals of 0.5, 1, 2, 4, 6, 12, and 24 h. The withdrawn samples were immediately replaced by equal volumes of the same medium. The drug content of the release samples was estimated spectrophotometrically at λ_{max} of 359 nm. The experiments were performed in triplicate.

2.9. Kinetic Evaluation of the Release Data

The data obtained from the in vitro release studies were analyzed using curve fitting to different kinetic models (zero order, first order, Higuchi diffusion model, and Korsmeyer–Peppas equation) [36]. The model that best described the data was selected based on the highest correlation coefficient (R^2).

2.10. Ex Vivo Skin Permeation Study

Skin permeation studies of RT were carried out for the selected RT-NC hydrogel formulation (RT-NC2) and free RT hydrogel using the abdominal skin of a male mouse according to previously described procedures [37]. The study protocol was approved by The Research Ethics Committee, Faculty of Pharmacy, South Valley University, Egypt (approval number P.S.V.U 125/22). The animals were sacrificed, the dorsal hair was removed and the skin was cleaned three times with phosphate buffer pH 7.4. Fresh skin specimens were stretched over one end of the open-ended glass tubes with a total base surface area of 3.14 cm^2 using an elastic rubber band. The tested gel formulations (1 g of free RT or RT-NC2 hydrogels equivalent to 5 mg of RT) were placed over the skin surface. The glass tubes were dipped in a glass beaker containing 100 mL of phosphate buffer (pH 6.5 with 0.25%, v/v ethanol). The beakers were shaken at 50 RPM and $37 \pm 0.5\ ^\circ$C for 24 h in a thermostatic shaker water bath. At different time intervals (0.5, 1, 2, 4, 6, 12, and 24 h), samples of 5.0 mL were withdrawn, replaced with an equal volume of the fresh release medium, and analyzed spectrophotometrically at λ_{max} of 359 nm for RT content. The measurements were carried out in triplicate. The cumulative amount of drug permeated per unit surface area was plotted as a function of time. The slope of the linear regression line was taken as the steady state flux (J_{ss}, μg·cm^{-2}·h^{-1}) [38]. The apparent permeability coefficient (P_{app}, cm·h^{-1}) was calculated using the following equation:

$$P_{app} = \frac{J_{ss}}{C_0} \tag{4}$$

where C_0 is the initial concentration of RT (μg/mL) in the donor compartment.

2.11. In Vivo Anti-Inflammatory Paw Edema Studies

The acute anti-inflammatory activity for the selected hydrogel formulation was performed using a carrageenan-induced rat paw edema model [39]. The study protocol was approved by The Research Ethics Committee, Faculty of Pharmacy, South Valley University, Egypt (approval number P.S.V.U 125/22). The approximate weight of each rat was 200 g. The rats were randomly divided into four groups, each of four rats. Carrageenan (1%, w/v) in saline solution was injected subcutaneously into the left hind paw of the rats for the induction of edema. Group 1 received a placebo HPMC hydrogel and was used as an untreated control. Groups 2 and 3 received free RT and RT-NC2 hydrogels, respectively. Group 4 received a marketed diclofenac sodium gel 1% (Olfen$^\circledR$, Medical Union Pharmaceuticals, Cairo, Egypt) as a reference anti-inflammatory agent. The tested formulations were applied on the edematous paw 30 min post induction which was considered as the zero time of treatment. The growth in the paw thickness was determined using a vernier caliper. The measurements were performed in triplicate. The percent edema and percent edema inhibition were calculated using the following equations:

$$\text{Edema (E, \%)} = \frac{V_t - V_0}{V_0} \times 100 \tag{5}$$

$$\text{Edema inhibition (\%)} = \frac{E_c - E_t}{E_c} \times 100 \qquad (6)$$

where V_0 and V_t are the mean paw volume before and after carrageenan injection at time t, respectively. E_c and E_t are the edema percentages of control and treated groups at the same time interval, respectively.

2.12. Statistical Analyses

The experiments were run in triplicate and the results were represented as mean $\pm$ SD. GraphPad Prism software version 8.0.1 (GraphPad Software Inc., La Jolla, CA, USA) was used to statistically analyze the data. One-way analysis of variance analysis (ANOVA) with Tukey's post-hoc test was used. A difference of $p < 0.05$ was predefined as statistically significant.

3. Results and Discussion

3.1. RT-NCs Preparation and Characterization

Despite the enormous advantages of nanocrystals including high drug loading and improved dissolution and saturation solubility, they suffer from poor physical stability that results from their small particle size and the associated increase in free energy leading to aggregation [40,41]. To enhance RT-NCs' stability, various stabilizers were used in this study including nonionic surfactants such as Tween 80 and nonionic polymers such as Pluronic F127, HP-β-CD, and PEG 6000. They are believed to stabilize nanocrystals through adsorption on their surface forming protective layers against particle aggregation and crystal growth [41].

3.2. Particle Size, Polydispersity Index, and Zeta Potential Measurements

Table 2 shows the particle size, polydispersity index, zeta potential, and percent drug entrapment efficiency (%EE) for RT-NCs prepared using various stabilizers. The particle sizes ranged from 270.5 ± 16.7 to 505.8 ± 20.5 nm. The smallest size was detected for HP-β-CD RT-NCs while Tween 80 formed the largest particles with the differences being statistically significant at $p < 0.05$ except RT-NC1 versus RT-NC2. The particle size of nanocrystals is controlled by several factors, such as the method of preparation, eventual presence of stabilizers, and the type of stabilizer. The generation of nanocrystals is associated with an enormous increase in surface area due to the production of a large number of small particles and a vast decrease in particle size. This is associated with increasing the system Gibb's free energy leading to thermodynamic instability [42]. These nanoparticles will eventually agglomerate in an attempt to minimize their total energy [43]. Stabilizers (e.g., surfactants and polymers) are thus required to minimize the system free energy and prevent agglomeration. A successful stabilizer should be able to control the particle growth during the production of uniform nanoparticles [42]. The larger size for Tween 80-stabilized nanocrystals might be related to their weak ability to sterically stabilize the nanoparticles, allowing them to grow in size during preparation. On the other hand, the smaller size detected for HP-β-CD-stabilized nanocrystals might be due to their ability to perfectly coat the newly formed nanoparticles which sterically stabilized them and prevented their aggregation and increase in size. These results are in agreement with previous work which showed a larger particle size of Tween 80-stabilized nanocrystals compared to those stabilized by HP-β-CD [41]. The PDI values were in the range of ~0.3 to 0.5, thus indicating the acceptable size distribution of the nanocrystals. A PDI value of 0.05 or smaller indicates a monodispersed population while heterogeneous nanoparticles have a PDI more than 0.7 [44].

The zeta potential was measured for all of the prepared rutin nanocrystal formulations due to its paramount importance for nanoparticle colloidal stability; it represents the electrostatic barrier that prevents nanoparticle aggregation and agglomeration [42,45]. Table 2 shows that the zeta potential of rutin nanocrystals ranged from -12.4 ± 1.0 to -28.8 ± 1.0 mV with the differences being statistically significant at $p < 0.05$ except RT-NC2 versus RT-NC3. RT-NCs had negative zeta potential values, probably due to the adsorption of water hydroxide ions at the nanocrystal surface [46,47]. It was previously shown that an absolute zeta potential value of around 30 mV is required for good colloidal stability [45]. However, this applies when the stabilization depends on pure electrostatic forces only without contributions from steric stabilization [42]. For instance, it was previously shown that nanosuspensions stabilized by non-ionic polymers and surfactants showed good colloidal stability while having zeta potential values much lower than the suggested value of 30 mV [28,41,48].

Table 2. Particle size, polydispersity index, zeta potential, and percent drug encapsulation efficiency of various RT-NCs formulations.

Formulation	Stabilizer	Size (nm)	PDI	Zeta Potential (mV)	%EE
RT-NC1	Pluronic F127	289.0 ± 13.5	0.50 ± 0.03	−17.8 ± 0.5	68.4 ± 0.8
RT-NC2	HP-β-CD	270.5 ± 16.7	0.32 ± 0.02	−28.8 ± 1.0	75.5 ± 0.9
RT-NC3	Tween 80	505.8 ± 20.5	0.56 ± 0.07	−27.62 ± 1.1	65.7 ± 0.7
RT-NC4	PEG 6000	370.5 ± 17.9	0.51 ± 0.09	−12.4 ± 1.0	66.2 ± 0.8

All data are presented as mean ± SD.

3.3. Percent Drug Entrapment Efficiency (%EE) Measurements

The percent drug entrapment efficiency (%EE) ranged from 65.7 ± 0.7% for RT-NCs prepared with Tween 80 to 75.5 ± 0.9% for those prepared with HP-β-CD. The differences were statistically significant at $p < 0.05$ except for those between RT-NC4 and either RT-NC1 or RT-NC3. The %EE was measured by an indirect method where the nanocrystals were separated by centrifugation and the drug content in the supernatant was measured. Thus, the highest %EE for HP-β-CD-stabilized nanocrystals is probably due to its ability to effectively coat and stabilize the nanoparticles which prevented their escape in the supernatant. On the other hand, the relatively lower %EE detected for Tween 80 (non-ionic surfactant) and Pluronic F127 (non-ionic polymer) might be due to their ability to partially solubilize the drug in water through micelle formation which might have facilitated its escape in the supernatant, thus decreasing the %EE [41,49].

In light of the above results RT-NCs with HP-β-CD as a stabilizer were selected for further studies since they showed the smallest particle size making them the most promising candidate to enhance rutin's anti-inflammatory properties and penetration into the skin [18,25]. In addition, these RT-NCs also had the highest %EE of 75.5 ± 0.9%, thus limiting the needed excipients and maximizing the drug/excipient ratio. They also had the highest zeta potential of −28.8 ± 1.0 mV which suggests better colloidal stability compared with other RT-NC preparations.

3.4. Stability Studies

3.4.1. Physical Stability of RT-NCs

The settlement volume ratio (F), the ratio between the volume or height of the nanocrystal suspension after and before sedimentation for a given period of time, is usually used as an indicator of nanocrystal suspension physical stability [31]. Table 3 shows that the F values were in the range of 0.15 to 0.95 with formulation RT-NC2 containing HP-β-CD showing the highest F values at all the studied time points. There was a general trend of a decrease in F values with time for all the studied preparations. At any time point, the F values followed this order: RT-NC2 (HP-β-CD) > RT-NC1 (Pluronic F127) > RT-NC3 (Tween 80) > RT-NC4 (PEG 6000). All the differences were statistically significant ($p < 0.05$). The lowest F values were detected for the nanocrystals with PEG 6000 as a stabilizer. Thus, a value of 0.21 ± 0.01 was measured for freshly prepared nanocrystals that gradually decreased to 0.15 ± 0.01 after three weeks indicating poor colloidal stability. In contrast, RT-NC2 containing HP-β-CD had the best colloidal stability as indicated by the highest F values among the tested preparations. A value of 0.95 ± 0.03 that was measured for freshly prepared samples decreased to 0.89 ± 0.05 after three weeks with no significant difference ($p > 0.05$). This high stability might be related to the relatively higher zeta potential of -28.8 ± 1.0 mV for HP-β-CD-stabilized nanocrystals in addition to their ability to sterically stabilize the nanocrystals. Similar high stability was previously observed for HP-β-CD-stabilized daidzein nanocrystals confirming its ability to efficiently coat the nanocrystals and prevent their agglomeration over time [41]. These results support the selection of formulation RT-NC2 for further studies.

Table 3. Settlement volume ratios for various RT nanocrystals after storage at room temperature for various time periods.

Time	Settlement Volume Ratio (*F*)			
	RT-NC1	**RT-NC2**	**RT-NC3**	**RT-NC4**
Freshly prepared	0.79 ± 0.02	0.95 ± 0.03	0.74 ± 0.03	0.21 ± 0.01
One week	0.77 ± 0.01	0.91 ± 0.01	0.72 ± 0.02	0.19 ± 0.02
Two weeks	0.75 ± 0.02	0.9 ± 0.03	0.69 ± 0.01	0.17 ± 0.01
Three weeks	0.72 ± 0.02	0.89 ± 0.05	0.68 ± 0.02	0.15 ± 0.01

3.4.2. Storage and Photostability

To further characterize the stability of RT nanocrystals, the selected formulation (RT-NC2 stabilized by HP-β-CD) was stored at room temperature (25 °C) and refrigerated conditions (4 °C) for 60 days and their particle size, polydispersity index, and percent drug entrapment efficiency were determined. Table 4 shows that there was a gradual decrease in the percent drug entrapment efficiency with time. Thus, after 60 days of storage the %EE decreased from 75.53 ± 0.91 to 71.23 ± 1.07 and 70.00 ± 1.00 for the samples stored at 4 °C and 25 °C, respectively. The %EE after 60 days was significantly smaller than that of either zero time or 30 days of storage ($p < 0.05$). Furthermore, the storage temperature had no important influence on the %EE as evidenced by a non-significant difference between the samples stored at 4 °C and 25 °C. The small decrease in %EE over time might be attributed to the drug solubilization over time by HP-β-CD which converts the drug from a nanocrystal to a solubilized form and facilitates its escape to the surrounding bulk medium.

Table 4. Effect of storage at room temperature (25 °C) and 4 °C on the percent drug entrapment efficiency (%EE), particle size (nm), and polydispersity index (PDI) of RT nanocrystals formulation RT-NC2.

Storage Temperature	Zero Time		30 Days		60 Days	
	4 °C	**25 °C**	**4 °C**	**25 °C**	**4 °C**	**25 °C**
%EE	75.5 ± 0.9	75.5 ± 0.9	73.6 ± 0.7	72.9 ± 1.2	71.2 ± 1.1	70.0 ± 1.0
Size (nm)	270.5 ± 16.7	270.5 ± 16.7	280.6 ± 5.4	275.6 ± 9.3	290.3 ± 7.8	320.4 ± 9.3
PDI	0.3 ± 0.02	0.3 ± 0.02	0.5 ± 0.02	0.4 ± 0.12	0.5 ± 0.13	0.4 ± 0.09

Regarding the particle size, there was a general size increase over time regardless of the storage temperature, albeit the increase at 25 °C was higher than at 4 °C. Thus, the size after 60 days of storage at 25 °C was significantly bigger compared to that of all other tested samples ($p < 0.05$). This indicates that storage in refrigerated conditions is advisable for these nanocrystals. A similar bigger particle size at a higher storage temperature was observed in other pieces of research and attributed to the increase in nanoparticle kinetic energy at higher temperatures leading to a higher probability of particle collisions and subsequently increasing the particle size [41,50,51]. Similarly, there was a general increase in the PDI values over time. However, the differences were not significant compared with the freshly prepared nanocrystals ($p > 0.05$).

Concerning RT photostability, previous studies have shown that RT is susceptible to photodegradation where exposure to UVB radiation for 120 min resulted in a decrease of 13.6% in RT content [52]. Our results show that light exposure caused progressive degradation of free RT (Figure 1). Thus, after 4 weeks of light exposure, only $42.7 \pm 0.7\%$ was remaining for free RT. In contrast, RT-NC2 had much better stability against light exposure. For instance, after the same time, the remaining RT for the nanocrystals was $95.1 \pm 3.4\%$. This confirms that the nanocrystals stabilized by HP-β-CD had around 2.3-fold better RT photostability. This much better stability for the nanocrystals might be attributed to the protection effect offered by HP-β-CD where it covers the drug nanocrystals. These results are in agreement with previous reports showing better photostability of RT when formulated into nanoparticles [53].

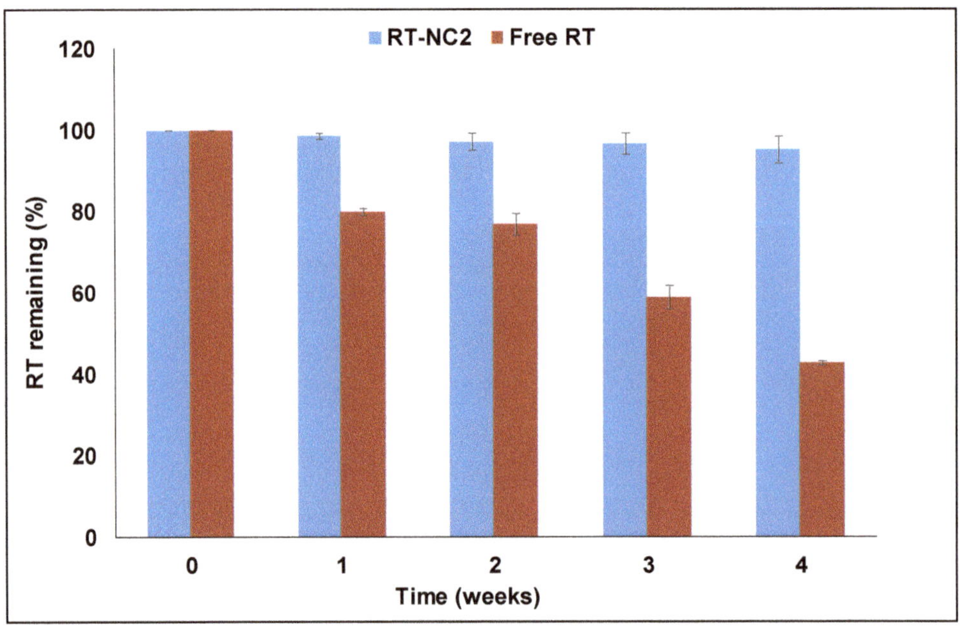

Figure 1. Percent remaining of RT as a function of exposure time to light for free RT and RT nanocrystals formulation RT-NC2.

3.5. FT-IR Spectroscopy Studies

The potential of chemical interactions between rutin and HP-β-CD in RT-NC2 was studied by recording the FT-IR spectra of rutin alone, HP-β-CD alone, their physical mixture (1:1, *w/w*), and the selected nanocrystal formulation (HP-β-CD-stabilized nanocrystals) and the results are shown in Figure 2. The spectrum of rutin alone shows a broad band centered at around 3430 cm^{-1} for OH bending, a sharp band at 1654 cm^{-1} due to C = O stretching, and a sharp band at 1594 cm^{-1} for C = C stretching of aromatic structures which is in agreement with published reports [54]. The spectrum of HP-β-CD alone shows a broad band centered at around 3400 cm^{-1} ascribed for vibration of free –OH groups and a band at 2927 cm^{-1} for vibration of bound -OH groups. The spectrum of the rutin/HP-β-CD physical mixture shows as sharp band at 1652 cm^{-1} ascribed to the stretching of rutin carbonyl groups while the stretching of rutin C = C of aromatic structures is slightly shifted to 1614 cm^{-1}. These bands appeared at the same wavenumbers in the spectrum of RT-NC2 nanocrystals (1652 and 1614 cm^{-1}, respectively) confirming the absence of chemical or physical interactions between rutin and HP-β-CD.

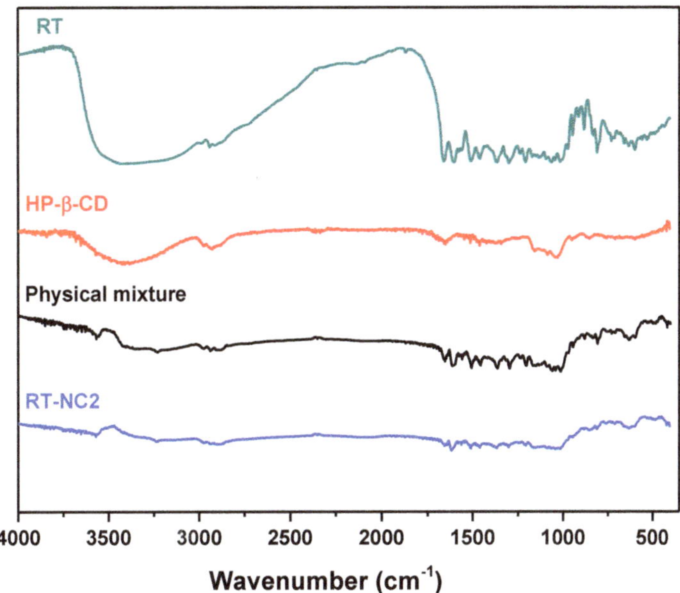

Figure 2. FT-IR spectra of rutin alone, HP-β-CD alone, their physical mixture (1:1, *w/w*), and HP-β-CD-stabilized nanocrystals (formulation RT-NC2).

3.6. Saturation Solubility Measurements

RT is a hydrophobic compound with poor aqueous solubility which limits its bioavailability and clinical benefits [55,56]. The results obtained (Figure 3) show that RT solubility in phosphate buffer pH 6.5 was 1.8 ± 0.7 µg/mL. Rutin is a weak acid with pKa in the range of 7.1 to 11.65 leading to a pH-dependent solubility profile [57]. Conversion of RT into NCs resulted in a significant increase in its aqueous solubility for all the tested stabilizers (Figure 3) ($p < 0.05$). For instance, NCs showed around a 102- to 202-fold increase in RT aqueous solubility that was dependent on the type of the stabilizer. The degree of solubility enhancement followed this descending order: HP-β-CD (RT-NC2) > Pluronic F127 (RT-NC1) > Tween 80 (RT-NC3) > PEG 6000 (RT-NC4). This might be related to the nanocrystal particle size where HP-β-CD-stabilized nanocrystals had the smallest particle size of 270.5 ± 16.7 nm among the tested stabilizers (Table 2). According to the Ostwald–Freundlich equation, the decrease in particle size results in increasing the particles' surface area which in turn leads to increasing rutin's aqueous solubility [58,59]. However, particle size is not the only factor influencing aqueous solubility. For example, Tween 80-stabilized nanocrystals had a bigger size than those stabilized by PEG 6000 but they had better solubility (Table 2). This is presumably attributed to the ability of Tween 80 to form micelles that encapsulate hydrophobic drugs such as rutin and increase their aqueous solubility [41,49].

The drug physical mixtures with the used stabilizers also achieved significantly higher drug aqueous solubility compared with the free drug hydrogel ($p < 0.05$) [60]. This might be attributed to the hydrophilicity of the used stabilizers which facilitates drug dissolution and solubility in water. In addition, the nanocrystals had significantly higher drug solubility compared with the corresponding physical mixture. This is probably due to the size reduction and increase in surface area achieved by the nanocrystals.

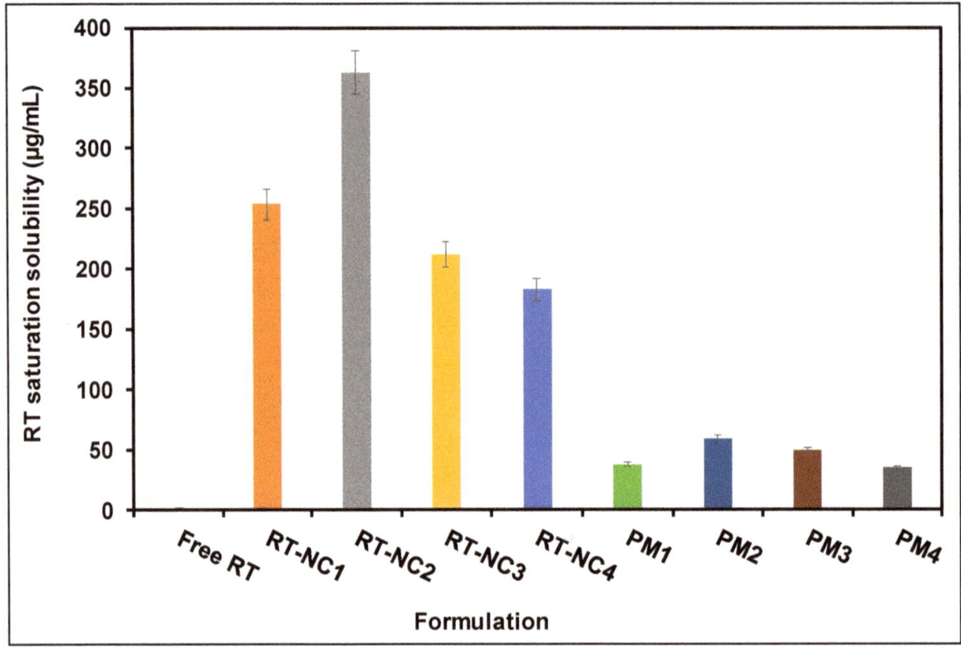

Figure 3. Saturation solubility of various RT-NCs in phosphate buffer pH 6.5 in comparison to free RT and various corresponding physical mixtures (PM). RT-NC1: rutin nanocrystals formulation 1, RT-NC2: rutin nanocrystals formulation 2, RT-NC3: rutin nanocrystals formulation 3, RT-NC4: rutin nanocrystals formulation 4.

3.7. Drug Dissolution Studies

Figure 4 shows the percent RT dissolved as a function of time for various nanocrystal formulations in comparison to the free drug. Free RT had the slowest dissolution rate among the tested preparations where around only 25% was dissolved after 120 min. RT is known as a hydrophobic compound with a slow dissolution rate which explains this slow dissolution [54]. Interestingly, the nanocrystal formulation RT-NC2 containing HP-β-CD as a stabilizer achieved 100% drug dissolution in 30 min compared with around only 15% for the free drug. Other nanocrystal formulations had significantly faster drug dissolution rates compared with the free drug ($p < 0.05$). However, except for RT-NC1, RT-NC2 had significantly faster drug dissolution compared with the other tested RT-NC formulations after 30 min ($p < 0.05$). After 30 min, the percent of drug dissolved followed this descending order: RT-NC2 > RT-NC1 > RT-NC3 > RT-NC4. Thus, they had 2.3-, 4.9-, 6-, and 6.7-fold higher dissolution rates compared with the free drug, respectively. This is the same order observed above for the saturation aqueous solubility and is probably attributed to the effect of particle size, surface area, and micelle formation on the drug dissolution rate. Previous studies have shown that the mechanism by which a given stabilizer enhanced the drug dissolution rate might have a more important influence compared with the particle size. Thus, etodolac nanocrystals' dissolution rate was affected by the particle size, as well as the type of stabilizer [42]. The % etodolac dissolved for β-cyclodextrin-stabilized nanocrystals with a particle size of 866 nm was higher than that observed for Tween 80-stabilized nanocrystals with a particle size of 393 nm. This observation was attributed to the ability of β-cyclodextrin to form a water-soluble inclusion complex with etodolac which increased its dissolution rate [61]. The viscosity of the dissolution medium and its ability to influence the drug ionization status were also found to affect the drug dissolution rate [42].

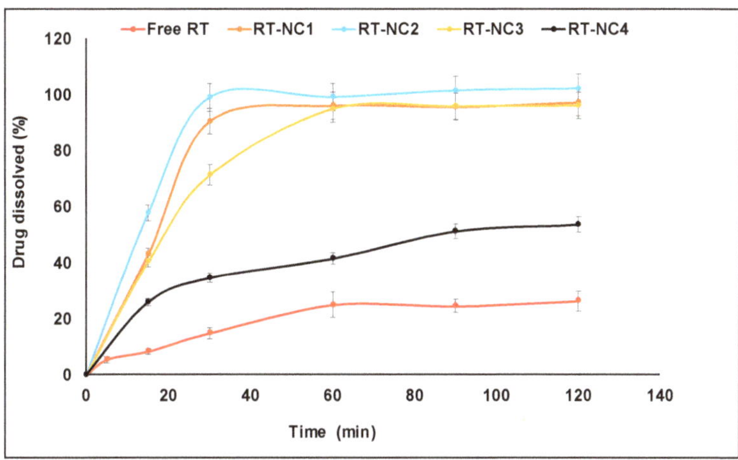

Figure 4. Dissolution profiles of various RT-NCs in phosphate buffer pH 6.5 at 37 °C in comparison to free RT.

3.8. SEM Observations

Figure 5 shows an SEM photomicrograph of RT-NCs prepared using HP-β-CD (RT-NC2). The nanocrystals appear as homogenously distributed spherical particles with distinctive boundaries and no aggregation. The size obtained from this measurement was 111.2 ± 29.5 nm. This size is smaller than that measured by DLS (270.5 ± 16.7 nm), probably due to the dry nature of the samples measured by SEM compared to the hydrated particles measured in DLS [62]. During sample measurement in SEM, the hydrated shell collapses during drying in the high-vacuum chamber of the SEM resulting in dried nanoparticles having a smaller particle size [63].

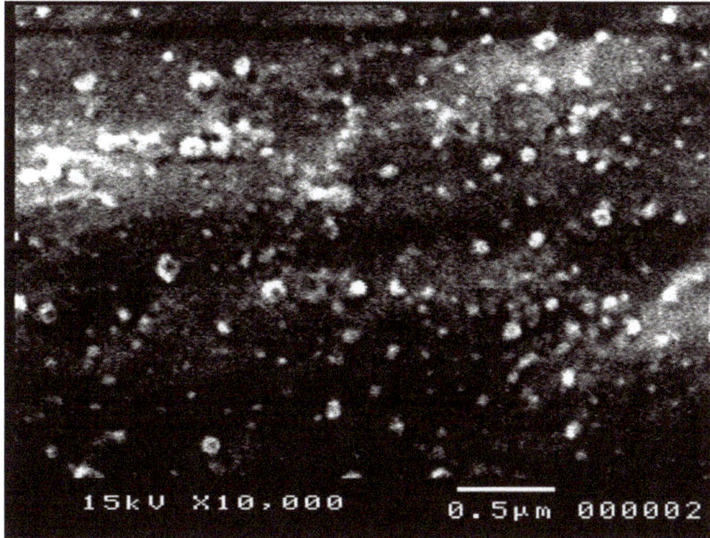

Figure 5. Scanning electron microscope photomicrograph of RT NCs prepared using HP-β-CD as a stabilizer (F2).

3.9. Characterization of Free RT and RT-NC2 Hydrogels

A hydrogel formulation was selected to facilitate RT-NC2 application on the skin since previous studies have shown that hydrogels were more efficacious than other vehicles in promoting drug–skin penetration from nanocrystal formulations [64,65]. Hydrogels have high water content, bioadhesive properties, and could serve as a depot system allowing for sustained drug delivery to the skin [66]. The properties of free RT and RT-NC2 hydrogels are shown in Table 5.

Table 5. Properties of free RT and RT-NC2 hydrogels.

Parameter	Drug Content	pH	Viscosity (cp)
Free RT hydrogel	95.92 ± 1.32%	6.8 ± 0.03	25,000 ± 45.1
RT-NC2 hydrogel	97.42 ± 1.1%	6.9 ± 0.01	45,263.33 ± 55.07

3.10. Drug Release Studies of Free RT and RT-NC2 Hydrogels

Figure 6A shows that drug release from the free drug hydrogel was slow whereby around only 30% of the drug was released after 24 h. This is presumably due to the hydrophobic nature of the drug which limits its dissolution rate and aqueous solubility. Release of a drug suspended in a hydrogel base is believed to include two steps: drug dissolution followed by diffusion of the solubilized drug through the hydrogel matrix. In contrast, a much faster drug release was observed for RT-NC2 hydrogel whereby almost complete drug release (~97%) was observed in 12 h. These results agree well with the enhanced dissolution rate and aqueous solubility described above for RT nanocrystals in comparison to the free drug. Similar behavior was also observed previously for nanocrystals suspended in a hydrogel base and was attributed to the small particle size of the nanocrystals leading to larger surface area and smaller diffusion distance and, hence, better drug dissolution and release [67]. The drug release medium was also reported to influence the drug release rate from nanocrystal formulations [42]. For instance, the pH of the release medium was found to affect the ionization status of ionic drugs leading to an important influence on their dissolution and release rate. Moreover, stabilizers that increase the release medium viscosity in the vicinity of a nanocrystal surface decreased the drug release rate from nanocrystal formulations [42].

The release data were analyzed using various mathematical models and the correlation coefficient (R^2) was calculated to obtain insights into the drug release mechanism (Figure 7) [68]. The R^2 values of free RT hydrogel were 0.74, 0.92, 0.87, and 0.97 for the zero order, first order, Higuchi, and Korsmeyer–Peppas models, respectively. In addition, the R^2 values of RT-NC2 hydrogel were, respectively 0.84, 0.86, 0.93, and 0.99 for the zero order, first order, Higuchi, and Korsmeyer–Peppas models (Table 6). This indicates that the drug release from both preparations followed the Korsmeyer–Peppas model. The release exponent (n) which indicates the release mechanism was 0.316 and 0.688 for the free drug hydrogel and RT-NC2 hydrogel, respectively. This confirms that the release from free drug hydrogel was governed by Fickian diffusion (case I diffusional) while that from the RT-NC2 hydrogel was governed by anomalous (non-Fickian) transport [69].

Table 6. Kinetic parameters of various models of RT release data from free RT and RT-NC2 hydrogels.

Kinetic Models	Zero Order		First Order		Higuchi Diffusion Model		Korsmeyer–Peppas		
	k_0	R^2	K_1	R^2	K_H	R^2	n	K_{kp}	R^2
Free RT hydrogel	0.777 ± 0.04	0.835 ± 0.06	0.010 ± 0.032	0.855 ± 0.07	4.918 ± 0.090	0.930 ± 0.012	0.316 ± 0.005	0.010 ± 0.0005	0.966 ± 0.002
RT-NC2 hydrogel	2.551 ± 0.125	0.741 ± 0.01	0.147 ± 0.028	0.915 ± 0.016	16.960 ± 0.786	0.867 ± 0.012	0.688 ± 0.063	0.106 ± 0.0004	0.999 ± 0.002

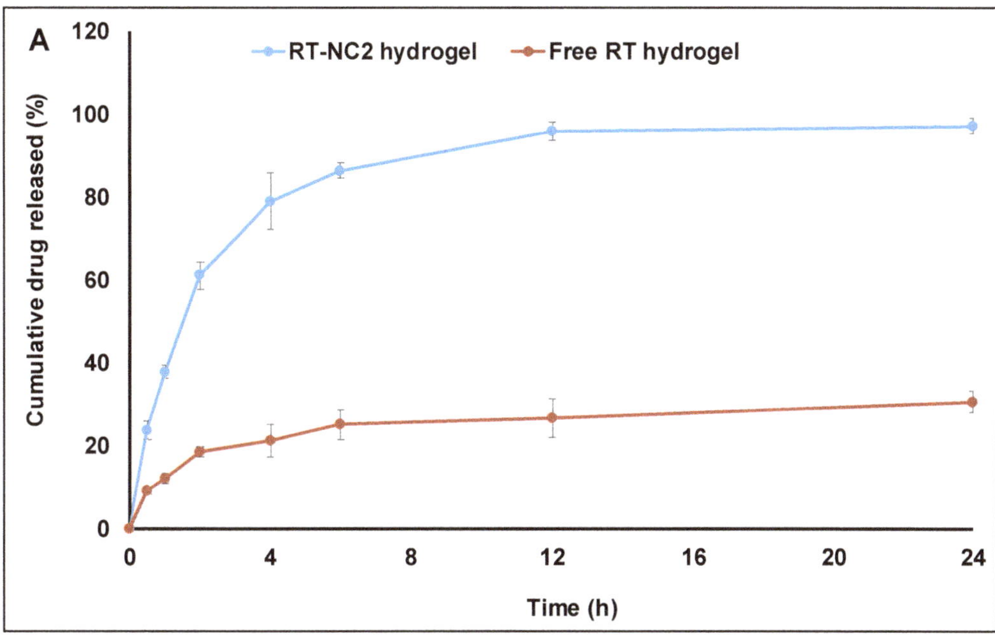

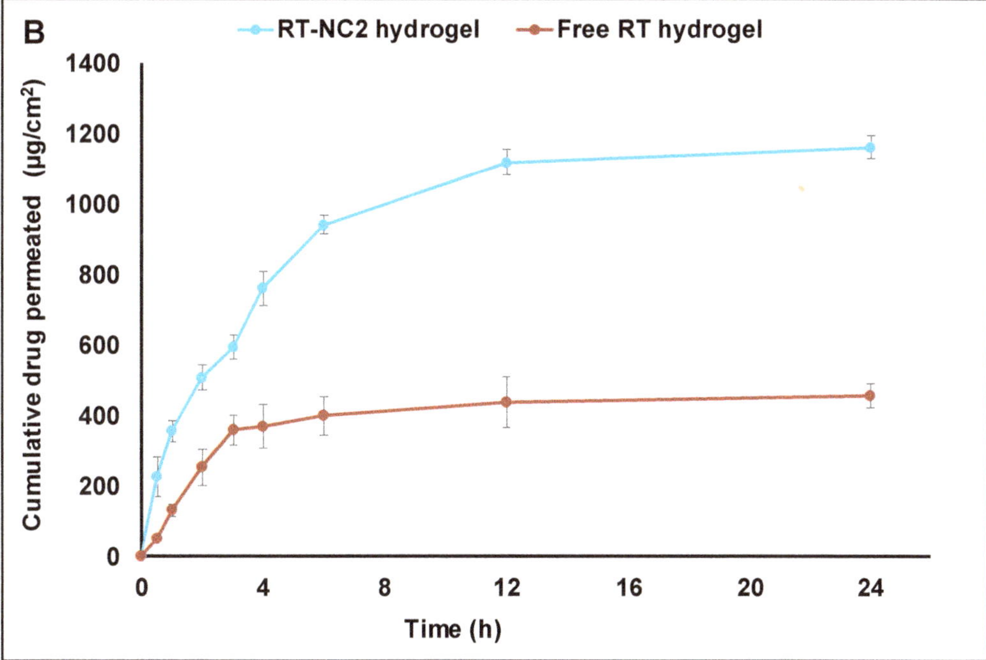

Figure 6. (A) Drug release profiles from RT-NC2 hydrogel in comparison to the free drug hydrogel in phosphate buffer pH 6.5 containing 0.25%, *v/v* ethanol at 37 °C. **(B)** Cumulative amount of RT permeated per unit surface area of mouse abdominal skin ($\mu g/cm^2$) for free RT hydrogel and RT-NC2 hydrogel.

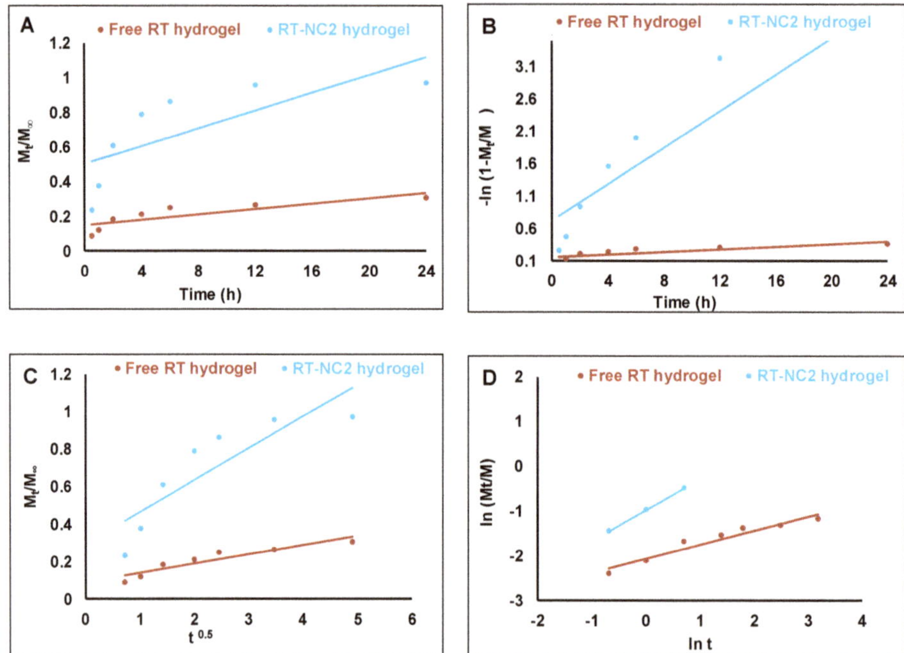

Figure 7. Plots of RT release data according to different kinetic models. (**A**) Zero order, (**B**) first order, (**C**) Higuchi diffusion model, (**D**) and Korsmeyer–Peppas equation.

3.11. Ex Vivo Skin Permeation Study

Figure 6B shows the cumulative amount of RT permeated through mouse abdominal skin for the selected RT-NC2 hydrogel in comparison to the free drug hydrogel. RT nanocrystal hydrogel had significantly higher drug skin permeation where the cumulative drug amounts permeated after 24 h were 456.7 ± 35.5 and 1163.9 ± 33.9 $\mu g \cdot cm^{-2}$ for the free drug hydrogel and nanocrystal hydrogel, respectively. This indicates that the nanocrystals achieved around a 2.5-fold enhancement in the amount of drug permeated through the skin. Furthermore, the flux (Jss) and apparent permeability coefficient (P_{app}) of the nanocrystal hydrogel were similarly enhanced by around 2.8- and 3.2-fold in comparison to the free drug hydrogel (Table 7), respectively. Similar enhancement in drug skin permeability properties was previously observed in other studies and attributed to the small particle size, enhanced dissolution, and solubility of the nanosized drug particles in comparison to the coarse drug particles. In addition, the nanocrystals might have better adhesion to the skin due to their small particle size and increased contact area with the skin which creates a positive concentration gradient between the nanocrystals and skin and ultimately leads to enhanced drug permeability [65,70,71].

Table 7. Ex vivo permeation parameters of RT from free drug hydrogel and RT-NC2 hydrogel through mouse abdominal skin.

Parameter	Q [a]	Jss [b]	P_{app} [c] $\times 10^3$
Free RT hydrogel	456.7 ± 35.5	12.9 ± 1.2	2.3 ± 0.2
RT-NC2 hydrogel	1163.9 ± 33.9	36.5 ± 1.7	7.3 ± 0.3

[a] Cumulative amount of RT permeated per unit area ($\mu g \cdot cm^{-2}$) after 24 h. [b] Flux (permeation rate constant) at steady state ($\mu g \cdot cm^{-2} \cdot h^{-1}$), obtained from the slope of the regression line after plotting the cumulative amount of RT permeated per unit area vs. time. [c] Apparent permeability coefficient ($cm \cdot s^{-1}$) calculated from Equation (4).

3.12. In Vivo Anti-Inflammatory Paw Edema Studies

The carrageenan-induced rat paw edema inflammatory model was used to assess the potential of HP-β-CD-stabilized RT-NCs hydrogel to enhance RT's anti-inflammatory properties in comparison to untreated control, free RT hydrogel and diclofenac sodium commercial gel (Olfen® gel) as a standard anti-inflammatory drug. The treatment was initiated 30 min post carrageenan injection and the percent edema was calculated (Figure 8A). The percent edema was highest at zero time for all of the tested preparations. Subsequently, there was a gradual decrease in the percent edema for all of the tested preparations. At any given time point, the percent edema followed this order: Control > free RT hydrogel > Olfen® (diclofenac sodium) commercial gel > RT-NC2 hydrogel. All the differences were statistically significant ($p < 0.05$).

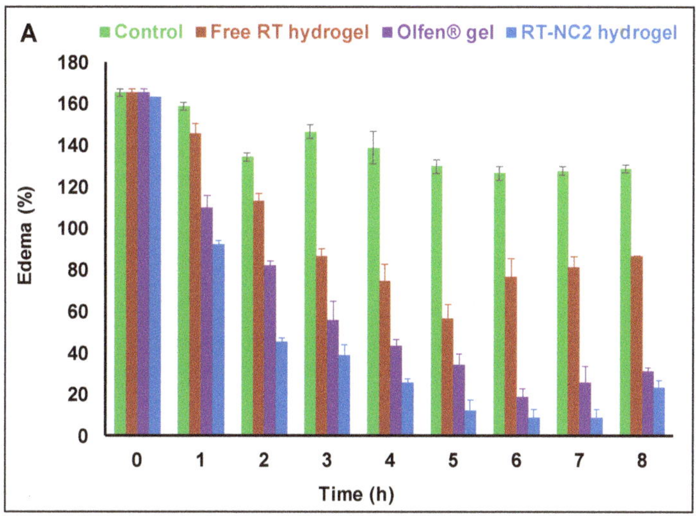

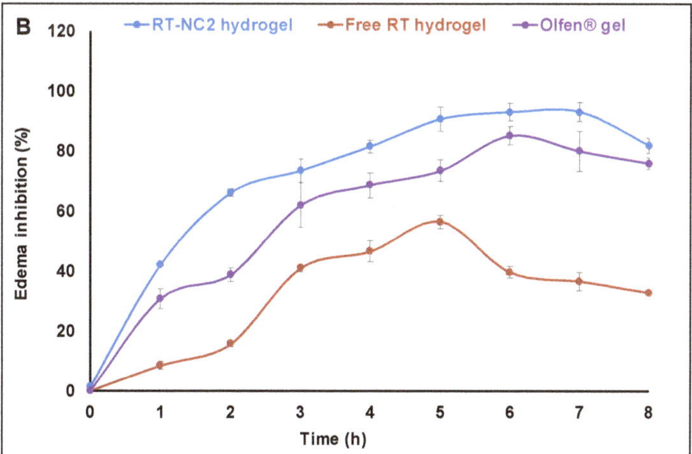

Figure 8. (**A**) Percent of paw edema as a function of time in rats after treatment with the selected rutin nanocrystal formulation (RT-NC2) hydrogel in comparison to rats treated with free RT hydrogel, Olfen® gel, and untreated rats. (**B**) Percent of paw edema inhibition as a function of time in rats after treatment with the selected rutin nanocrystal formulation (RT-NC2) hydrogel in comparison to rats treated with free RT hydrogel and commercial Olfen® gel.

The percent edema inhibition was also calculated for the tested preparations and taken as a measure of their anti-inflammatory activity (Figure 8B). Both Olfen® gel and RT-NC2 hydrogel achieved significantly higher percent edema inhibition compared with the free RT hydrogel at all the studied time points ($p < 0.05$). Peak edema inhibition was achieved at 7 h post administration for both Olfen® gel and RT-NC2 hydrogel. At this time point, Olfen® gel and RT-NC2 hydrogel had around 2.2- and 2.5-fold higher edema inhibition compared with free RT hydrogel, respectively. In addition, RT-NC2 hydrogel had significantly higher percent edema inhibition at all the studied time points except at 3 and 6 h compared with Olfen® gel. The enhanced anti-inflammatory activity observed for RT-NC2 hydrogel could be explained on the basis of enhanced drug release and skin permeability (Figure 6) compared with the free drug hydrogel which might facilitate drug delivery to the inflammation site. Interestingly, the RT-NC2 hydrogel also had a better anti-inflammatory effect compared with the commercial diclofenac sodium gel (Olfen® gel) which is presumably due to the nanometric particle size of the RT crystals which enhances drug dissolution and augments its penetration through deep skin layers and eventually results in a better anti-inflammatory effect. This finding is promising since a better anti-inflammatory effect is achieved by the RT-NC2 hydrogel than by the standard non-steroidal, anti-inflammatory drug diclofenac sodium without its notorious side effects which might increase patient compliance. In addition, site-specific drug delivery through topical application is expected to further improve drug safety and efficacy. Previous studies have shown that nanocrystal preparations were able to increase the anti-inflammatory properties of several other drugs [72–74].

4. Conclusions

Rutin nanocrystals were successfully prepared by the anti-solvent nanoprecipitation–ultrasonication method using various stabilizers such as non-ionic surfactants and non-ionic polymers. The type of stabilizer had a great influence on the nanocrystal properties. Thus, HP-β-CD gave the most favorable nanocrystal properties in terms of small particle size, high drug entrapment efficiency, high zeta potential, good colloidal stability, and the highest drug photostability. In addition, HP-β-CD-stabilized nanocrystals had around a 202- and 6.7-fold enhancement in drug aqueous saturation solubility and dissolution rate, respectively. HP-β-CD also affected the drug release rate and permeability through skin. Thus, HP-β-CD-stabilized rutin nanocrystals dispersed in HPMC hydrogel had around a 2.5-fold higher skin permeability than the free drug hydrogel. This better permeability resulted in an enhanced in vivo anti-inflammatory effect compared to the free drug hydrogel and commercial diclofenac sodium gel. Collectively, these results show the importance of the careful selection of nanocrystal stabilizers to optimize drugs' physicochemical properties and maximize their in vivo efficacy.

Author Contributions: Conceptualization, A.S.H. and G.M.S.; methodology, A.S.H.; software, A.S.H. and G.M.S.; data curation, A.S.H. and G.M.S.; formal analysis, A.S.H. and G.M.S.; writing—original draft, A.S.H. and G.M.S.; writing—review and editing, A.S.H. and G.M.S. All authors have read and agreed to the published version of the manuscript.

Funding: This research received no external funding.

Institutional Review Board Statement: The study protocol was approved by The Research Ethics Committee, Faculty of Pharmacy, South Valley University, Egypt (approval number P.S.V.U 125/22).

Informed Consent Statement: Not applicable.

Data Availability Statement: Not applicable.

Conflicts of Interest: The authors declare no conflict of interest.

Abbreviations

Hydroxypropyl beta-cyclodextrin (HP-β-CD), liquid crystalline nanoparticles (LCNs), nanocrystals (NCs), rutin (RT), hydroxypropyl methyl cellulose (HPMC), rutin nanocrystals (RT-NCs), polydispersity index (PDI), percent drug entrapment efficiency (%EE), settlement volume ratio (*F*), Fourier transform infrared spectroscopy (FT-IR), scanning electron microscopy (SEM), one-way analysis of variance (ANOVA), physical mixtures (PM), correlation coefficient (R^2), release exponent (n), flux (*Jss*), and apparent permeability coefficient (P_{app}).

References

1. Lukšič, L.; Bonafaccia, G.; Timoracka, M.; Vollmannova, A.; Trček, J.; Nyambe, T.K.; Melini, V.; Acquistucci, R.; Germ, M.; Kreft, I. Rutin and quercetin transformation during preparation of buckwheat sourdough bread. *J. Cereal Sci.* **2016**, *69*, 71–76. [CrossRef]
2. Mauludin, R.; Müller, R.H.; Keck, C.M. Development of an oral rutin nanocrystal formulation. *Int. J. Pharm.* **2009**, *370*, 202–209. [CrossRef]
3. Imani, A.; Maleki, N.; Bohlouli, S.; Kouhsoltani, M.; Sharifi, S.; Dizaj, S.M. Molecular mechanisms of anticancer effect of rutin. *Phytother. Res.* **2020**, *35*, 2500–2513. [CrossRef]
4. Ghorbani, A. Mechanisms of antidiabetic effects of flavonoid rutin. *Biomed. Pharmacother.* **2017**, *96*, 305–312. [CrossRef]
5. Negahdari, R.; Bohlouli, S.; Sharifi, S.; Dizaj, S.M.; Saadat, Y.R.; Khezri, K.; Jafari, S.; Ahmadian, E.; Jahandizi, N.G.; Raeesi, S. Therapeutic benefits of rutin and its nanoformulations. *Phytother. Res.* **2021**, *35*, 1719–1738. [CrossRef]
6. Satari, A.; Ghasemi, S.; Habtemariam, S.; Asgharian, S.; Lorigooini, Z. Rutin: A Flavonoid as an Effective Sensitizer for Anticancer Therapy; Insights into Multifaceted Mechanisms and Applicability for Combination Therapy. *Evid. -Based Complement. Altern. Med.* **2021**, *2021*, 9913179. [CrossRef]
7. Memar, M.Y.; Yekani, M.; Sharifi, S.; Dizaj, S.M. Antibacterial and biofilm inhibitory effects of rutin nanocrystals. *Biointerface Res. Appl. Chem.* **2022**, *13*, 132.
8. Gul, A.; Kunwar, B.; Mazhar, M.; Faizi, S.; Ahmed, D.; Shah, M.R.; Simjee, S.U. Rutin and rutin-conjugated gold nanoparticles ameliorate collagen-induced arthritis in rats through inhibition of NF-κB and iNOS activation. *Int. Immunopharmacol.* **2018**, *59*, 310–317. [CrossRef]
9. Oyagbemi, A.A.; Bolaji-Alabi, F.B.; Ajibade, T.O.; Adejumobi, O.A.; Ajani, O.S.; Jarikre, T.A.; Omobowale, T.O.; Ola-Davies, O.E.; Soetan, K.O.; Aro, A.O.; et al. Novel antihypertensive action of rutin is mediated via inhibition of angiotensin converting enzyme/mineralocorticoid receptor/angiotensin 2 type 1 receptor (ATR1) signaling pathways in uninephrectomized hypertensive rats. *J. Food Biochem.* **2020**, *44*, e13534. [CrossRef]
10. Ziaee, A.; Zamansoltani, F.; Nassiri-Asl, M.; Abbasi, E. Effects of Rutin on Lipid Profile in Hypercholesterolaemic Rats. *Basic Clin. Pharmacol. Toxicol.* **2009**, *104*, 253–258. [CrossRef]
11. Wang, Y.-D.; Zhang, Y.; Sun, B.; Leng, X.-W.; Li, Y.-J.; Ren, L.-Q. Cardioprotective effects of rutin in rats exposed to pirarubicin toxicity. *J. Asian Nat. Prod. Res.* **2018**, *20*, 361–373. [CrossRef]
12. Wu, H.; Su, M.; Jin, H.; Li, X.; Wang, P.; Chen, J.; Chen, J. Rutin-Loaded Silver Nanoparticles with Antithrombotic Function. *Front. Bioeng. Biotechnol.* **2020**, *8*, 598977. [CrossRef] [PubMed]
13. Gullón, B.; Lu-Chau, T.A.; Moreira, M.T.; Lema, J.M.; Eibes, G. Rutin: A review on extraction, identification and purification methods, biological activities and approaches to enhance its bioavailability. *Trends Food Sci. Technol.* **2017**, *67*, 220–235. [CrossRef]
14. Miyake, K.; Arima, H.; Hirayama, F.; Yamamoto, M.; Horikawa, T.; Sumiyoshi, H.; Noda, S.; Uekama, K. Improvement of Solubility and Oral Bioavailability of Rutin by Complexation with 2-Hydroxypropyl-β-cyclodextrin. *Pharm. Dev. Technol.* **2000**, *5*, 399–407. [CrossRef] [PubMed]
15. Remanan, M.K.; Zhu, F. Encapsulation of rutin using quinoa and maize starch nanoparticles. *Food Chem.* **2021**, *353*, 128534. [CrossRef] [PubMed]
16. Paudel, K.R.; Wadhwa, R.; Tew, X.N.; Lau, N.J.X.; Madheswaran, T.; Panneerselvam, J.; Zeeshan, F.; Kumar, P.; Gupta, G.; Anand, K.; et al. Rutin loaded liquid crystalline nanoparticles inhibit non-small cell lung cancer proliferation and migration in vitro. *Life Sci.* **2021**, *276*, 119436. [CrossRef] [PubMed]
17. Liversidge, G.G.; Cundy, K.C. Particle size reduction for improvement of oral bioavailability of hydrophobic drugs: I. Absolute oral bioavailability of nanocrystalline danazol in beagle dogs. *Int. J. Pharm.* **1995**, *125*, 91–97. [CrossRef]
18. Pelikh, O.; Stahr, P.-L.; Huang, J.; Gerst, M.; Scholz, P.; Dietrich, H.; Geisel, N.; Keck, C.M. Nanocrystals for improved dermal drug delivery. *Eur. J. Pharm. Biopharm.* **2018**, *128*, 170–178. [CrossRef]
19. Zhou, Y.; Du, J.; Wang, L.; Wang, Y. Nanocrystals Technology for Improving Bioavailability of Poorly Soluble Drugs: A Mini-Review. *J. Nanosci. Nanotechnol.* **2017**, *17*, 18–28. [CrossRef]
20. Noyes, A.A.; Whitney, W.R. The rate of solution of solid substances in their own solutions. *J. Am. Chem. Soc.* **1897**, *19*, 930–934. [CrossRef]
21. Yue, P.; Zhou, W.; Huang, G.; Lei, F.; Chen, Y.; Ma, Z.; Chen, L.; Yang, M. Nanocrystals based pulmonary inhalation delivery system: Advance and challenge. *Drug Deliv.* **2022**, *29*, 637–651. [CrossRef] [PubMed]

22. Gülsün, T.; Gürsoy, R.N.; Öner, L. Nanocrystal technology for oral delivery of poorly water-soluble drugs. *FABAD J. Pharm. Sci.* **2009**, *34*, 55.
23. Parmar, P.K.; Wadhawan, J.; Bansal, A.K. Pharmaceutical nanocrystals: A promising approach for improved topical drug delivery. *Drug Discov. Today* **2021**, *26*, 2329–2349. [CrossRef]
24. Mishra, P.R.; Al Shaal, L.; Müller, R.H.; Keck, C.M. Production and characterization of Hesperetin nanosuspensions for dermal delivery. *Int. J. Pharm.* **2009**, *371*, 182–189. [CrossRef] [PubMed]
25. Pyo, S.M.; Meinke, M.C.; Keck, C.M.; Müller, R.H. Rutin—Increased Antioxidant Activity and Skin Penetration by Nanocrystal Technology (smartCrystals). *Cosmetics* **2016**, *3*, 9. [CrossRef]
26. Li, J.; Ni, W.; Aisha, M.; Zhang, J.; Sun, M. A rutin nanocrystal gel as an effective dermal delivery system for enhanced anti-photoaging application. *Drug Dev. Ind. Pharm.* **2021**, *47*, 429–439. [CrossRef]
27. Zhai, X.; Lademann, J.; Keck, C.M.; Müller, R.H. Nanocrystals of medium soluble actives—Novel concept for improved dermal delivery and production strategy. *Int. J. Pharm.* **2014**, *470*, 141–150. [CrossRef]
28. Shaikh, F.; Patel, M.; Patel, V.; Patel, A.; Shinde, G.; Shelke, S.; Pathan, I. Formulation and optimization of cilnidipine loaded nanosuspension for the enhancement of solubility, dissolution and bioavailability. *J. Drug Deliv. Sci. Technol.* **2022**, *69*, 103066. [CrossRef]
29. Xia, D.; Quan, P.; Piao, H.; Piao, H.; Sun, S.; Yin, Y.; Cui, F. Preparation of stable nitrendipine nanosuspensions using the precipitation–ultrasonication method for enhancement of dissolution and oral bioavailability. *Eur. J. Pharm. Sci.* **2010**, *40*, 325–334. [CrossRef]
30. Cristiano, M.C.; Barone, A.; Mancuso, A.; Torella, D.; Paolino, D. Rutin-Loaded Nanovesicles for Improved Stability and Enhanced Topical Efficacy of Natural Compound. *J. Funct. Biomater.* **2021**, *12*, 74. [CrossRef]
31. Sun, M.; Gao, Y.; Pei, Y.; Guo, C.; Li, H.; Cao, F.; Yu, A.; Zhai, G. Development of Nanosuspension Formulation for Oral Delivery of Quercetin. *J. Biomed. Nanotechnol.* **2010**, *6*, 325–332. [CrossRef] [PubMed]
32. Allouni, Z.E.; Cimpan, M.R.; Høl, P.J.; Skodvin, T.; Gjerdet, N.R. Agglomeration and sedimentation of TiO_2 nanoparticles in cell culture medium. *Colloids Surf. B Biointerfaces* **2009**, *68*, 83–87. [CrossRef]
33. Tai, K.; Rappolt, M.; Mao, L.; Gao, Y.; Li, X.; Yuan, F. The stabilization and release performances of curcumin-loaded liposomes coated by high and low molecular weight chitosan. *Food Hydrocoll.* **2020**, *99*, 105355. [CrossRef]
34. Gulsun, T.; Borna, S.E.; Vural, I.; Sahin, S. Preparation and characterization of furosemide nanosuspensions. *J. Drug Deliv. Sci. Technol.* **2018**, *45*, 93–100. [CrossRef]
35. El-Mahdy, M.M.; Hassan, A.S.; El-Badry, M.; El-Gindy, G.E.-D.A. Performance of curcumin in nanosized carriers niosomes and ethosomes as potential anti-inflammatory delivery system for topical application. *Bull. Pharm. Sci. Assiut* **2020**, *43*, 105–122. [CrossRef]
36. Mircioiu, C.; Voicu, V.; Anuta, V.; Tudose, A.; Celia, C.; Paolino, D.; Fresta, M.; Sandulovici, R.; Mircioiu, I. Mathematical Modeling of Release Kinetics from Supramolecular Drug Delivery Systems. *Pharmaceutics* **2019**, *11*, 140. [CrossRef]
37. Pillai, O.; Panchagnula, R. Transdermal delivery of insulin from poloxamer gel: Ex vivo and in vivo skin permeation studies in rat using iontophoresis and chemical enhancers. *J. Control. Release* **2003**, *89*, 127–140. [CrossRef]
38. Sloan, K.B.; Beall, H.D.; Weimar, W.R.; Villanueva, R. The effect of receptor phase composition on the permeability of hairless mouse skin in diffusion cell experiments. *Int. J. Pharm.* **1991**, *73*, 97–104. [CrossRef]
39. Escribano, E.; Calpena, A.C.; Queralt, J.; Obach, R.; Doménech, J. Assessment of diclofenac permeation with different formulations: Anti-inflammatory study of a selected formula. *Eur. J. Pharm. Sci.* **2003**, *19*, 203–210. [CrossRef]
40. Bujňáková, Z.; Dutkova, E.; Baláž, M.; Turianicova, E.; Baláž, P. Stability studies of As 4 S 4 nanosuspension prepared by wet milling in Poloxamer 407. *Int. J. Pharm.* **2015**, *478*, 187–192. [CrossRef]
41. Wang, H.; Xiao, Y.; Wang, H.; Sang, Z.; Han, X.; Ren, S.; Du, R.; Shi, X.; Xie, Y. Development of daidzein nanosuspensions: Preparation, characterization, in vitro evaluation, and pharmacokinetic analysis. *Int. J. Pharm.* **2019**, *566*, 67–76. [CrossRef]
42. Afifi, S.A.; Hassan, M.A.; Abdelhameed, A.S.; Elkhodairy, K.A. Nanosuspension: An Emerging Trend for Bioavailability Enhancement of Etodolac. *Int. J. Polym. Sci.* **2015**, *2015*, 938594. [CrossRef]
43. Wang, Y.; Zheng, Y.; Zhang, L.; Wang, Q.; Zhang, D. Stability of nanosuspensions in drug delivery. *J. Control. Release* **2013**, *172*, 1126–1141. [CrossRef]
44. Danaei, M.; Dehghankhold, M.; Ataei, S.; Hasanzadeh Davarani, F.; Javanmard, R.; Dokhani, A.; Khorasani, S.; Mozafari, M.R. Impact of Particle Size and Polydispersity Index on the Clinical Applications of Lipidic Nanocarrier Systems. *Pharmaceutics* **2018**, *10*, 57. [CrossRef]
45. Bhattacharjee, S. DLS and zeta potential—What they are and what they are not? *J. Control. Release* **2016**, *235*, 337–351. [CrossRef]
46. Zimmermann, R.; Freudenberg, U.; Schweiß, R.; Küttner, D.; Werner, C. Hydroxide and hydronium ion adsorption—A survey. *Curr. Opin. Colloid Interface Sci.* **2010**, *15*, 196–202. [CrossRef]
47. Tian, Y.; Chen, L.; Zhang, W. Influence of Ionic Surfactants on the Properties of Nanoemulsions Emulsified by Nonionic Surfactants Span 80/Tween 80. *J. Dispers. Sci. Technol.* **2016**, *37*, 1511–1517. [CrossRef]
48. Singh, M.K.; Pooja, D.; Ravuri, H.G.; Gunukula, A.; Kulhari, H.; Sistla, R. Fabrication of surfactant-stabilized nanosuspension of naringenin to surpass its poor physiochemical properties and low oral bioavailability. *Phytomedicine* **2018**, *40*, 48–54. [CrossRef]

49. Homayouni, A.; Sadeghi, F.; Varshosaz, J.; Garekani, H.A.; Nokhodchi, A. Comparing various techniques to produce micro/nanoparticles for enhancing the dissolution of celecoxib containing PVP. *Eur. J. Pharm. Biopharm.* **2014**, *88*, 261–274. [CrossRef]

50. Hong, C.; Dang, Y.; Lin, G.; Yao, Y.; Li, G.; Ji, G.; Shen, H.; Xie, Y. Effects of stabilizing agents on the development of myricetin nanosuspension and its characterization: An in vitro and in vivo evaluation. *Int. J. Pharm.* **2014**, *477*, 251–260. [CrossRef]

51. Al-Gebory, L.; Mengüç, M.P.; Koşar, A.; Şendur, K. Effect of electrostatic stabilization on thermal radiation transfer in nanosuspensions: Photo-thermal energy conversion applications. *Renew. Energy* **2018**, *119*, 625–640. [CrossRef]

52. Savic, I.M.; Savic-Gajic, I.M.; Nikolic, V.D.; Nikolic, L.B.; Radovanovic, B.C.; Milenkovic-Andjelkovic, A. Enhencemnet of solubility and photostability of rutin by complexation with β-cyclodextrin and (2-hydroxypropyl)-β-cyclodextrin. *J. Incl. Phenom. Macrocycl. Chem.* **2016**, *86*, 33–43. [CrossRef]

53. Almeida, J.S.; Lima, F.; Ros, S.D.; Bulhões, L.O.S.; de Carvalho, L.M.; Beck, R.C.R. Nanostructured Systems Containing Rutin: In Vitro Antioxidant Activity and Photostability Studies. *Nanoscale Res. Lett.* **2010**, *5*, 1603–1610. [CrossRef] [PubMed]

54. Gera, S.; Pooladanda, V.; Godugu, C.; Challa, V.S.; Wankar, J.; Dodoala, S.; Sampathi, S. Rutin nanosuspension for potential management of osteoporosis: Effect of particle size reduction on oral bioavailability, in vitro and in vivo activity. *Pharm. Dev. Technol.* **2020**, *25*, 971–988. [CrossRef] [PubMed]

55. Frutos, M.J.; Rincón-Frutos, L.; Valero-Cases, E. Chapter 2.14—Rutin. In *Nonvitamin and Nonmineral Nutritional Supplements*; Nabavi, S.M., Silva, A.S., Eds.; Academic Press: Cambridge, MA, USA, 2019; pp. 111–117. [CrossRef]

56. Krewson, C.; Naghski, J. Some Physical Properties of Rutin. *J. Am. Pharm. Assoc.* **1952**, *41*, 582–587. [CrossRef] [PubMed]

57. Rashidinejad, A.; Jameson, G.B.; Singh, H. The Effect of pH and Sodium Caseinate on the Aqueous Solubility, Stability, and Crystallinity of Rutin towards Concentrated Colloidally Stable Particles for the Incorporation into Functional Foods. *Molecules* **2022**, *27*, 534. [CrossRef] [PubMed]

58. Tian, X.; Li, H.; Zhang, D.; Liu, G.; Jia, L.; Zheng, D.; Shen, J.; Shen, Y.; Zhang, Q. Nanosuspension for parenteral delivery of a p-terphenyl derivative: Preparation, characteristics and pharmacokinetic studies. *Colloids Surf. B Biointerfaces* **2013**, *108*, 29–33. [CrossRef] [PubMed]

59. Sun, J.; Wang, F.; Sui, Y.; She, Z.; Zhai, W.; Wang, C.; Deng, Y. Effect of particle size on solubility, dissolution rate, and oral bioavailability: Evaluation using coenzyme Q_{10} as naked nanocrystals. *Int. J. Nanomed.* **2012**, *7*, 5733–5744. [CrossRef]

60. Akbari, J.; Saeedi, M.; Morteza-Semnani, K.; Kelidari, H.R.; Moghanlou, F.S.; Zareh, G.; Rostamkalaei, S. The Effect of Tween 20, 60, and 80 on Dissolution Behavior of Sprionolactone in Solid Dispersions Prepared by PEG 6000. *Adv. Pharm. Bull.* **2015**, *5*, 435–441. [CrossRef] [PubMed]

61. Shan-Yang, L.; Yuh-Horng, K. Solid particulates of drug-β-cyclodextrin inclusion complexes directly prepared by a spray-drying technique. *Int. J. Pharm.* **1989**, *56*, 249–259. [CrossRef]

62. Alaaeldin, E.; Mostafa, M.; Mansour, H.F.; Soliman, G.M. Spanlastics as an efficient delivery system for the enhancement of thymoquinone anticancer efficacy: Fabrication and cytotoxic studies against breast cancer cell lines. *J. Drug Deliv. Sci. Technol.* **2021**, *65*, 102725. [CrossRef]

63. Fissan, H.; Ristig, S.; Kaminski, H.; Asbach, C.; Epple, M. Comparison of different characterization methods for nanoparticle dispersions before and after aerosolization. *Anal. Methods* **2014**, *6*, 7324–7334. [CrossRef]

64. Jin, N.J.N.; Pyo, S.M.P.S.M.; Keck, C.M.; Müller, R.H. Azithromycin nanocrystals for dermal prevention of tick bite infections. *Pharmazie* **2019**, *74*, 277–285. [CrossRef]

65. Kumar, M.; Shanthi, N.; Mahato, A.K.; Soni, S.; Rajnikanth, P.S. Preparation of luliconazole nanocrystals loaded hydrogel for improvement of dissolution and antifungal activity. *Heliyon* **2019**, *5*, e01688. [CrossRef]

66. Hoare, T.R.; Kohane, D.S. Hydrogels in drug delivery: Progress and challenges. *Polymer* **2008**, *49*, 1993–2007. [CrossRef]

67. Wei, S.; Xie, J.; Luo, Y.; Ma, Y.; Tang, S.; Yue, P.; Yang, M. Hyaluronic acid based nanocrystals hydrogels for enhanced topical delivery of drug: A case study. *Carbohydr. Polym.* **2018**, *202*, 64–71. [CrossRef]

68. Sang, Y.; Miao, P.; Chen, T.; Zhao, Y.; Chen, L.; Tian, Y.; Han, X.; Gao, J. Fabrication and Evaluation of Graphene Oxide/Hydroxypropyl Cellulose/Chitosan Hybrid Aerogel for 5-Fluorouracil Release. *Gels* **2022**, *8*, 649. [CrossRef]

69. Ritger, P.L.; Peppas, N.A. A simple equation for description of solute release II. Fickian and anomalous release from swellable devices. *J. Control. Release* **1987**, *5*, 37–42. [CrossRef]

70. Patel, V.; Sharma, O.P.; Mehta, T. Nanocrystal: A novel approach to overcome skin barriers for improved topical drug delivery. *Expert Opin. Drug Deliv.* **2018**, *15*, 351–368. [CrossRef]

71. Wang, W.P.; Hul, J.; Sui, H.; Zhao, Y.S.; Feng, J.; Liu, C. Glabridin nanosuspension for enhanced skin penetration: Formulation optimization, in vitro and in vivo evaluation. *Pharmazie* **2016**, *71*, 252–257. [CrossRef]

72. Macedo, L.D.O.; Barbosa, E.J.; Löbenberg, R.; Bou-Chacra, N.A. Anti-inflammatory drug nanocrystals: State of art and regulatory perspective. *Eur. J. Pharm. Sci.* **2021**, *158*, 105654. [CrossRef] [PubMed]

73. Fang, Y.; Li, S.; Ye, L.; Yi, J.; Li, X.; Gao, C.; Wu, F.; Guo, B. Increased bioaffinity and anti-inflammatory activity of florfenicol nanocrystals by wet grinding method. *J. Microencapsul.* **2020**, *37*, 109–120. [CrossRef] [PubMed]

74. Gujar, K.; Wairkar, S. Nanocrystal technology for improving therapeutic efficacy of flavonoids. *Phytomedicine* **2020**, *71*, 153240. [CrossRef] [PubMed]

pharmaceutics

Article

Targeted Fisetin-Encapsulated β-Cyclodextrin Nanosponges for Breast Cancer

Alaa R. Aboushanab [1], Riham M. El-Moslemany [1], Amal H. El-Kamel [1,*], Radwa A. Mehanna [2,3], Basant A. Bakr [4] and Asmaa A. Ashour [1]

[1] Department of Pharmaceutics, Faculty of Pharmacy, Alexandria University, Alexandria 21525, Egypt
[2] Department of Medical Physiology, Faculty of Medicine, Alexandria University, Alexandria 21525, Egypt
[3] Center of Excellence for Research in Regenerative Medicine and Applications (CERRMA), Faculty of Medicine, Alexandria University, Alexandria 21525, Egypt
[4] Department of Zoology, Faculty of Science, Alexandria University, Alexandria 21525, Egypt
*   Correspondence: amalelkamel@alexu.edu.eg; Tel.: +20-010050-80510

Abstract: Fisetin (FS) is considered a safer phytomedicine alternative to conventional chemotherapeutics for breast cancer treatment. Despite its surpassing therapeutic potential, its clinical utility is hampered by its low systemic bioavailability. Accordingly, as far as we are aware, this is the first study to develop lactoferrin-coated FS-loaded β-cyclodextrin nanosponges (LF-FS-NS) for targeted FS delivery to breast cancer. NS formation through cross-linking of β-cyclodextrin by diphenyl carbonate was confirmed by FTIR and XRD. The selected LF-FS-NS showed good colloidal properties (size 52.7 ± 7.2 nm, PDI < 0.3, and ζ-potential 24 mV), high loading efficiency (96 ± 0.3%), and sustained drug release of 26 % after 24 h. Morphological examination using SEM revealed the mesoporous spherical structure of the prepared nanosponges with a pore diameter of ~30 nm, which was further confirmed by surface area measurement. Additionally, LF-FS-NS enhanced FS oral and IP bioavailability (2.5- and 3.2-fold, respectively) compared to FS suspension in rats. Antitumor efficacy evaluation in vitro on MDA-MB-231 cells and in vivo on an Ehrlich ascites mouse model demonstrated significantly higher activity and targetability of LF-FS-NS (30 mg/kg) compared to the free drug and uncoated formulation. Consequently, LF-FS-NS could be addressed as a promising formulation for the effective management of breast cancer.

Keywords: phytomedicine; nanosponges; lactoferrin; bioavailability; MDA-MB-231 cells; caspase-3; cyclin-D1

Citation: Aboushanab, A.R.; El-Moslemany, R.M.; El-Kamel, A.H.; Mehanna, R.A.; Bakr, B.A.; Ashour, A.A. Targeted Fisetin-Encapsulated β-Cyclodextrin Nanosponges for Breast Cancer. *Pharmaceutics* **2023**, *15*, 1480. https://doi.org/10.3390/pharmaceutics15051480

Academic Editors: Franco Dosio and Ana Isabel Fernandes

Received: 30 March 2023
Revised: 25 April 2023
Accepted: 10 May 2023
Published: 12 May 2023

1. Introduction

Nanosponges (NS) are promising polymeric colloidal systems. As the name implies, NS are nano-sized porous structures offering ideal properties for drug delivery. NS can be synthesized from various polymers and copolymers, such as hyper cross-linked polystyrenes [1], ethyl cellulose [2], or cyclodextrins [3]. Cyclodextrins (CDs) are generally recognized as safe (GRAS), as listed by the Food and Drug Administration (FDA) [4]. Including β-CD as the most studied and most frequently used, CDs are cyclic oligosaccharides with a cage-like structure and a distinct cone-shaped lipophilic cavity [5]. This unique structure contributes to their ability to form inclusion complexes with various molecules, resulting in enhanced aqueous solubility and protection against degradation [6]. However, the ease of dissociation of the formed complexes upon dilution, in addition to the inability to complex certain molecules, are considered major drawbacks [6]. To overcome these limitations, structural modifications of CDs have been suggested to increase the inclusion capacity and allow for a wider scope of biological applications [7]. Of these, cyclodextrin-based nanosponges (CDNS) have recently attracted attention and been synthesized by polymer cross-linking to form a highly porous branched matrix [3]. The obtained amphiphilic spongy structure confers the ability to accommodate hydrophobic molecules in the CD cavities and fewer lipophilic molecules in the more hydrophilic outer polymeric network with high loading capacity and controlled drug release [8]. Furthermore, CDNS offer other advantages, such as being highly biocompatible, biodegradable, and of low cytotoxicity [9]. In addition, this nanosystem was shown to improve permeation across biological barriers and enhance the bioavailability of active substances [6]. As a controlled drug delivery system, CDNS has been explored for oral, topical, and parenteral drug delivery for numerous applications, such as anticancer, antiviral, antihypertensive and antiplatelet therapy [5,10]. One of the most relevant fields is drug delivery for cancer therapy [9,11], including breast cancer. In this aspect, CDNS showed improved anticancer drug effects both in vitro [12,13] and in vivo [14].

The breast cancer (BC) burden has been rising sharply over the past decades. Having replaced lung cancer, it is now the most diagnosed cancer worldwide, representing a quarter of all cancer cases in females. Despite the significant advances in BC treatment, a continuous rapid increase in the number of new cases and deaths is recorded, with a 40% projection expected especially in low human developing index countries by 2040 compared to 2020 [15]. This could be imputed to the limitations of chemotherapy, including non-selectivity to cancer cells, multidrug resistance (MDR), and ineffective inhibition of tumor growth, metastasis, and recurrence. Thus, recent strategies for the prevention and treatment of BC focused on the use of herbal medicine as a safe, effective, and low-cost alternative to cytotoxic drugs. Considerable anticancer activity in natural phytochemicals has been confirmed against different cancers via numerous in vitro and in vivo studies [16]. Flavonoids are plant polyphenols exhibiting various beneficial properties for breast cancer therapy [17]. They were shown to affect growth and proliferation via cell cycle arrest, necrosis, and apoptosis. Furthermore, they exhibit antioxidant, anti-inflammatory, and anti-mutagenic properties [18]. A potent flavonoid found in various fruits and vegetables and exhibiting cytotoxic activity is fisetin (3,3′,4′,7-tetrahydroxy flavone, FS) [19]. The anticancer effect of FS on different breast cancer cell lines including triple-negative breast cancer (TNBC) cells was manifested [20,21]. Among several mechanistic studies on different breast cancer cell lines, FS was shown to inhibit cell proliferation and metastasis, prevent cell cycle progression, induce apoptosis, cause cleavage of poly ADP ribose polymerase (PARP), and modulate Bcl-2 family protein expression [22]. Moreover, FS was capable of suppressing PKCα/ROS/ERK1/2 and p38 MAPK signaling pathway activation, reducing NF-κB activation, and lowering TET1 expression in a concentration- and time-dependent manner [22,23]. It also reverses the epithelial to mesenchymal transition (EMT) process mediated by the PTEN/Akt/GSK-3β signaling pathway [20].

Despite its bioactive potential in the prevention and treatment of different cancer conditions, the clinical applications of FS have been impeded by its highly lipophilic nature, and, hence, limited aqueous solubility [24]. Furthermore, it is rapidly metabolized, enzymatically degraded, and liable to p-glycoprotein efflux following oral administration, which leads to a short half-life and poor bioavailability [24,25]. These curtailments interfere with FS bioaccessibility, necessitating the development of novel oral drug delivery approaches, such as loading into nanosystems [26,27]. Furthermore, the nanoparticles' surface could be modified with active ligands, such as folic acid, hyaluronic acid, chondroitin sulfate, and lactoferrin, for selective and efficient accumulation at tumor sites via targeting specific receptors overexpressed on BC cells [26]. Lactoferrin (LF) is an iron-binding glycoprotein possessing a strong affinity to transferrin receptors overexpressed on breast cancer cells [28]. It is capable of either promoting or inhibiting cell proliferation and migration depending on whether the cell it acts upon is normal or cancerous, respectively [29]. Moreover, being a part of the innate immune system, LF boosts adaptive immune response, thus, preventing or inhibiting cancer development [29].

In this regard, the current study aimed to develop FS-loaded cyclodextrin nanosponges (FS-NS) coated with the bioactive protein LF to enhance FS bioavailability and anticancer activity. This novel nanosystem would provide multiple advantages, including high drug loading and entrapment efficiency with sustainment of release and an improvement in the anticancer activity of FS via both passive targeting by enhanced permeability and retention (EPR) effect and active targeting by the LF coating. Following formulation optimization and in vitro characterization, the change in FS pharmacokinetic parameters upon loading into proposed LF-FS-NS relative to FS suspension and uncoated FS-NS was assessed. Moreover, the cytotoxicity and cellular uptake of the test formulations were assessed on MDA-MB-231, a human TNBC cell line. A murine Ehrlich ascites breast cancer mouse model was used to assess in vivo anticancer efficacy.

2. Materials and Methods

2.1. Materials

β-cyclodextrin, diphenyl carbonate (DPC), coumarin 6 (C6), and quercetin were purchased from Sigma-Aldrich (St. Louis, MO, USA). Lactoferrin (LF) was obtained from Lactoferrin.co (Frankfurt, Germany). Fisetin (FS) was from Arctom Scientific (Agoura Hills, CA, USA). HPLC-grade acetonitrile, methanol, dimethylformamide (DMF), and formic acid were purchased from Fischer Scientific (Loughborough, UK). Hoechst 33342 stain, an annexin V FITC/propidium iodide (PI) kit, fetal bovine serum (FBS), and Dulbecco's modified Eagle's medium (DMEM) were purchased from Sigma-Aldrich (St. Louis, MO, USA). Penicillin and streptomycin solution (100 U/mL each) were from BioWhittaker® (Lonza, Belgium). 3-[4, 5-dimethylthiazol-2-yl]-2,5-diphenyl tetrazolium bromide (MTT) was purchased from Serva (Heidelberg, Germany). PCR primers were obtained from Eurofins Scientific (Luxembourg). A one-step RT qPCR kit (SYBR Green with low ROX) was from Enzynomics Co. Ltd., Yuseonggu, Daejeon, Korea. All the other reagents were of analytical grade and were used without further purification.

2.2. Preparation of Blank Nanosponges

NS formulations were prepared using diphenyl carbonate (DPC) cross-linker as previously described, with some modifications [30]. In brief, β-cyclodextrin (CD) was dissolved in N, N-dimethylformamide (DMF), and DPC was added in a molar ratio of 1:6. The reaction mixture was mixed under magnetic stirring at 450 rpm for 20 min in a water bath at 80 °C. For optimization, the effects of heating the reaction mixture at 90, 120, or 150 °C for different time intervals (2 or 5 h) while using different volumes of DMF (3 or 6 mL) were investigated. The product was subjected to 5 cycles of washing with deionized water and acetone and then left to dry in a desiccator for 48 h. The purified powdered formulation was ground in a mortar and accurately weighed to calculate the percentage yield using the following equation:

$$\% \text{ Yield} = \frac{\text{Weight of the nanosponge}}{\text{Weight of } \beta - \text{CD} + \text{Weight of DPC}} \times 100 \tag{1}$$

The obtained powder (14 mg) was dispersed in 2 mL deionized water and sonicated using a probe sonicator (Bandelin Sonoplus, Germany) at 60% amplitude for 10 min. Homogenization at 10,000 rpm for 10 min was subsequently carried out to obtain a nano-dispersion (NS).

2.3. Ferric Chloride Test

To qualitatively verify NS formation, the $FeCl_3$ test was performed to detect the presence of phenols formed as a by-product during the cross-linking esterification reaction between β-CD and DPC. Unwashed NS (10 mg) were dispersed in 3 mL of deionized water followed by the addition of 1 mL $FeCl_3$ solution. The appearance of a deep violet color indicated the presence of phenol, proving NS formation [30].

2.4. Preparation of Fisetin-Loaded Nanosponges

For the preparation of fisetin-loaded nanosponges (FS-NS), the drug was dissolved in ethanol and added to the NS dispersion in a ratio of 1:4 to obtain a final drug concentration of 1.75 mg/mL. For complete drug loading, the mixture was subjected to 5 min sonication in a bath sonicator followed by overnight stirring [10].

2.5. Preparation of Lactoferrin (LF)-Coated FS-NS

FS-NS formulation dispersion (2 mL) was dropped into 100 μL of LF solution in PBS pH 6.5 and stirred at 500 rpm for 30 min. Different LF concentrations (25–100 mg/mL) were tested. Efficient LF coating was verified by size and ζ-potential measurements.

2.6. Physicochemical Characterization

2.6.1. Surface Area and Porosity Analysis

Specific surface area and porosity of the prepared NS were measured using a nitrogen absorption–desorption isotherm (Belsorp-Mini II analyzer, Japan). The sample (0.2 g) was degassed for 3 h before analysis. The surface area and porosity were determined using Brunauer–Emmett–Teller (BET) and Barrett–Joiner–Halenda analyses [31].

2.6.2. Microscopical Examination

Scanning Electron Microscopy (SEM)

The surface morphology and porosity of blank NS were evaluated using a scanning electron microscope (SM-IT200; JEOL, Tokyo, Japan). The NS dispersion was mounted on a metal stub, air-dried, and sputter-coated with gold before the examination.

Transmission Electron Microscopy (TEM)

Transmission electron microscopy (TEM) (model JEM-100CX, JEOL, Japan) was used to further investigate the morphology as well as to determine the average size of NS, FS-NS, and LF-FS-NS [32]. Dispersions were dropped on carbon-coated copper grids, stained with uranyl acetate (1% w/v), and air-dried before the examination. For the determination of particle size (PS), 50 measurements from different fields were carried out using image-analysis software [31] (Fiji 1.52p; National Institutes of Health, Bethesda, MD, USA) and the polydispersity index (PDI), calculated by the following equation:

$$\text{PDI} = \left(\frac{SD}{d}\right)^2 \tag{2}$$

where SD is the standard deviation and d is the average diameter.

2.6.3. ζ-Potential Measurement

ζ-potential of different formulations (NS, FS-NS, LF-FS-NS) was determined using a Malvern Zetasizer (Nano-ZS Series DTS 1060, Malvern Instruments, UK). Measurements were performed at a fixed angle of 173° at 25 °C. Formulations were adequately diluted with deionized distilled water in the ratio 1:100 and measured in triplicate.

2.6.4. Entrapment Efficiency and Drug Loading Determination

Fisetin entrapment efficiency% (EE%) was carried out using the dialysis technique [33]. FS-NS and LF-FS-NS dispersions (0.5 mL) were placed in a dialysis bag (Visking 36/32, 28 mm, MWCO 12–14 KDa; Serva, Heidelberg, Germany) and then immersed in 26 mL PBS (pH 7.4) containing 0.1% Tween® 80 to maintain sink conditions, before being centrifuged at 25 °C and 500 rpm for 30 min using a high-speed cooling centrifuge (Model 3K-30; Sigma Laborzentrifugen GmbH, Osterode, Germany). The free unentrapped drug in the eluent was quantified spectrophotometrically at 360 nm using a UV–visible spectrophotometer (Cary 60 UV–visible spectrophotometer, Agilent, Santa Clara, CA, USA). Linearity was checked in the range 5–15 μg/mL with a coefficient of determination $R^2 = 0.99$. For % drug loading (DL%) determination, NS was redispersed in ethanol following centrifugation. The amount of FS analyzed denoted the weight of FS in NS formulations [10]. EE% and DL% were calculated using the following equations:

$$EE\ \% = \frac{\text{Total drug amount } - \text{ unentraped drug amount}}{\text{Total drug amount}} \times 100 \qquad (3)$$

$$DL\ \% = \frac{\text{Drug amount in NS}}{\text{Total weight of NS}} \times 100 \qquad (4)$$

2.6.5. In Vitro Drug Release

In vitro release of FS from both FS-NS and LF-FS-NS in comparison to the drug solution was carried out using the dialysis bag method. The drug solution was prepared in an aqueous PEG 400 solution (50% v/v). NS dispersions (0.5 mL equivalent to 0.9 mg FS) were added into dialysis bags and then placed into 30 mL PBS (pH 7.4) with 0.1% w/v Tween® 80, maintaining sink conditions. The experiment was carried out in a thermostatically controlled shaking water bath at 37 °C and 100 rpm. At different time intervals (1–24 h), samples (1 mL) were withdrawn and replaced with fresh medium. The drug concentration was determined spectrophotometrically at 360 nm. Then, the percentage of cumulative FS released was calculated in triplicate.

Drug release kinetics from NS dispersions were assessed using model-dependent methods [34] calculated by the Excel add-in DDsolver [35].

2.6.6. Fourier Transform Infrared Spectroscopy (FTIR)

Prior to analysis, FS-NS dispersion was dried using Aerosil® 200, as previously described [36]. Briefly, FS-NS aqueous dispersion was mixed with Aerosil® 200 in a ratio of 4:1, and then the mixture was allowed to dry overnight in a desiccator. The FTIR spectra of Aerosil® 200, β-CD, DPC, NS, FS, and FS-NS were obtained using an FTIR spectrometer (PerkinElmer Inc., Waltham, MA, USA). Tested samples were mixed with KBr (1:100 w/w) and compressed to form a disc that was scanned in the range 4000–500 cm^{-1}.

2.6.7. X-ray Powder Diffractometry (XRD)

The crystallinity of Aerosil® 200, NS, FS, and FS-NS was assessed using XRD (XRD-7000 X-ray diffractometer, Bruker D2-Phaser; Madison, WI, USA). The diffraction pattern was performed in a step scan model of 30 kV and 30 mA with a scanning region for the diffraction angle, 2θ, from 0 to 100°, with step size 0.02°.

2.7. In Vitro Cell Culture Studies

In vitro cell culture studies were performed on the human breast cancer cell line (MDA-MB-231) obtained from ATCC (HTB-26™) and cultured at CERRMA (Center of Excellence for Research in Regenerative Medicine and its Applications), Faculty of Medicine, Alexandria University. Cells were grown in Dulbecco's modified Eagle's medium (DMEM)-high glucose, enriched with (10 % v/v) fetal bovine serum (FBS) and antibiotics (100 U/mL penicillin, 100 μg/mL streptomycin). Cells were maintained in a 5% CO_2 incubator at 37 °C.

2.7.1. MTT Cytotoxicity Assay

MDA-MB-231 cells were grown as monolayer cultures in the culture media mentioned above and monitored daily using the phase-contrast inverted microscope (CKX41SF; Olympus) for their growth and morphology. Media were changed every 2–3 days. When confluent, cells were plated into 96-well plates with a uniform seeding density of 5×10^3 cells/well and allowed to adhere for 24 h. The medium was then replaced with medium containing different concentrations of FS, FS-NS, and LF-FS-NS ranging from 10 to 100 μg/mL and the similarly diluted blank formulations NS and LF-NS, and then incubated for 48 h at 37 °C and 5% CO_2. The media were then aspirated, replaced by MTT solution, and incubated for another 4 h. Finally, the formazan blue crystals were dissolved in DMSO, and the absorbance was measured at 570 nm by an ELISA well-plate reader (Tecan, Infinite F50, Männedorf, Switzerland). The values obtained were compared with the control, which was regarded as 100% living cells. Experiments were performed in triplicate [26].

2.7.2. Apoptosis Assay (Annexin V-FTIC/Propidium Iodide Assay)

Annexin V assay was used to assess the apoptotic effect of FS, FS-NS, and LF-FS-NS [26]. Cells were incubated in 6-well plates in a density of 2×10^5 cells/well and left to adhere for 24 h at 37 °C and 5% CO_2. Cells were treated with samples equivalent to half the IC_{50} value calculated from the cytotoxicity experiment for 24 h. Cells were then trypsinized, collected by centrifugation at 2000 rpm, and stained with annexin V-fluorescein isothiocyanate (FITC) (50 µg/mL, 5 µL) and propidium iodide (PI) (50 µg/mL, 5 µL) as per the manufacturer's protocol. Analysis of apoptotic cells was performed by 20,000 cells gating by flow cytometry (BD FACSCalibur™ flow cytometer, San Jose, CA, USA). The experiment was performed in triplicate with representative images provided.

2.7.3. Cell Migration by Wound Healing Assay

The effect of FS, FS-NS, and LF-FS-NS on the migration ability of MDA-MB-231 cells was examined by the wound healing assay [37]. Cells were seeded in 12-well plates and incubated at 37 °C with 5% CO_2 until confluent. A 100 µL pipette tip was used to make a single scratch in the cell monolayer, and an image of the scratch was captured using the phase-contrast inverted microscope (CKX41SF; Olympus). The cells were then treated with concentrations corresponding to half IC_{50} values and incubated for 24 h. Following incubation, the scratch width was imaged. Images were analyzed using NIH ImageJ software, and the ratio of cell migration was calculated as the percentage of the remaining cell-free area in comparison to the area of the initial scratch.

2.7.4. Cellular Uptake

Fluorescently labeled NS and LF-NS were prepared using coumarin 6 (C6). A similar loading procedure was adopted with the addition of the dye to ethanol instead of FS. Briefly, 100 µL of freshly prepared C6 solution (250 µg/mL in ethanol) was added to 2 mL of NS and stirred overnight. MDA-MB-231 cells were seeded in 6-well plates at an initial density of 8×10^4 cells/well on coverslips. Then, 24 h later, C6-labeled NS formulations, both uncoated and LF-coated, and the free dye were added to the corresponding wells and incubated at 37 °C and 5% CO_2 for 4 h. The cells were then washed three times with PBS and fixed with 4% paraformaldehyde at room temperature for 30 min in the dark. Cell nuclei were stained with Hoechst 33,342 for another 20 min. The cells were observed using laser scanning confocal microscopy (LEICA, DMi8, Mannheim/Wetzlar, Germany), and the fluorescent signals of the conjugated formulations were analyzed and compared with the control and free C6. Image processing was conducted using the Leica Application Suite X (LAS X) software. The uptake study was conducted in triplicates. Finally, quantification of fluorescence intensity in the obtained images was performed using ImageJ software, National Institute of Health, NIH (version 1.45s) [38].

2.8. In Vivo Studies

2.8.1. Pharmacokinetic Study

Study Design

The study was conducted on female Wistar rats (200 ± 20 g) that were kept under standard conditions of light, temperature, and humidity, with free access to food and water during the study.

Rats were randomly divided into 6 groups (each of 6 animals). Animals were fasted overnight prior to the experiment. A single dose of FS suspension, FS-NS, and LF-FS-NS equivalent to 30 mg FS/kg was administered either orally by gastric gavage or by IP injection. FS suspension was prepared in 0.5% *w/v* carboxymethyl cellulose sodium for oral administration and in 10% PEG 400 saline solution for IP injection. Blood samples were collected via the orbital plexus under anesthesia at predetermined time points (0.25, 0.5, 1, 3, 5, 7, and 24 h) in EDTA-containing tubes. Blood samples were centrifuged at 4000 rpm for 10 min. Plasma samples were frozen and maintained at -80 °C pending analysis.

Quantification of FS in Plasma Samples

FS in plasma samples was analyzed using a reported HPLC-UV method for the quantification of FS in plasma samples, with slight modification [39]. Plasma samples (300 μL) were vortex mixed with 40 μL quercetin (1 μg/mL) as an internal standard and an equal volume of methanol acidified with 0.5% v/v formic acid for plasma protein precipitation for 1 min, and then centrifuged at 10,000 rpm for 10 min. The obtained supernatants were filtered through 0.22 μm syringe PTFE filters. The HPLC instrument (Agilent 1260 infinity, Germany) equipped with a quaternary pump (G1311C) and a UV variable wavelength detector (G1314F) was used. Samples (100 μL) were manually injected in a reversed-phase C_{18} column (Agilent HC-C_{18}(2), 4.6 × 150 mm, 5 μm). The mobile phase was a mixture of acetonitrile and water acidified with 0.5% v/v formic acid in the ratio of 40:60. The flow rate was maintained at 1 mL/min, and detection was performed at 360 nm. Quantification was achieved using a calibration curve for peak area ratios of FS/quercetin in spiked plasma obtained under the same conditions.

Pharmacokinetic Data Analysis

The plasma concentrations vs. time data were analyzed by a non-compartmental pharmacokinetic model using the Excel pharmacokinetic solver add-in [40], and pharmacokinetic parameters were calculated. Results were expressed as mean ± standard deviation (SD) (n = 6).

2.8.2. Antitumor Efficacy Evaluation

Experimental Design

Efficacy and toxicity studies were performed on female Swiss albino mice (7–8 weeks of age, about 20–25 g). Animals were maintained under controlled temperature and humidity with a 12 h dark/light cycle, with free access to food and water.

Mammary tumors were induced in mice using the Ehrlich ascites tumor (EAT) model [26]. EAT cells, obtained from the National Cancer Institute, Cairo, Egypt, were properly diluted ($\sim 2 \times 10^6$) and subcutaneously injected into the left mammary fat pad of mice. The tumor volume was evaluated until it reached ~ 100 mm^3 (12 days post-inoculation). Tumor-bearing animals were then randomly assigned into four groups (n = 6) as follows: positive control group (untreated), FS suspension in 10% PEG 400 saline solution, FS-NS, and LF-FS-NS administered by IP injection. Treated groups received 7 doses on an alternate-day dosing regimen of either FS suspension or NS formulations equivalent to 30 mg FS/kg. A negative control group was included for comparison. At the end of the study, mice were sacrificed, and tumors were excised and washed with ice-cold phosphate buffer. Tumor weight was determined, and then tumors were divided into portions for further assessment.

Assessment of Tumor Growth

The percentage change in tumor volume compared to baseline volume was determined as an indicator of tumor growth. Tumor volume was assessed twice weekly. A vernier caliper was used to measure the tumor length (major axis) and width (minor axis), and then tumor volume was calculated according to the following equation [26]:

$$\text{Tumor volume} = \text{length} \times (\text{width})^2 \times 0.5 \tag{5}$$

Tumor Biomarkers

Excised tumors were homogenized in phosphate buffer saline (pH 7.4) to make a final 10% tissue homogenate that was used for the quantitative determination of tumor growth biomarkers. ELISA kits were used to assess cyclin-D1 level (Mouse cyclin-D1 ELISA Kit, EIAal™, Waltham, MA, USA) as an indicator of proliferation rate and the anti-apoptotic BCl-2 protein (catalog no. CSB-E08855m). Quantification was performed according to the manufacturer's protocol.

The relative expression of Bax and caspase-3 genes in tumor tissues, using quantitative real-time reverse transcriptase polymerase chain reactions (qRT-PCR), was performed. Forward and reverse primer sequences for PCR amplification are shown in Table 1. The RT-PCR reactions were run in triplicate with signal collection at the end of each cycle. An internal housekeeping gene (GAPDH) was applied to normalize relative transcript levels. The comparative threshold cycle ($\Delta\Delta Ct$) method was used to determine sample differences.

Table 1. qRT-PCR primers used to evaluate gene expression levels of Bax and caspase-3 in tumor tissues.

Gene	Forward	Reverse
Bax	GCTGACATGTTTGCTGATGG	GATCAGCTCGGGCACTTTAG
Caspase-3	AGGGGTCATTTATGGGACA	TACACGGGATCTGTTTCTTTG
GAPDH	TCACCACCATGGAGAAGGC	GCTAAGCAGTTGGTGGTGCA

Histopathological Examination

Excised tumor specimens were fixed in 10% formalin for 24 h. Sections (5 μm thickness) were cut, stained with hematoxylin and eosin (H&E) stain for 5 min, dehydrated in alcohol, and mounted in Canada balsam prior to microscopical examination.

2.8.3. In Vivo Toxicity Study

The effect of NS formulations vs. FS suspension on general animal health during the treatment course was evaluated. Change in body weight was recorded on a weekly basis. Following sacrifice, the liver, kidney, and spleen were excised and fixed in 10% formalin for histopathological examination. Furthermore, liver and kidney functions were evaluated by measuring serum alanine aminotransferase (ALT), aspartate aminotransferase (AST), urea, and creatinine in comparison to healthy animals.

2.9. Statistical Analysis

All experiments were conducted in triplicate, and results were represented as mean ± SD. Statistical analyses were performed using an unpaired Student's *t*-test and one-way analysis of variance (ANOVA) followed by a post-hoc Tukey's test for multiple comparisons using GraphPad Prism (Version 7.04, San Diego, CA, USA). The level of significance was set at $p \leq 0.05$.

3. Results and Discussion

3.1. Preparation and Optimization of β-CD-NS

NS formulations were prepared using DPC as a cross-linker for β-CD hydroxyl groups with the formation of new carbonate bonds [30]. A molar ratio 1:6 of β-CD:DPC was selected based on previous reports showing a higher percentage yield compared to other ratios tested [10,30]. Optimization of the cross-linking reaction conditions was carried out by changing the temperature, time, and volume of the reaction vehicle (DMF) (Table 2). The successful formation of NS was primarily confirmed by the observation of a gel-like mass during the reaction (gelification) together with the appearance of a deep violet color upon adding ferric chloride [30]. As shown in Table 2, using a small volume of DMF (3 mL) resulted in an incomplete reaction at different reaction conditions (F1–F4). Increasing the DMF volume to 6 mL resulted in the successful formation of NS, except for F5 (90 °C, 2 h). Furthermore, increasing the reaction temperature to 120 and 150 °C allowed for NS formation in the shorter time interval tested (2 h, F7, and F9). Based on the percentage yield (49.85 ± 1.2%), 150 °C was selected as the optimum reaction temperature (F9).

Table 2. Optimization of cross-linking reaction conditions.

Formulation Code	DMF Volume (mL)	Reaction Temperature (°C)	Reaction Time (h)	Ferric Chloride Test	Gelification
F1	3	90	2	Negative	No
F2	3	90	5	Negative	No
F3	3	120	2	Negative	No
F4	3	120	5	Negative	No
F5	6	90	2	Negative	No
F6	6	90	5	Positive	Yes
F7	6	120	2	Positive	Yes
F8	6	120	5	Positive	Yes
F9	6	150	2	Positive	Yes
F10	6	150	5	Positive	Yes

FS loading (FS-NS) was achieved by simple overnight stirring of drug ethanolic solution and the aqueous nano-dispersion of blank NS (at a weight ratio of 1:4) [10].

3.2. Preparation of LF-Coated FS-NS

LF has been extensively studied as a natural tumor-targeting ligand owing to its ability to bind specific receptors overexpressed on the surface of cancerous cells, in addition to its biocompatibility and biodegradability [29]. In the current work, LF coating on NS was investigated for the first time. This was based on the electrostatic interaction between cationic LF and the negatively charged surface of FS-NS (-26 mV). The effect of the LF concentration added (25–100 mg/mL) on the ζ-potential of FS-NS was studied (Figure 1). Upon increasing the LF concentration from 0 to 75 mg/mL, ζ-potential shifted from -26 ± 6.5 to 24 ± 1.1 mV, indicating the deposition of LF on the surface of NS. Further increases in LF beyond 75 mg/mL only brought about a slight insignificant ($p \geq 0.05$) change in ζ-potential (25.1 ± 2 mV for 100 mg/mL LF), reflecting complete FS-NS surface coverage with an LF layer. Hence, 75 mg/mL was chosen as the optimum LF concentration for coating FS-NS.

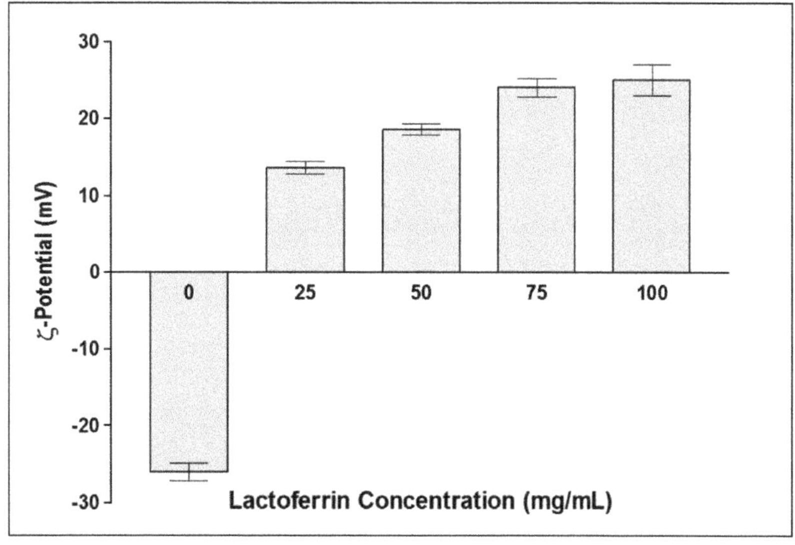

Figure 1. Effect of lactoferrin concentration on FS-NS ζ-potential (mV). Data presented as mean $\pm$ SD (n = 3).

3.3. Physicochemical Characterization

3.3.1. Colloidal Properties and Entrapment Efficiency

The colloidal properties of selected NS formulations are shown in Table 3. The optimized blank NS formulation (F9) chosen for further development showed an average PS, PDI, and ζ-potential of 46.1 ± 6.2 nm, 0.13, and −22 ± 0.8 mV, respectively. The negative charge could be attributed to the presence of free β-Cyclodextrin hydroxyl groups and DPC carbonyl groups [10,41]. FS loading into NS resulted in a slight decrease in PS (38.2 ± 3.8 nm) with an insignificant ($p \geq 0.05$) change in ζ-potential (−26 ± 6.5 mV). These results reflect efficient drug entrapment in the porous NS. LF-coated FS-NS showed a significant increase in PS ($p \leq 0.05$) compared to the uncoated formulation. The increase in size, together with the shift in ζ-potential from negative to positive, suggests successful coating and deposition of LF on the surface of NS [26]. The high ζ-potential value of the three NS formulations tested allows for sufficient particle repulsion, indicating good colloidal stability [41].

Table 3. Physicochemical properties of the selected nanosponge formulations (n = 3).

Formulation	Size (nm)	PDI	ζ-Potential (mV)	EE%	DL%
NS	46.1 ± 6.2	0.13	−22 ± 1.8	-	-
FS-NS	38.2 ± 3.8	0.19	−26 ± 6.5	96.1 ± 0.3	23.8 ± 0.2
LF-FS-NS	52.7 ± 7.2	0.09	24 ± 1.1	95.9 ± 0.2	-

A high entrapment efficiency of FS (>95%) that was unaffected by LF coating, and DL% ~24%, were calculated. (Table 2). High EE% of lipophilic drugs in ß-CD nanosponge formulations have been previously reported for curcumin [42] and quercetin [10]. The high EE% could be ascribed to the porous nature of NS, in which the drug is deposited [43].

3.3.2. Analysis of Surface Area and Porosity of NS

Figure 2 illustrates the nitrogen adsorption–desorption isotherm (Figure 2A) and the pore size distribution (Figure 2B) of the selected NS formulation (F9). A type IV isotherm with a hysteresis loop and an average pore diameter of ~26 nm were obtained, reflecting the mesoporous nature of NS [44]. The total surface area and total pore volume were 15.8 m^2/g and 0.1 cm^3/g, respectively. The large surface area and pore volume observed allow for efficient drug loading [31], subsequently supporting the high FS EE% observed.

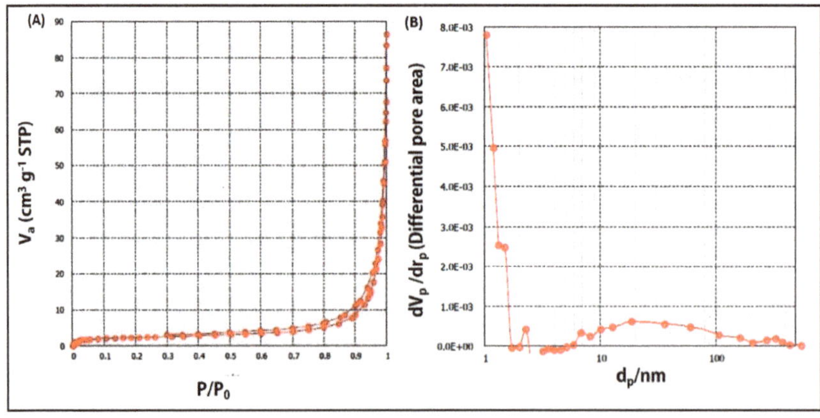

Figure 2. (**A**) Nitrogen adsorption–desorption isotherm; (**B**) pore size distribution of selected NS formulation (F9).

3.3.3. Microscopical Examination

SEM imaging showed a rough, porous surface (Figure 3a) with a pore diameter of ~30 nm, confirming the mesoporous structure of NS. TEM micrographs revealed spherical non-aggregated nanoparticles for blank NS (Figure 3b), which was not affected by FS loading except for the decrease in particle size (Figure 3c). A dense coat surrounding the nanosponges could be observed for LF-FS-NS (Figure 3d).

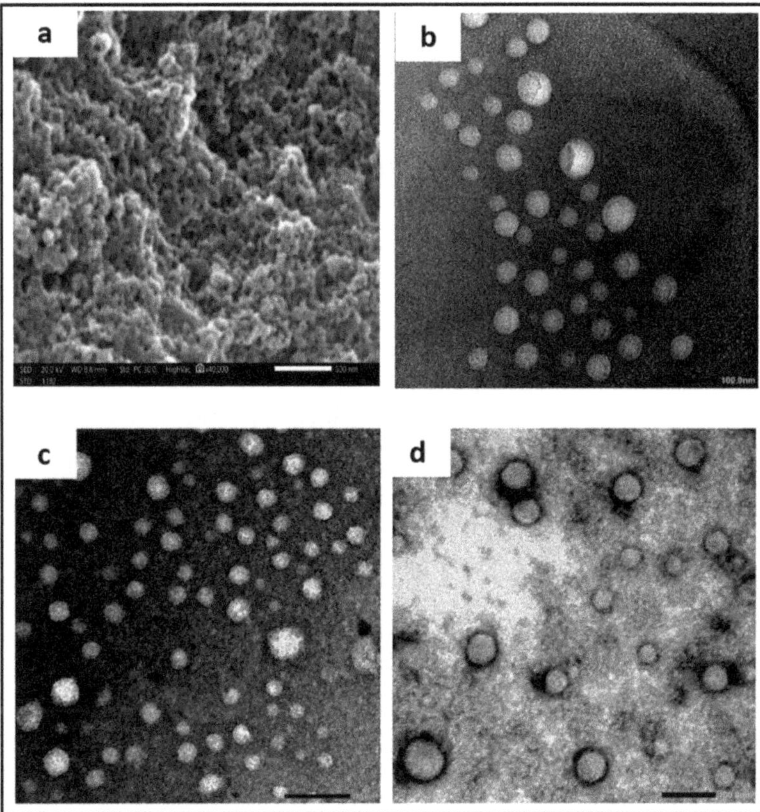

Figure 3. (**a**) SEM of NS with measurement of pore size. Magnification × 40 K, scale bar represents 500 nm. TEM images (**b–d**) showing the morphology of (**b**) NS, (**c**) FS-NS, and (**d**) LF-FS-NS. Magnification × 30 K, scale bar represents 100 nm.

3.3.4. In Vitro Drug Release

In vitro release profiles of FS solution, FS-NS, and LF-FS-NS are shown in Figure 4. FS solution exhibited complete release, reaching 100% after 3 h. On the other hand, FS release from FS-NS and LF-FS-NS was slow and steady, reaching only 35 and 26% after 24 h, respectively. In addition to the prolonged sustained release profile, the low initial burst observed (less than 10%) strongly confirms the presence of FS inside the pores of the nanosponge, as previously stated for other drugs loaded in ß-CD-NS [12,13]. The decrease in the slope of the drug release curve with time reflects the gradual increment of diffusion distance in the polymeric matrix. Moreover, the lower percentage of FS released for LF-FS-NS as compared to FS-NS could be attributed to the additional barrier created by the coating layer restricting release medium diffusion into the NS matrix [45].

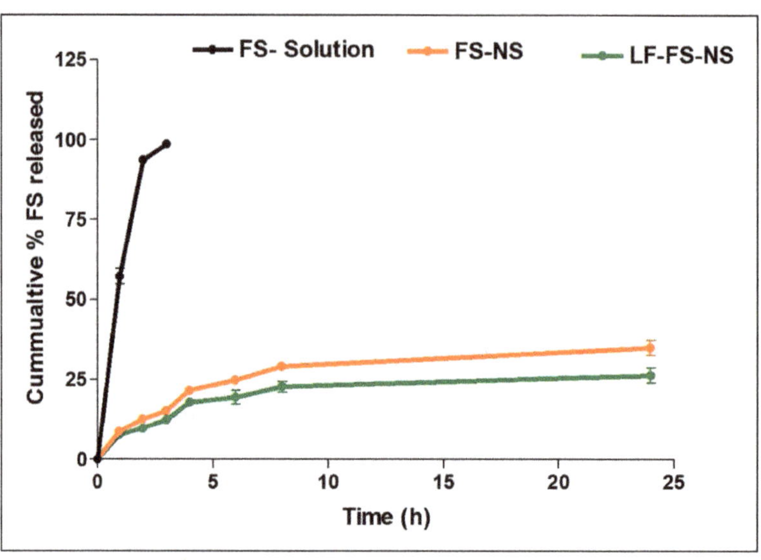

Figure 4. In vitro release profile of FS from FS solution, FS-NS, and LF-FS-NS at 37 °C in PBS (7.4) with 0.1% *w/v* Tween® 80. Data represent mean ± SD (n = 3).

The drug release mechanism of FS-NS and LF-FS-NS was determined by fitting to different release kinetics models, namely the zero-order, first-order, Higuchi, Korsmeyer–Peppas, and Hixson–Crowell models. To designate the data best fit, the largest correlation coefficient (r) and smallest mean standard error (MSE) were used as statistical parameters. The results indicated diffusion-controlled FS release from both FS-NS and LF-FS-NS as the greatest r value (0.94), and the minimum MSE value were observed for the Korsmeyer–Peppas model. A release exponent value (n) $\leq$ 0.5 was observed as being indicative of Fickian diffusion.

3.3.5. Fourier Transform Infrared Spectroscopy (FTIR)

FTIR spectra of β-CD, DPC, FS, blank NS, and FS-NS are shown in Figure 5A. β-CD displayed characteristic bands at 3288 for O–H stretching, 2921 for C-H stretching, 1416 for C-H bending, and 1047 cm^{-1} for C–O stretching. The most characteristic band in the FTIR spectrum of DPC is 1771 cm^{-1}, corresponding to C=O. In the NS spectrum, the appearance of a new sharp absorption band at 1759 cm^{-1} corresponding to carbonyl (C=O) stretching vibration, which was absent in the FTIR spectrum of pure β-CD, together with the reduction in O–H stretching vibration, indicates carbonate linkage formation with OH groups of β-CD, proving efficient cross-linking with DPC. This is consistent with previous reports [10,12].

The FS spectrum showed absorption bands at 3519 and 3351 cm^{-1} corresponding to aromatic ring attached O–H stretching, 1607 cm^{-1} for C=O stretching, 1571 cm^{-1} for C=C stretching, 1477 cm^{-1} for C–O stretching, and 1279 cm^{-1} for C–O–H bending vibrations [26].

The FTIR spectrum of FS-NS exhibited some changes in the fingerprint region (400–1400 cm^{-1}) compared to the FS spectrum. Peak broadening and shifting in the spectrum of FS-NS compared to blank NS suggest definite interactions between FS and β-CD NS, further confirming the effective internalization of the drug into the pores of the formed NS [10].

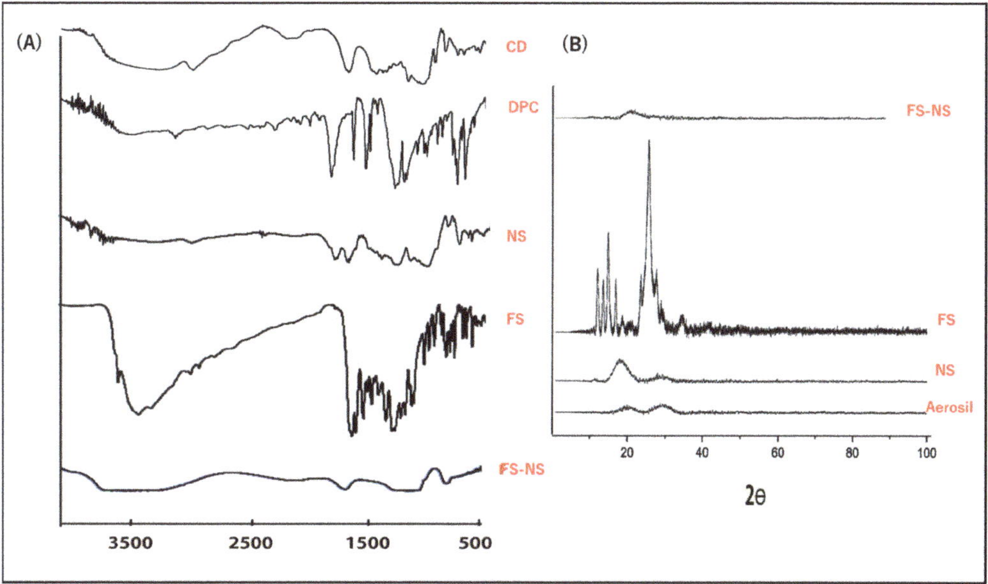

Figure 5. (**A**) FTIR spectra for FS-NS and its components; (**B**) X-ray diffraction pattern of FS, NS, and FS-NS.

3.3.6. X-ray Diffractometry (XRD)

To assess drug crystallinity in the developed FS-NS, nondestructive XRD was conducted. Figure 5B shows the XRD patterns of FS, NS, and FS-NS, as well as Aerosil® 200, which was used for drying the aqueous dispersions of tested NS. The Aerosil® 200 diffractogram showed its amorphous structure, as only a broad peak at 22.5° was observed [36]. The FS diffractogram revealed prominent sharp diffraction peaks at 2θ angles 12.5, 15.2, 17.1, 25.7, and 28.5°, reflecting FS's crystalline nature [46]. These characteristic sharp peaks of FS disappeared upon loading into FS-NS, indicating its presence in an amorphous state which could be attributed to the supramolecular complex formation of FS with NS [47].

3.4. Cell Line Studies
3.4.1. Cytotoxicity Evaluation

The antiproliferative effect of FS on TNBC has been previously reported [37,48,49]. In this study, the cytotoxicity of FS-NS and LF-FS-NS in comparison with FS solution was evaluated on MDA-MB-231 cells using MTT assay in the concentration range of 10–100 µg/mL. To compare the antitumor activity of FS solution and NS formulations, %viability and, consequently, IC$_{50}$ values were calculated. FS showed concentration-dependent cytotoxicity with an IC$_{50}$ value of 59.18 µg/mL. Interestingly, FS encapsulation into NS (FS-NS) resulted in a 1.3-fold decrease in IC$_{50}$ (45.06 µg/mL). The obtained results are in agreement with previous reports on the promotion of cytotoxicity of different biomolecules following loading into ß-CD-NS formulations [12,13]. Further coating of FS-NS with LF boosted the antiproliferative effect of FS with a reduction in IC$_{50}$ to 27.67 µg/mL (a 2.1- and 1.6-fold decrease compared to FS and FS-NS, respectively). Similar effects of LF modification on cytotoxicity enhancement were shown for PLGA nanoparticles [28], bilosomes [33], and nanostructured lipid carriers (NLC) [50]. This could be partly a result of the innate anticancer effect of LF [29]. Blank formulations (NS and LF-NS) were included in the study and showed percentage MDA-MB-231 cell viability exceeding 80% for all concentrations tested. This proves the cytocompatibility of the blank nano-formulations and provides evidence that the observed cytotoxicity of FS-loaded nanosponge formulations is due to the enhanced cellular interactions of FS following loading into NS [26].

3.4.2. In Vitro Apoptosis Assay

The apoptosis (programmed cell death)-mediated anticancer effect of FS and FS-loaded nanosponge formulations was investigated using flow cytometry of annexin V-stained apoptotic cells (Figure 6A). FS resulted in a significant ($p \leq 0.05$) apoptotic activity compared to the control group (23.9 ± 0.17% and 7.3 ± 0.28 %, respectively). Loading into NS formulation brought about a significant increase in FS apoptotic activity (29.9 ± 0.3%) which was further augmented ($p \leq 0.05$) following LF coating of FS-NS (36.6 ± 0.61%). The significant apoptotic effect of LF-FS-NS compared to the uncoated formulation could somewhat be attributed to the inhibitory effect of LF on plasmalemmal V-H+-ATPase [51]. The higher cellular apoptosis for FS-NS and LF-FS-NS formulations compared to the FS solution elucidates in part the observed enhancement of cytotoxicity.

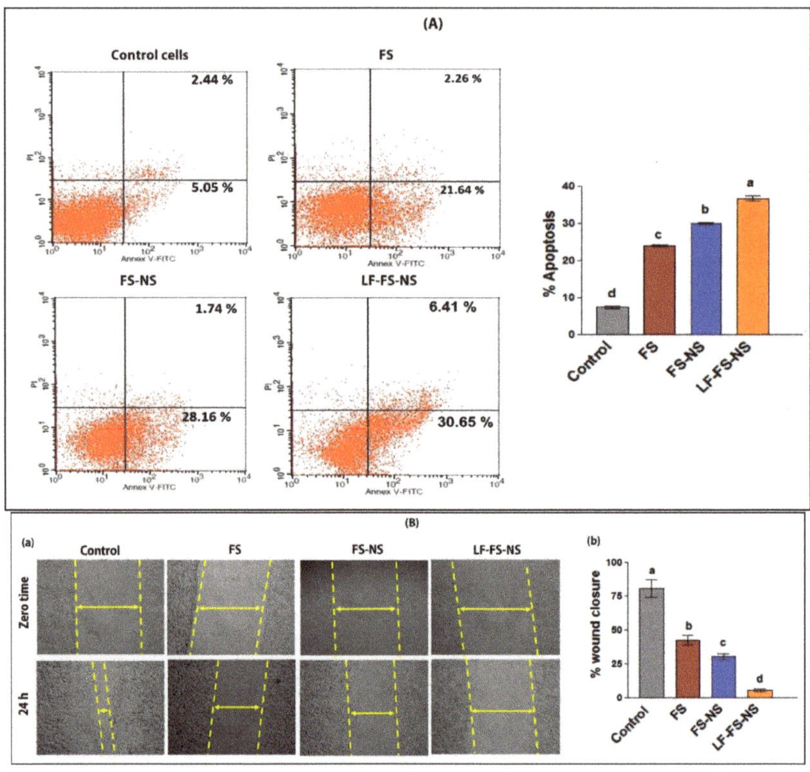

Figure 6. (**A**) Percentage apoptosis for FS, FS-NS, and LF-FS-NS by flow cytometry using annexin V FITC/propidium iodide assay after incubation for 24 h with MDA-MB-231 cells and (**B**) scratch wound assay: (a) migration inhibitory activity of free FS, FS-NS, and LF-FS-NS on MDA-MB-231 cells (magnification ×20) and (b) percentage wound closure. Data were expressed as means ± SD (n = 3). Data were analyzed using one-way ANOVA followed by Tukey's post-hoc test for group comparisons. Means of similar symbols were statistically insignificant: a > b > c > d ($p \leq 0.05$).

3.4.3. Cell Migration

FS has been previously shown to suppress migration and metastasis of TNBC through epithelial-to-mesenchymal transition reversal via the PTEN/Akt/GSK3β signaling pathway [20]. In the current study, the ability of FS-NS and LF-FS-NS vs. FS solution to inhibit MDA-MB-321 cell migration was studied by the wound healing assay (Figure 6B). Compared to control cells, FS significantly reduced wound closure by two-fold. This is in agreement with previous reports on the ability of FS to effectively reduce the migration of TNBC cells in the range of 20 to 76% [52]. Interestingly, FS-NS succeeded in further significant ($p \leq 0.05$) inhibition of wound closure (42.5 ± 3.5 and 30.1 ± 2.1% for FS solution and FS-NS, respectively). LF-FS-NS was the most suppressing formulation, achieving an approximately 15-, 8-, and 6-fold decrease in wound closure compared to the control, FS solution, and FS-NS, respectively.

3.4.4. Cellular Uptake

The extent of the internalization of C6-labeled NS and LF-NS by MDA-MB-231 cells was evaluated by confocal microscopy (Figure 7). Following 4 h exposure, the free dye was internalized by the cells; however, C6 loading in NS resulted in a 3- and 6-fold increase in cellular uptake for the uncoated and LF-coated formulations, respectively. The improvement in cellular uptake following FS loading into NS and LF-NS could partially explain the enhanced antiproliferative, apoptotic, and migration-inhibitory effects observed. Enhanced cellular uptake of the nano-formulations could be attributed to their small size (below 100 nm) which allows for endocytosis and not just simple diffusion as previously reported [33]. The superior cellular uptake observed for LF-C6-NS compared to C6-NS reflects faster and more efficient internalization of the formulation into breast cancer cells due to LF interaction with its target receptors overexpressed on metabolically active cancer cells [28]. Moreover, LF's positive charge allows for cell entry via electrostatic interaction with the negatively charged cell membrane glycosaminoglycans [53]. Surface charge was shown to not only affect cellular uptake rates but also intracellular trafficking [50].

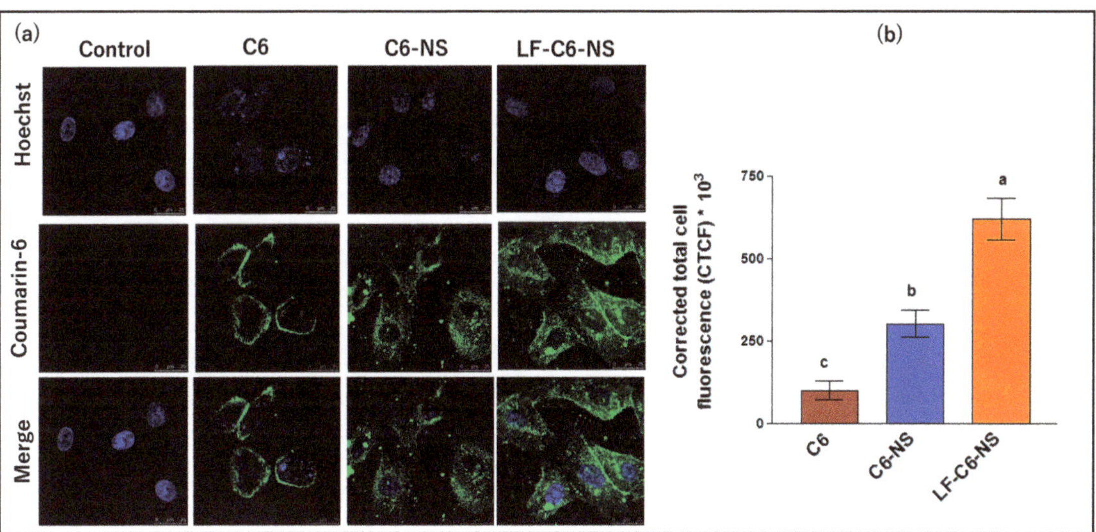

Figure 7. (**a**) Confocal laser scanning microscope images showing cellular uptake of free coumarin 6 solution and its NS formulations after 4 h incubation with MDA-MB-231 cells and (**b**) corrected total fluorescence intensity. Data were expressed as means ± SD (n = 3). Data were analyzed using one-way ANOVA followed by Tukey's post-hoc test for group comparisons. Means of similar symbols were statistically insignificant: a > b > c ($p \leq 0.05$).

3.5. In Vivo Studies

3.5.1. Pharmacokinetic Study

The bioavailability of FS administered to rats either orally or by IP injection was studied for free FS, FS-NS, and LF-FS-NS (Table 4, Figure 8). The drug plasma concentration–time profiles after oral administration of a single FS dose (30 mg/kg) either free or loaded into NS formulations are demonstrated in Figure 8a, and the calculated pharmacokinetic parameters are listed in Table 4. Results revealed marked changes in the pharmacokinetic behavior of FS after loading into both uncoated and LF-coated NS. A significant decrease in T_{max} was observed for LF-FS-NS (0.25 h) compared to FS suspension and FS-NS (1 h) indicating a faster rate of absorption. This might be attributed to the positive charge on LF-FS-NS, which allows for favorable distribution in the small intestine and uptake via multiple endocytosis pathways compared to negatively charged nanoparticles [54]. Furthermore, positively charged nanoparticles were shown to electrokinetically interact with mucus, thus, opening epithelial cells' tight junctions and promoting absorption via the paracellular pathway [55].

Table 4. Pharmacokinetic parameters of FS after oral and IP administration of a single dose (30 mg/kg) of FS suspension and NS formulations in rats.

Parameter	Oral			Intraperitoneal		
	FS Suspension	FS-NS	LF-FS-NS	FS Suspension	FS-NS	LF-FS-NS
$t_{1/2}$ (h)	7.2 [c] ± 1.5	15.4 [a] ± 3.5	11.2 [b] ± 2.1	8.5 [c] ± 2.1	17.7 [a] ± 2.5	12.2 [b] ± 1.2
T_{max} (h)	1	1	0.25	0.25	1	0.25
C_{max} (ng/mL)	24.3 [c] ± 4.6	41.3 [b] ± 7.4	56.2 [a] ± 5.9	509.8 [c] ± 31.6	1622.4 [a] ± 242.9	862.9 [b] ± 60.1
AUC_{0-24} (ng/mL × h)	135.1 [b] ± 4.2	312.1 [a] ± 116.8	274.1 [a] ± 18.5	2538.8 [b] ± 121.6	8474.5 [a] ± 1834.9	6160.1 [a] ± 121.9
$AUC_{0-\infty}$ (ng/mL x h)	149.5 [b] ± 2.7	457.6 [a] ± 251.5	368.7 [a] ± 28.5	2874.6 [c] ± 254.1	12,371.5 [a] ± 2143.5	9126.1 [b] ± 565.1
* Relative bioavailability		3.1	2.5		4.3	3.2

The study was conducted on female Wistar rats with six animals in each group. Values were expressed as mean ± SD. Data were analyzed using one-way (ANOVA) followed by post-hoc test (Tukey's). The level of significance was set at $p \leq 0.05$. Means of similar symbols are statistically insignificant, a > b > c. * Relative bioavailability was calculated by dividing $AUC_{0-\infty}$ of different NS formulations to that of FS suspension.

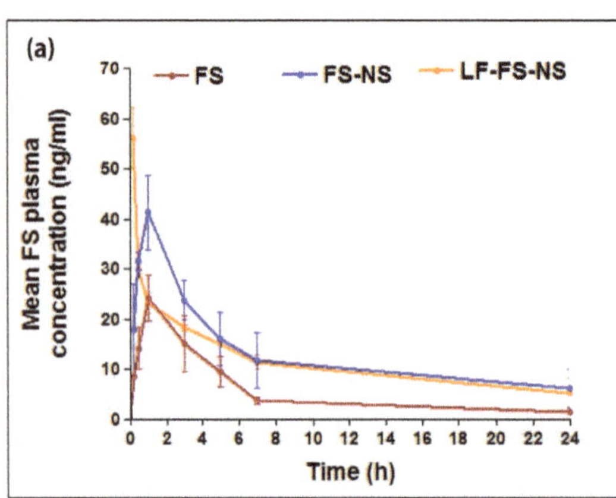

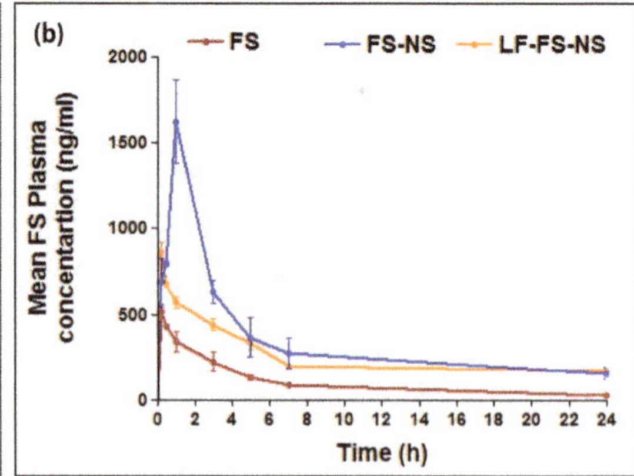

Figure 8. Mean FS plasma concentration–time profiles of FS solution and NS formulations following (a) oral and (b) intraperitoneal administration of a single dose (35 mg/kg) to rats. Data represent mean ± SD (n = 6).

Whereas FS loading into uncoated NS did not affect T_{max}, a significant increase in C_{max} by 1.7-fold (41.3 ± 7.4 ng/mL) compared to FS suspension (24.3 ± 4.6 ng/mL) ($p \leq 0.001$) was observed. Moreover, LF-FS-NS resulted in a 2.3-fold increase in C_{max} (56.2 ± 5.9 ng/mL) compared to FS suspension, which was also significantly higher than the uncoated formulation, ($p \leq 0.01$). Again, both NS formulations achieved a significant increase in $AUC_{0-\infty}$, with 3.1- and 2.5-fold for the uncoated and LF-coated NS, respectively, compared to drug suspension ($p \leq 0.05$), with a significant increase in $t_{1/2}$ ($p \leq 0.05$). It is noteworthy that LF-FS-NS exhibited a significantly lower $t_{1/2}$ compared to FS-NS ($p \leq 0.05$), owing to its cationic nature, which could facilitate serum protein aggregation, opsonization, and, thus, systemic clearance by macrophages [56]. Despite their difference in $t_{1/2}$, the differences between AUC for FS-NS and LF-FS-NS did not reach statistical significance.

The pharmacokinetic behavior of FS-NS and LF-FS-NS vs. FS suspension was also investigated following IP administration. Figure 8b shows the drug plasma concentration–time profiles and Table 4 demonstrates the main pharmacokinetic parameters. Following IP administration, an obvious increase in the amount and extent of FS reaching the circulation and, hence, bioavailability, was observed for NS formulations compared to FS suspension, as indicated by the significantly higher C_{max} (3-fold for FS-NS and 1.7-fold for LF-FS-NS) and $AUC_{0-\infty}$ (4.3-fold for FS-NS and 3.2-fold for LF-FS-NS) ($p \leq 0.05$). Furthermore, FS half-life was significantly prolonged following its loading into either coated or uncoated NS formulations, ($p \leq 0.05$). Nevertheless, $t_{1/2}$ for FS-NS was significantly higher than that of LF-FS-NS ($p \leq 0.001$) in a similar manner to that observed following its oral administration, suggesting a possible systemic clearance by macrophages being positively charged [56].

The improved drug bioavailability following administration of FS-NS formulations via either oral or IP routes compared to FS suspension matches previous reports on the potential of NS to enhance erlotinib bioavailability as a result of supramolecular complex formation between the drug and porous NS [57]. Drug inclusion within the nanocavities reduced its particle size and, accordingly, increased solubility and dissolution rate, hence, facilitating absorption [57]. Furthermore, being deeply incorporated in the nanopores could possibly allow for the avoidance of drug pre-systemic intestinal and first-pass hepatic metabolism [57].

It is worth mentioning that FS bioavailability was significantly higher via the IP route compared to the oral one when administered either as a suspension or loaded into NS formulations. This could be attributed to by-passing the drug pre-systemic intestinal metabolism following IP injection [58]. Furthermore, IP administration allows for a higher and faster drug absorption rate due to the rapid uptake of drug from the peritoneal cavity, resulting in a more rapid saturation of the drug-metabolizing enzymes than following oral administration. Consequently, higher concentrations of the unmetabolized drug will be available in systemic circulation, resulting in higher drug bioavailability [59].

3.5.2. In Vivo Evaluation of Anticancer Potential

The anticancer efficacy of different NS formulations was assessed on female mice bearing Ehrlich ascites tumors, a well-established model of spontaneous murine mammary adenocarcinoma [60]. Efficacy assessment started when tumor size reached ~100 mm^3. Treatment was administered by IP injection over 14 days.

Tumor Growth Inhibition

Figure 9A(a) demonstrates the percentage increase in tumor size for different study groups during the study period. Representative photographs of excised tumors following sacrifice are shown in Figure 9A(b). The positive control mice exhibited a significant increase in tumor size throughout the study period, with ~600% after 16 days ($p \leq 0.01$). Tumor growth inhibitory effect was achieved following treatment with FS suspension, FS-NS, and LF-FS-NS, with a percentage increase in tumor size of 209.4, 128.9, and 116.7 %, respectively, after 16 days, which is consistent with the excised tumor images (Figure 9A(b)).

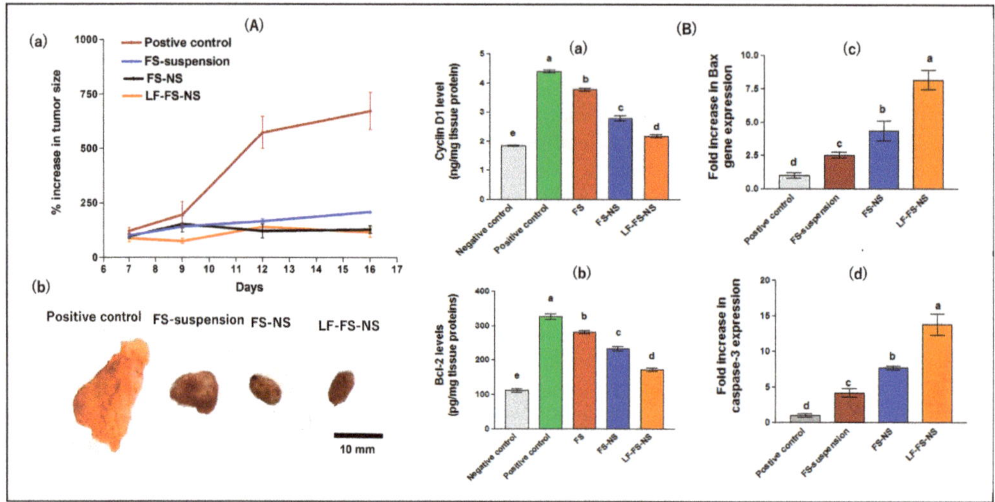

Figure 9. (**A**) Antitumor effect of FS and NS formulations in comparison with untreated control 16 days post-IP treatment of Ehrlich ascites mammary tumor in mice showing: (**a**) percentage change in tumor size relative to pretreatment volume and (**b**) digital images of excised tumors. (**B**) Tumor biomarker levels following 16-day treatment with FS and FS-NS formulations compared to untreated control: (**a**) cyclin D1, (**b**) Bcl-2, (**c**) Bax, and (**d**) caspase-3. Data were expressed as means ± SD (n = 3). Data were analyzed using one-way ANOVA followed by post-hoc Tukey's test for group comparisons. Means of similar symbols were statistically insignificant: a > b > c > d > e ($p \leq 0.05$).

Assessment of Tumor Biomarkers

Cyclins are a family of cell proteins controlling cell cycle progression and promoting tumor proliferation with cyclin D1 (CD1) being a cell-cycle regulator essential for G1-phase progression. It is overexpressed in more than 50% of breast tumors [61]. In the current study, the positive control group demonstrated a significant overexpression in CD1 level by 2.4-fold compared to the negative control group ($p \leq 0.001$) (Figure 9B(a)). Conversely, all treated groups showed a significant reduction in CD1 level compared to the positive control ($p \leq 0.001$) in the following order: FS suspension < FS-NS < LF-FS-NS. The FS effect on lowering CD1 transcription in breast cancer is consistent with previous reports [26]. This effect was significantly augmented for FS-NS, which could be explained by the FS bioavailability enhancing effect of NS and the improved cell penetration via the EPR effect [62,63]. The LF-FS-NS formulation achieved the lowest expression of CD1, confirming its superior cytotoxic potential.

Intrinsic apoptosis, or programmed cell death, includes initial mitochondrial perturbation arising from cytotoxicity and is mainly regulated by the Bcl-2 protein family [64]. The latter is subdivided into pro-apoptotic proteins, such as Bax, and anti-apoptotic ones, such as Bcl-2. The affinities and relative abundance of different Bcl-2 proteins control whether anti-apoptotic or pro-apoptotic reactions predominate [64]. Overexpression of Bcl-2, present mainly on the outer membrane of the mitochondria, protects against apoptosis induced by many cytotoxic agents [65]. On the contrary, the increased expression of the pro-apoptotic proteins, such as Bax, will stimulate the release of mitochondrial cytochrome C into the cytoplasm and, subsequently, the activation of caspase-3 [65]. Caspase-3, a key mediator of apoptosis, is a frequently activated death protease that catalyzes the specific cleavage of several vital cellular proteins [66]. In this context, the levels of Bcl-2, Bax, and caspase-3 biomarkers were quantitively determined in tumor tissues to investigate the possible molecular pathway for the anticancer activity of FS. As shown in Figure 9B(b,c), the positive control group showed a significant up-regulation in the levels of the anti-apoptotic protein Bcl-2 by ~3-fold compared to the negative control group ($p \leq 0.001$). Contrarily, FS suspension significantly decreased the levels of Bcl-2 by 13.5% and increased the expression of the Bax gene by ~2.5-fold compared to the positive control group ($p \leq 0.05$), which is in agreement with previously reported data [22,49]. Furthermore, FS loading into NS resulted in a significant down-regulation in Bcl-2 levels by 17.6% with an up-regulation in Bax gene expression by 1.7-fold compared to FS suspension ($p \leq 0.05$), elucidating the role of supramolecular complex formation between FS and NS in enhancing FS anticancer activity. Interestingly, LF-FS-NS significantly achieved a higher reduction in Bcl-2 levels by 26.2% and an increase in Bax gene expression (1.9-fold) vs. uncoated FS-NS ($p \leq 0.05$).

To further confirm the apoptotic capability of different treatment groups, analysis of the caspase-3 gene was carried out. As previously reported, FS-induced apoptosis acts through mitochondrial- and caspase-3-dependent pathways [67]. As shown in Figure 9B(d), all treatments significantly increased the expression of the caspase-3 gene with variable degrees vs. the positive control group ($p \leq 0.05$). This could be due to the proven ability of FS to up-regulate Bax gene expression and, subsequently, the induction of apoptosis via caspase-3 activation [67]. Again, the LF-coated formulation exhibited the highest elevation in caspase-3 expression, reflecting superior efficacy. This could be explained by its ability to actively target LF receptors overexpressed on breast cancer cells [28], thus, allowing for efficient FS cellular uptake and internalization, as demonstrated by the cellular uptake study (Section 3.4.4). Moreover, on the subcellular level, LF can increase drug nuclear localization, hence, achieving optimum efficacy, as the nucleus is the main site of action for most anticancer drugs [53]. This highlights the potential of NS surface modification with LF to improve the anticancer activity of FS via an active targeting mechanism. Indeed, LF-targeted formulations have previously shown higher anticancer activity compared to both free drug and untargeted nanotherapy [33,56].

Histopathological Evaluation

Findings of the tumor growth inhibition study and tumor biomarkers were additionally verified via histopathological examination of excised tumor tissue (Figure 10). Examined sections from normal control groups (Figure 10a,b) showed normal breast tissue architecture. Glands were organized into lobules of complex branching alveolar glands with extensive interlobular connective tissue and fat between them. The stromal compartment predominated and is packed with adipocytes, which offered insulation and aided in the protection of the fragile mammary gland tissue. The ductal lobular system's cellular lining was bilayered and composed of inner (luminal) epithelial cells, which were cuboidal to columnar in shape and had a pale eosinophilic cytoplasm. Outer (basal) myoepithelium cells were variable in form, ranging from flattened cells with compressed nuclei to prominent epithelioid cells with a copious transparent cytoplasm, and could occasionally have a myoid appearance.

Conversely, the typical lobular and ductal architecture was lost in the positive control group (Figure 10c,d), which revealed the highest intraductal proliferation in the gland, as evidenced by the formation of irregular dark proliferation sites with aberrant nuclei and a mild appearance of lymphoid tissue. Moreover, there was little evidence of fibrous interductal stroma, indicating medullary cancer. Furthermore, myoepithelium loss was seen as an indication of invasion.

Treatment with FS suspension influenced the histopathological characteristics by the formation of some intralobular connective tissue but still condensed in a non-fibrous aspect with obvious proliferating ductal glands showing lumen degradation, reflecting minor efficacy (Figure 10e,f). On the other hand, treatment with different FS-loaded NS formulations presented variable effectiveness. In this respect, FS-NS accomplished lower efficacy, as indicated by the decrease in the number of proliferative cells, but aberrant stroma and many lymphocytes were still seen (Figure 10g,h). Interestingly, LF-FS-NS demonstrated the greatest anticancer potential, as indicated by the absence of any proliferative cells, but still showed disorganized cells (Figure 10i,j) owing to the short treatment time. Some adipose tissue started to be retrieved. Nearly normal breast tissue architecture was seen. Additionally, the interlobular stroma appeared normal with many lymphocytes.

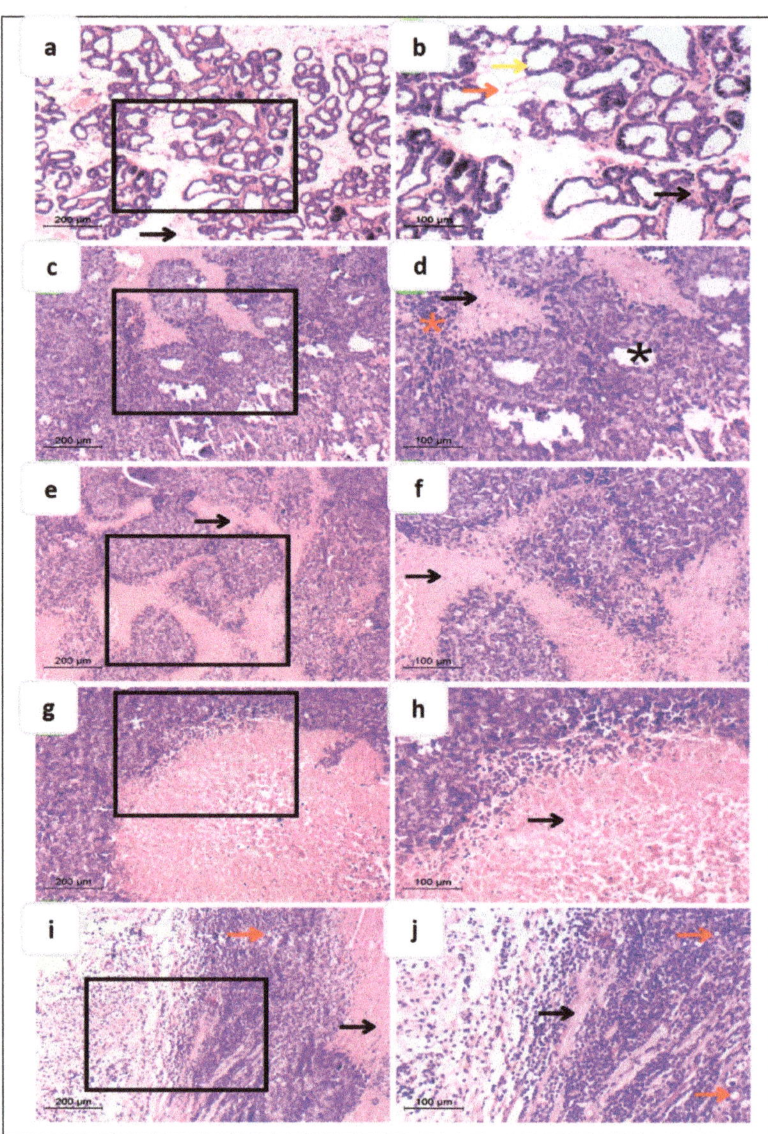

Figure 10. Photomicrographs illustrating H&E-stained breast tissue specimens of (**a**,**b**) normal breast and (**c**,**d**) positive control group and groups treated with (**e**,**f**) FS suspension, (**g**,**h**) FS-NS, and (**i**,**j**) LF-FS-NS. Yellow arrow: ductal system. Red arrow: adipose connective tissue. Black arrow: connective tissue stroma. Black star: degeneration. Red star: lymphocytes. The right panel represents a higher magnification of the area selected in the left panel. Magnification ×100, scale bar 200 μm (low)- ×200, scale bar 100 μm (high).

To sum up, combined results of biochemical and histopathological evaluation verified the superiority of LF-FS-NS in the treatment of breast cancer compared to both FS suspension and uncoated formulation, suggesting successful tumor targeting and drug accumulation in cancer cells.

3.5.3. In Vivo Toxicity

All mice treated with FS test formulations survived and appeared healthy throughout the study. An insignificant change in body weight ($p \geq 0.05$) was observed. Evaluation of liver and kidney functions was performed, and the results are described in Table 5. Liver enzymes, namely ALT and AST, were assessed as indicators for proper liver performance, while serum levels of both urea and creatinine were measured as indicators of renal function. Results demonstrated insignificant changes in the measured parameters for all tested groups compared to normal healthy control mice ($p \geq 0.05$), suggesting the safety of the developed formulations. Our results were in accordance with many reported articles describing the safety and biocompatibility of ß-CD-NS [68,69].

Table 5. Effect of free FS suspension and NS formulations on liver and kidney function tests.

Group	ALT (U/mL)	AST (U/mL)	Urea (mg/dL)	Creatinine (mg/dL)
Negative control	85.5 ± 2.9	60.1 ± 5.5	43 ± 2.4	0.94 ± 0.1
Positive control	92.1 ± 5.1	60.5 ± 2.8	48 ± 3.8	1 ± 0.3
FS suspension	91.5 ± 4.3	61.6 ± 5.2	44 ± 3.5	1.07 ± 0.2
FS-NS	90.3 ± 6.2	62.2 ± 3.6	49.2 ± 4.6	1.05 ± 0.2
LF-FS-NS	91.8 ± 3.4	61.2 ± 6.1	47 ± 2.2	1.2 ± 0.2

The study was conducted on female albino mice with six animals in each group (n = 6). Values were expressed as mean ± SD. Data were analyzed using one-way ANOVA followed by a post-hoc test (Tukey's) for group comparisons.

To further assess the in vivo toxicity, histopathological examination of the most affected organs (liver, kidney, and spleen) was performed at the end of treatment (Figure 11A–C). The spleen sections (Figure 11A) isolated from different study groups showed almost normal architecture, with a well-delineated white and red pulp with continuous trabecular throughout the tissues encountered by a capsule. The white pulp included lymphoid follicles as well. Nonetheless, the FS suspension group showed some dilatation and the development of bleeding (Figure 11A(c)).

Liver sections for FS-loaded NS groups demonstrated typical hepatocyte appearance with normal cells in the center having polyhedral shape, vacuolated acidophilic cytoplasm, and rounded vesicular nuclei (Figure 11B(d,e)). However, examined specimens from the FS suspension group exhibited mild inflammation near the central vein (Figure 11B(c)).

Despite the extensive reports on the nephroprotective role of FS [70], the renal tissue was the most affected by FS suspension (Figure 11C(c)). Since this was only obvious based on histopathological but not biochemical analysis, the slight changes observed could be attributed to different factors, including mice physiological state and other effects on nutrients uptake, among others [68]. Interestingly, renal tissues showed no symptoms of toxicity for FS-NS formulations (Figure 11C(d,e)). Normal appearance of (Malpighian) corpuscles was evident, which were comprised of glomerular capillaries and Bowman's capsules with subcapsular space. Many proximal convoluted tubules were lined with simple truncated cubical (pyramidal) cells with basal spherical nuclei and had narrow lumina. The lumina in distal convoluted tubules were large.

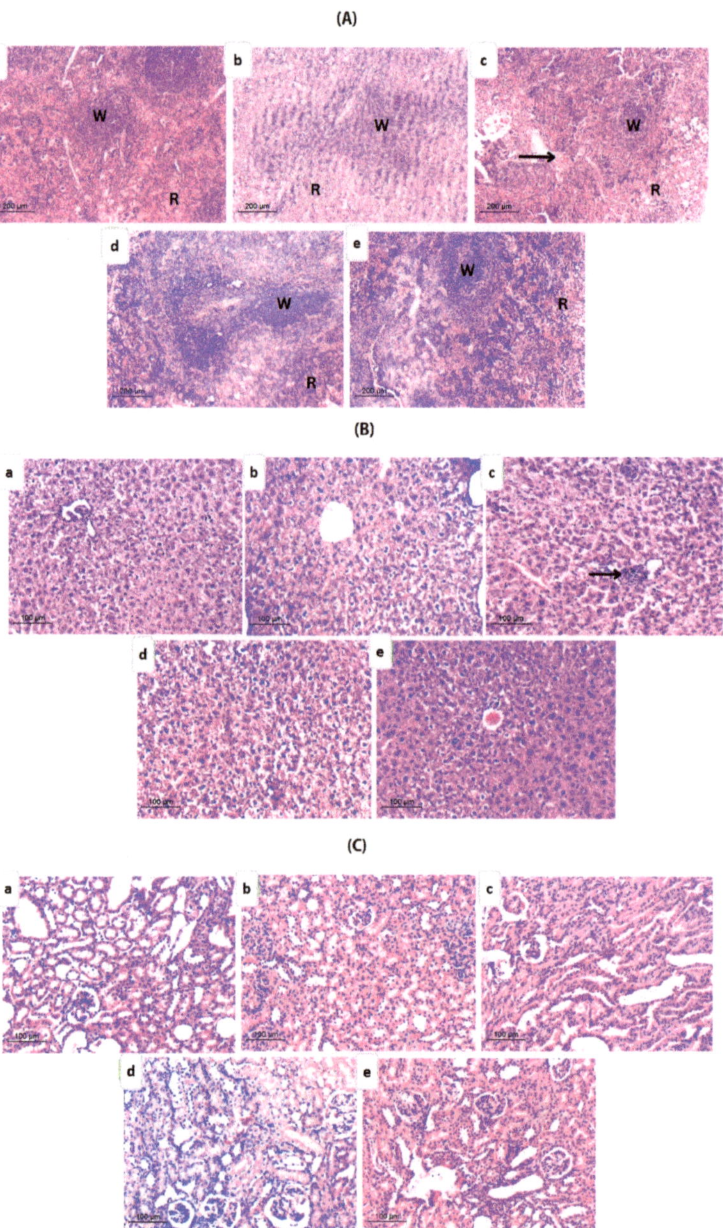

Figure 11. Photomicrographs illustrating H&E-stained (**A**) spleen tissue specimens of (**a**) normal group, (**b**) positive control group, (**c**) FS suspension-treated group, (**d**) FS-NS-treated group, and (**e**) LF-FS-NS-treated group. Black arrow: hemorrhage. W: white pulp. R: red Pulp. (**B**) Liver tissue specimens of (**a**) normal group, (**b**) positive control group, (**c**) FS suspension-treated group, (**d**) FS-NS-treated group, and (**e**) LF-FS-NS-treated group. Black arrow: inflammation. (**C**) Kidney tissue specimens of (**a**) normal group, (**b**) positive control group, (**c**) FS suspension-treated group, (**d**) FS-NS-treated group, and (**e**) LF-FS-NS-treated group. Magnification ×100 spleen-×200 liver and kidney.

4. Conclusions

In the current study, ß-CD nanosponges cross-linked with diphenyl carbonate were selected for improving the bioavailability and anticancer potential of fisetin. For the first time, to the best of our knowledge, FS-NS was coated with the active targeting ligand LF (LF-FS-NS). The optimized formulations of FS-NS and LF-FS-NS were shown to enhance FS bioavailability both orally and via IP injection. NS formulations administered by IP significantly enhanced the efficacy of FS against breast cancer. This was verified at both the cellular level on MDA-MB-231 cells and the molecular level in an Ehrlich ascites tumor compared to FS suspension, with superior effects achieved by the actively targeted formulation (LF-FS-NS). Accordingly, LF-FS-NS offers a highly adaptable strategy for bioactive targeted nanotherapy for the delivery of FS and possibly other phytomedicines to treat breast cancer.

Author Contributions: Conceptualization, R.M.E.-M., A.H.E.-K. and A.A.A.; methodology, A.R.A., R.A.M., A.H.E.-K. and A.A.A.; software, A.R.A. and A.A.A.; validation, R.M.E.-M., A.H.E.-K. and A.A.A.; formal analysis, A.R.A., R.M.E.-M., R.A.M., B.A.B. and A.A.A.; investigation, A.R.A., R.A.M., A.H.E.-K. and B.A.B.; resources, A.R.A. and R.M.E.-M.; data curation, A.R.A., R.M.E.-M and A.A.A.; writing original draft, A.R.A., R.M.E.-M., B.A.B. and A.A.A.; writing—review and editing, R.M.E.-M., A.H.E.-K., R.A.M., B.A.B. and A.A.A.; visualization, R.M.E.-M. and A.A.A.; supervision, R.M.E.-M., A.H.E.-K. and A.A.A.; project administration, A.H.E.-K.; funding acquisition, A.R.A. All authors have read and agreed to the published version of the manuscript.

Funding: This research received no external funding.

Institutional Review Board Statement: Animal studies were conducted according to the ethical guidelines and approved by the Institutional Ethics Committee of the Faculty of Pharmacy, Alexandria University, Alexandria, Egypt (protocol code AU-062023-1-141, and date of approval 24/1/2023). The MDA-MB-231 cell line was obtained from ATCC (HTB-26™) and cultured in CERRMA (Center of Excellence for Research in Regenerative Medicine and its Applications), Faculty of Medicine, Alexandria University.

Informed Consent Statement: Not applicable.

Data Availability Statement: The data presented in this study are available in the article.

Conflicts of Interest: The authors declare no conflict of interest.

References

1. Davankov, V.A.; Ilyin, M.M.; Tsyurupa, M.P.; Timofeeva, G.I.; Dubrovina, L.V. From a Dissolved Polystyrene Coil to an Intramolecularly-Hyper-Cross-Linked "Nanosponge". *Macromolecules* **1996**, *29*, 8398–8403. [CrossRef]
2. Anwer, M.K.; Fatima, F.; Ahmed, M.M.; Aldawsari, M.F.; Alali, A.S.; Kalam, M.A.; Alshamsan, A.; Alkholief, M.; Malik, A.; Az, A.; et al. Abemaciclib-loaded ethylcellulose based nanosponges for sustained cytotoxicity against MCF-7 and MDA-MB-231 human breast cancer cells lines. *Saudi Pharm. J.* **2022**, *30*, 726–734. [CrossRef] [PubMed]
3. Utzeri, G.; Matias, P.M.C.; Murtinho, D.; Valente, A.J.M. Cyclodextrin-Based Nanosponges: Overview and Opportunities. *Front. Chem.* **2022**, *10*, 859406. [CrossRef] [PubMed]
4. Sherje, A.P.; Dravyakar, B.R.; Kadam, D.; Jadhav, M. Cyclodextrin-based nanosponges: A critical review. *Carbohydr. Polym.* **2017**, *173*, 37–49. [CrossRef]
5. Pawar, S.; Shende, P.; Trotta, F. Diversity of beta-cyclodextrin-based nanosponges for transformation of actives. *Int. J. Pharm.* **2019**, *565*, 333–350. [CrossRef] [PubMed]
6. Gabr, M.M.; Mortada, S.M.; Sallam, M.A. Carboxylate cross-linked cyclodextrin: A nanoporous scaffold for enhancement of rosuvastatin oral bioavailability. *Eur. J. Pharm. Sci.* **2018**, *111*, 1–12. [CrossRef]
7. Crini, G.; Fourmentin, S.; Fenyvesi, É.; Torri, G.; Fourmentin, M.; Morin-Crini, N. Cyclodextrins, from molecules to applications. *Environ. Chem. Lett.* **2018**, *16*, 1361–1375. [CrossRef]
8. Asela, I.; Donoso-Gonzalez, O.; Yutronic, N.; Sierpe, R. beta-Cyclodextrin-Based Nanosponges Functionalized with Drugs and Gold Nanoparticles. *Pharmaceutics* **2021**, *13*, 513. [CrossRef]
9. Iravani, S.; Varma, R.S. Nanosponges for Drug Delivery and Cancer Therapy: Recent Advances. *Nanomaterials* **2022**, *12*, 2440. [CrossRef]
10. Abou Taleb, S.; Moatasim, Y.; GabAllah, M.; Asfour, M.H. Quercitrin loaded cyclodextrin based nanosponge as a promising approach for management of lung cancer and COVID-19. *J. Drug Deliv. Sci. Technol.* **2022**, *77*, 103921. [CrossRef]

11. Kapil, G.; Sankha, B.; Bhupendra, P. Recent Pharmaceutical Developments in the Treatment of Cancer Using Nanosponges. In *Advanced Drug Delivery Systems*; Chapter 4; Bhupendra, P., Ed.; IntechOpen: Rijeka, Croatia, 2022.

12. Anwer, M.K.; Ahmed, M.M.; Aldawsari, M.F.; Iqbal, M.; Kumar, V. Preparation and Evaluation of Diosmin-Loaded Diphenylcarbonate-Cross-Linked Cyclodextrin Nanosponges for Breast Cancer Therapy. *Pharmaceuticals* **2022**, *16*, 19. [CrossRef] [PubMed]

13. Khazaei Monfared, Y.; Mahmoudian, M.; Cecone, C.; Caldera, F.; Zakeri-Milani, P.; Matencio, A.; Trotta, F. Stabilization and Anticancer Enhancing Activity of the Peptide Nisin by Cyclodextrin-Based Nanosponges against Colon and Breast Cancer Cells. *Polymers* **2022**, *14*, 594. [CrossRef]

14. Argenziano, M.; Gigliotti, C.L.; Clemente, N.; Boggio, E.; Ferrara, B.; Trotta, F.; Pizzimenti, S.; Barrera, G.; Boldorini, R.; Bessone, F.; et al. Improvement in the Anti-Tumor Efficacy of Doxorubicin Nanosponges in In Vitro and in Mice Bearing Breast Tumor Models. *Cancers* **2020**, *12*, 162. [CrossRef] [PubMed]

15. Arnold, M.; Morgan, E.; Rumgay, H.; Mafra, A.; Singh, D.; Laversanne, M.; Vignat, J.; Gralow, J.R.; Cardoso, F.; Siesling, S.; et al. Current and future burden of breast cancer: Global statistics for 2020 and 2040. *Breast* **2022**, *66*, 15–23. [CrossRef]

16. Benoni, G.; Cuzzolin, L. Safety and Efficacy of Phytomedicines in Cancer Prevention and Treatment. In *Herbal Drugs: Ethnomedicine to Modern Medicine*; Ramawat, K.G., Ed.; Springer: Berlin/Heidelberg, Germany, 2009; pp. 207–220.

17. Vachetta, V.S.; Marder, M.; Troncoso, M.F.; Elola, M.T. Opportunities, obstacles and current challenges of flavonoids for luminal and triple-negative breast cancer therapy. *Eur. J. Med. Chem. Rep.* **2022**, *6*, 100077. [CrossRef]

18. Hazafa, A.; Rehman, K.U.; Jahan, N.; Jabeen, Z. The Role of Polyphenol (Flavonoids) Compounds in the Treatment of Cancer Cells. *Nutr. Cancer* **2020**, *72*, 386–397. [CrossRef]

19. Kubina, R.; Krzykawski, K.; Kabala-Dzik, A.; Wojtyczka, R.D.; Chodurek, E.; Dziedzic, A. Fisetin, a Potent Anticancer Flavonol Exhibiting Cytotoxic Activity against Neoplastic Malignant Cells and Cancerous Conditions: A Scoping, Comprehensive Review. *Nutrients* **2022**, *14*, 2604. [CrossRef]

20. Li, J.; Gong, X.; Jiang, R.; Lin, D.; Zhou, T.; Zhang, A.; Li, H.; Zhang, X.; Wan, J.; Kuang, G.; et al. Fisetin Inhibited Growth and Metastasis of Triple-Negative Breast Cancer by Reversing Epithelial-to-Mesenchymal Transition via PTEN/Akt/GSK3β Signal Pathway. *Front. Pharmacol.* **2018**, *9*, 772. [CrossRef]

21. Khozooei, S.; Lettau, K.; Barletta, F.; Jost, T.; Rebholz, S.; Veerappan, S.; Franz-Wachtel, M.; Macek, B.; Iliakis, G.; Distel, L.V.; et al. Fisetin induces DNA double-strand break and interferes with the repair of radiation-induced damage to radiosensitize triple negative breast cancer cells. *J. Exp. Clin. Cancer Res.* **2022**, *41*, 256. [CrossRef]

22. Imran, M.; Saeed, F.; Gilani, S.A.; Shariati, M.A.; Imran, A.; Afzaal, M.; Atif, M.; Tufail, T.; Anjum, F.M. Fisetin: An anticancer perspective. *Food Sci. Nutr.* **2021**, *9*, 3–16. [CrossRef]

23. Rahmani, A.H.; Almatroudi, A.; Allemailem, K.S.; Khan, A.A.; Almatroodi, S.A. The Potential Role of Fisetin, a Flavonoid in Cancer Prevention and Treatment. *Molecules* **2022**, *27*, 9009. [CrossRef]

24. Krishnakumar, I.M.; Jaja-Chimedza, A.; Joseph, A.; Balakrishnan, A.; Maliakel, B.; Swick, A. Enhanced bioavailability and pharmacokinetics of a novel hybrid-hydrogel formulation of fisetin orally administered in healthy individuals: A randomised double-blinded comparative crossover study. *J. Nutr. Sci.* **2022**, *11*, e74. [CrossRef] [PubMed]

25. Kadari, A.; Gudem, S.; Kulhari, H.; Bhandi, M.M.; Borkar, R.M.; Kolapalli, V.R.; Sistla, R. Enhanced oral bioavailability and anticancer efficacy of fisetin by encapsulating as inclusion complex with HPβCD in polymeric nanoparticles. *Drug Deliv.* **2017**, *24*, 224–232. [CrossRef] [PubMed]

26. Talaat, S.M.; Elnaggar, Y.S.R.; El-Ganainy, S.O.; Gowayed, M.A.; Abdel-Bary, A.; Abdallah, O.Y. Novel bio-inspired lipid nanoparticles for improving the anti-tumoral efficacy of fisetin against breast cancer. *Int. J. Pharm.* **2022**, *628*, 122184. [CrossRef] [PubMed]

27. Kumar, R.M.; Kumar, H.; Bhatt, T.; Jain, R.; Panchal, K.; Chaurasiya, A.; Jain, V. Fisetin in Cancer: Attributes, Developmental Aspects, and Nanotherapeutics. *Pharmaceuticals* **2023**, *16*, 196. [CrossRef]

28. Halder, A.; Jethwa, M.; Mukherjee, P.; Ghosh, S.; Das, S.; Helal Uddin, A.B.M.; Mukherjee, A.; Chatterji, U.; Roy, P. Lactoferrin-tethered betulinic acid nanoparticles promote rapid delivery and cell death in triple negative breast and laryngeal cancer cells. *Artif. Cells Nanomed. Biotechnol.* **2020**, *48*, 1362–1371. [CrossRef]

29. Cutone, A.; Rosa, L.; Ianiro, G.; Lepanto, M.S.; Bonaccorsi di Patti, M.C.; Valenti, P.; Musci, G. Lactoferrin's Anti-Cancer Properties: Safety, Selectivity, and Wide Range of Action. *Biomolecules* **2020**, *10*, 456. [CrossRef]

30. Garrido, B.; González, S.; Hermosilla, J.; Millao, S.; Quilaqueo, M.; Guineo, J.; Acevedo, F.; Pesenti, H.; Rolleri, A.; Shene, C.; et al. Carbonate-β-Cyclodextrin-Based Nanosponge as a Nanoencapsulation System for Piperine: Physicochemical Characterization. *J. Soil Sci. Plant Nutr.* **2019**, *19*, 620–630. [CrossRef]

31. El-Habashy, S.; Eltaher, H.; Gaballah, A.; Mehanna, R.; El-Kamel, A.H. Biomaterial-Based Nanocomposite for Osteogenic Repurposing of Doxycycline. *Int. J. Nanomed.* **2021**, *16*, 1103–1126. [CrossRef]

32. Shende, P.K.; Trotta, F.; Gaud, R.S.; Deshmukh, K.; Cavalli, R.; Biasizzo, M. Influence of different techniques on formulation and comparative characterization of inclusion complexes of ASA with β-cyclodextrin and inclusion complexes of ASA with PMDA cross-linked β-cyclodextrin nanosponges. *J. Incl. Phenom. Macrocycl. Chem.* **2012**, *74*, 447–454. [CrossRef]

33. Kharouba, M.; El-Kamel, A.; Mehanna, R.; Thabet, E.; Heikal, L. Pitavastatin-loaded bilosomes for oral treatment of hepatocellular carcinoma: A repurposing approach. *Drug Deliv.* **2022**, *29*, 2925–2944. [CrossRef] [PubMed]

34. Costa, P.; Sousa Lobo, J.M. Modeling and comparison of dissolution profiles. *Eur. J. Pharm. Sci.* **2001**, *13*, 123–133. [CrossRef]

35. Zhang, Y.; Huo, M.; Zhou, J.; Zou, A.; Li, W.; Yao, C.; Xie, S. DDSolver: An add-in program for modeling and comparison of drug dissolution profiles. *AAPS J.* **2010**, *12*, 263–271. [CrossRef] [PubMed]

36. Shehata, E.M.M.; Gowayed, M.A.; El-Ganainy, S.O.; Sheta, E.; Elnaggar, Y.S.R.; Abdallah, O.Y. Pectin coated nanostructured lipid carriers for targeted piperine delivery to hepatocellular carcinoma. *Int. J. Pharm.* **2022**, *619*, 121712. [CrossRef]

37. Lu, Q.; Lai, Y.; Zhang, H.; Ren, K.; Liu, W.; An, Y.; Yao, J.; Fan, H. Hesperetin Inhibits TGF-β1-Induced Migration and Invasion of Triple Negative Breast Cancer MDA-MB-231 Cells via Suppressing Fyn/Paxillin/RhoA Pathway. *Integr. Cancer Ther.* **2022**, *21*, 15347354221086900. [CrossRef] [PubMed]

38. Radwan, S.E.-S.; El-Moslemany, R.M.; Mehanna, R.A.; Thabet, E.H.; Abdelfattah, E.-Z.A.; El-Kamel, A. Chitosan-coated bovine serum albumin nanoparticles for topical tetrandrine delivery in glaucoma: In vitro and in vivo assessment. *Drug Deliv.* **2022**, *29*, 1150–1163. [CrossRef]

39. Kumar, R.; Kumar, R.; Khurana, N.; Singh, S.K.; Khurana, S.; Verma, S.; Sharma, N.; Kapoor, B.; Vyas, M.; Khursheed, R.; et al. Enhanced oral bioavailability and neuroprotective effect of fisetin through its SNEDDS against rotenone-induced Parkinson's disease rat model. *Food Chem. Toxicol.* **2020**, *144*, 111590. [CrossRef]

40. Zhang, Y.; Huo, M.; Zhou, J.; Xie, S. PKSolver: An add-in program for pharmacokinetic and pharmacodynamic data analysis in Microsoft Excel. *Comput. Methods Programs Biomed.* **2010**, *99*, 306–314. [CrossRef]

41. Zidan, M.F.; Ibrahim, H.M.; Afouna, M.I.; Ibrahim, E.A. In vitro and in vivo evaluation of cyclodextrin-based nanosponges for enhancing oral bioavailability of atorvastatin calcium. *Drug Dev. Ind. Pharm.* **2018**, *44*, 1243–1253. [CrossRef]

42. Pushpalatha, R.; Selvamuthukumar, S.; Kilimozhi, D. Cross-linked, cyclodextrin-based nanosponges for curcumin delivery— Physicochemical characterization, drug release, stability and cytotoxicity. *J. Drug Deliv. Sci. Technol.* **2018**, *45*, 45–53. [CrossRef]

43. Sharma, K.; Kadian, V.; Kumar, A.; Mahant, S.; Rao, R. Evaluation of solubility, photostability and antioxidant activity of ellagic acid cyclodextrin nanosponges fabricated by melt method and microwave-assisted synthesis. *J. Food Sci. Technol.* **2022**, *59*, 898–908. [CrossRef] [PubMed]

44. Gholibegloo, E.; Mortezazadeh, T.; Salehian, F.; Ramazani, A.; Amanlou, M.; Khoobi, M. Improved curcumin loading, release, solubility and toxicity by tuning the molar ratio of cross-linker to β-cyclodextrin. *Carbohydr. Polym.* **2019**, *213*, 70–78. [CrossRef]

45. Shilpi, S.; Vimal, V.D.; Soni, V. Assessment of lactoferrin-conjugated solid lipid nanoparticles for efficient targeting to the lung. *Prog. Biomater.* **2015**, *4*, 55–63. [CrossRef]

46. Zhang, J.-q.; Jiang, K.-m.; An, K.; Ren, S.-H.; Xie, X.-g.; Jin, Y.; Lin, J. Novel water-soluble fisetin/cyclodextrins inclusion complexes: Preparation, characterization, molecular docking and bioavailability. *Carbohydr. Res.* **2015**, *418*, 20–28. [CrossRef] [PubMed]

47. Khalid, Q.; Ahmad, M.; Minhas, M.U.; Batool, F.; Malik, N.S.; Rehman, M. Novel β-cyclodextrin nanosponges by chain growth condensation for solubility enhancement of dexibuprofen: Characterization and acute oral toxicity studies. *J. Drug Deliv. Sci. Technol.* **2021**, *61*, 102089. [CrossRef]

48. Smith, M.L.; Murphy, K.; Doucette, C.D.; Greenshields, A.L.; Hoskin, D.W. The Dietary Flavonoid Fisetin Causes Cell Cycle Arrest, Caspase-Dependent Apoptosis, and Enhanced Cytotoxicity of Chemotherapeutic Drugs in Triple-Negative Breast Cancer Cells. *J. Cell Biochem.* **2016**, *117*, 1913–1925. [CrossRef] [PubMed]

49. Sun, X.; Ma, X.; Li, Q.; Yang, Y.; Xu, X.; Sun, J.; Yu, M.; Cao, K.; Yang, L.; Yang, G.; et al. Anti-cancer effects of fisetin on mammary carcinoma cells via regulation of the PI3K/Akt/mTOR pathway: In vitro and in vivo studies. *Int. J. Mol. Med.* **2018**, *42*, 811–820. [CrossRef]

50. Zhang, J.; Xiao, X.; Zhu, J.; Gao, Z.; Lai, X.; Zhu, X.; Mao, G. Lactoferrin- and RGD-comodified, temozolomide and vincristine-coloaded nanostructured lipid carriers for gliomatosis cerebri combination therapy. *Int. J. Nanomed.* **2018**, *13*, 3039–3051. [CrossRef]

51. Pereira, C.S.; Guedes, J.P.; Gonçalves, M.; Loureiro, L.; Castro, L.; Gerós, H.; Rodrigues, L.R.; Côrte-Real, M. Lactoferrin selectively triggers apoptosis in highly metastatic breast cancer cells through inhibition of plasmalemmal V-H+-ATPase. *Oncotarget* **2016**, *7*, 62144–62158. [CrossRef]

52. Shahi Thakuri, P.; Gupta, M.; Singh, S.; Joshi, R.; Glasgow, E.; Lekan, A.; Agarwal, S.; Luker, G.D.; Tavana, H. Phytochemicals inhibit migration of triple negative breast cancer cells by targeting kinase signaling. *BMC Cancer* **2020**, *20*, 4. [CrossRef]

53. Abdelmoneem, M.A.; Abd Elwakil, M.M.; Khattab, S.N.; Helmy, M.W.; Bekhit, A.A.; Abdulkader, M.A.; Zaky, A.; Teleb, M.; Elkhodairy, K.A.; Albericio, F.; et al. Lactoferrin-dual drug nanoconjugate: Synergistic anti-tumor efficacy of docetaxel and the NF-κB inhibitor celastrol. *Mater. Sci. Eng. C* **2021**, *118*, 111422. [CrossRef]

54. Du, X.J.; Wang, J.L.; Iqbal, S.; Li, H.J.; Cao, Z.T.; Wang, Y.C.; Du, J.Z.; Wang, J. The effect of surface charge on oral absorption of polymeric nanoparticles. *Biomater. Sci.* **2018**, *6*, 642–650. [CrossRef] [PubMed]

55. Chellampillai, B.; Pawar, A.P. Improved bioavailability of orally administered andrographolide from pH-sensitive nanoparticles. *Eur. J. Drug Metab. Pharmacokinet.* **2011**, *35*, 123–129. [CrossRef] [PubMed]

56. El-Lakany, S.A.; Elgindy, N.A.; Helmy, M.W.; Abu-Serie, M.M.; Elzoghby, A.O. Lactoferrin-decorated vs PEGylated zein nanospheres for combined aromatase inhibitor and herbal therapy of breast cancer. *Expert Opin. Drug Deliv.* **2018**, *15*, 835–850. [CrossRef] [PubMed]

57. Dora, C.P.; Trotta, F.; Kushwah, V.; Devasari, N.; Singh, C.; Suresh, S.; Jain, S. Potential of erlotinib cyclodextrin nanosponge complex to enhance solubility, dissolution rate, in vitro cytotoxicity and oral bioavailability. *Carbohydr. Polym.* **2016**, *137*, 339–349. [CrossRef]

58. Al Shoyaib, A.; Archie, S.R.; Karamyan, V.T. Intraperitoneal Route of Drug Administration: Should it Be Used in Experimental Animal Studies? *Pharm. Res.* **2019**, *37*, 12. [CrossRef]
59. Claassen, V. *Neglected Factors in Pharmacology and Neuroscience Research: Biopharmaceutics, Animal Characteristics, Maintenance, Testing Conditions*, 1st ed.; Elsevier: Amsterdam, The Netherlands, 1994; Volume 12.
60. Mishra, S.; Tamta, A.K.; Sarikhani, M.; Desingu, P.A.; Kizkekra, S.M.; Pandit, A.S.; Kumar, S.; Khan, D.; Raghavan, S.C.; Sundaresan, N.R. Subcutaneous Ehrlich Ascites Carcinoma mice model for studying cancer-induced cardiomyopathy. *Sci. Rep.* **2018**, *8*, 5599. [CrossRef] [PubMed]
61. El-Far, S.W.; Helmy, M.W.; Khattab, S.N.; Bekhit, A.A.; Hussein, A.A.; Elzoghby, A.O. Phytosomal bilayer-enveloped casein micelles for codelivery of monascus yellow pigments and resveratrol to breast cancer. *Nanomedicine* **2018**, *13*, 481–499. [CrossRef]
62. Allahyari, S.; Zahednezhad, F.; Khatami, M.; Hashemzadeh, N.; Zakeri-Milani, P.; Trotta, F. Cyclodextrin nanosponges as potential anticancer drug delivery systems to be introduced into the market, compared with liposomes. *J. Drug Deliv. Sci. Technol.* **2022**, *67*, 102931. [CrossRef]
63. Jasim, I.K.; Abd Alhammid, S.N.; Abdulrasool, A.A. Synthesis and evaluation of B-cyclodextrin Based Nanosponges of 5-Fluorouracil by Using Ultrasound Assisted Method. *Iraqi J. Pharm. Sci.* **2020**, *29*, 88–98. [CrossRef]
64. Kale, J.; Osterlund, E.J.; Andrews, D.W. BCL-2 family proteins: Changing partners in the dance towards death. *Cell Death Differ.* **2018**, *25*, 65–80. [CrossRef] [PubMed]
65. Pfeffer, C.M.; Singh, A.T.K. Apoptosis: A Target for Anticancer Therapy. *Int. J. Mol. Sci.* **2018**, *19*, 448. [CrossRef] [PubMed]
66. Huang, J.; Cui, H.; Peng, X.; Fang, J.; Zuo, Z.; Deng, J.; Wu, B. The Association between Splenocyte Apoptosis and Alterations of Bax, Bcl-2 and Caspase-3 mRNA Expression, and Oxidative Stress Induced by Dietary Nickel Chloride in Broilers. *Int. J. Environ. Res. Public Health* **2013**, *10*, 7310–7326. [CrossRef]
67. Chen, Y.C.; Shen, S.C.; Lee, W.R.; Lin, H.Y.; Ko, C.H.; Shih, C.M.; Yang, L.L. Wogonin and fisetin induction of apoptosis through activation of caspase 3 cascade and alternative expression of p21 protein in hepatocellular carcinoma cells SK-HEP-1. *Arch. Toxicol.* **2002**, *76*, 351–359. [CrossRef]
68. Monfared, Y.K.; Pedrazzo, A.R.; Mahmoudian, M.; Caldera, F.; Zakeri-Milani, P.; Valizadeh, H.; Cavalli, R.; Matencio, A.; Trotta, F. Oral supplementation of solvent-free kynurenic acid/cyclodextrin nanosponges complexes increased its bioavailability. *Colloids Surf. B Biointerfaces* **2023**, *222*, 113101. [CrossRef]
69. Khazaei Monfared, Y.; Mahmoudian, M.; Caldera, F.; Pedrazzo, A.R.; Zakeri-Milani, P.; Matencio, A.; Trotta, F. Nisin delivery by nanosponges increases its anticancer activity against in-vivo melanoma model. *J. Drug Deliv. Sci. Technol.* **2023**, *79*, 104065. [CrossRef]
70. Alsawaf, S.; Alnuaimi, F.; Afzal, S.; Thomas, R.M.; Chelakkot, A.L.; Ramadan, W.S.; Hodeify, R.; Matar, R.; Merheb, M.; Siddiqui, S.S.; et al. Plant Flavonoids on Oxidative Stress-Mediated Kidney Inflammation. *Biology* **2022**, *11*, 1717. [CrossRef] [PubMed]

pharmaceutics

MDPI

Article

Influence of the Topology of Poly(L-Cysteine) on the Self-Assembly, Encapsulation and Release Profile of Doxorubicin on Dual-Responsive Hybrid Polypeptides

Dimitra Stavroulaki [1], Iro Kyroglou [1], Dimitrios Skourtis [1], Varvara Athanasiou [1], Pandora Thimi [1], Sosanna Sofianopoulou [2], Diana Kazaryan [1], Panagiota G. Fragouli [3], Andromahi Labrianidou [4], Konstantinos Dimas [4], Georgios Patias [5], David M. Haddleton [5] and Hermis Iatrou [1,*]

[1] Industrial Chemistry Laboratory, Department of Chemistry, National and Kapodistrian University of Athens, Panepistimiopolis, Zografou, GR-15771 Athens, Greece
[2] Hellenic Police Headquarters, Forensic Science Division, Chemical and Physical Examinations Department, GR-10442 Athens, Greece
[3] DIDPE, Dyeing, Finishing, Dyestuffs and Advanced Polymers Laboratory, University of West Attica, 250 Thevon Street, GR-12241 Athens, Greece
[4] Laboratory of Pharmacology, Faculty of Medicine, University of Thessaly, Viopolis, GR-41500 Larissa, Greece
[5] Department of Chemistry, University of Warwick, Gibbet Hill, Coventry CV4 7AL, UK
* Correspondence: iatrou@chem.uoa.gr

Citation: Stavroulaki, D.; Kyroglou, I.; Skourtis, D.; Athanasiou, V.; Thimi, P.; Sofianopoulou, S.; Kazaryan, D.; Fragouli, P.G.; Labrianidou, A.; Dimas, K.; et al. Influence of the Topology of Poly(L-Cysteine) on the Self-Assembly, Encapsulation and Release Profile of Doxorubicin on Dual-Responsive Hybrid Polypeptides. *Pharmaceutics* **2023**, *15*, 790. https://doi.org/10.3390/pharmaceutics15030790

Academic Editor: Ana Isabel Fernandes

Received: 20 January 2023
Revised: 16 February 2023
Accepted: 23 February 2023
Published: 27 February 2023

Abstract: The synthesis of a series of novel hybrid block copolypeptides based on poly(ethylene oxide) (PEO), poly(L-histidine) (PHis) and poly(L-cysteine) (PCys) is presented. The synthesis of the terpolymers was achieved through a ring-opening polymerization (ROP) of the corresponding protected N-carboxy anhydrides of N^{im}-*Trityl*-L-histidine and S-*tert*-butyl-L-cysteine, using an end-amine-functionalized poly(ethylene oxide) (mPEO-NH$_2$) as macroinitiator, followed by the deprotection of the polypeptidic blocks. The topology of PCys was either the middle block, the end block or was randomly distributed along the PHis chain. These amphiphilic hybrid copolypeptides assemble in aqueous media to form micellar structures, comprised of an outer hydrophilic corona of PEO chains, and a pH- and redox-responsive hydrophobic layer based on PHis and PCys. Due to the presence of the thiol groups of PCys, a crosslinking process was achieved further stabilizing the nanoparticles (NPs) formed. Dynamic light scattering (DLS), static light scattering (SLS) and transmission electron microscopy (TEM) were utilized to obtain the structure of the NPs. Moreover, the pH and redox responsiveness in the presence of the reductive tripeptide of glutathione (GSH) was investigated at the empty as well as the loaded NPs. The ability of the synthesized polymers to mimic natural proteins was examined by Circular Dichroism (CD), while the study of zeta potential revealed the "stealth" properties of NPs. The anticancer drug doxorubicin (DOX) was efficiently encapsulated in the hydrophobic core of the nanostructures and released under pH and redox conditions that simulate the healthy and cancer tissue environment. It was found that the topology of PCys significantly altered the structure as well as the release profile of the NPs. Finally, in vitro cytotoxicity assay of the DOX-loaded NPs against three different breast cancer cell lines showed that the nanocarriers exhibited similar or slightly better activity as compared to the free drug, rendering these novel NPs very promising materials for drug delivery applications.

Keywords: polypeptides; drug delivery; nanoparticles; doxorubicin; cancer; topology of poly-L-cystein

1. Introduction

Cancer remains one of the leading causes of death worldwide. Due to its complex nature, multiple metabolic pathways and ability to resist numerous drugs, so far, selective elimination of cancer cells without influencing healthy tissues has not yet been achieved [1,2].

Most of the anticancer drugs used are poorly water soluble, leading to poor absorption and low bioavailability. Therefore, it is necessary to design and synthesize smart drug delivery systems which can transport therapeutic agents in a timely and spatially controlled manner. This need can be satisfied by using nanotechnology for the synthesis of nanostructured materials as drug carriers, the most common being micelles [3–5], liposomes [6], polymersomes, magnetic NPs and mesoporous silica NPs as well as hydrogels [7–9]. Among the materials used, polymers play a pivotal role due to their high functionalization. Drugs can be encapsulated or chemically conjugated to polymers, lowering their toxicity and increasing their solubility as well as their circulation time in the blood and preventing renal clearance for more efficient accumulation in the solid tumor either through the EPR effect or an active targeting mechanism [10].

Ideally, smart drug delivery systems to target cancer cells should have the ability to bypass numerous biological obstacles such as vasculature and non-vasculature barriers and tumor microenvironment as well as intracellular barriers [11]. The tumor microenvironment is dynamic and is characterized by acidity, hypoxia and ischemia. Usually, the pH value of a tumor tissue is 6.3–6.8, while higher concentrations of various biological substances are detected in cancer cells such as GSH and matrix metalloproteinase 2 (MMP2) as well as active oxidative species (ROS) [12,13]. The blood serum has a pH value of 7.4, while the endosomal and extracellular pH of the cancer tissues exhibits a pH range from 6.3–6.8, and that of the lysosomal compartments of the cell can be between pH 5.5–5.0 [14,15]. Therefore, pH-responsive systems can induce controlled drug release and penetrate deeper into cancer cells, due to pH variations between the intracellular organelles of the cell and the extracellular matrix. In the case of polypeptides bearing an amine or carboxylic acid in their side chains, pH changes activate the protonation/deprotonation mechanism, leading to the disassembly of their conformation and the desired triggered release of their cargo.

Therefore, the NPs that will be used to treat cancer should exhibit the following main characteristics: good biocompatibility and minimal cytotoxicity without systemic side effects, highly selective accumulation in pathological tissues, accurate stimulus responses to result in selective release of the cargo and long-term stability in blood circulation as well as minimal cargo loss before arrival at the target. In addition, drug delivery systems should be inert and stable in the aggressive environment of the blood compartment. We envisioned synthetic NPs that fulfill the above properties and can change their structure when they reach the pathological environment, releasing the cargo in a controlled way, leading to efficient and highly selective elimination of cancer cells.

PHis is a unique polypeptide with the ability to respond to physiological pH variations as its imidazole ring with a pKa around 6.5 can be protonated and deprotonated within the physiological values, altering the hydrophilicity of the polypeptide [16]. Bilalis et al. [17] have described the synthesis of PHis-grafted mesoporous silica NPs (MSNs) which can efficiently encapsulate the anticancer drug DOX, and can release it in a pH-controlled manner. Our group has also referred to the synthesis of linear and 3-miktoarm star hybrid polypeptides based on PHis, poly(L-glutamic acid), poly(L-lysine) and PEO. These novel structures could load the anticancer drug Everolimus and release it in response to pH changes [18].

Redox responsive nano-structures, whose action is determined by the microenvironment of cancer tissue, also induce controlled drug release [19,20]. The redox reaction depends on the concentration of active substances in the cell organelles. For example, the concentration of GSH is four times higher in cancerous tumors compared to its concentration in healthy tissues. Different concentrations of GSH were also found in intracellular (~2–10 mM) and extracellular (~2–10 µM) fractions. In particular, intracellular compartments comprising the cytoplasm, mitochondria, and nucleus contain higher concentrations of GSH compared to extracellular fluids [21–23]. It is well known that disulfide bonds are sensitive to GSH, as GSH can cause the rapid cleavage of disulfide bonds, leading to cytosolic delivery of anticancer agents [24–27].

Wang et al. [28] managed to synthesize redox-responsive SCL micelles based on poly(ethylene glycol)-*b*-poly(L-cysteine)-*b*-poly(L-phenylalanine) triblock copolymers, which could load DOX. A sustained release profile of DOX was observed from these NPs where the PCys was in a block form. Moreover, Wu et al. [29,30] reported the synthesis of DOX-loaded and gold-embedded micelles based on poly(L-cysteine) which exhibit synergistic chemo- and photothermal therapy of cancer cells. In both works, PCys was in a block form and a similar sustained release of the drug was observed. In a previous publication by our group [31], we presented the synthesis of disulfide crosslinked polypeptide nanogels consisting of PHis and PCys, which show satisfying pH, redox and thermo-responsiveness to the external stimuli. In that work, the disulfide bonds were randomly distributed along the polymeric chains. We have shown that the macromolecular architecture and topology of the blocks can play a critical role on the self-assembly of amphiphilic polymers [32–36]. Although the topology of cysteine along the polypeptide chain can play a critical role at the release profile of the drug, to our knowledge, there is no publication so far which studies this parameter on the structure and drug release profile of NPs.

Herein, we present the synthesis of three series of hybrid polypeptide copolymers composed of PEO, PCys and PHis. PEO was always the first block in all these series. At the first series, PCys was the second block, while PHis was the last block. At the second series, PHis was the second block while PCys was the last. Finally, at the third series, PHis and PCys were randomly distributed along the chain. Two different compositions of PHis and PCys were used, keeping the total number of the monomeric units of PHis and PCys the same, while the PEO block was the same for all hybrid polypeptides. It was found that the empty hybrid terpolymers can self-assemble into micellar nanostructures and exhibit pH and redox responsiveness. Moreover, they can effectively load the anticancer drug DOX in differently structured NPs compared to the empty one and can release in a controlled manner, in response to pH and redox variations. The release profiles depended on the structure of the NPs. In order to elucidate the influence of PCys topology at the release profile of DOX, the release was studied in media with various concentrations of GSH as well as different pH values. Finally, in vitro studies of the efficacy of the NPs in breast cancer cell lines prove that the DOX-loaded NPs could be potentially used for cancer treatment. In order to elucidate the impact of the PCys topology on encapsulation efficiency and release profiles, the results were compared to the NPs obtained under similar conditions with similar polymers such as PEO-*b*-PHis or PEO-*b*-Poly(sarcosine)-*b*-PCys (PEO-PSAR-PCys).

2. Experiment

2.1. Materials and Methods

2.1.1. Materials

Ethyl acetate (>99.9%, Carlo Erba, Val de Reuil, France) was fractionally distilled over phosphorous pentoxide. Tetrahydrofuran (THF) (>99.9%, Carlo Erba) was purified over Na-K alloy, using standard high vacuum techniques. The purification of *N,N*-Dimethylformamide (DMF) (99.9%, Alfa Aesar, Waltham, MA, USA anhydrous, amine free) was performed by short-path fractional distillation under high vacuum in a custom-made apparatus, and the middle fraction was used. The final product was stored in a vacuum flask at 3 °C. Benzene (99%, thiophen-free grade, Sigma Aldrich, Saint Louis, MO, USA) was treated with calcium hydride and was allowed to be stirred overnight for moisture removal. It was then distilled under high vacuum and stored in a different flask containing *n*-BuLi. Diethyl ether (>99.5%), Dichloromethane (99.8%) and Chloroform (>99.8%) were purchased from Fluka, Charlotte, NC, USA. *n*-Hexane (>95%) was obtained from Carlo Erba, Val de Reuil, France. Methyl Sulfoxide (DMSO) (99.8+%, for peptide synthesis) was supplied from Acros Organics, Waltham, MA, USA. BOC-His(*Trt*)-OH (>99%) was acquired from Christof Senn Laboratories AG, Dielsdorf, Switzerland. Sarcosine (98%) was purchased from Alfa Aesar. *S*-tert-Butylmercapto-L-Cysteine (99%) was provided by Sigma Aldrich. Methoxypolyethylene glycol amine (*m*PEO-NH$_2$) with average M_n = 10,000 g mol^{-1} was obtained from Sigma Aldrich. Triethylamine (Et_3N) (99.83%, Fluka) was dried over calcium hydride for one day and then distilled and stored under vacuum over sodium. Triphosgene (99%), Thionyl chloride (99.5+%), Hydrogen peroxide (ACS reagent, 30 wt.%, solution in water, non-stabilized) and DL-1,4-Dithiothreitol (99%) were provided by Acros Organics, Waltham, MA, USA. (R)-(+)-Limonene (97%) was purchased from Alfa Aesar. Trifluoroacetic acid (TFA) (>99%) was obtained from Fluka. Triisopropylsilane (98%) was supplied from Sigma Aldrich. L-Glutathione (98%, reduced form) was obtained from FluoroChem, Hadfield, UK. Sodium Chloride (99.9%) was purchased by Penta chemicals, Prague, Check Republic. Sodium Hydroxide pearls (99.4%) was acquired by Lachner, Neratovice, Check Republic. Hydrochloric Acid $1\ mol\ L^{-1}$ and Acetic Acid glacial (99.8%) were obtained from Chem-Lab, Zedelgem, Belgium. Tris base ultrapure (99.9%) was purchased from Duchefa Biochemie, RV Haarlem, Netherlands. Sodium Phosphate Monobasic (98–100.5%) was supplied from Riedel-de Haen, Charlotte, NC, USA. Doxorubicin Hydrochloride (>99%) was obtained from Selleckchem, Planegg, Germany. Distilled water was further purified by a Milli-Q Direct Water purification system (18.2 MΩ·cm, Merck Millipore, Darmstadt, Germany).

2.1.2. NMR Spectroscopy

^{1}H-NMR measurements were carried out on a 400 MHz Bruker Avance Neo instrument, Billerica, MA, USA. A mixture of deuterium oxide (D_2O) and deuterium chloride (DCl 1%) was used as the solvent for the polymers, while deuterated chloroform ($CDCl_3$) was employed as the solvent for NCAs, at room temperature.

2.1.3. FT-IR Spectroscopy

Fourier transform infrared (FT-IR) spectroscopy measurements were conducted using a Perkin Elmer Spectrum One instrument (Waltham, MA, USA), in KBr pellets at room temperature, in the 450–4000 cm^{-1} range.

2.1.4. Size Exclusion Chromatography

Size exclusion chromatography (SEC) was employed to determine the M_n and $\text{Đ}M = M_w/M_n$ values. The analysis was performed using two different SEC sets. The first one was composed of a Waters Breeze instrument (Milford, MA, USA) equipped with a 2410 differential refractometer and a Precision PD 2020 two angles (15°, 90°) light scattering detector. The carrier solvent was a 0.10% TFA (v/v) solution of water/acetonitrile (80/20 v/v) at a flow rate of 0.8 mL min^{-1} at 35 °C. Three linear Waters hydrogel columns were used as a stationary phase. The second system was composed of a Waters 600 high-performance liquid chromatographic pump, Waters Ultrastyragel columns (HT-2, HT-4, HT-5E and HT-6E), a Waters 410 differential refractometer and a Precision PD 2020 two angles (15°, 90°) light scattering detector at 60 °C. A 0.1 M LiBr in DMF solution was used as an eluent at a rate of 1 mL min^{-1}.

2.1.5. Circular Dichroism

Circular Dichroism measurements were conducted via a JASCO J–815 model in a 0.1 cm cell. The aqueous solutions of polymers had a concentration of 10^{-5} g mL^{-1} and the desired pH was adjusted with the addition of droplets either of 0.01 N HCl, or 0.01 N NaOH. The temperature was stabilized to 25 °C with the use of a dedicated digital thermostat. The nitrogen flow was adjusted to 6.0 L min^{-1}.

2.1.6. UV Spectroscopy

UV spectroscopy was carried out using a Perkin Elmer Lamda 650 spectrometer, (Waltham, MA, USA) in the range of 250–800 nm, at room temperature, with cells requiring 3 mL. A Waters Diode-Array 690 detector (Milford, MA, USA) was used for the calibration and on-line determination of the DOX drug loading efficiency at λ = 485 nm.

2.1.7. Dynamic Light Scattering

DLS measurements were conducted with a Brookhaven Instruments BI-200SM Research Goniometer system (Holtsville, NY, USA) operating at $\lambda = 640$ nm and with 40 mW laser power. Correlation functions were analyzed by the cumulant method and the Contin software. The correlation function was measured at 90°, at 25 °C. All measurements were performed in either an isotonic PBS or Tris buffer (10 mM, 150 mM NaCl) at pH = 7.4, PBS buffer (10 mM, 150 mM NaCl) at pH = 6.5, and an isotonic acetate buffer (10 mM, 150 mM NaCl) at pH = 5.0. The concentration range measured was between 2×10^{-3}–1×10^{-5} g mL^{-1}.

2.1.8. Static Light Scattering

SLS measurements were carried out on an ALV/CGS-3 Compact Goniometer System (ALV GmbH, Langen, Germany), equipped with an ALV-5000/EPP multi-tau digital correlator with 288 channels and an ALV/LSE-5003 light scattering electronics unit for stepper motor drive and limit switch control. A JDS Uniphase 22 mW He-Ne laser was used as the light source. The instrument was connected to a Polyscience model 9102 bath for temperature control, allowing measurements at variable temperature.

2.1.9. Electrophoretic Mobility

The electrophoretic mobility measurements of the empty and drug-loaded nanoparticle dispersions were conducted using a Brookhaven Instruments Nanobrook Omni system (Holtsville, NY, USA) operating at $\lambda = 640$ nm and with 40 mW power, operating in PALLS mode. All the measurements were performed in isotonic Tris buffer (10 mM, 150 mM NaCl) at pH = 7.4 at 37 °C and were the average of at least three runs.

2.1.10. Transmission Electron Microscopy

Transmission electron microscopy images were obtained using a Jeol 2100 TEM, operated at 200 kV and fitted with a Gatan Ultrascan 1000 camera (Pleasanton, CA, USA). Samples for TEM analysis were prepared via drop-casting a few milliliters of sample dispersions after ultrasonication onto holey carbon grids, allowing the solvent to evaporate and leaving the sample to rest for 24 h at ambient temperature.

2.1.11. Cell Culture

The MCF-7, MDA-MB 231 and T47D cell lines were cultured in RPMI 1640 growth medium, supplemented with 5% fetal bovine serum, 2 mM L-Glutamine, 100 U mL^{-1} penicillin and 100 µg mL^{-1} streptomycin. Cells were maintained at 37 °C in a humidified 5% CO$_2$ incubator. ER(+) human breast cancer cell line MCF7, ER/PR (+) human breast cancer cell line T47D and triple negative human breast cancer (TNBC) cell line MB231 were purchased from NCI (NCI, NIH, Frederick, MD, USA).

2.2. Synthesis of the Monomers

2.2.1. Synthesis of N^{im}-Trityl-L-Histidine N-Carboxy Anhydride (N^{im}-Trityl-L-His-NCA)

For the synthesis of *N^{im}-Trityl*-L-His-NCA a previously reported method by our group [16] was mainly followed with some modifications. The synthesis was conducted in two steps. In the first step, the hydrochloric salt of *N^{im}-Trityl*-L-His-NCA was formed, and finally the desired pure *N^{im}-Trityl*-L-His-NCA was synthesized after the removal of HCl. The whole synthetic procedure is presented in detail in Supplementary Materials.

2.2.2. Synthesis of S-tert-Butyl-mercapto-L-Cysteine N-Carboxy Anhydride (tBM-L-Cys-NCA)

The *N*-carboxy anhydride of *t*BM-L-Cys-NCA was synthesized in a similar way to a previously described process [37]. The synthetic steps are provided in Supplementary Materials.

2.2.3. Synthesis of Sarcosine N-Carboxy Anhydride (Sar-NCA)

The synthesis of the Sarcosine *N*-Carboxy Anhydride was conducted following previously reported procedures [16,38]. The total synthetic route is described thoroughly in Supplementary Materials.

2.3. Synthesis of the Hybrid-Polypeptides

The synthesized polymers are illustrated in Scheme 1. The synthetic procedure is described in detail in Supplementary Materials. Briefly, the synthesis of the hybrid polypeptide terpolymers was achieved through a ring-opening polymerization (ROP) process [39–41] of the corresponding *N*-carboxy anhydrides, using an amino end-functionalized poly(ethylene oxide) (*m*-PEO-NH$_2$) macroinitiator, with molecular weight 10.0×10^3 g mol^{-1}. Highly purified DMF was the solvent at all polymerizations. In case of the *m*PEO-*b*-PHis-*b*-PCys as well as the *m*PEO-*b*-PCys-*b*-PHis hybrid terpolymers, the sequential addition synthetic approach of the corresponding anhydrides of the amino acids was used, after the completion of the polymerization of each monomer. In case of *m*PEO-*b*-[PCys-*co*-PHis] terpolymers, the macroinitiator polymerized the mixture of the two anhydrides. Then, the hybrid polypeptides were precipitated followed by deprotection of the trityl group of His by TFA. Finally, the deprotection of cysteine was achieved by using 1,4 dithiothreitol (DTT). A general reaction sequence for the synthesis of hybrid terpolymers of the general type mPEO-*b*-P(Cys)-*b*-P(His) (by the term general type we mean the three different structures with different PCys topology) is given in Scheme 2.

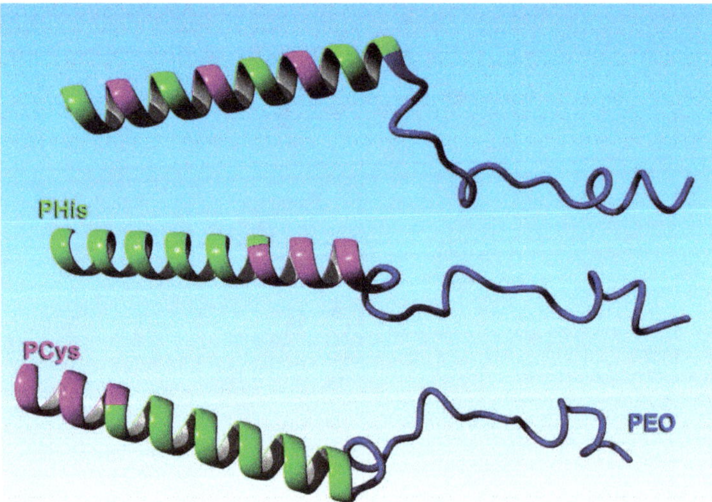

Scheme 1. Schematic representation of the synthesized hybrid terpolymers. PEO is depicted with blue color. PCys is colored magenta, while PHis is green. At the top terpolymer, the polypeptidic monomers are randomly distributed along the chain, at the middle, the PCys is located between the PEO and PHis chains, while at the lower polymer, the PCys is located at the edge of the polymeric chain.

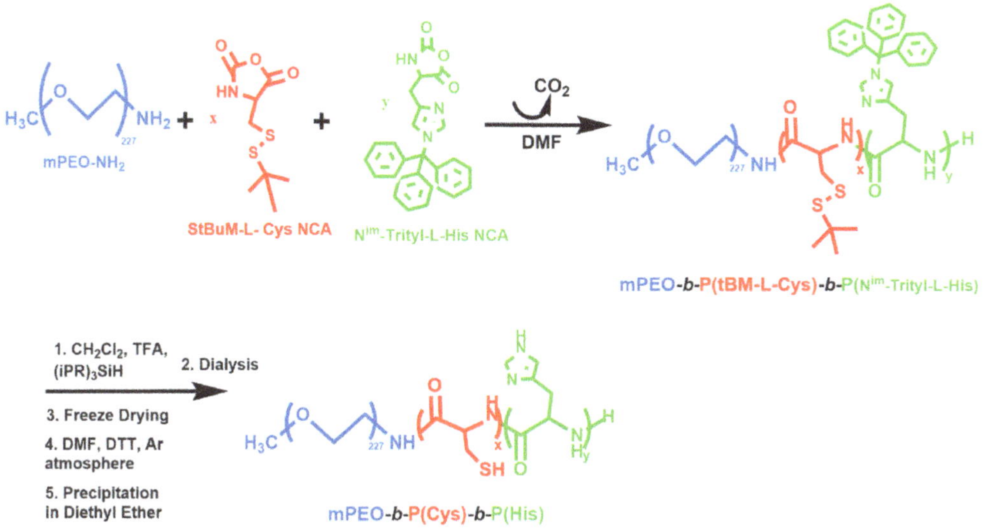

Scheme 2. Reactions used for the synthesis of the fully protected polymers of the general type *m*PEO-*b*-P(Cys)-*b*-P(His). PHis has a green color, PCys is red, while PEO is blue.

Since the PEO blocks were equal for all the polymers, the code of the hybrid polypeptides was defined by the order of the blocks as well as the monomeric units of L-cysteine; therefore, the abbreviation PHis-PCys5 refers to the triblock $m\text{PEO}_{227}$-*b*-P(His)$_{40}$-*b*-P(Cys)$_5$ and PCys10-PHis refers to $m\text{PEO}_{227}$-*b*-P(Cys)$_{10}$-*b*-P(His)$_{35}$, while in case of the $m\text{PEO}_{227}$-*b*-[P(Cys)$_5$-*co*-P(His)$_{40}$], where the polypeptidic block is composed of randomly distributed peptides, PCys5COPHis will be mentioned.

2.4. Self-Assembly of Empty NPs via Solvent Switch Method

The ability of the synthesized polymers to self-assemble in aqueous media was examined at five different isotonic buffers, with different pH values and GSH concentrations. The pH values of 7.4 and 6.5 were adjusted with a PBS buffer solution (10 mM, 150 mM NaCl), while the pH 5.0 was achieved with an acetate buffer solution (10 mM, 150 mM NaCl). At pH 6.5 and 5.0, another two buffers were prepared containing 10 mM of GSH, in order to study the influence of this reducing agent in the self-assembly behavior of the NPs. In a typical procedure, 10 mg of the hybrid polypeptides, as well as 0.02 g of DTT were dissolved in 2 mL of DMSO. After the complete dissolution, 18 mL of MilliQ water were added dropwise, and the whole mixture was left under stirring overnight. The next day, the solution was placed in a dialysis membrane (Spectrapor MWCO 3500 Da) and was dialyzed against 2 L of PBS buffer pH = 7.4 for 3 h. Then, the dialysis membrane was transferred to a fresh media of the same buffer and dialyzed for another 3 h with the presence of 10 mL H_2O_2. The last dialysis was lasted 12 h and then the solution of the NPs was collected and divided in 5 equal parts. The first part was kept for DLS measurements, while the remaining solution was transferred equally in four different dialysis membranes (Spectrapor MWCO 3500 Da) and was dialyzed against 2 L of the following buffers for 24 h with frequent changes of the external media: PBS pH = 6.5, PBS pH = 6.5 and 10 mM GSH, pH = 5.0, pH = 5.0 and 10 mM GSH. Finally, the solution of each membrane was collected and measured with DLS, after filtration with a 0.45 µm hydrophilic filter.

2.5. Loading of Anticancer DOX

In a typical experiment, 10 mg of the fully deprotected polymers was dissolved in 2 mL of DMSO and left under stirring overnight, to afford clear solutions. In case of the polymers containing PCys, 0.02 g (0.13 mmol, 9:1 mol DTT/mol Cys) DTT was added, in order to avoid the undesirable crosslinking reactions. Subsequently, a special treatment of DOX (HCl-salt) was conducted, according to a standard procedure described by Kataoka et al. [42]. In line with this protocol, 5 mg of DOX hydrochloride was dissolved in 100 mL of MilliQ water, and the resulting red solution was added in a separatory funnel containing 100 mL of chloroform. Then, 3.0 equivalent of triethylamine (TEA) (mol Et$_3$N: mol Dox $\times$ HCl = 3:1) was added in the aqueous phase and the color immediately turned to purple. After shaking the solution, the color became red again and the DOX was distributed in the organic phase. The concentration of DOX in the aqueous phase was estimated photometrically at 485 nm and the pH measured was close to neutral. Then, the hydrophobic DOX-free base dissolved in chloroform was separated and collected in a flask. The organic solvent was distilled off and the solid DOX was obtained. Afterwards, the solution of each polymer in DMSO was added in the flask containing the dried DOX and was left for half an hour to be dissolved. Then, 8 mL of PBS buffer (pH = 7.4) was added dropwise to the above mixture over a period of 10 min. The solution was then placed in a dialysis bag (Spectrapor, MWCO: 3500 Da, 25 °C) and was dialyzed against 4 L of isotonic PBS buffer at pH = 7.4 (150 mM NaCl, 10 mM PBS), in order to remove the excess drug. After 3 h of dialysis, the external buffer was renewed, and 30 mL of H$_2$O$_2$ was added in the fresh buffer, in the case of the polymers containing poly(L-cysteine), in order to induce the crosslinking reaction. The dialysis lasted another 3 hours and then the same procedure was repeated, without the addition of H$_2$O$_2$, for 12 h in total. The next day, the solution inside the membrane was obtained and the volume measured was about 12 mL. Then, about 4 mL of the NP solution was preserved for further analysis and the rest of the solution was divided into five equal parts of 1.5 mL and each part was added in a new dialysis membrane (Spectrapor, MWCO: 6000–8000 Da) and was immediately immersed in 35 mL of buffers with different characteristics, as far as the pH, the temperature and the concentration of GSH are concerned, in order to study the in vitro DOX release profile. The encapsulation efficiency (EE) and the loading capacity (LC) of the different NPs were calculated by UV absorption spectroscopy at 485 nm, as the polymer did not absorb at this wavelength, while free DOX does. Quantification was achieved by calibrating the instrument with dissolved DOX in the corresponding PBS buffer.

The encapsulation efficiency and the loading capacity were calculated according to the following equations:

$$\text{EE (\%)} = (\text{mass of Dox in NPs/ mass of Dox in the initial solution}) \times 100$$

$$\text{LC (\%)} = (\text{mass of Dox in NPs/ polymer mass}) \times 100$$

2.6. In Vitro Drug Release Studies

In vitro DOX release experiments were conducted at three different pH values (pH = 7.4, 6.5 and 5.0), at two temperatures (37 °C and 40 °C) and as far as the polymers with poly(L-cysteine) in their polypeptidic block are concerned, the factor of the addition of GSH was studied. More precisely, after the completion of the dialysis procedure, the remaining solution of NPs was divided into five equal parts of 1.5 mL, as mentioned above, was transferred into a new dialysis bag (Spectrapor, MWCO: 6000–8000 Da) and was immediately immersed in 35 mL of each of the in vitro release medium. The first membrane was ingrained in a PBS buffer at pH = 7.4 and at 37 °C (0.010 M PBS, 0.150 M NaCl) under stirring at 200 rpm. For the release studies at the acidic pH (6.5 and 5.0), two different samples were employed, for each pH value. The first dialysis bag was immersed in a PBS buffer at pH = 6.5, at 40 °C, (0.010 M PBS, 0.150 M NaCl) and the other was introduced into the same release medium containing 10 mM of GSH. Similarly, in the case of the pH = 5.0, the first membrane was added in an acetate buffer at pH = 5.0, at 40 °C, (0.010 M acetate, 0.150 M NaCl), and the last was sank into the same buffer, at the same conditions with further addition of 10 mM GSH. The cumulative release of the drug was measured at the exterior solution at defined time intervals. The dialysis membrane was transferred into a fresh buffer solution at every interval, in order to avoid saturation of the solution from the hydrophobic drug. The DOX concentration was calculated by UV spectroscopy at $\lambda = 485$ nm, using a calibration curve obtained with solutions of known DOX concentration measured using the same instrument.

2.7. In Vitro Cytotoxic Activity: Sulforhodamine B (SRB) Assay

The established human cell lines from breast cancer MCF-7 (estrogen and progesterone receptor positive invasive ductal carcinoma), T-47D (progesterone receptor positive invasive ductal carcinoma) and MDA-MB231 (triple negative breast cancer) were used and provided by the pharmacology laboratory of NCI (National Cancer Institute, NIH, Frederick, MD, USA).

Cell culture was performed in RPMI 1640 medium (Gibco®, Code: 31870025) supplemented with 5% fetal bovine serum (FBS: fetal bovine serum, (Biosera, Code: 1001G)), 2 mM L-glutamine (Biosera, Code: XO-T1715), 100 U mL^{-1} penicillin and 100 µg mL^{-1} streptomycin (Biosera, Code: XO-A4122). The cell cultures were kept in an incubation oven, at 37 °C, in an atmosphere of 5% CO_2 and 95% humidity.

The antiproliferative activity of the NPs was tested by the colorimetric method of SRB [43,44]. SRB is an anionic micromolecular compound that is stoichiometrically attached to the basic amino acid residues of protein chains, under slightly acidic conditions, and then extracted, under basic conditions.

This process involves the following steps. At the beginning of each experiment, the viability of the cells is checked with the trypan blue method so that it is always greater than 96%. The cells are added to 96-well flat-bottom cell culture plates (density 5000–10,000 cells per position) and incubated for 24 hours in an incubation oven at 37 °C, 5% CO_2 and 95% humidity to return to the logarithmic development phase (adjustment period). After 24 h, the NP solutions are added. In some cells, only culture material is added to provide the control cells (control, C). Each NP solution was tested in four logarithmic concentrations with a maximum concentration of 10 µM. The final concentration of DMSO in each cell culture was not higher than 0.1%. A number of sites from each cell line in each experiment are fixed with 50% v/v TCA (Trichloroacetic Acid) (Applichem, Code: A1431) cold solution for 1 h at 4 °C, after 24 hours of the adjustment period, aiming the representation of cell culture in the phase of addition at NPs (Tz). After 48 hours of incubating the cells with the NPs, the cells are fixed by gently adding 50% v/v TCA to each site of the cell culture plate, for 1 hour, at 4 °C. The cells are then carefully washed, 3 times, with deionized water, the excess water is removed and the plates are allowed to dry at room temperature. The cells are stained with a solution of 0.04% w/v SRB (from SIGMA, Code: S9012) in 1% acetic acid (from Fluka, Code: 45731), for 10 minutes, at room temperature. After incubation, the excess dye is removed by repeated rinsing with 1% v/v acetic acid and the cell monolayers are allowed to dry at room temperature. A 10 mM Tris base solution is then added and the cells are incubated for 10 minutes at 37 °C. Under these conditions, the protein-bound dye is released into the slightly basic Tris base solution. For each concentration of the studied NP solution, the optical absorption at 540 nm (Ti) is measured on a BioTek microplate reader (Biotek, EI-311).

Using the optical absorption measurements of the cells at the time of addition of the NPs (Tz), the control cells (C), as well as the cells under the influence of the examined NPs, the percentage growth of the cells (% growth rate) can be calculated with the use of the following equations:

$$[(Ti - Tz)/(C - Tz)] \times 100, \text{ for concentrations where Ti} \geq \text{Tz and}$$

$$[(Ti - Tz)/Tz] \times 100, \text{ for concentrations where Ti} < \text{Tz}$$

From the resulting dose–response curves (response, the cell growth rate, % growth rate) the parameters GI50, TGI and LC50 are determined, where:

GI50, Growth Inhibition 50% = the concentration of the drug through which cell growth is inhibited by 50%.

TGI, Total Growth Inhibition = the concentration of the drug through which total inhibition of cell growth is achieved.

LC50, Lethal Concentration 50% = the concentration of the drug that causes death in 50% of the cell population [45,46].

3. Results and Discussion

3.1. Synthesis and Characterization of the N-Carboxy Anhydrides (NCAs)

The synthesis of the *N*-carboxy anhydrides of α-amino acids was monitored by FT-IR spectroscopy, while the successful synthesis and the high purity of the final monomers were confirmed by ^{1}H-NMR spectroscopy. The results from the characterization of the *N*-carboxy anhydrides are summarized in Supplementary Materials (Figures S1–S6, Schemes S1–S3).

3.2. Synthesis and Characterization of the Polymers

Initially, the novel fully protected polymers of the general type of PEO-*b*-P(N^{im}-*Trityl*-L-His)-*b*-P(*t*BM-L-Cys) were synthesized followed by the selective deprotection of each polypeptide block, to afford the fully deprotected polymers of the general type of *m*PEO-*b*-P(Cys)-*b*-P(His). The synthetic procedure was monitored by FT-IR spectroscopy, and the molecular weights were obtained by using SEC-TALLS while the controlled cleavage of the protective groups was confirmed by ^{1}H-NMR. The polymers were excessively characterized and the characterization results are shown in Table 1. It can be seen that the novel hybrid terpolymers exhibited a high degree of molecular and compositional homogeneity, while the experimentally obtained molecular characteristics were within 10% close to the stoichiometric one. In addition, the total molecular weight of the polypeptidic blocks were close in all polymers although the ratios between the PHis and PCys were different, while PEO was the same.

Table 1. Molecular characteristics of the hybrid terpolymers of the general type of *m*PEO-*b*-P(Cys)-*b*-P(His).

Polymer	M_n PEO	M_n P(Cys)$_x$-P(His)$_y$	I
	$\times 10^{-3}$ (g mol^{-1}) [a]	$\times 10^{-3}$ (g mol^{-1}) [b]	($\frac{M_w}{M_n}$) [b]
PCys5-PHis	10.0	5.9	1.16
PCys10-PHis	10.0	5.9	1.11
PHis-PCys5	10.0	6.0	1.17
PHis-PCys10	10.0	5.7	1.21
PCys5COPHis	10.0	6.1	1.18
PCys10COPHis	10.0	5.8	1.15

[a] M_n PEO. [b] Experimental M_n of deprotected P(Cys)$_x$-P(His)$_y$ blocks obtained by SEC-TALLS subtracting the MW of the PEO block. Measurements were conducted using 0.10% TFA (*v/v*) solution of H$_2$O/ACN (80/20 *v/v*) as the eluent at 35 °C.

As an example, we will present the characterization of the PCys5-PHis. The FT-IR spectra of the block copolymer *m*PEO$_{227}$-*b*-P(*t*BM-L-Cys)$_5$-*b*-P(N^{im}-*Trityl*-L-His)$_{40}$ is presented in Figure S22. Spectrum A corresponds to the protected copolymer *m*PEO$_{227}$-*b*-P(*t*BM-L-Cys)$_5$. The vibration at 1637 cm^{-1} is attributed to the C=O bond of the amide bond. Other characteristic peaks appear at 1100 cm^{-1} and 2890 cm^{-1}, which are due to the amplitude vibration of the ether bond C–O–C of PEO and C–H bonds respectively. In addition, the vibration at 1740 cm^{-1} corresponds to the vibration of one carbonyl group of L-cysteine *N*-carboxy anhydride, which indicates that the polymerization of the first monomer has not been completed at the time of the measurement. Spectrum B (Figure S22) corresponds to the copolymer *m*PEO$_{227}$-*b*-P(*t*BM-L-Cys)$_5$-*b*-P(N^{im}-*Trityl*-L-His)$_{40}$ and was obtained approximately 14 days after the addition of the second monomer (N^{im}-*Trityl*-L-His NCA). In this spectrum, the characteristic peak at 1679 cm^{-1} is observed, which corresponds to the amide bond, as well as the absorption bands at 1106 cm^{-1} of PEO. Additional peaks that appear in this spectrum are the vibration at 1784 cm^{-1}, which is due to the one carbonyl group of N^{im}-*Trityl*-L-His NCA, and indicates that the polymerization of the PHis block has not yet been completed. Additionally, the peaks at 745 cm^{-1} and 703 cm^{-1} are attributed to the bending vibrations of the -CH=CH- bonds of the aromatic rings of the *trityl* protecting groups of PHis. Spectrum C corresponds to the final fully protected block copolymer, which was isolated after precipitation in diethyl ether. This spectrum shows exactly the same absorption bands as spectrum B, with the difference that the vibration at 1784 cm^{-1} of the anhydride is absent, as the histidine monomer is completely consumed. Finally, the spectrum D corresponds to the desired fully deprotected PCys5PHis. In this spectrum the vibrations at 746 cm^{-1} and 702 cm^{-1}, which correspond to the *trityl* protecting groups of the poly(L-histidine), are absent, proving the successful deprotection of this polypeptide block. However, there are no accurate data from the FT-IR spectrum to verify the successful deprotection of poly(L-cysteine) building blocks, as the vibration signals of -SS-, -CH$_2$-S-, -SH, -CH (*t*-butyl) bonds are very weak. The successful synthesis and the purity of the terpolymer was confirmed by ^{1}H-NMR spectroscopy in D$_2$O/DCl 1% solvent, after each deprotection step. It is observed that all peaks in both spectra (Figure S23) are attributed to the hydrogens of the polymer. In Figure S23, the upper spectrum corresponds to the histidine-deprotected mPEO$_{227}$-b-P(tBM-L-Cys)$_5$-b-P(His)$_{40}$, while the second is attributed to the final fully deprotected PCys5PHis. Figure S23, Spectrum A: ^{1}H-NMR (600 MHz, D$_2$O/DCl 1%, δ, ppm): 1.34 (i: 9H, (CH$_3$)$_3$–C–), 3.18–3.24 (f + g: 4H, –CH$_2$–), 3.41 (h: 3H, CH$_3$–O–), 3.35–3.90 (e: 4H, –CH$_2$–CH$_2$–O–), 4.40 (d: 1H, NH–CH(CH$_2$–S–S)–C=O), 4.77 (c: 1H, NH–CH(CH$_2$–Im)–C=O), 7.35 (b: 1H, –C=CH–N–), 8.71 (a: 1H, –N=CH–N–). Figure S23, Spectrum B: ^{1}H-NMR (600 MHz, D$_2$O/DCl 1%, δ, ppm): 1.35 (i: 9H, (CH$_3$)$_3$–C–), 3.20–3.24 (f + g: 4H, –CH$_2$–), 3.42 (h: 3H, CH$_3$–O–), 3.35–3.90 (e: 4H, –CH$_2$–CH$_2$–O–), 4.53 (d: 1H, NH–CH(CH$_2$–S–S)–C=O), 4.75 (c: 1H, NH–CH(CH$_2$–Im)–C=O), 7.36 (b: 1H, –C=CH–N–), 8.71 (a: 1H, –N=CH–N). Finally, the histidine-deprotected terpolymer mPEO$_{227}$-b-P(tBM-L-Cys)$_5$-b-P(His)$_{40}$ was characterized by SEC chromatography in H$_2$O/TFA solvent (Figure S24).

A similar procedure was followed for the synthesis and characterization of all hybrid terpolymers. The characterization results from all polymers obtained by FT-IR, ^{1}H-NMR and SEC are presented in Supplementary Materials (Figures S7–S30).

3.3. Secondary Structure through Cyclic Dichroism

It is well known that polypeptides have the ability to mimic natural proteins by adopting secondary structures in response to various external stimuli (temperature, pH, etc.).

In order to investigate their structural and conformational changes by pH and temperature, we studied the synthesized polymers by CD. More precisely, CD measurements were conducted at four different pH values: pH = 7.4 (pH of the healthy tissue), pH = 6.5 (pH of the extracellular environment of the tumor tissue as well as early endosome pH within the cells), pH = 5.0 (pH of the lysosomes within the cell) and pH =3.0, and at three different temperatures: 25 °C (room temperature), 37 °C (temperature of the healthy tissue) and 40 °C (temperature of cancer tissue). We studied the PCys-protected polymers, while only PHis was deprotected, in order to avoid crosslinking.

In most cases, the results revealed a similar conformational transition of the secondary structure from a beta turn at higher pH values (pH = 7.4 and pH = 6.5) to a random coil conformation at lower pH values (pH = 5.0 and pH = 3.0), as shown in Figure 1 as well as Figures S31–S34. The negative peaks at 190 in combination with the positive peak at 205 nm and a slight negative peak at 218 nm are indicative of the β-turn type 2 conformation [47], the negative peak at 218 and a positive at 195 nm reveal a β-sheet conformation, while the negative peak at 225 nm is indicative of an α-helix conformation. At lower pH, the negative peak at 196 nm in combination with the positive peak at 218 nm are characteristic of the random coil structure.

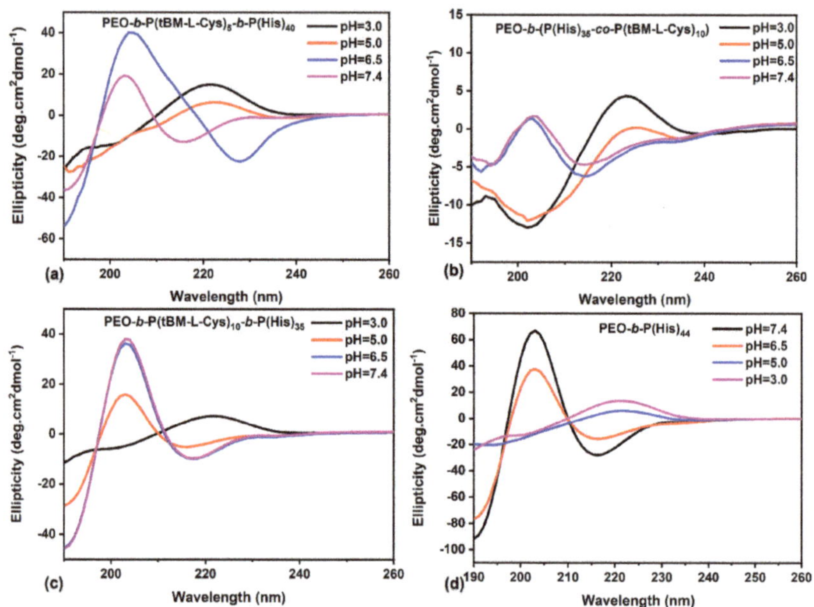

Figure 1. CD spectra at different pH values, at 25 °C: (**a**) $mPEO_{227}$-*b*-P(*t*BM-L-Cys)$_5$-*b*-P(His)$_{40}$, (**b**) $mPEO$-*b*-(P(His)$_{35}$-*co*-P(*t*BM-L-Cys)$_{10}$), (**c**) $mPEO_{227}$-*b*-P(*t*BM-L-Cys)$_{10}$-*b*-P(His)$_{35}$, (**d**) $mPEO$-*b*-P(His)$_{44}$.

As we showed in our previous work, at higher pH, the β-turn is a conformation that enthalpically favors the structure of PHis homopolymer [48]. In this conformation, the imidazole rings come close, developing the maximum hydrogen bonds. In this structure, a loop is created every three amino acids, since the nitrogen of the imidazole ring of an amino acid forms hydrogen bonds with the carbonyl group of the following amino acid and at the same time, forms hydrogen bonds with the hydrogen of the imidazole ring of the following amino acid. At lower pH, the secondary structure changes from β-turn to random coil conformation.

The conformational transitions obtained by altering the pH ($mPEO_{227}$-*b*-P(*t*BM-L-Cys)$_{10}$-*b*-P(His)$_{35}$ (Figure 1c), $mPEO_{227}$-*b*-P(His)$_{40}$-*b*-P(*t*BM-L-Cys)$_5$, $mPEO_{227}$-*b*-P(His)$_{35}$-*b*-P(*t*BM-L-Cys)$_{10}$ and $mPEO_{227}$-*b*-[P(*t*BM-L-Cys)$_5$-*co*-P(His)$_{40}$]) (Figure S32–S34, see supporting information) are similar to the one obtained by PEO-*b*-PHis diblock copolymer (Figure 1d). At the triblocks, a lower pH is required as compared to the one required for PEO-*b*-PHis to achieve the transition to the random coil conformation, due to the higher amount of hydrophobic blocks. At these terpolymers, the absorption of the PHis block dominates and overlaps the absorbance of the β-sheet conformation of the protected PCys block. However, in some cases, the secondary structure of the β-sheet is evident, which is the typical conformation of the free and protected PCys [49]. The terpolymer PEO-*b*-P(*t*BM-L-Cys)$_5$-*b*-P(His)$_{40}$ (Figure 1a) at pH = 7.4 exhibits a mixed structure of β-turn and β-sheet, which is attributed to both the PHis and PCys moieties. At pH = 6.5, a mixed structure of β-turn and α-helix is observed, as we can see a negative peak at 190 and a positive at 205 nm, while we also observe a negative peak at 230 nm indicative of the α-helix conformation. Finally, at more acidic pH (pH = 5.0 and pH = 3.0), only the conformation of the random coil is observed.

In case of $mPEO$-*b*-(P(His)$_{35}$-*co*-P(*t*BM-L-Cys)$_{10}$) (Figure 1b), the presence of a larger amount of PCys randomly distributed along the PHis chain induces a larger amount of β-sheet conformation at neutral pH. This is more pronounced at this copolypeptide due to the higher amount of PCys (Figure 1b) rather the one with lower amount and the same structure (random distribution of PCys) (Figure S34). It seems that the small amount of PCys did not have significant impact on the secondary structure. The difference of the secondary structure obtained from the random as compared to the block copolypeptides is proof of the random distribution of PCys along the PHis chain on the PCys5COPHis and PCys10COPHis terpolymers. Finally, it was found that by increasing the temperature maintaining a constant pH, from 25 °C, to 37 °C and then to 40 °C, the conformation is not altered (Figure S31), which shows that the polymers do not show a temperature responsiveness.

3.4. Self-Assembly of the Empty Hybrid Polymers

The ability of the synthesized polymers of the general type *m*PEO-*b*-P(Cys)-*b*-P(His) to form nanostructures was achieved via a solvent switch method, by applying the dialysis technique, with the use of DMSO as the common good solvent and aqueous solution at pH = 7.4, as the final media. During this procedure, a simultaneous crosslinking reaction was conducted, using H_2O_2 as the oxidative agent to form the disulfide bonds which stabilize the NPs. The ability of the polymers to self-assemble as well as the structural characteristics of the NPs were examined by DLS, SLS and TEM. At pH = 7.4, in which the self-assembly and crosslinking takes place, two populations are always observed by DLS (Figures S35–S39). The average size of the small population is about 30 nm, while the larger population is about 250 nm. The appearance of two populations is also observed by TEM microscopy. More specifically, for the crosslinked polymer PCys5-PHis the TEM image depicted at Figure 2g shows small spherical and elliptical vesicles within the core of a larger spherical nanostructure. The large NPs are composed of a large core containing multiple small vesicles. The matrix of the core is a mixture of PHis and PCys. Therefore, the dimensions of the NPs obtained by DLS (~31 nm, Figure S38) are the small vesicles within the core of the large NPs shown as the second population. The core of the NPs obtained by TEM for these NPs is almost 210 nm but if we add the PEO corona, we will achieve dimensions close to 220 nm which are smaller than the 250 nm obtained by DLS (Figure S35). This is probably due to the different processes followed for the sample preparation for DLS and TEM. For DLS, the nanoparticles are dissolved in PBS buffer pH = 7.4, while for TEM imaging, the treatment includes the removal of the salts by dialysis, freeze drying and finally redissolution in MilliQ water to be placed on the grid and the final evaporation of water to dryness. This difference in dimensions obtained between the two methods is common in many works [28], where TEM gives smaller dimensions, and can be attributed to the shrinkage caused by the evaporation of water. No TEM measurement was performed for the crosslinked polymer PCys10-PHis, as it precipitated during the process of self-assembly and crosslinking at PBS buffer pH = 7.4. The precipitation is due to the increased hydrophobicity and crosslinking, which is a consequence of the higher percentage of PCys in the polypeptide block and the close packing of PCys since they are obliged to be organized and located at the interphase between the PEO and PHis phases. In all cases, TEM measurements (Figure 2) revealed spherical micellar structures, with a multivesicular core comprised of PHis and PCys polypeptides and a hydrophilic corona of PEO. This kind of self-organization is consistent with the results from DLS and SLS, as mentioned.

SLS measurements confirm the self-assembly of the NPs in structures containing a multivesicular core, as the ratio R_g/R_h is close or slightly larger than 1. Table 2 summarizes the polymers of the present work and the corresponding values of the sizes R_g, R_h and R_g/R_h, at pH = 7.4, at 25 °C.

Finally, a general observation concerning all NPs is that in TEM images, around the gray core, a faint, white crown can be seen, which is attributed to the PEO block, as it does not create a strong contrast. This phenomenon comes in agreement with the results from z-potential measurements (Table 2), which reveal that at pH = 7.4, the mean value of the z-potential is in the range [−6.8 mV, + 3.3 mV], therefore, all the synthesized NPs have in most cases a neutral surface charge, indicating that the PEO block consists of the outer periphery of the nanostructures.

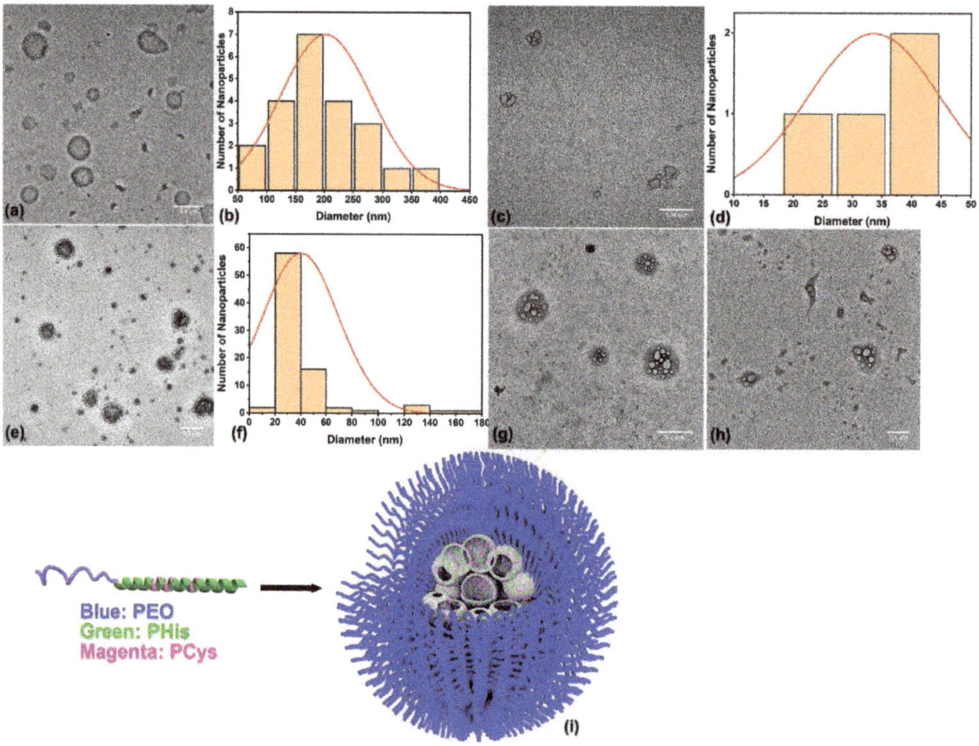

Figure 2. TEM image of the empty crosslinked NPs of (**a**) PCys5COPHis (scale bar is 0.5 μm); (**b**) histogram of size distribution of NPs of (**a**); (**c**) PHis-PCys10 (scale bar is 0.10 μm); (**d**) histogram of size distribution of NPs of (**c**); (**e**) PCys10COPHis (scale bar is 0.2 μm); (**f**) histogram of size distribution of NPs of (**e**); (**g**) PCys5-Phis (scale bar is 0.2 μm); (**h**) PHis-PCys5 (scale bar is 0.1 μm); (**i**) illustration of a NP featuring a multivesicular core.

Table 2. Molecular characteristics of the empty NPs by DLS and TEM.

POLYMER	R_g (nm)	R_h (D_h) [a] (nm)	R_g/R_h	Diameter by DLS (nm)	PDI by DLS	Average Core Diameter by TEM (nm)	Zeta Potential (mV) pH/GSH				
							7.4	6.5	5	6.5/GSH 10 mM	5/GSH 10 mM
PCys5-PHis	128	125 (250)	1.1	A.31 B.250	0.294	205	+2.4	+3.2	−3.0	−0.3	+7.8
PCys10-PHis	-	-	-	-	-	-	-	−0.6	+5.1	+0.4	+1.4
PHis-PCys5	134	122 (245)	1.1	A.32 B.215	0.344	52	−0.1	0.0	−	+3.3	−
PHis-PCys10	150	107 (215)	1.4	A.30 B.200	0.284	45	+3.3	−3.5	+0.4	−0.3	+0.5
PCys5COPHis	131	100 (200)	1.3	A.18 B.256	0.268	194	+3.1	−0.3	+1.3	+3.4	+3.8
PCys10COPHis	87	128 (256)	1.2	A.31 B.250	0.294	35, 158	+2.4	+3.2	−3.0	−0.3	+7.8

[a] The population with the largest dimension used was obtained by DLS.

3.5. pH and Redox Responsiveness of the Empty NPs

In order to investigate the pH and redox responsiveness of the synthesized NPs at both healthy and cancerous tissue conditions, different aqueous solutions of the polymers were prepared and measurements took place at different pH values and GSH concentrations. More specifically, DLS measurements were conducted to the solutions of the empty crosslinked polymers, resulting from the dialysis process, at pH = 7.4 (pH of human blood and healthy tissues), at pH = 6.5 (extracellular pH of cancer cells and early endosomes inside the cells) and at pH = 5.0 (lysosomal pH inside the cells), as well as at pH = 6.5 and pH = 5.0 with the addition of 10 mM GSH (intracellular GSH of cancer cells).

DLS measurements revealed responsiveness towards pH and GSH concentration. In the case of PCys5-PHis NPs (Figure S35), at pH = 7.4, the results indicate the existence of two populations, one small of about 31 nm and a larger one of about 250 nm. With a decrease in pH from 7.4 to 6.5 and 5.0 (25 °C, 90°), a slight increase in the diameter is observed, due to swelling of PHis through interaction with water, since at this pH PHis is protonated, which in turn leads to an increase in its hydrophilicity.

In the presence of 10 mM GSH at the acidic pH, the NPs exhibit a further redox response. GSH acts as a reducing agent and causes the cleavage of the disulfide bonds. The concentration of GSH is about 10–20 mM in cancer cells, while it is about 2 μM in healthy tissues. It is observed that at pH = 6.5, in the presence of GSH, a third population appears at 7 nm, most likely due to the rupture of the NPs to smaller particles or even single chains. In addition, at pH = 5.0 in the presence of GSH, except for the third population that appears at 7 nm, there is an additional increase in the size of the larger population from 285 nm (pH = 5.0, without GSH) to 427 nm (pH = 5.0, with GSH). Both of these results prove the synergistic response of PHis and PCys, through the variation in pH and GSH under healthy and cancerous conditions. The same trend is observed for all the polymers and the results from DLS measurements are summarized in Supplementary Materials (Figures S35–S39).

3.6. Self-Assembly of the DOX-Loaded NPs

The encapsulation of DOX was performed on all series of the polymers, as well as on another two polymers PEO$_{227}$-*b*-P(His)$_{44}$ and PEO$_{227}$-*b*-P(Sar)$_{98}$-*b*-P(Cys)$_{30}$. DLS and TEM techniques were employed to obtain the structure and the morphological characteristics of NPs, while z-potential measurements were conducted to determine their surface charge. All the results are summarized in Table 3 and Figure 3, while DLS results are presented in Supplementary Materials (Figures S40–S43).

Table 3. Molecular characteristics by DLS and TEM, encapsulation efficiency and loading capacity of the DOX-loaded NPs.

POLYMER	Diameter by DLS (nm)	PDI by DLS	Average Core Diameter by TEM (nm)	Zeta Potential (mV)	EE (%)	LC (%)
PCys5-PHis	204	0.227	120	−7.6	13.6	8.5
PCys10-PHis	193	0.136	108	−8.1	19.8	9.7
PHis-PCys5	159	0.137	15	1.1	19.4	9.5
PHis-PCys10	208	0.196	23	0.88	15.2	7.9
PCys5COPHis	118	0.146	100	4.3	20.3	10.0
PCys10COPHis	154	0.164	130	1.4	23.0	12.0

Surprisingly, the results from DLS and TEM revealed that in some cases, the DOX-loaded NPs can self-assemble into homogeneous core–shell micellar structures, instead of polydisperse micellar structures with a multivesicular core obtained by the empty one. It can be seen (Table 3) that the diameter of the NPs obtained by TEM is smaller than the R$_h$, due to the different preparation methods followed, as referred to previously. As an example, the TEM images and the size distribution of the core of the NPs obtained by the PCysX-PHis, are shown in Figure 3b–d. It is obvious that the NPs formed by these terpolymers are core–shell micelles, composed of a core containing the hydrophobic polypeptides along with encapsulated DOX, with a corona composed of a PEO chain and water. The dimension of the NPs obtained by DLS is close to 190 nm, while the one obtained by TEM is close to 110 nm. In case of the TEM images, we see only the core of the NPs, while by DLS we see the outer dimensions of the NPs. Still, the addition of the PEO which is smaller than 10 nm on the dimensions obtained by TEM cannot match the dimensions of the DLS. The structure of the micelles is illustrated in Figure 3n.

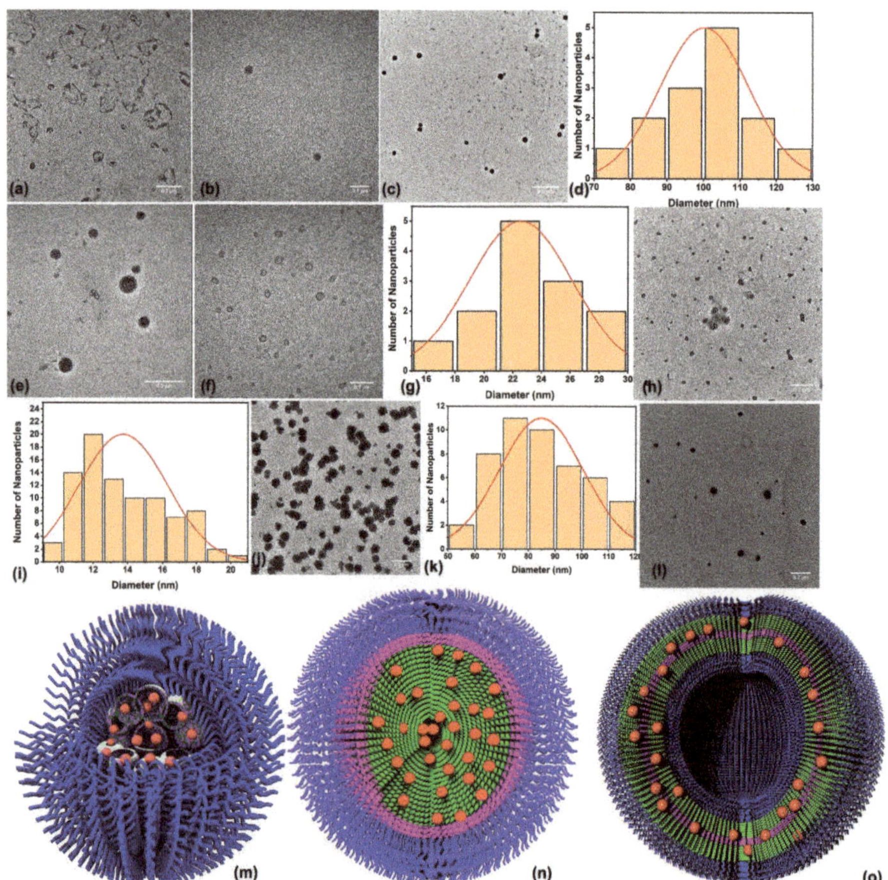

Figure 3. TEM image of the DOX-loaded NPs of (**a**) PCys5COPHis (scale bar is 0.5 μm); (**b**) PCys10-PHis (scale bar is 0.1 μm); (**c**) PCys5-Phis (scale bar is 0.5 μm); (**d**) histogram of size distribution of NPs of (**c**); (**e**) PCys10COPHis (scale bar is 0.2 μm); (**f**) PHis-PCys10 (scale bar is 0.1 μm); (**g**) histogram of size distribution of NPs of (**f**); (**h**) PHis-PCys5 (scale bar is 0.2 μm); (**i**) histogram of size distribution of NPs of (**h**); (**j**) $mPEO_{22}$-b-P(His)$_{44}$ (scale bar is 0.2 μm); (**k**) histogram of size distribution of NPs of (**k**); (**l**) $mPEO_{227}$-b-P(Sar)$_{98}$-b-P(Cys)$_{30}$ (scale bar is 0.2 μm); (**m**) cartoon of a micellar structure composed of a multivesicular core loaded with DOX; (**n**) cartoon of a core–shell micelle, with the core loaded with DOX; (**o**) cartoon of a vesicular structure with the polymeric monolayer loaded with DOX. PHis has a green color, PCys is magenta, while PEO is blue. The vesicles at the core at Figure 3 (**m**) is a mixture of PHis (green) with PCys (magenta).

The TEM images of the DOX-loaded NPs of the PHis-PCysX as well as the size distribution of the core of the NPs are shown in Figure 3f–i. The NPs are vesicular structures of rather small dimensions. In these images, a faint white diffuse cloud is observed that surrounds the vesicle and is due to the hydrophilic block of PEO. The possible structure of the vesicles is illustrated in Figure 3o. Particularly at these two polymers, the significant difference between the DLS and TEM dimensions is probably due to the formation of vesicular structures, where the elimination of the solvent is expected to result in a significant reduction in dimensions (as observed by TEM) due to the shrinkage of the NPs.

The TEM images of the DOX-loaded NPs of PCysXCOPHis are depicted on Figure 3a,e. The TEM image of PCys5COPHis in Figure 3ashows ruptured aggregates, probably ruptured vesicular structures. It seems that the small amounts of PCys at this polymer did not result in efficient crosslinking that would stabilize the aggregate. It seems that the random distribution of the small amount of PCys did not result in efficient crosslinking. On the contrary, in the case of the PCys10COPHis with higher amount of Pcys (Figure 3e), the NPs were more robust and aggregates with a core containing smaller vesicles were formed. The vesicular structures within the core were very small and could not be distinguished by DLS. In that series of NPs, the dimensions obtained by DLS and TEM are close.

The TEM image and the core size distribution of the DOX-loaded NPs of $mPEO_{227}$-b-$P(His)_{44}$ are shown in Figure 3j,k, respectively, while the TEM of the $mPEO_{227}$-b-$P(Sar)_{98}$-b-$P(Cys)_{30}$ is shown in Figure 3l. Both NPs self-assemble into spherical core–shell micellar structures. The hydrophobic core of the micelle, where DOX is encapsulated, consists of the blocks of PHis and PCys for the polymers $mPEO_{227}$-b-$P(His)_{44}$ and $mPEO_{227}$-b-$P(Sar)_{98}$-b-$P(Cys)_{30}$, respectively. The outer hydrophilic corona of the micelles is attributed to the PEO block in the case of $mPEO_{227}$-b-$P(His)_{44}$ and to the PEO and poly(sarcosine) blocks for the polymer $mPEO_{227}$-b-$P(Sar)_{98}$-b-$P(Cys)_{30}$. It is obvious that the presence of PCys at the terpolymers significantly altered the structure of the NPs as compared to the one that lacked the PCys layer.

Finally, Z-potential measurements were conducted in all DOX-loaded NPs, in order to determine their surface charge. The results are presented in Table 3 and show that at pH = 7.4, the mean value of the z-potential is in the range [−8.1 mV, + 1.1 mV], concluding that all synthesized NPs have a neutral surface charge. These results come in accordance with the observations from TEM images, which prove that the uncharged and hydrophilic block of PEO is located at the outer periphery of the nanoparticle. In summary, TEM imaging revealed how the PCys topology as well as the encapsulation of DOX affects the morphology of the DOX-loaded NPs. Thus, it is expected that the topology of the polypeptidic blocks will influence the kinetics of drug release under healthy and cancer cell conditions.

3.7. Drug Loading and In Vitro Release Studies

Drug loading was performed at pH = 7.4, using PBS isotonic buffer (150 mM NaCl, 10 mM PBS). The encapsulation efficiency (EE) of the drug and the loading capacity (LC) of the various NPs were determined by UV–Vis spectrophotometry at 485 nm, since only DOX absorbs in this wavelength. Quantification was performed using a standard DOX calibration curve in the corresponding PBS buffer pH = 7.4, presented in Supplementary Materials (Figure S44). Table 3 summarizes the results from UV–Vis spectroscopy measurements.

The drug release profile was examined at various pHs, temperatures, and in the presence of GSH, in order to simulate the release conditions in both healthy (pH = 7.4, 37 °C) and cancer tissue (pH = 6.5, 40 °C, 10 mM GSH) as well as late lysosomes environment of the cancer cells (pH = 5.0, 40 °C, 10 mM GSH). The amount of DOX released was determined by UV–Vis spectrophotometry at 485 nm and quantification was performed using standard calibration curves of the drug in the respective buffers, presented in Supplementary Materials (Figures S45 and S46).

It can be seen that the NPs consisting of the aggregated polymer PCys5-PHis in Figure 4a are pH-stimuli responsive, since after 144 hours, 34% of the drug has been released at pH 7.4, 56% at pH 6.5 and 60% at pH 5.0. It is obvious that as the pH of the release medium decreases, the percentage of released DOX increases. This effect is expected, since at acidic pH, the imidazole ring of histidine is protonated, rendering the PHis blocks hydrophobic, leading to the swelling or the rupture of NPs. This is in agreement with the DLS results. Finally, the rupture of the nanoparticles leads to the release of DOX in a pH-controlled manner. It seems that PCys is not contributing significantly to the release of the drug, since the lowering of only the pH results in a significant increase at the release rate of DOX. Therefore, 5 monomeric units of Cys is not enough to create a strong crosslinked layer that will direct the release of the drug.

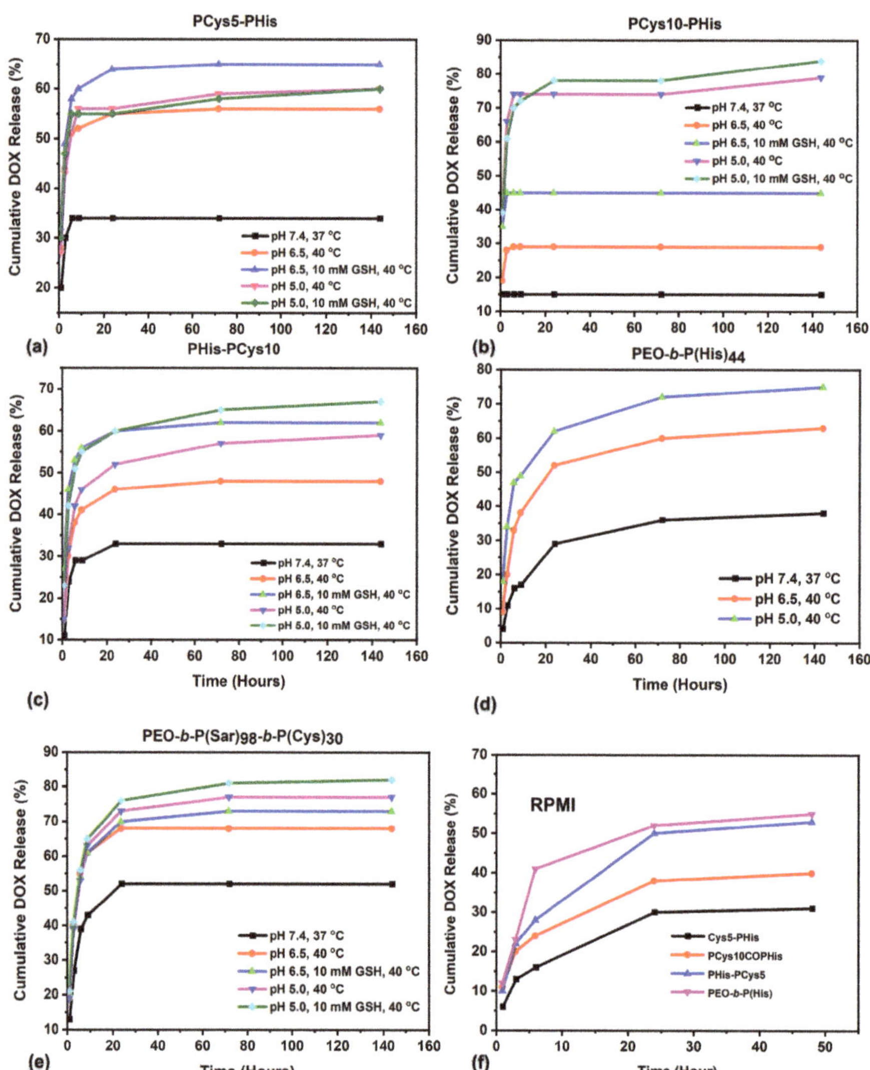

Figure 4. Cumulative release of the DOX-loaded NPs: (**a**) PCys5-PHis, (**b**) PCys10-PHis, (**c**) PHis-PCys10, (**d**) PEO-*b*-PHis, (**e**) PEO-*b*-(PSAR)$_{98}$-*b*-(PCys)$_{30}$, (**f**) RPMI.

Contrary to the NPs formed by PCys5-PHis with the lower amount of PCys, the NPs formed by the polymers exhibited the same architecture but higher PCys amount, i.e., PCys10-PHis, at 144 hours, only 15% of the drug was released at pH = 7.4, 29% at pH = 6.5, while at pH = 5.0, 79% was released (Figure 4b).

It is worth noting that the percentages of the drug released at pH = 7.4 and 6.5 from the PCys10-PHis NPs are lower obtained at all the NPs. This may be due to the presence of the crosslinked PCys layer at the interphase of PEO that maintain the cargo within the core until it is heavily ruptured by an increased concentration of GSH (see Figure 3n). At the same time, the greater stability of the NPs due to the more extensive crosslinking leads to a more pronounced response to the GSH concentration, as at pH = 6.5 without GSH, the release reaches 29%, and at the same pH in the presence of GSH, the percentage increases to 45%. At pH = 5.0, without GSH the release reaches 75%, while at pH = 5.0 in the presence of GSH, it reaches 84%, demonstrating a strong synergistic response to these stimuli (pH and GSH). It seems that the strong swelling of the core at pH = 5.0 even without GSH leads almost to the rupture of the PCys crosslinked layer.

For the DOX-loaded NPs of PHis-PCys5 hybrid terpolymer, pH plays a more critical role than GSH. The NPs formed are micellar structures with a multivesicular core that are expected to be less robust than the core–shell micelles. As in the case of the PCys5-PHis, by lowering the pH we have a significant release of the drug (Figure S47). Due to the formation of a PCys monolayer within the bilayer of the hybrid copolypeptide (see Figure 3o), the release of the drug at pH = 7.4 remains rather low (about 25%) but increases significantly by the protonation of PHis when the pH is lowered. This can be attributed to the transition of PHis from hydrophobic to hydrophilic, and since most of the drug exists within the PHis layer, this switch leads to increased release.

In case of PHis-PCys10 NPs, it was found that the vesicles formed release a significant amount of drug at pH = 7.4. The release become gradual by lowering the pH as well as the increased concentration of GSH. In this case both parameters contribute equally. The maximum cumulative release was 65% (Figure 4c).

For the DOX-loaded PCys5COPHis NPs, core–shell structures composed of a multi-vesicular core were probably formed that were not robust as indicated by the rupturing under vacuum to dryness (Figure 3a). We have a significant release of drug even at pH = 7.4 which increases gradually by the lowering of pH (Figure S48) and the addition of GSH. The maximum cumulative release obtained was 75%. This release profile is similar to the one of PHis-PCys10, where both stimuli, pH and redox, contribute equally.

Slower release was obtained by the NPs formed by PCys10COPHis hybrid copolypep-tides (random structure) (Figure S49), due to the presence of a larger amount of PCys that hinders the release of the drug. The NPs have a similar core–shell structures exhibiting a multivesicular core similar to the one with the lower amount of PCys and the same structure. Although the release at pH = 7.4 is rather high, there is a gradual increase in the release by lowering the pH and addition of GSH. Similar gradual release profile was obtained by the vesicular structures of PHis-PCys10 as well as PCys5COPHis NPs.

In order to elucidate the influence of the PCys on the release profile of the loaded NPs, we studied the encapsulation and release of DOX of the NPs obtained by the hybrid copolypeptides $mPEO_{227}$-b-P(His)$_{44}$ and $mPEO_{227}$-b-P(Sar)$_{98}$-b-P(Cys)$_{30}$ shown in Figure 3d,e, respectively. The DOX loaded NPs of the polymer $mPEO_{227}$-b-P(His)$_{44}$ are core–shell micelles. They show responsiveness only to pH due to the PHis block. Thus, at 144 hours 35% of the drug has been released at pH = 7.4, 63% at pH = 6.5 and 75% at pH = 5.0. It is obvious that although the cumulative release at pH = 7.4 is comparable with most of the NPs in this work, at pH = 6.5 and 5.0 the release is higher. The lower release rates of the DOX at the terpolymers is due to the contribution of the hydrophobic PCys and its crosslinking. At pH = 7.4 where both polypeptides are hydrophobic, the release profile do not depend on the presence of PCys significantly, unless PCys is located at the interphase of PEO. At lower pH, the NPs from $mPEO_{227}$-b-P(His)$_{44}$ lose their structure faster than the one containing PCys due to the crosslinks formed by this amino acid that stabilize the structure and a lower pH is required to reach the same release rate.

In order to compare the release profile of the terpolymers with the one containing only a hydrophilic polymer and PCys, we synthesized many block polypeptides, initially PEO-*b*-PCys$_{44}$. However, the polymer was not soluble, and we found that the best solubility was on the triblock terpolymer mPEO$_{227}$-*b*-P(Sar)$_{98}$-*b*-P(Cys)$_{30}$. The increase in PEO did not result in significant enhancement of the solubility of the NPs, and we incorporated PSar for that purpose. It was found that the NPs formed by the mPEO$_{227}$-*b*-P(Sar)$_{98}$-*b*-P(Cys)$_{30}$ hybrid copolypeptide showed a significant release even at pH = 7.4 and inability to maintain the cargo even at neutral pH. In addition, the response to GSH was very strong, while to pH, it was minimal. The weak dependence of the release to pH is due to the protonation of DOX at lower pH which renders it more hydrophilic and not to the polymer.

In order to further examine the influence of the complex media of RPMI + FBS that mimic an even closer environment at the blood compartment, we performed drug release profiles in this media. To our knowledge, this is the first time that release curves have been performed at the cell culture medium and not only in buffers (Figure 4f). It is obvious that the release profiles are similar to that of the isotonic PBS buffer at pH = 7.4. The results show that even after 2 days, most of the drug is still encapsulated into the NPs and thus, the delivery of the drug is directed by the carriers.

3.8. In Vitro Cytotoxic Activity

The antiproliferative activity of the various NP solutions was tested by the colorimetric method of sulforhodamine B (SRB, Sulfurhodamine B).

From the three cell lines tested, the most sensitive to both DOX and nanoformulations were found to be MCF-7 cells, followed by T-47Ds, while MDA-MB231 were found to be the least sensitive under the experimental conditions used (Figure 5 and Table 4). Interestingly, DOXIL (or CAELYX) does not work well in these experimental conditions, which is probably explained by its composition. In contrast to DOXIL, the four nanocarriers tested showed similar activity to DOX, as shown by both the three cell line growth curves (Figure 5) and the GI50, TGI and LC50 parameters (Table 4). The mPEO$_{227}$-*b*-P(His)$_{44}$ NPs showed a slightly better effect on MB231 cells at a concentration of 1 μM (Figure 5); however, the other four nanocarriers did not exhibit any specificity in the cytotoxic activity, similarly to free DOX (Table 4). Finally, none of the empty nanocarriers tested in the same cell lines and experimental conditions showed toxicity.

Table 4. GI50 (Growth Inhibiting concentration 50), TGI (Total Growth Inhibition) and LC50 (Lethal Concentration 50) against three established human breast cancer cell lines. All values are in μM.

Cancer Cell Line	D$_{OXO}$_PEO-*b*-P(Cys)$_5$-*b*-P(His)$_{40}$			D$_{OXO}$_PEO-*b*-P(His)			D$_{OXO}$_PEO-*b*-P(His)$_{40}$-*b*-P(Cys)$_5$			D$_{OXO}$_PEO-*b*-[P(Cys)$_{10}$-*co*-P(His)$_{35}$			DOXIL			DOX		
	GI$_{50}$	TGI	LC$_{50}$	GI$_{50}$	TGI	LC$_{50}$	GI$_{50}$	TGI	LC$_{50}$	GI$_{50}$	TGI	LC$_{50}$	GI$_{50}$	TGI	LC$_{50}$	GI$_{50}$	TGI	LC$_{50}$
T47D	0.3	0.9	>10	0.4	0.8	>10	0.4	0.8	>10	0.2	0.8	>10	>10	>10	>10	0.4	0.9	>10
MCF7	0.5	4.5	>10	0.3	1.0	>10	0.7	5.4	>10	0.8	5.4	>10	>10	>10	>10	0.5	4.7	>10
MB231	0.6	6.5	>10	0.4	1.0	>10	0.4	7.3	>10	0.6	7.2	>10	>10	>10	>10	0.5	6.3	>10

3.9. Influence of the PCys Topology on Self-Assembly, DOX Loading, In Vitro Release Profile as Well as In Vitro Cytotoxic Activity

From the systematic study of the series of hybrid polymers, it is obvious that the topology of PCys plays a critical role in the structure of the NPs formed and thus, the release profile of the drug.

Concerning the empty NPs, the self-assembly resulted in the formation of micellar NPs composed of a multivesicular core and a shell from PEO chains.

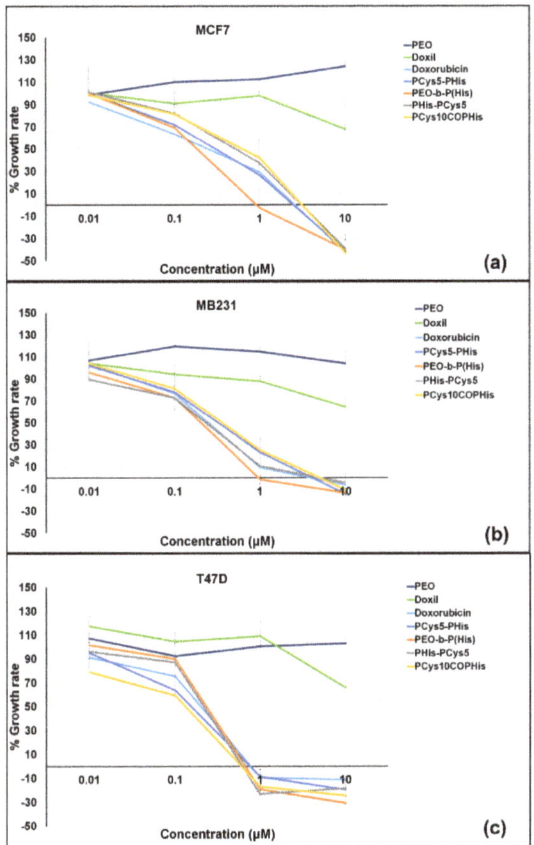

Figure 5. Growth rate curves for (**a**) the breast cancer cell lines MCF-7, (**b**) the breast cancer cell lines MB231, (**c**) the breast cancer cell lines T47D. Negative values of growth rate denote cytotoxic activity (see Section 2.7 under Materials and Methods for the calculation of the growth rate).

The loaded NPs with DOX presented differently structured NPs as compared to the empty one, which differs depending on the topology of PCys. In the case that PCys is at the interphase between the PEO and PHis, the formed loaded NPs are core–shell micelles. The crosslinked PCys interphase tightly close the drug within the core preventing its leakage at healthy tissue conditions. The drug is released slowly at extracellular cancer pH conditions, while it is released fast and efficiently under intracellular cancer cell conditions, where we have a combination of low pH and high concentration of GSH. Under intracellular healthy conditions, the release is slower than intracellular cancer cell conditions, but slightly higher than under extracellular healthy conditions.

When PCys is at the edge of the polymeric chain, PCys which is more hydrophobic than PHis at neutral pH, interacts with DOX directing the aggregation. So, the chains aggregate first through the PCys end block creating a bilayer with two PEG hydrophilic layers at the outer part leading to the formation of vesicular structures. As illustrated in Figure 3o, the PCys (magenta) layer aggregates in an antiparallel manner to form the bilayer, leading to the formation of vesicular structures. However, due to the encapsulation of DOX also at the PHis layer, we have a significant initial release even at higher pH values, since the crosslinked PCys layer does not hinder the leakage of the drug as in the case when

PCys was at the interphase between PEO and PHis (Figure 3n). Still, the release is gradual and responds at both stimuli (pH and redox) when the amount of PCys is higher.

When the PCys is randomly distributed along the PHis chain, the formed loaded NPs are core–shell structures with a multivesicular core. They present a significant initial release at higher pH as in the case of the NPs where the PCys was at the edge of the polymeric chain and the amount of PCys was large. In that case, PCys do not form a crosslinked tight layer, since the monomeric units of Cys are not close together.

This work shows that it is possible to select a drug release profile and the structure of the NPs formed by altering the topology of PCys. In most cases, both stimuli were participating at the release profile of the NPs. Slow release can be achieved by placing PCys at the interphase of the NPs and will be released fast when the NPs reach an intracellular cancer cell environment. When PCys is located at the edge of the polymeric chain, the NPs will form vesicular structures and it will be possible to encapsulate both a hydrophobic drug within the bilayer and a hydrophilic drug at the empty interior. The release will be performed in a gradual way, by lowering the pH and increasing GSH concentration. In that case a significant amount of PCys has to be incorporated at the NPs.

A similar gradual release can be achieved when PCys is randomly distributed along the PHis chain. This release profile can be achieved even for low amounts of PCys and through a core–shell micelle structure exhibiting a multivesicular core.

Usually, the nanoparticulate drug delivery results in a higher cancer cell growth rate as compared to the corresponding growth rate of the free drug [50]. In our work, the cell culture results showed that the GI50 of the NPs is comparable or better to the free drug after two days, although at least half of the drug is still encapsulated within the NPs. In the case of the PEO-*b*-Phis hybrid copolymer, the efficacy against all cancer cell lines was even better than free DOX. This shows that the presence of PHis favors the efficient accumulation of the drugs within the cancer cells through the rupturing of the endosomes by the "proton sponge mechanism", improving the delivery of the drug within the cells. These results are very encouraging for these materials to be used as drug delivery carriers for anticancer agents.

4. Conclusions

In this work, three series of novel hybrid amphiphilic terpolymers have been synthesized from the general type *m*PEO-*b*-P(Cys)-*b*-P(His) exhibiting different PCys topology, i.e., either between the PEO and PHis blocks, at the end of the polymeric chain or randomly distributed along the PHis chain. The terpolymers self-assemble to afford empty NPs mainly exhibiting the core–shell micellar structured NPs with multivesicular core. The polymeric materials can encapsulate the anticancer drug DOX to result in NPs exhibiting pH and redox responsiveness due to the PHis and PCys moieties, respectively, while PEO is always at the outer periphery presenting "stealth" properties, as z-potential measurements revealed. The encapsulated DOX was released in a controlled manner upon both stimuli, pH and GSH concentration. Depending on the PCys topology, NPs with different structures as well as release profiles were achieved. When the PCys is in the middle of the polymeric chain, core–shell micelles are formed, while the crosslinked PCys layer do not allow the leakage of the drug under healthy pH and GSH conditions. When the PCys is at the edge of the chain, vesicular structures are formed with gradual release of DOX depending on both stimuli. Finally, when PCys is randomly distributed, less robust core–shell micellar structures with a multivesicular structured core are formed that present a gradual release of the drug concerning both stimuli. The antiproliferative activity of these "smart" DOX-loaded NPs was tested in three breast cancer cell lines (MCF-7, T-47D and MDA-MB231) and the results revealed similar activity to DOX. Doxorubicin continues to be a cornerstone of anticancer chemotherapy being first line drug in different types of cancers and is probably the most commonly prescribed anticancer drug. However, doxorubicin suffers from severe side effects, the development of drug-induced toxicity, mainly cardiotoxicity, and the development of drug resistance being the most significant [51,52]. The cardiotoxicity is dose-dependent and the dose-limiting side effect of the drug may even result in the withdrawal of doxorubicin from the chemotherapeutic regimen. Drug resistance against doxorubicin and anthracyclines in general often occurs via the upregulation of MDR (Multi Drug Resistant) genes that control three different types of efflux proteins, pumping out of the cells, thus the drug reducing its intracellular concentration and ultimately diminishing its anticancer efficacy. To address these important clinical drawbacks of doxorubicin and because of the importance of this drug in oncology, drug delivery systems have been employed. The efforts so far have led to the development and clinical use of only two such systems: the liposomal doxorubicin and the pegylated liposomal formulation, while several other efforts have failed to advance such systems in the clinics. Thus, in this context, and if the subsequent evaluation of these "smart" systems show promise in addressing these pitfalls of the drug in vitro and most importantly in vivo in animal models of cancer, these materials could be very promising candidates in cancer treatment.

Supplementary Materials: The following supporting information can be downloaded at: https://www.mdpi.com/article/10.3390/pharmaceutics15030790/s1. Materials used and details on the synthesis of N-carboxy anhydrides are given in Schemes S1–S3, Figures S1–S6; details on the synthesis and characterization of the polymers are given in Scheme S4–S5, Figures S7–S30; Circular Dichroism results are shown in Figures S31–S34, DLS results are shown in Figures S35–S43; in vitro DOX release studies are shown in Figures S44–S46; release curves of the loaded nanoparticles are given in Figures S47–S49. References [53–57] are cited in the supplementary materials.

Author Contributions: Conceptualization, H.I.; bibliographic search, H.I., D.S. (Dimitra Stavroulaki), V.A., I.K., D.S. (Dimitrios Skourtis), P.G.F., A.L., D.K. and K.D.; methodology, D.S. (Dimitra Stavroulaki), V.A., I.K., D.S. (Dimitrios Skourtis), P.G.F., A.L., S.S., G.P. and P.T.; formal analysis, D.S. (Dimitra Stavroulaki), V.A., I.K., D.S. (Dimitrios Skourtis), P.G.F., A.L., S.S., G.P. and D.K.; supervision, H.I., K.D. and D.M.H.; writing—review and editing, H.I., P.G.F., K.D. and D.M.H.; resources, H.I., K.D. and D.M.H.; data curation, H.I., K.D. and D.M.H. All authors have read and agreed to the published version of the manuscript.

Funding: The research project was supported by the Hellenic Foundation for Research and Innovation (H.F.R.I.) under the "2nd Call for H.F.R.I. Research Projects to support Faculty Members & Researchers" (Project Number: 2762).

Institutional Review Board Statement: Not applicable.

Informed Consent Statement: Not applicable.

Data Availability Statement: Not applicable.

Conflicts of Interest: None of the authors have any conflicts of interest including any financial, personal or other relationships with other people or organizations.

Abbreviations

BOC, *tert*-butyloxycarbonyl; CD, circular dichroism; DCl, deuterium chloride; DMF, dimethylformamide; DMSO, methyl sulfoxide; DOX, doxorubicin; DLS, dynamic light scattering; DTT, dithiothreitol; EE, encapsulation efficiency; FBS, fetal bovine serum; FT-IR, Fourier transform infrared spectroscopy; GSH, glutathione; HVL, high vacuum line; LC, loading capacity; MMP2, matrix metalloproteinase 2; MSNs, mesoporous silica nanoparticles; MWCO, molecular weight-cutoff; MilliQ®, water purification systems from Merck Millipore; NCAs, *N*-carboxy anhydrides; NMR, nuclear magnetic resonance; NPs, nanoparticles; PBS, phosphate buffered saline; P(Cys), poly(L-cysteine); PEO, poly(ethylene oxide); P(His), poly(L-histidine); PPhe, poly(L-phenylalanine); PTFE, poly(tetrafluoroethylene); R_g, radius of gyration; R_h, hydrodynamic radius; ROP, ring-opening polymerization; ROS, reactive oxidative species; SCL, shell cross-linked; SEC, size exclusion chromatography; SLS, static light scattering; SRB, sulforhodamine B; TCA, trichloroacetic acid; TEA, triethylamine; TEM, transmission electron microscopy; TFA, trifluoroacetic acid; THF, tetrahydrofuran.

References

1. Norouzi, M.; Amerian, M.; Amerian, M.; Atyabi, F. Clinical applications of nanomedicine in cancer therapy. *Drug Discov. Today* **2020**, *25*, 107–125. [CrossRef] [PubMed]
2. Liao, J.; Jia, Y.; Wu, Y.; Shi, K.; Yang, D.; Li, P.; Qian, Z. Physical-, chemical-, and biological-responsive nanomedicine for cancer therapy. *Wiley Interdiscip. Rev. Nanomed. Nanobiotechnology* **2020**, *12*, e1581. [CrossRef] [PubMed]
3. Blanco, E.; Kessinger, C.W.; Sumer, B.D.; Gao, J. Multifunctional micellar nanomedicine for cancer therapy. *Exp. Biol. Med.* **2009**, *234*, 123–131. [CrossRef] [PubMed]
4. Bae, Y.; Nishiyama, N.; Fukushima, S.; Koyama, H.; Yasuhiro, M.; Kataoka, K. Preparation and biological characterization of polymeric micelle drug carriers with intracellular pH-triggered drug release property: Tumor permeability, controlled subcellular drug distribution, and enhanced in vivo antitumor efficacy. *Bioconjugate Chem.* **2005**, *16*, 122–130. [CrossRef]
5. Yoo, H.S.; Park, T.G. Folate receptor targeted biodegradable polymeric doxorubicin micelles. *J. Control. Release* **2004**, *96*, 273–283. [CrossRef]
6. Torchilin, V.P. Recent advances with liposomes as pharmaceutical carriers. *Nat. Rev. Drug Discov.* **2005**, *4*, 145–160. [CrossRef]
7. Liarou, E.; Varlas, S.; Skoulas, D.; Tsimblouli, C.; Sereti, E.; Dimas, K.; Iatrou, H. Smart polymersomes and hydrogels from polypeptide-based polymer systems through α-amino acid N-carboxyanhydride ring-opening polymerization. From chemistry to biomedical applications. *Prog. Polym. Sci.* **2018**, *83*, 28–78. [CrossRef]
8. Skoulas, D.; Mangiapia, G.; Parisi, D.; Kasimatis, M.; Glynos, E.; Stratikos, E.; Vlassopoulos, D.; Frielinghaus, H.; Iatrou, H. Tunable Hydrogels with Improved Viscoelastic Properties from Hybrid Polypeptides. *Macromolecules* **2021**, *54*, 10786–10800. [CrossRef]
9. Zhang, S.; Alvarez, J.D.; Sofroniew, V.M.; Deming, J.T. Design and synthesis of nonionic copolypeptide hydrogels with reversible thermoresponsive and tunable physical properties. *Biomacromolecules* **2015**, *16*, 1331–1340. [CrossRef]
10. Yhee, J.Y.; Son, S.; Son, S.; Joo, M.K.; Kwon, I.C. The EPR effect in cancer therapy. In *Cancer Targeted Drug Delivery*; Springer: Berlin/Heidelberg, Germany, 2013; pp. 621–632.
11. Li, J.; Kataoka, K. Chemo-physical Strategies to Advance the in Vivo Functionality of Targeted Nanomedicine: The Next Generation. *J. Am. Chem. Soc.* **2021**, *143*, 538–559. [CrossRef]
12. Mürdter, T.E.; Friedel, G.; Backman, J.T.; McClellan, M.; Schick, M.; Gerken, M.; Bosslet, K.; Fritz, P.; Toomes, H.; Kroemer, H.K.; et al. Dose Optimization of a Doxorubicin Prodrug (HMR 1826) in Isolated Perfused Human Lungs: Low Tumor pH Promotes Prodrug Activation by β-Glucuronidase. *J. Pharmacol. Exp. Ther.* **2002**, *301*, 223–228. [CrossRef]
13. Thistlethwaite, A.J.; Leeper, D.B.; Moylan, D.J.; Nerlinger, R.E. pH distribution in human tumors. *Int. J. Radiat. Oncol. Biol. Phys.* **1985**, *11*, 1647–1652. [CrossRef]
14. Najafi, M.; Goradel, N.H.; Farhood, B.; Salehi, E.; Solhjoo, S.; Toolee, H.; Kharazinejad, E.; Mortezaee, K. Tumor microenvironment: Interactions and therapy. *J. Cell. Physiol.* **2019**, *234*, 5700–5721. [CrossRef]
15. Feng, L.; Dong, Z.; Tao, D.; Zhang, Y.; Liu, Z. The acidic tumor microenvironment: A target for smart cancer nano-theranostics. *Natl. Sci. Rev.* **2018**, *5*, 269–286. [CrossRef]
16. Mavrogiorgis, D.; Bilalis, P.; Karatzas, A.; Skoulas, D.; Fotinogiannopoulou, G.; Iatrou, H. Controlled polymerization of histidine and synthesis of well-defined stimuli responsive polymers. Elucidation of the structure–aggregation relationship of this highly multifunctional material. *Polym. Chem.* **2014**, *5*, 6256–6278. [CrossRef]
17. Bilalis, P.; Tziveleka, L.-A.; Varlas, S.; Iatrou, H. pH-Sensitive nanogates based on poly (l-histidine) for controlled drug release from mesoporous silica nanoparticles. *Polym. Chem.* **2016**, *7*, 1475–1485. [CrossRef]
18. Karatzas, A.; Haataja, J.S.; Skoulas, D.; Bilalis, P.; Varlas, S.; Apostolidi, P.; Sofianopoulou, S.; Stratikos, E.; Houbenov, N.; Ikkala, O. Marcromolecular Architecture and Encapsulation of the Anticancer Drug Everolimus Control the Self-Assembly of Amphiphilic Polypeptide-Containing Hybrids. *Biomacromolecules* **2019**, *20*, 4546–4562. [CrossRef]
19. Li, B.; Xu, Q.; Li, X.; Zhang, P.; Zhao, X.; Wang, Y. Redox-responsive hyaluronic acid nanogels for hyperthermia-assisted chemotherapy to overcome multidrug resistance. *Carbohydr. Polym.* **2019**, *203*, 378–385. [CrossRef]
20. Elzes, M.R.; Akeroyd, N.; Engbersen, J.F.; Paulusse, J.M. Disulfide-functional poly (amido amine) s with tunable degradability for gene delivery. *J. Control. Release* **2016**, *244*, 357–365. [CrossRef]
21. Gamcsik, M.P.; Kasibhatla, M.S.; Teeter, S.D.; Colvin, O.M. Glutathione levels in human tumors. *Biomarkers* **2012**, *17*, 671–691. [CrossRef]
22. Perry, R.R.; Mazetta, J.; Levin, M.; Barranco, S.C. Glutathione levels and variability in breast tumors and normal tissue. *Cancer* **1993**, *72*, 783–787. [CrossRef] [PubMed]
23. Cheng, R.; Feng, F.; Meng, F.; Deng, C.; Feijen, J.; Zhong, Z. Glutathione-responsive nano-vehicles as a promising platform for targeted intracellular drug and gene delivery. *J. Control. Release* **2011**, *152*, 2–12. [CrossRef] [PubMed]
24. Wu, J.; Zhao, L.; Xu, X.; Bertrand, N.; Choi, W.I.; Yameen, B.; Shi, J.; Shah, V.; Mulvale, M.; MacLean, J.L. Hydrophobic cysteine poly (disulfide)-based redox-hypersensitive nanoparticle platform for cancer theranostics. *Angew. Chem.* **2015**, *127*, 9350–9355. [CrossRef]
25. Wang, L.; You, X.; Lou, Q.; He, S.; Zhang, J.; Dai, C.; Zhao, M.; Zhao, M.; Hu, H.; Wu, J. Cysteine-based redox-responsive nanoparticles for small-molecule agent delivery. *Biomater. Sci.* **2019**, *7*, 4218–4229. [CrossRef] [PubMed]
26. Huo, M.; Yuan, J.; Tao, L.; Wei, Y. Redox-responsive polymers for drug delivery: From molecular design to applications. *Polym. Chem.* **2014**, *5*, 1519–1528. [CrossRef]

27. Gyarmati, B.; Némethy, Á.; Szilágyi, A. Reversible disulphide formation in polymer networks: A versatile functional group from synthesis to applications. *Eur. Polym. J.* **2013**, *49*, 1268–1286. [CrossRef]

28. Wang, K.; Luo, G.-F.; Liu, Y.; Li, C.; Cheng, S.-X.; Zhuo, R.-X.; Zhang, X.-Z. Redox-sensitive shell cross-linked PEG–polypeptide hybrid micelles for controlled drug release. *Polym. Chem.* **2012**, *3*, 1084–1090. [CrossRef]

29. Wu, X.; Zhou, L.; Su, Y.; Dong, C.-M. Plasmonic, targeted, and dual drugs-loaded polypeptide composite nanoparticles for synergistic cocktail chemotherapy with photothermal therapy. *Biomacromolecules* **2016**, *17*, 2489–2501. [CrossRef]

30. Wu, X.; Zhou, L.; Su, Y.; Dong, C.-M. An autoreduction method to prepare plasmonic gold-embedded polypeptide micelles for synergistic chemo-photothermal therapy. *J. Mater. Chem. B* **2016**, *4*, 2142–2152. [CrossRef]

31. Bilalis, P.; Varlas, S.; Kiafa, A.; Velentzas, A.; Stravopodis, D.; Iatrou, H. Preparation of hybrid triple-stimuli responsive nanogels based on poly (L-histidine). *J. Polym. Sci. Part A Polym. Chem.* **2016**, *54*, 1278–1288. [CrossRef]

32. Lee, J.H.; Orfanou, K.; Driva, P.; Iatrou, H.; Hadjichristidis, N.; Lohse, D.J. Linear and Nonlinear Rheology of Dendritic Star Polymers: Experiment. *Macromolecules* **2008**, *41*, 9165–9178. [CrossRef]

33. Junnila, S.; Houbenov, N.; Karatzas, A.; Hadjichristidis, N.; Hirao, A.; Iatrou, H.; Ikkala, O. Side-Chain-Controlled Self-Assembly of Polystyrene-Polypeptide Miktoarm Star Copolymers. *Macromolecules* **2012**, *45*, 2850–2856. [CrossRef]

34. Yamauchi, K.; Akasaka, S.; Hasegawa, H.; Iatrou, H.; Hadjichristidis, N. Blends of a 3-miktoarm star terpolymer (3 mu-ISD) of isoprene (I), styrene (S), and dimethylsiloxane (D) with PS and PDMS. Effect on microdomain morphology and grain size. *Macromolecules* **2005**, *38*, 8022–8027. [CrossRef]

35. Gitsas, A.; Floudas, G.; Mondeshki, M.; Lieberwirth, I.; Spiess, H.W.; Iatrou, H.; Hadjichristidis, N.; Hirao, A. Hierarchical Self-Assembly and Dynamics of a Miktoarm Star chimera Composed of Poly(gamma-benzyl-L-glutamate), Polystyrene, and Polyisoprene. *Macromolecules* **2010**, *43*, 1874–1881. [CrossRef]

36. Houbenov, N.; Haataja, J.S.; Iatrou, H.; Hadjichristidis, N.; Ruokolainen, J.; Faul, C.F.J.; Ikkala, O. Self-Assembled Polymeric Supramolecular Frameworks. *Angew. Chem. Int. Ed.* **2011**, *50*, 2516–2520. [CrossRef]

37. Habraken, G.J.; Koning, C.E.; Heuts, J.P.; Heise, A. Thiol chemistry on well-defined synthetic polypeptides. *Chem. Commun.* **2009**, *24*, 3612–3614. [CrossRef]

38. Fetsch, C.; Grossmann, A.; Holz, L.; Nawroth, J.F.; Luxenhofer, R. Polypeptoids from N-substituted glycine N-carboxyanhydrides: Hydrophilic, hydrophobic, and amphiphilic polymers with poisson distribution. *Macromolecules* **2011**, *44*, 6746–6758. [CrossRef]

39. Hadjichristidis, N.; Iatrou, H.; Pitsikalis, M.; Sakellariou, G. Synthesis of well-defined polypeptide-based materials via the ring-opening polymerization of α-amino acid N-carboxyanhydrides. *Chem. Rev.* **2009**, *109*, 5528–5578. [CrossRef]

40. Deming, T.J. Living polymerization of α-amino acid-N-carboxyanhydrides. *J. Polym. Sci. Part A Polym. Chem.* **2000**, *38*, 3011–3018. [CrossRef]

41. Kricheldorf, H.R. Polypeptides and 100 years of chemistry of α-amino acid N-carboxyanhydrides. *Angew. Chem. Int. Ed.* **2006**, *45*, 5752–5784. [CrossRef]

42. Kataoka, K.; Matsumoto, T.; Yokoyama, M.; Okano, T.; Sakurai, Y.; Fukushima, S.; Okamoto, K.; Kwon, G.S. Doxorubicin-loaded poly (ethylene glycol)–poly (β-benzyl-l-aspartate) copolymer micelles: Their pharmaceutical characteristics and biological significance. *J. Control. Release* **2000**, *64*, 143–153. [CrossRef]

43. Skehan, P.; Storeng, R.; Scudiero, D.; Monks, A.; McMahon, J.; Vistica, D.; Warren, J.T.; Bokesch, H.; Kenney, S.; Boyd, M.R. New colorimetric cytotoxicity assay for anticancer-drug screening. *JNCI J. Natl. Cancer Inst.* **1990**, *82*, 1107–1112. [CrossRef]

44. Sereti, E.; Tsimplouli, C.; Kalaitsidou, E.; Sakellaridis, N.; Dimas, K. Study of the Relationship between Sigma Receptor Expression Levels and Some Common Sigma Ligand Activity in Cancer Using Human Cancer Cell Lines of the NCI-60 Cell Line Panel. *Biomedicines* **2021**, *9*, 38. [CrossRef]

45. Kamposioras, K.; Tsimplouli, C.; Verbeke, C.; Anthoney, A.; Daoukopoulou, A.; Papandreou, C.N.; Sakellaridis, N.; Vassilopoulos, G.; Potamianos, S.P.; Liakouli, V.; et al. Silencing of caveolin-1 in fibroblasts as opposed to epithelial tumor cells results in increased tumor growth rate and chemoresistance in a human pancreatic cancer model. *Int. J. Oncol.* **2019**, *54*, 537–549. [CrossRef]

46. Iatrou, H.; Dimas, K.; Gkikas, M.; Tsimblouli, C.; Sofianopoulou, S. Polymersomes from Polypeptide Containing Triblock Co- and Terpolymers for Drug Delivery against Pancreatic Cancer: Asymmetry of the External Hydrophilic Blocks. *Macromol. Biosci.* **2014**, *14*, 1222–1238. [CrossRef]

47. Greenfield, N.J. Analysis of Circular Dichroism Data. In *Methods in Enzymology*; Academic Press: Cambridge, MA, USA, 2004; Volume 383, pp. 282–317.

48. Bilalis, P.; Skoulas, D.; Karatzas, A.; Marakis, J.; Stamogiannos, A.; Tsimblouli, C.; Sereti, E.; Stratikos, E.; Dimas, K.; Vlassopoulos, D. Self-healing pH-and enzyme stimuli-responsive hydrogels for targeted delivery of gemcitabine to treat pancreatic cancer. *Biomacromolecules* **2018**, *19*, 3840–3852. [CrossRef]

49. Ulkoski, D.; Scholz, C. Synthesis and application of aurophilic poly (cysteine) and poly (cysteine)-containing copolymers. *Polymers* **2017**, *9*, 500. [CrossRef]

50. D'Angelo, N.A.; Noronha, M.A.; Câmara, M.C.C.; Kurnik, I.S.; Feng, C.; Araujo, V.H.S.; Santos, J.H.P.M.; Feitosa, V.; Molino, J.V.D.; Rangel-Yagui, C.O.; et al. Doxorubicin nanoformulations on therapy against cancer: An overview from the last 10 years. *Mater. Sci. Eng. C* **2021**, *133*, 112623. [CrossRef]

51. Al-malky, H.S.; Al Harthi, S.E.; Osman, A.-M.M. Major obstacles to doxorubicin therapy: Cardiotoxicity and drug resistance. *J. Oncol. Pharm. Pract.* **2020**, *26*, 434–444. [CrossRef]

52. Shafei, A.; El-Bakly, W.; Sobhy, A.; Wagdy, O.; Reda, A.; Aboelenin, O.; Marzouk, A.; El Habak, K.; Mostafa, R.; Ali, M.A.; et al. A review on the efficacy and toxicity of different doxorubicin nanoparticles for targeted therapy in metastatic breast cancer. *Biomed. Pharmacother.* **2017**, *95*, 1209–1218. [CrossRef]
53. Aliferis, T.; Iatrou, H.; Hadjichristidis, N. Living polypeptides. *Biomacromolecules* **2004**, *5*, 1653–1656. [CrossRef] [PubMed]
54. Pickel, D.L.; Politakos, N.; Avgeropoulos, A.; Messman, J.M. A mechanistic study of α-(amino acid)-N-carboxyanhydride polymerization: Comparing initiation and termination events in high-vacuum and traditional polymerization techniques. *Macromolecules* **2009**, *42*, 7781–7788. [CrossRef]
55. Hadjichristidis, N.; Iatrou, H.; Pispas, S.; Pitsikalis, M. Anionic polymerization: High vacuum techniques. *J. Polym. Sci. A Polym. Chem.* **2000**, *38*, 3211–3234. [CrossRef]
56. Uhrig, D.; Mays, J.W. Experimental techniques in high–vacuum anionic polymerization. *J. Polym. Sci. A Polym. Chem.* **2005**, *43*, 6179–6222. [CrossRef]
57. Hadjichristidis, N.; Hirao, A. *Anionic Polymerization*; Springer: Berlin/Heidelberg, Germany, 2015.

pharmaceutics

Article

Design and Evaluation of Pegylated Large 3D Pore Ferrisilicate as a Potential Insulin Protein Therapy to Treat Diabetic Mellitus

B. Rabindran Jermy [1,*], Mohammed Salahuddin [2], Gazali Tanimu [3], Hatim Dafalla [4], Sarah Almofty [5] and Vijaya Ravinayagam [6,*]

[1] Department of NanoMedicine Research, Institute for Research and Medical Consultations (IRMC), Imam Abdulrahman Bin Faisal University, P.O. Box 1982, Dammam 31441, Saudi Arabia

[2] Department of Clinical Pharmacy, Institute for Research and Medical Consultations (IRMC), Imam Abdulrahman Bin Faisal University, P.O. Box 1982, Dammam 31441, Saudi Arabia

[3] Center for Refining and Advanced Chemicals, Research Institute, King Fahd University of Petroleum and Minerals, P.O. Box 5040, Dhahran 31261, Saudi Arabia

[4] Core Research Facilities (CRF), King Fahd University of Petroleum and Minerals, P.O. Box 613, Dhahran 31261, Saudi Arabia

[5] Department of Stem Cell Research, Institute for Research and Medical Consultations (IRMC), Imam Abdulrahman Bin Faisal University, P.O. Box 1982, Dammam 31441, Saudi Arabia

[6] Deanship of Scientific Research & Department of NanoMedicine Research, Institute for Research and Medical Consultations (IRMC), Imam Abdulrahman Bin Faisal University, P.O. Box 1982, Dammam 31441, Saudi Arabia

* Correspondence: rjermy@iau.edu.sa (B.R.J.); vrnayagam@iau.edu.sa (V.R.)

Citation: Jermy, B.R.; Salahuddin, M.; Tanimu, G.; Dafalla, H.; Almofty, S.; Ravinayagam, V. Design and Evaluation of Pegylated Large 3D Pore Ferrisilicate as a Potential Insulin Protein Therapy to Treat Diabetic Mellitus. *Pharmaceutics* **2023**, *15*, 593. https://doi.org/10.3390/pharmaceutics15020593

Academic Editors: Koyo Nishida and Tomáš Etrych

Received: 22 October 2022
Revised: 27 January 2023
Accepted: 1 February 2023
Published: 9 February 2023

Abstract: An iron-based SBA-16 mesoporous silica (ferrisilicate) with a large surface area and three-dimensional (3D) pores is explored as a potential insulin delivery vehicle with improved encapsulation and loading efficiency. Fe was incorporated into a framework of ferrisilicate using the isomorphous substitution technique for direct synthesis. Fe^{3+} species were identified using diffuse reflectance spectroscopy. The large surface area (804 m^2/g), cubic pores (3.2 nm) and insulin loading were characterized using XRD, BET surface area, FTIR and TEM analyses. For pH sensitivity, the ferrisilicate was wrapped with polyethylene glycol (MW = 400 Daltons) (PEG). For comparison, Fe (10 wt%) was impregnated on a Korea Advanced Institute of Science and Technology Number 6 (KIT-6) sieve and Mesocellular Silica Foam (MSU-F). Insulin loading was optimized, and its release mechanism was studied using the dialysis membrane technique (MWCO = 14,000 Da) at physiological pH = 7.4, 6.8 and 1.2. The kinetics of the drug's release was studied using different structured/insulin nanoformulations, including Santa Barbara Amorphous materials (SBA-15, SBA-16), MSU-F, ultra-large-pore FDU-12 (ULPFDU-12) and ferrisilicates. A different insulin adsorption times (0.08–1 h), insulin/ferrisilicate ratios (0.125–1.0) and drug release rates at different pH were examined using the Korsmeyer–Peppas model. The rate of drug release and the diffusion mechanisms were obtained based on the release constant (k) and release exponent (n). The cytotoxicity of the nanoformulation was evaluated by 3-(4,5-dimethylthiazol-2-yl)-2,5-diphenyltetrazolium bromide (MTT) assay using human foreskin fibroblast (HFF-1) cells. A low cytotoxicity was observed for this nanoformulation starting at the highest concentrations used, namely, 400 and 800 µg. The hypoglycemic activity of insulin/ferrisilicate/PEG on acute administration in Wistar rats was studied using doses of 2, 5 and 10 mg/kg body weight. The developed facile ferrisilicate/PEG nanoformulation showed a high insulin encapsulation and loading capacity with pH-sensitive insulin release for potential delivery through the oral route.

Keywords: ferrisilicate; PEG; insulin; encapsulation; diabetic mellitus

1. Introduction

Diabetic mellitus (DM) is a metabolic disorder categorized by hyperglycemia due to inopportune insulin secretion. DM is differentiated as type I (insulin-reliant) and type II diabetes (insulin resistance) [1,2]. Diabetic mellitus, a disease termed as a life style disease, is quickly turning into a global epidemic. The prime reason for this is attributed due to change in life style, unhealthy diets and lack of awareness. In 2019, diabetic severity resulted in 1.5 million mortalities, and notably, 48% of these deaths occurred before 70 years of age. Among the two types of diabetes, type II diabetes is dominant, accounting for 95% (WHO). Recent data show that worldwide, about 537 million people have diabetes, and this number is expected to reach 783 million by 2045 [3]. Such a high percentage and increasing rate of DM is primarily attributed to obesity and changes in life style. The healthcare expenditure spent on diabetic treatment was estimated to be USD 966 billion in 2021.

Insulin was discovered by Frederick Banting in 1921, while Charles Best developed the clinical use of insulin in 1922 [4]. Insulin is administered to control the blood glucose level. Insulin helps to uptake glucose by binding with the insulin receptor and intitiating several protein activations cascades (e.g., Glut-4 transporter to plasma member, influx of glucose, synthesis of glycogen, glycolysis and triglyceride production). Type 1 DM (due to defective pancreatic β cells) depends on the lifelong supply of insulin. In the case of type 2 DM patients, the peripheral cells resist the administration of insulin, while some patients at the lateral life stage also require insulin. In order to treat type 1 DM, a common mode of insulin administration is through the subcutaneous route with about four injections per day. The treatment affects patient compliance and induces several side effects (lipoatrophy). Still, the subcutaneous administration route is preferred due to insulin's low bioavailability and challenges in developing an effective drug delivery system due to low membrane permeability and molecular size constraints.

In order to limit the number of injection cycles, a controlled insulin release strategy was followed using zinc and protamine. The used formulation showed poor reproducible kinetic parameters and the effect between meal periods was minimal and led to hypoglycemic events [5]. Accordingly, similar to insulin, insulin detemir and insulin glargine were shown to subvert the hyproglycemic action, but the required dose level was almost double compared to the normal dose of insulin for one day [6]. It has been reported that diabetic mediators similar to insulin could also alter the mitogenic pathway and can be a potential carcinogen in the long run [7]. The construction of a stimuli-responsive smart drug release system is the most recent attractive research direction involving interdisciplinary research between material science and medical science. An appropriately constructed nanovehicle with controlled insulin delivery using biocompatible nanosilica is proposed to overcome the deficiencies in subcutaneous therapy, improve therapeutic efficiency, enhance the stability of drug release and make ease diet control and exercise regiments. Several glucose-sensitive smart drug delivery system based on phenylboronic acid (PBA) and proteins such as concanavalin and glucose oxidase have been reported [8]. However, such a glucose-sensitive, sensor-based nanovehicle requires a multi-step synthesis procedure, the use of solvents and an advanced chemical set-up.

Recently, a biocompatible drug delivery system based on structured silica/polymeric nanocomposites are shown to be a promising nanovehicle to carry insulin [9]. A micronee-dle design based on mesoporous silica capped with zinc oxide in the form of an insulin reservoir has been reported to effectively control insulin delivery for prolonged periods of time [10]. Several studies are ongoing to improve the efficacy of protein delivery using mesoporous silica/chitosan and poly(lactic-co-glycolic) acid nanoformulations and improve their permeability [11]. The isomorphous substitution of biocompatible metals such as Fe, Zn, Ti, etc., into the silica framework is gaining importance in biomedical applications [12,13]. The use of Fe cations (Fe^{3+} and Fe^{2+}) with particle sizes ranging between 3 and 15 nm is gaining attention in multifunctional therapeutics as contrasting agents for magnetic resonance imaging, in hyperthermia treatments and as drug delivery agents [14]. The presence of iron oxide nanoparticles (FeNPs) favors biocompatibility [15], and as such, they are applied in hyperthermia for their anticancer [16,17] and antibacterial activity [18], as well as in tissue engineering [19]. Previously, we reported a direct hydrothermal synthesis of Iron-incorporated Santa Barbara Amorphous 16 (FeSBA-16) [20]. The presence of large

3D cubic pores of ferrisilicate could be exploited for insulin entrapment/loading capacity and insulin release. The wrapping of the nanocarrier with polyethylene glycol is reported to improve the bioavailability and drug stabilization and facilitate the transport of protein across human gastrointestinal fluid [21]. In this study, we investigated the effect of a pegylated, large 3D porous ferrisilicate/insulin nanoformulation for diabetes management. The textural characteristics are investigated using different physico-chemical characterization techniques. The insulin encapsulation/loading capacity and the pH-based, smart kinetic release behavior in response to stimuli were studied for insulin release. Furthermore, the nanoformulation toxicity in vitro and hypoglycemic effect in vivo were assessed.

2. Material and Methods

The silica source tetraethylorthosilicate (reagent grade, 98%, Sigma Aldrich, Darmstadt, Germany) and non-ionic template Pluronic F127 (BioReagent, suitable for cell culture, BASF, Wyandotte, MI, USA), iron(III) nitrate nonahydrate ($\geq$99.95%, BioReagent, suitable for cell culture, Sigma Aldrich, Saint Louis, MO, USA), n-butanol ($\geq$99%, anhydrous, Sigma Aldrich, Saint Louis, MO, USA), human recombinant insulin (rHu, dry powder, Sigma-Aldrich Chemie Holding GmbH, Taufkirchen, Germany) and polyethylene glycol (BioUltra, MW = 400 Daltons, Sigma-Aldrich Chemie Holding GmbH, Taufkirchen, Germany) were obtained from Sigma Aldrich. All chemicals were used as received without any further purification.

2.1. Synthesis

2.1.1. Synthesis of Ferrisilicate Using Hydrothermal Technique

Fe-SBA-16, termed as ferrisilicate, was prepared using sol–gel technique. The ferrisilicate containing Fe species can be tuned between SiO_2/Fe_2O_3 ratios of 50 and 250. In the present study, the Fe content can reach a SiO_2/Fe_2O_3 ratio of 50. In brief, 5 g of F127 was dissolved in acidic HCl solution (2 M) and stirred for 1 h. Then, 16 g of n-butanol (co-solvent) was added along with 24 g of tetraethylorthosilicate and the iron source (0.186 g of iron nitrate nonahydrate (Si/Fe ratio 250)) and stirred for 24 h. The mixture was stored in a polypropylene bottle (Nalgene, Rochester, NY, USA, Thermo Fisher Scientific, Bohemia, NY, USA) and transferred to an oven to be hydrothermally aged at 100 °C for 24 h. The precipitate was filtered, washed several times with excess water and dried at 100 °C for 12 h. The as-synthesized sample was finally calcined at 550 °C for 6 h.

2.1.2. Synthesis of Iron-Impregnated Structured Silica (10 wt% Fe/KIT-6 and 10 wt% Fe/MSU-F) Using Impregnation Technique

Firstly, 0.7235 g of iron nitrate nonahydrate was dissolved in 80 mL of distilled water. Then, 1.0 g of KIT-6, mesosilicalite, or Mesocellular Silica Foam (MSU-F) was added and stirred for 24 h at room temperature (RT). The solution was dried at 120 °C for 3 h and calcined at 500 °C for 2 h.

2.1.3. Insulin/Ferrisilicate

For insulin loading, 80 mg of insulin was added to 8 mL of 0.01 M HCl solution and stirred for 20 min. Then, 160 mg of ferrisilicate was added and stirred at 300 rpm overnight in an ice-cold environment. After that, the mixture was filtered, washed with 5 mL of distilled water and dried at RT (5 h) and stored at 4 °C.

2.1.4. Insulin/Ferrisilicate/PEGylation

For PEGylation, 14 µL of PEG (Molecular weight = 400) was added in 3 mL of deionized water, stirred for 20 min under argon atmosphere and then 150 mg of Insulin/Ferrisilicate was added and stirred under an ice-cold environment for 24 h. Then, the mixture was freeze-dried using the lyophilization technique.

2.2. Characterization Techniques

The phase of insulin, insulin/ferrisilicate/PEG, was identified using benchtop XRD (Miniflex 600, Rigaku, Tokyo, Japan). The textural features, including BET surface area, pore size and pore volume, were measured using the nitrogen adsorption technique (ASAP-2020 plus, Micromeritics, Norcross, GA, USA). The ferrite nanoparticles' chemical coordination was analyzed using DRS-UV-visible spectroscopy analysis (V-750, JASCO, Tokyo, Japan). The insulin functional groups of our nanoformulation were determined using FT-IR spectroscopy (L160000A, Perkin Elmer, Waltham, MA, USA). The morphological variations of insulin/ferrisilicate/PEG were investigated using transmission electron microscopy (TEM, JEM2100F, JEOL, Tokyo, Japan).

2.3. Insulin Entrapment Efficiency and Loading Capacity

Insulin entrapment efficiency (EE %) and loading capacity (LC %) over ferrisilicate were estimated using UV-visible spectroscopy at the specific wavelength of 275 nm.

$$EE(\%) = \left[\left(\frac{\text{Amount of insulin in ferrisilicate}}{\text{Initial amount of insulin}}\right)\right] \times 100$$

$$LC(\%) = \left[\left(\frac{\text{Initial amount of insulin} - \text{Insulin in supernatant}}{\text{Amount of ferrisilicate} + \text{insulin}}\right)\right] \times 100$$

2.4. Insulin Release Study

The release trends of different nanoformulations were studied using the dialysis membrane technique. First, 30 mg of the nanoformulation was placed inside 3 mL of PBS solution inside the dialysis membrane. The release of insulin was monitored under different pH solutions (7.4, 6.8 and 1.2) at 37 °C. At regular time intervals, 10 mL of solution was withdrawn and replaced with an equal volume of fresh solution. The amount of released insulin was identified at a specific wavelength of 275 nm. In order to measure the insulin release, a standard curve for insulin was established through calibration. An initial stock solution of 1000 μg/mL of insulin using phosphate buffer solution was prepared. We prepared 10 mL of 6 different concentrations of 5, 10, 15, 20, 25 and 30 μg/mL from stock solution using the working PBS solution (pH = 1.2 or 6.8 or 7.4), and then a calibration plot was established at the maximum absorption wavelength (λ_{max} = 275 nm). The linear regression was found to be $y = 0.0069x + 0.0071$, where y corresponds to absorbance and x to the concentration of the released drug (μg/mL). A linear calibration plot with a correlation coefficient of 0.993 was used to quantify insulin release from the nanoformulations. Each experiment was performed in triplicates.

2.5. Cytotoxicity of Insulin/Ferrisilicate Nanoformulation against HFF-1 Cells

Human foreskin fibroblast (HFF-1) cells were obtained as (SCRC-1041TM, ATCC, Manassas, VA, USA) and maintained in DMEM supplied with 10% fetal bovine serum, 1% L-glutamine and 1% penicillin-streptomycin (Gibco, Thermo Fisher Scientific, Waltham, MA, USA) in a humidified 5% CO_2 incubator (Galaxy® 170S, Eppendorf, Stevenage, UK) at 37 °C. The cells were seeded in a 96-well plate at (10^4 cells/well) and treated with group (a), insulin/ferrisilicate/PEG, insulin/10 wt% Fe/KIT-6/PEG, ferrisilicate/PEG and 10 wt% Fe/KIT-6/PEG using 25, 50, 100, 200, 400 and 800 μg/mL, and group (b), free insulin at 12.5, 25, 50, 100, 200 and 400 μg/mL, for 24, 48 and 72 h, including (c) untreated cells as control cells in each experiment. Both groups of concentrations are labeled as 1, 2, 3, 4 and 5 and were applied corresponding to the functionalized encapsulated insulin. MTT assay was performed to determine the cytotoxicity of the ferrisilicate nanoformulations by using a colorimetric-based reaction using MTT reagent 3-(4,5-dimethylthiazol-2-yl)-2,5-diphenyltetrazolium bromide to assess the metabolic activity of the cells. For the assay, 10 μL of MTT reagent (98%, 2128-1G, Sigma-Aldrich Chemie Holding GmbH, Taufkirchen, Germany) was added to obtain a final dilution of 1:10 and incubated for 4 h in a CO_2 incubator at 37 °C. The formed formazan blue dye was solubilized by adding 100 μL of DMSO (Dimethyl sulfoxide) and read at 570 nm wavelength by a SYNERGY Neo2 multi-mode microplate reader (BioTek Instruments, Winooski, VT, USA). Cell viability was calculated as:

$$\text{Cell viability}(\%) = \left(\frac{\text{Absorbance of Sample}}{\text{Absorbance of Control}}\right) \times 100$$

2.6. Statistics

The cytotoxicity assay was performed in three independent experiments. Statistical analysis was performed using Prism 9 software (GraphPad, La Jolla, CA, USA). The analysis was performed using two-way ANOVA with Dunnett's multiple comparison test. Statistically non-significant p values were indicated as (ns). The data analysis of drug delivery was conducted using Prism 8 software and SPSS software version 20.0 (IBM Corp., Armonk, NY, USA).

2.7. In Vivo Study

Wistar rats of either sex weighing 180–200 g were used for the experiments. The animals are obtained from the IRMC animal house and were maintained and treated according to the IACUC policy. The study was approved by Imam Abdulrahman Bin Faisal University IRB through IRB number IRB-2020-13-278 with approval date of 30 September 2020. A single-dose study was performed to learn the effect of testing samples in diabetic rats using acute administration. The diabetic rats were divided into three groups, as follows:

Experimental Grouping:

Group 1: Diabetic rats orally administrated with normal saline (marketed dehydration fluid and electrolyte replenishment of sodium chloride) without the nanoformulation;

Group 2, 3 and 4: Diabetic rats treated with nanoformulations (orally administered) at three gradient doses of 2, 5 and 10 IU/kg body weight, respectively.

Blood glucose levels were measured at the start of the study (will be considered as initial blood glucose level) for all the animals included in the study. The glucose solution (2 g/kg) was administered orally 30 min after the administration of the testing samples, and blood samples were collected at 1, 2, 3, 4, 5 and 6 h after glucose administration to estimate the blood glucose levels.

3. Results and Discussion

3.1. Characterization of Ferrisilicate/Insulin/PEG

Figure 1 depicts the XRD patterns of (A) insulin, (B) ferrisilicate and the (C) insulin/ferrisilicate/PEG nanoformulation. The recombinant insulin powder exhibited a crystalline peaks (2θ range 5–15°), characteristic of macrostructured protein. The ferrisilicate exhibited a broad amorphous silica peak between 10 and 40°. In the insulin/ferrisilicate/PEG nanoformulation, the crystalline peak of insulin is absent, which clearly indicates the molecular dispersion or amorphous transformation of ferrisilicate.

A nitrogen adsorption isotherm was used to characterize the surface texture and pore diameter of mesoporous materials in the range of 2–50 nm. In our case, the changes in the ferrisilicate's surface texture and three-dimensional cubic pores before and after insulin and PEG modifications were analyzed (Figure 2A,B). Parent ferrisilicate with an *Im3m* space group in the calcined form exhibited a type IV isotherm pattern with surface area of 804 m^2/g, pore volume of 0.65 cm^3/g and pore size centered at about 3.2 nm. After insulin loading and PEG wrapping, the quantity of nitrogen adsorbed reflected in the peak height that steeply decreased with capillary condensation (Figure 2A(a,b)). The hysteresis loop slightly decreases, indicating the occupation of pores by insulin. In parallel, the surface area (335 m^2/g) and pore volume (0.28 cm^3/g) decreases while the pore size stays at about 3.3 nm (Figure 2B(c,d)).

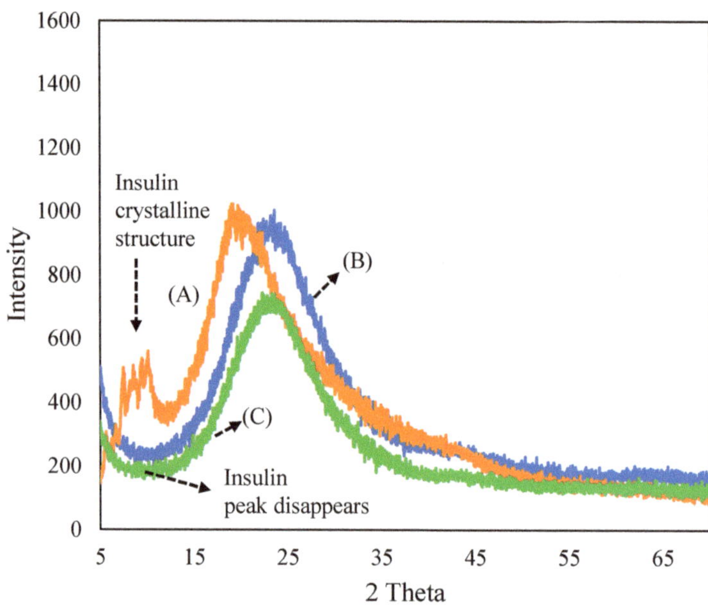

Figure 1. X-ray diffraction pattern of (**A**) insulin, (**B**) ferrisilicate and (**C**) insulin/ferrisilicate/PEG.

The textural changes in the samples before and after insulin/PEG wrapping were analyzed using nitrogen adsorption isotherm, and the results are presented in Table 1. A systematic change in the surface and pore volume indicates a successful insulin inclusion into the cubic pores of the ferrisilicate, while PEG wrapping occupies about 42% of the area around the 3D ferrisilicate. FeKIT-6 and mesosilicalite exhibited a similar high surface area and porous architecture. Overall, the textural modification indicates the insulin deposition in cubic cage pores of ferrisilicate (Figure 2A,B). The diffuse reflectance UV-visible spectra of ferrisilicate, insulin, and insulin/ferrisilicate/PEG are shown in Figure 2C(e–g). Ferrisilicate indicates a broad adsorption band in the 200–300 nm range, with peak maxima at 216 and 245 nm. The presence of a high-energy band is ascribed to the ligand-to-metal charge transfer due to tetrahedral Fe^{3+} species [22]. The band at 216 nm and 245 nm is ascribed to th(e electronic transition of O^{2-} to t_{2g} and e_g orbitals of Fe^{3+} in the iron oxide cluster. Unlike iron oxide-loaded mesoporous silica, ferrisilicate also shows a broad unresolved absorption band expanding between 400 and 500 nm. The absorption at 400 and 500 nm are ascribed to the quantum size effect of alpha-Fe_2O_3 species. Ferrisilicate shows the presence of Fe^{3+} species and the presence of a few hexacoordinated alpha-Fe_2O_3 species (Figure 2C(e)). Such iron oxide species occur in aggregated form, with octahedral or distorted octahedral coordination [23], inside the cubic pore channels of ferrisilicate. The insulin binding ability of ferrisilicate was measured using DRS-UV-Visible spectra. Insulin showed a strong absorption band at about 285 nm. The administering of ferrisilicate loaded with insulin (50 wt/wt%) followed by washing showed a similar peak to that of insulin at about 281 nm. The PEG wrapping shows effective conjugation with insulin through hydrogen bonding (Figure 2C(f,g)). Hinds et al. (2002) [24] stated that the PEG hydroxyl group interact with the amine functional moiety of the aminoacids of insulin. The FTIR spectra of ferrisilicate, insulin/ferrisilicate/PEG and insulin are shown in Figure 2D(h–j). Ferrisilicate exhibited characteristic peaks at the hydroxyl region between 3730 cm^{-1} and 3610 cm^{-1}. The presence of a broad band at about 3610 cm^{-1} is attributed to the hydroxyl group bridging the signalling between the Bronsted sites related to the isomorphous substitution of Fe for Si in the ferrisilicate framework [25]. Insulin showed characteristic carbonyl (C=O) stretching vibration bands due to the presence of amide at 1646 cm^{-1} and 1533 cm^{-1}. In case of insulin/ferrisilicate/PEG, the characteristic insulin vibration bands appear at about 1639 cm^{-1} and 1518 cm^{-1}. An increase in such vibrational peaks clearly shows the coupling of insulin with ferrisilicate. Moreover, there is a shift in the main vibration bands of insulin occurring from 1646 cm^{-1} to 1639 cm^{-1}. Such a shift in bands after the functionalization of insulin indicates the effective interaction between the insulin and ferrisilicate [26]. The loading capacity and entrapment of insulin inside the pore channels play a critical role in its release and anti-diabetic activity. Elongation in the broad hydroxyl (3000 cm^{-1}) and amine stretching (3291 cm^{-1}) bands compared to ferrisilicate indicates the effective functionalization of insulin in the insulin/ferrisilicate/PEG nanoformulation.

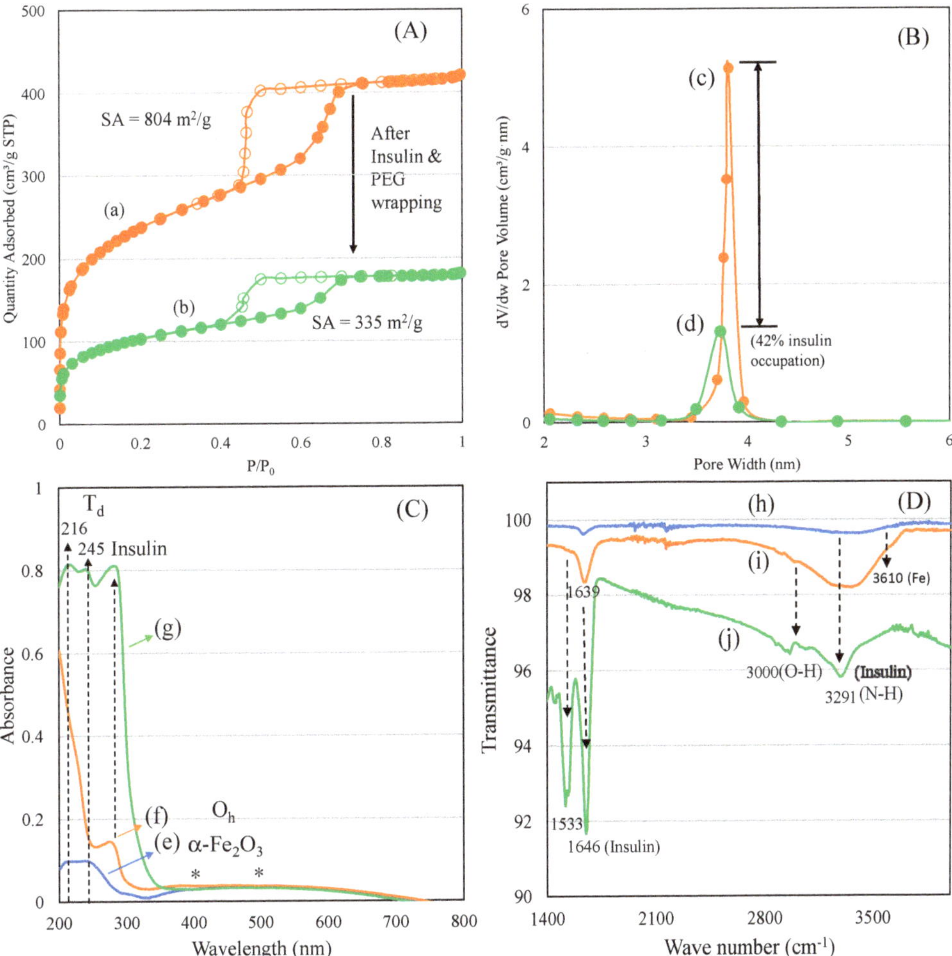

Figure 2. (**A**) Nitrogen adsorption isotherm: (a) ferrisilicate and (b) insulin/ferrisilicate/PEG. (**B**) Pore size distributions: (c) ferrisilicate and (d) insulin/ferrisilicate/PEG. (**C**) Diffuse reflectance spectra of (e) ferrisilicate, (f) insulin/ferrisilicate/PEG and (g) insulin. (**D**) FTIR spectra of (h) ferrisilicate, (i) insulin/ferrisilicate/PEG and (j) insulin.

Table 1. Textural properties of insulin and PEG-wrapped nanoformulations.

Nanocarriers	BET Surface Area (m^2/g)	Pore Volume (cm^3/g)	Average Pore Size (nm)
Ferrisilicate	804	0.65	3.2
Insulin/Ferrisilicate/PEG	335	0.28	3.3
Insulin/Fe-Mesosilicalite/PEG	360	0.38	4.2
Insulin/Fe-KIT-6/PEG	560	0.84	5.9

Figure 3A–D shows the morphological analyses of ferrisilicate and insulin-loaded, PEG-coated ferrisilicate. Ferrisilicate exhibits clear pore channels running on the parallel axis (Figure 3A). The analysis of the insulin-loaded, PEG-wrapped sample clearly shows the occurrence of structural transformation and enveloping by the polymeric layers. This indicates successful insulin loading and wrapping by PEG. In addition, magnification to 10 nm shows the presence of pore channels without any major changes inside the ferrisilicate (Figure 3C,D).

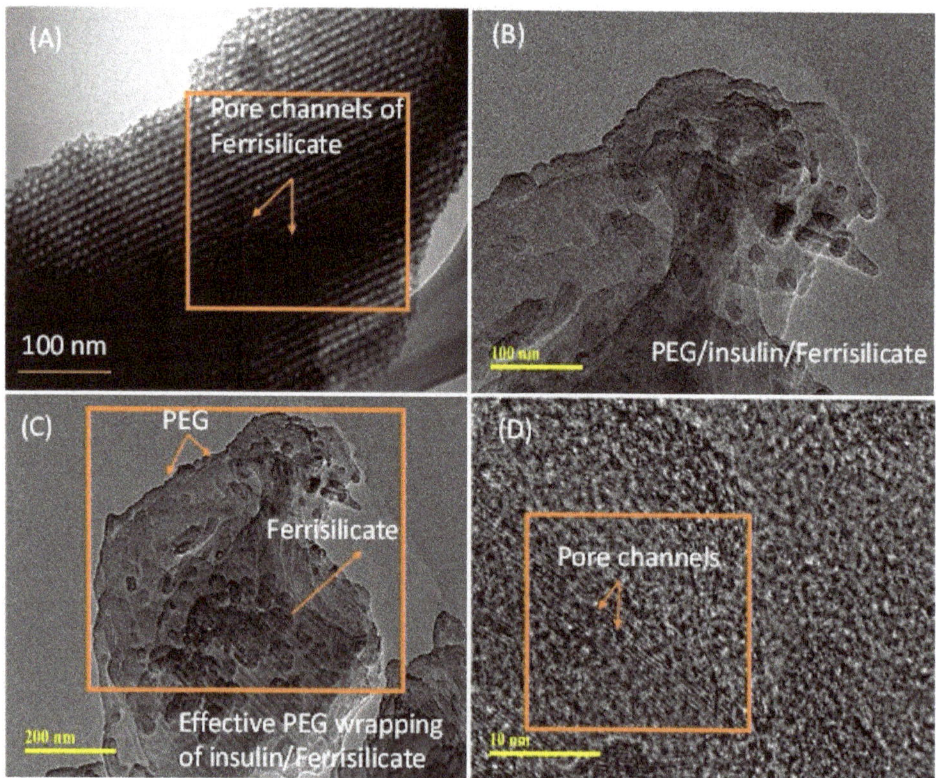

Figure 3. Transmission electron microscope images of (**A**) ferrisilicate and (**B–D**) insulin-loaded, PEG-wrapped ferrisilicate.

The effect of different porous-structured nanoformulations, entrapment efficiencies, loading capacities and insulin adsorption times (t = 0.08 h, 0.5 h, 0.75 h and 1.0 h) on insulin release are shown in Figure 4A–D. The insulin entrapment efficiency of hexagon-shaped SBA-15, cubic-shaped SBA-16, mesocellular forms with large pore windows, ultralarge-pore FDU-12 and ferrisilicate were studied. MSU-F and ULPFDU-12 contain large pores of

22 nm and 25 nm, respectively. SBA-15, SBA-16 and ferrisilicate have pore sizes of 8 nm, 5 nm and 3.2 nm, respectively. The entrapment efficiency was in the range of 82–93%, while the loading capacity was in the range of 38–46%. Interestingly, ferrisilicate showed high entrapment (92%) efficiency compared to large-pore nanomaterials MSU-F (91%) and ULPFDU-12 (93%). SBA-15 showed a slightly lower entrapment efficiency of 82% (Figure 4A). The insulin release profiles of encapsulated samples were studied using the dialysis membrane technique (Figure 4B). As expected, in the absence of polymer wrapping, the samples showed a high percentage cumulative insulin release. SBA-16, in the absence of iron species, showed a lower release of 68% at 72 h. The introduction of iron into SBA-16 silica (ferrisilicate) improves the insulin release by 95% compared to its silica counterpart, SiSBA-16. The presence of large pores favors the high release profile for MSU-F (91%), ULPFDU-12 (98%) and SBA-15 (91%).

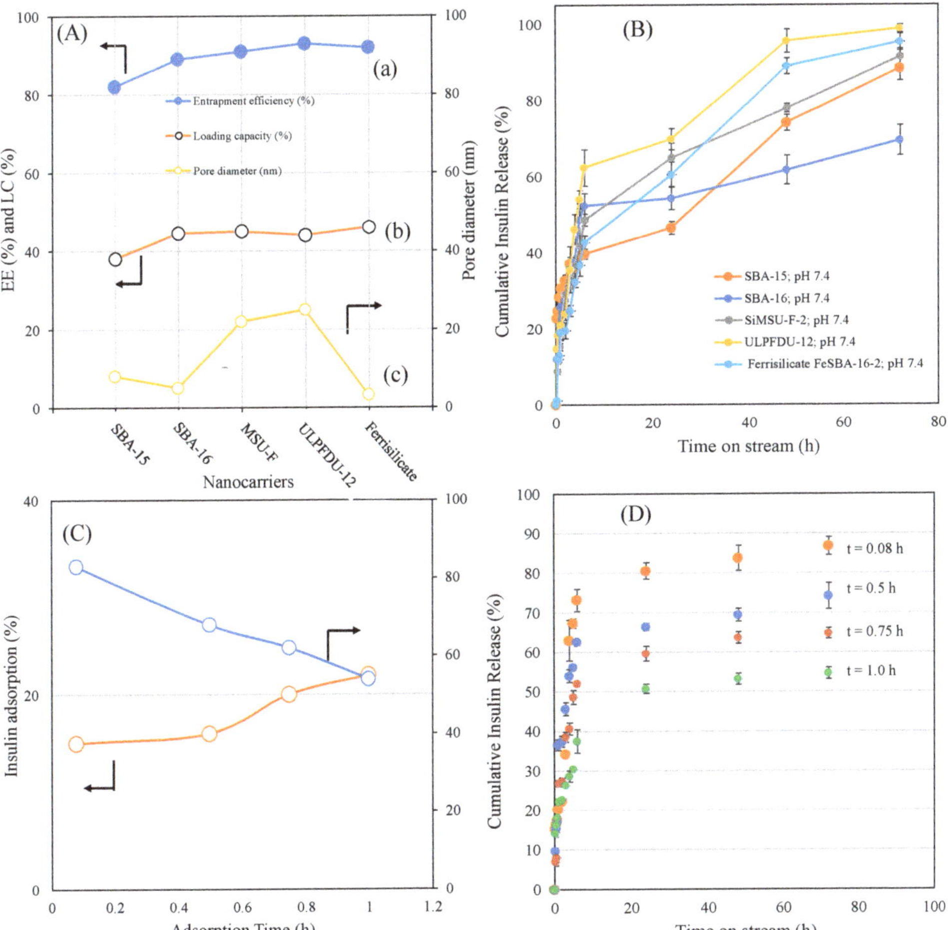

Figure 4. (**A**) Effect of different porous structured nanoformulations on (a) entrapment efficiency, (b) loading capacity and (c) pore diameter. (**B**) Percentage cumulative insulin delivery from different structured silicas. (**C**) Effect of different insulin adsorption times (t = 0.08 h, 0.5 h, 0.75 h and 1.0 h) on ferrisilicates versus percentage cumulative insulin release measured at 72 h. (**D**) Effect of insulin release (0.25–72 h) on ferrisilicate using different insulin adsorption times (0.08–1.0 h).

The effect of insulin adsorption over different adsorption times (t = 0.08 h, 0.5 h, 0.75 h and 1.0 h) on insulin release is shown in Figure 4C,D. Ferrisilicate shows an increase in insulin adsorption with time (15.7% in 0.08 h, 18.6% in 0.5 h, 20.5% in 0.75 h and 22% within 1 h). This result demonstrates that a higher adsorption time reduces insulin release. It indicates an effective entrapment with higher adsorption time. The entrapment efficiency and loading capacity reveals the efficiency of the drug-loaded nanoformulation [27]. A glucose-responsive system has been reported from PBA containing structured silica coated with diol-based copolymers (*N*-acryloyl glucosamine and *N*-isopropyl acrylamide). The formulation showed a high loading capacity (14.7%) and encapsulation efficiency (85.9%) with glucose responsive release at pH = 7.4 [28]. It has been reported that PBA- and diol-based block copolymers along with post-modification improved the glucose-sensitive release of insulin [29]. Replacing PBA with fluorophenylboronic acid has been reported to improve insulin loading, resulting in a high encapsulation efficiency and better glucose responses in physiological conditions [30]. Glucose-responsive sulfonamide–PBA, with its temperature-responsive properties, was reported to improve both loading capacity and encapsulation efficiency. These nanoparticles are reported to be safe and effective for the subcutaneous injection of insulin [31]. In our case, the entrapment efficiency was 92%, while the drug loading capacity reached up to 46%.

The insulin release over pegylated ferrisilicate, 10 wt% Fe/mesosilicalite and Fe/KIT-6 at different pH conditions (7.4, 6.8 and 1.2) are studied for 530 h (Figure 5). Ferrisilicate displays a high percentage of cumulative release of insulin of about 40–50%. The 10 wt% Fe/mesosilicalite and Fe/KIT-6, which contain 3D pore architectures, showed a release of about 20–30% over 530 h. This indicates the ink-shaped pores of SBA-16 (about 3.3 nm) are slightly restricted with Fe impregnation, showing a sustained release behavior with respect to insulin. The presence of 3D cage-type pores with Ia3d structures in KIT-6 was found to favor the slow release of insulin (20% for 530 h), while mesosilicalite with hexagonal pores of MCM-41 showed a release of about 30% over 530 h. This suggests that insulin tends to functionalize on the external micropores of mesosilicalite, while cage-type pores are able to accommodate the insulin inside the mesopores. Polysaccharide pullulan hydrogel in the form of carboxylation was reacted with concanavalin A (Con A) using amidization reaction. The nanoformulation was shown to control the release of insulin due to the specific bonding occurring between protein andglucose binding and to the crosslinked structure with uniform pores that accommodates insulin [32]. A detailed study on the glucose-responsive insulin release behavior of hydrogel/microgels-Con A nanoformulations shows that the release trend is controlled by the bolus and basal insulin release and network composition [33]. Konjac Mannan (heteropolysaccharide) fabricated with Con A through the crosslinking technique exhibited a targeted insulin release. The nanoformulation with a particle size of about 500 nm facilitated a glucose-responsive insulin release with a reversible pattern with different levels of glucose. Furthermore, the in vivo study reveals that the nanoparticles are non-toxic and able to control the blood sugar level for 6 h [34]. An immobilization of glucose oxidase on a biocompatible linear polysaccharide alginate/phenylboronic acid-derived composite exhibited an improved glucose sensitivity with a potential option for subcutaneous insulin delivery [35]. However, the protein stability, antigenicity and synthesis cost limit the design of such protein-based nanoformulations for clinical translation [36,37]. In case of phenylboronic acid functionalization, the cytotoxicity and lower solubility due to higher pKa (~9.0) are some limitations [38]. In the present study, the facile and simple nanocomposite formation between ferrisilicate and PEG can perform similar pH-sensitive insulin release.

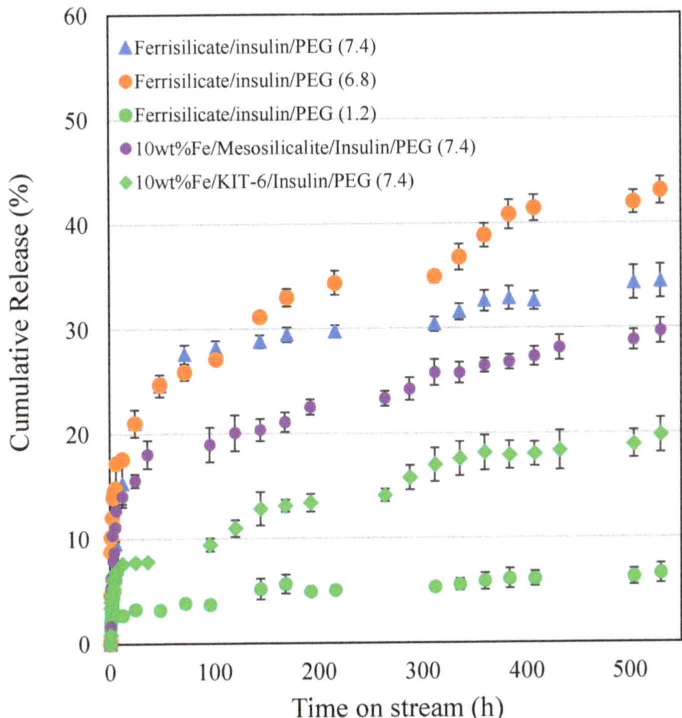

Figure 5. Insulin release by pegylated ferrisilicate, 10 wt% Fe/mesosilicalite and Fe/KIT-6 at different pH conditions (7.4, 6.8 and 1.2). The drug release experiment was performed in triplicates.

3.2. Kinetics of Different Ferrisilicate/Insulin Nanoformulation Drug Release Using the Korsmeyer–Peppas Model

The different ferrisilicate/insulin nanoformulation drug release profiles at different pHs were examined using the Korsmeyer–Peppas model, expressed using the equation:

$$R\% = kt^n$$

where R% is the different ferrisilicalite/insulin drug percentage release at time (t) and k and n are the kinetic release rate constant and the release exponent, respectively.

The kinetic parameters with their 95% confidence intervals are presented in Table 2.

For all the base materials; SiSBA-15, SiSBA-16, SiMSU-F and SiULPFDU-12, the rate of drug release is enhanced as observed from the higher release constants which are all higher than that of the modified FeSBA-16 base material. Similarly, the release mechanisms of the drugs nanoformulation obtained from the base materials followed the fickian diffusion mechanism (n < 0.45), while the modified FeSBA-16 drug release followed the non-fickian (0.45 < 0.488 < 0.89) diffusion mechanism. These indicate that modifying the base materials affect both the rate of release and the release diffusion mechanism.

The effect of four different insulin adsorption times (5, 30, 45 and 60 min) on the rate of release and diffusion mechanism of ferrisilicate nanoformulation drug shown that, the drug release mechanism at all the insulin adsorption times followed the fickian (n < 0.45) diffusion. However, the rate of drug release as signified by the released constant revealed that, the rate of drug release is highest after 30 min of insulin adsorption. Also, there is no direct correlation between the adsorption times studied and the rate of drug release, because the rate increased up to a maximum at 30 min adsorption, then continued to decline at higher insulin adsorption. This indicates that, there is a saturation limit of insulin adsorption, above which the release rate of the drug is affected.

Table 2. Kinetic parameters for insulin release from different porous-structured nanoformulations. Influence of adsorption time, insulin-to-ferrisilicate ratios and release at different pH conditions.

Drug Formulations	k (h^{-n})	n	$n-0.45$	R^2
SBA-15; pH 7.4	25.2086 ± 3.4391	0.2671 ± 0.0596	0.1829	0.9088
SBA-16; pH 7.4	21.9335 ± 3.2175	0.2882 ± 0.0638	0.1618	0.9102
MSU-F; pH 7.4	18.0398 ± 2.4209	0.4153 ± 0.0587	0.0347	0.9613
ULPFDU-12; pH 7.4	18.0954 ± 2.9981	0.4337 ± 0.0714	0.0163	0.9482
Ferrisilicate FeSBA-16; pH 7.4	13.7013 ± 2.1190	0.4879 ± 0.0670	0.0379	0.9634
Ferrisilicate; Adsorp time = 5 min	23.5404 ± 4.6002	0.3746 ± 0.0832	0.0754	0.9097
Ferrisilicate; Adsorp time = 30 min	27.6661 ± 5.4414	0.2840 ± 0.0837	0.1660	0.8512
Ferrisilicate; Adsorp time = 45 min	21.1912 ± 3.7614	0.3351 ± 0.0761	0.1149	0.9058
Ferrisilicate; Adsorp time = 60 min	20.1574 ± 1.2871	0.2555 ± 0.0288	0.1945	0.9750
Insulin/Ferrisilicate ratio 0.125-11-7.4	21.6909 ± 2.1147	0.2558 ± 0.0507	0.1942	0.9353
Insulin/Ferrisilicate ratio 0.25-12-7.4	26.0871 ± 3.4667	0.2814 ± 0.0680	0.1686	0.9067
Insulin/Ferrisilicate ratio 0.75-13-7.4	21.9251 ± 4.1966	0.4021 ± 0.0955	0.0479	0.9097
Insulin/Ferrisilicate ratio 1.0-14-7.4	36.3453 ± 3.6093	0.2102 ± 0.0516	0.2398	0.9040
Ferrisilicate/insulin/PEG (7.4)	1.4153 ± 0.5882	0.5722 ± 0.0837	0.1222	0.9013
Ferrisilicate/insulin/PEG (6.8)	7.8563 ± 1.3468	0.2728 ± 0.0381	0.1772	0.9093
Ferrisilicate/insulin/PEG (1.2)	2.6017 ± 0.1705	0.1791 ± 0.0151	0.2709	0.9631
10 wt% Fe/Mesosilicalite/insulin/PEG (7.4)	5.4089 ± 0.9689	0.2771 ± 0.0385	0.1729	0.9016
10 wt % Fe/KIT-6/insulin/PEG (7.4)	3.7004 ± 0.4118	0.2533 ± 0.0241	0.1967	0.9490

Different insulin/ferrisilicate ratios of 0.125, 0.25, 0.75 and 1.0 were utilized in the nanoformulation of drugs. The rate of drug release is highest at higher ratio of 1.0, indicating that the higher the insulin/ferrisilicate ratio, the greater the rate of drug release, except at the ratio of 0.75. The nanoformulation at all ratios followed the fickian diffusion mechanism. This confirmed that, the kinetic release profile of the nanoformed insulin/ferrisilicate drug is affected by the ratio of the insulin to the ferrisilicate base material.

The effect of different pH of ferrisilicate on the diffusion mechanisms and rate of ferrisilicate/insulin/PEG drug release revealed that, at close to neutral pH of 6.8, the rate of drug release is highest relative to highly acidic (pH = 1.2) and slightly basic (pH = 7.4) conditions. However, at both pH of 1.2 and 6.8, the release exponent signified fickian diffusion mechanism, while at pH of 7.4, the release mechanism changed to non-fickian diffusion. For these nanoformulation, it can be inferred that the pH of ferrisilicate affects both the rate of drug release and the diffusion mechanism.

Four drug nanoformulation with different base materials modification of ferrisilicate but having the same PEG coating at the same pH revealed that, the rate of drug release is enhanced with the modification, because of their higher release constant compared to the unmodified ferrisilicate. Also, the diffusion mechanism is affected by the modification as follows; 10 wt% Fe/KIT-6 and 10 wt% Fe/Mesosilicalite based nanoformulation drugs release followed the fickian (n < 0.45) diffusion mechanism while unmodified ferrisilicate-followed the non-fickian (0.45 < n < 0.89) diffusion mechanism.

For the null hypothesis: there is no significant differences between the various release exponent and 0.45, one-paired t-test was carried out and the calculated *p*-value (0.00003736) which is <<0.05 showed that the null hypothesis is invalid, rather the alternative hypothesis: that there are significant differences between the release exponents (n) and 0.45 is valid.

3.3. In Vitro Study

The cytotoxicity of ferrisilicate nanoformulations were studied on Human foreskin fibroblast (HFF-1) cells (Figure 6). Insulin/Ferrisilicate/PEG and insulin/10 wt% Fe/KIT-6/PEG were obliviously showing no cytotoxicity at 25, 50, 100, 200 μg/mL after each timepoint even after 72 h, and started to show cytotoxic effect by less than 50% cell viability at the highest concentration 800 μg/mL. Ferrisilicate, insulin/10 wt% Fe/KIT-6 and insulin were used as control groups to assess the cytotoxicity of empty vectors or NP material, as shown ferrisilicate and 10 wt% Fe/KIT-6 were not toxic even at the highest concentration 800 μg/mL post 72 h of treatment. Expectedly, insulin has stimulated cell growth to reach to 130% in comparison to DMEM- treated cells which have a 100% cell viability and used as control values in this experiment.

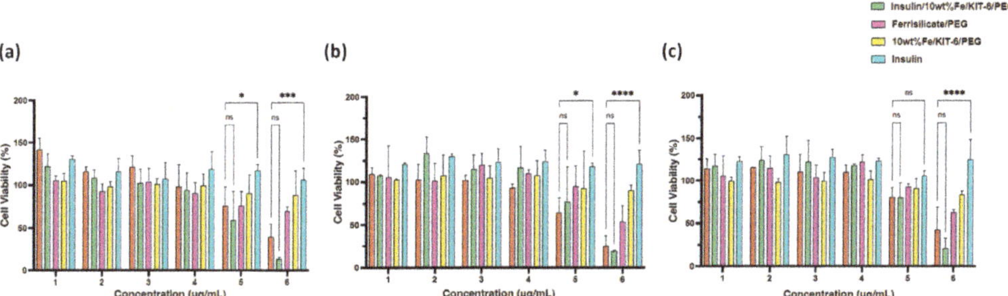

Figure 6. Cytotoxicity of Insulin/Ferrisilicate/PEG nanoformulations against Human foreskin fibroblast (HFF-1) cells. The cells were treated with Insulin/Ferrislicate/PEG, Insulin/10 wt% Fe/KIT-6/PEG, Ferrisilicate/PEG, 10 wt% FeKIT-6/PEG and insulin for (**a**) 24, (**b**) 48 and (**c**) 72 h. Error bars ± SD. (ns) non-significant *p* value, * $p < 0.05$; *** $p < 0.001$; **** $p < 0.0001$ as Insulin/Ferrisilicate/PEG versus Insulin/10 wt% Fe/KIT-6/PEG and insulin using two-way ANOVA with Dunnett's multiple comparison test. The cytotoxicity assay was performed in three independent experiments.

3.4. In Vivo Study

To find out the hypoglycemic effect of the prepared insulin nanoformulation, it was administered orally to the diabetic animals at three gradient doses 2, 5 and 10 (Insulin Unit) IU/kg body weight (Figure 7). The blood glucose level was measured at different intervals up to 6 h. The results of the study indicate the hypoglycemic effect of insulin nanoformulation at a dose of 5 and 10 IU/kg, and the formulation was found to be effective after 2 h of administration at a dose of 5 mg/kg. Furthermore, it reduced the blood glucose level significantly ($p < 0.0001$) until 3 h, after which it starts to move towards the initial blood glucose values. At 5 IU/kg, the insulin/ferrisilicate/PEG nanoformulation decreases the blood glucose level from 501 to 375 mg/dL (reduction of 25%). Moreover, the formulation at 10 IU/kg body weight showed a significant reduction ($p < 0.0001$) in blood glucose level after 1 h of administration and was found to be significant until 3 h of administration. It reduced the blood glucose level from 417 to 268 mg/dL (35% reduction). These effects of the nanoformulation at a dose of 5 and 10 IU/kg body weight were found to be significant

($p < 0.0001$), as compared to diabetic control. No significant effect of insulin formulation was found at 2 IU/kg body weight.

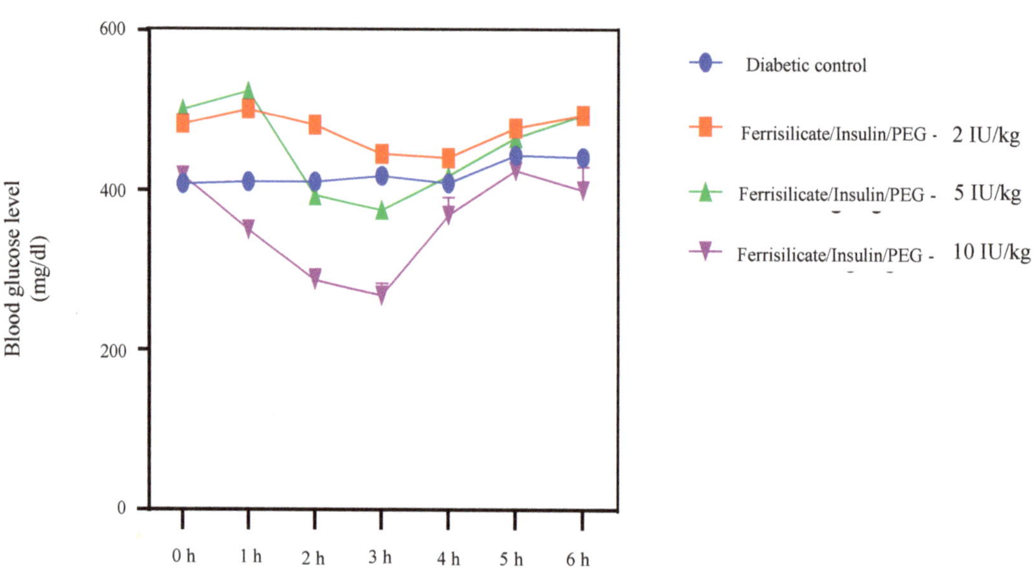

Figure 7. Hypoglycemic effect of insulin/ferrisilicate nanoformulation Insulin/Ferrisilicate/PEG administered to the diabetic animals at three gradient doses 2, 5, and 10 IU/kg body weight (n = 6).

4. Conclusions

In the present investigation, pegylated 3D cubic porous ferrisilicate was explored for diabetic management. The physico-chemical characterization reveals the incorporation of Fe^{3+} species into the framework (via isomorphous substitution) of SBA-16. A large surface area and pore size were found to be sufficient for a high encapsulation of insulin (46%) at 1.5 h. The adsorption increases with time to about 57% at 1.5 h. Ferrisilicate exhibited a high percentage of cumulative insulin release of about 32% in 530 h at pH 7.4. Insulin release was notably improved by PEG wrapping (0.08 µL/mg) to about 41% in 530 h at pH 6.8. As a comparison, Fe/Mesocellular foam, Fe/Mesosilicalite and Fe/KIT-6 showed about 40%, 28% and 19.2% insulin release in 530 h. The kinetic studies using the Korsmeyer–Peppas model revealed that the nature of the base materials, different insulin adsorption times, varying insulin/ferrisilicate ratios, different pHs and different base material modifications all affect the rate of insulin release and its diffusion mechanism. The nanoformulation showed very low toxicity in in vitro study and a hypoglycemic effect in in vivo study. Doses of 5 and 10 mg/kg body weight of pegylated ferrisilicate were found to be significant ($p < 0.0001$) compared to the diabetic control. Overall, the developed 3D porous ferrisilicate wrapped with polyethylene glycol mimics smart stimuli-responsive behavior with a pH-sensitive insulin release with high cell viability for potential application in diabetic management.

Author Contributions: B.R.J.: conceptualization, methodology, investigation, supervision; M.S.: in vivo study, investigation, methodology; S.A.: cell line investigation, original draft preparation; G.T.: kinetic study; H.D.: microscopic analysis, investigation, V.R.: conceptualization, validation, supervision, draft preparation, review. All authors have read and agreed to the published version of the manuscript.

Funding: B. Rabindran Jermy would like to acknowledge the funding obtained from the Deanship of Scientific Research, Imam Abdulrahman Bin Faisal University, grant number 2017-091-IRMC.

Institutional Review Board Statement: The in vivo study has been approved by Imam Abdulrahman Bin Faisal University IRB through Institutional Review Board number IRB-2020-13-278 with approval date of 30 September 2020 and conducted according to the guidelines of the Declaration of Helsinki.

Informed Consent Statement: Not applicable.

Data Availability Statement: The data sets, figures and analyses are provided in this manuscript. Further additional data will be available from the corresponding author upon request.

Acknowledgments: The authors would like to acknowledge Dorothy Joy H. Huelar for her assistance in the in vitro studies. The technical assistance of Janaica Yu Logan in the nanomaterial characterization is acknowledged. B.R.J would like to acknowledge the state of art facilities provided by Institute for Research and Medical Consultations, Imam Abdulrahman BinFaisal University.

Conflicts of Interest: The authors declare no conflict of interest.

References

1. Von Scholten, B.J.; Kreiner, F.F.; Gough, S.C.L.; von Herrath, M. Current and future therapies for type 1 diabetes. *Diabetologia* **2021**, *64*, 1037–1048. [CrossRef] [PubMed]
2. Reed, J.; Bain, S.; Kanamarlapudi, V. A review of current trends with Type 2 diabetes epidemiology, aetiology, pathogenesis, treatments and future perspectives. *Diabetes Metab. Syndr. Obes.* **2021**, *14*, 3567. [CrossRef] [PubMed]
3. IDF Diabetes Atlas. Available online: https://diabetesatlas.org (accessed on 20 January 2023).
4. Vecchio, I.; Tornali, C.; Bragazzi, N.L.; Martini, M. The discovery of insulin: An important milestone in the history of medicine. *Front. Endocrinol.* **2018**, *9*, 613. [CrossRef]
5. Holleman, F.; Gale, E.A. Nice insulins, pity about the evidence. *Diabetologia* **2007**, *50*, 1783–1790. [CrossRef] [PubMed]
6. Girish, C.; Manikandan, S.; Jayanthi, M. Practitioners section-Newer insulin analogues and inhaled insulin. *Indian J. Med. Sci.* **2006**, *60*, 117–123. [CrossRef]
7. Jonasson, J.; Ljung, R.; Talbäck, M.; Haglund, B.; Gudbjörnsdòttir, S.; Steineck, G. Insulin glargine use and short-term incidence of malignancies—A population-based follow-up study in Sweden. *Diabetologia* **2009**, *52*, 1745–1754. [CrossRef]
8. Najmeddine, A.A.; Saeed, M.; Beadham, I.G.; ElShaer, A. Efficacy and safety of glucose sensors for delivery of insulin: A Systematic Review. *PharmaNutrition* **2021**, *18*, 100280. [CrossRef]
9. Kuang, Y.; Zhai, J.; Xiao, Q.; Zhao, S.; Li, C. Polysaccharide/mesoporous silica nanoparticle-based drug delivery systems: A review. *Int. J. Biol. Macromol.* **2021**, *193*, 457–473. [CrossRef]
10. Fu, Y.; Liu, P.; Chen, M.; Jin, T.; Wu, H.; Hei, M.; Wang, C.; Xu, Y.; Qian, X.; Zhu, W. On-demand transdermal insulin delivery system for type 1 diabetes therapy with no hypoglycemia risks. *J. Colloid Interface Sci.* **2022**, *605*, 582–591. [CrossRef]
11. Wang, M.; Wang, C.; Ren, S.; Pan, J.; Wang, Y.; Shen, Y.; Zeng, Z.; Cui, H.; Zhao, X. Versatile Oral Insulin Delivery Nanosystems: From Materials to Nanostructures. *Int. J. Mol. Sci.* **2022**, *23*, 3362. [CrossRef]
12. Tong, X.; Li, Z.; Chen, W.; Wang, J.; Li, X.; Mu, J.; Tang, Y.; Li, L. Efficient catalytic ozonation of diclofenac by three-dimensional iron (Fe)-doped SBA-16 mesoporous structures. *J. Colloid Interface Sci.* **2020**, *578*, 461–470. [CrossRef]
13. Kunde, G.B.; Yadav, G.D. Green strategy for the synthesis of mesoporous, free-standing MAl2O4 (M = Fe, Co, Ni, Cu) spinel films by sol–gel method. *Mater. Sci. Eng. B* **2021**, *271*, 115244. [CrossRef]
14. Samrot, A.V.; Sahithya, C.S.; Selvarani, J.; Purayil, S.K.; Ponnaiah, P. A review on synthesis, characterization and potential biological applications of superparamagnetic iron oxide nanoparticles. *Curr. Res. Green Sustain. Chem.* **2021**, *4*, 100042. [CrossRef]
15. Materón, E.M.; Miyazaki, C.M.; Carr, O.; Joshi, N.; Picciani, P.H.; Dalmaschio, C.J.; Davis, F.; Shimizu, F.M. Magnetic nanoparticles in biomedical applications: A review. *Appl. Surf. Sci. Adv.* **2021**, *6*, 100163. [CrossRef]
16. Jermy, B.R.; Alomari, M.; Ravinayagam, V.; Almofty, S.A.; Akhtar, S.; Borgio, J.F.; AbdulAzeez, S. SPIONs/3D SiSBA-16 based Multifunctional Nanoformulation for target specific cisplatin release in colon and cervical cancer cell lines. *Sci. Rep.* **2019**, *9*, 14523. [CrossRef]
17. Balasamy, R.J.; Ravinayagam, V.; Alomari, M.; Ansari, M.A.; Almofty, S.A.; Rehman, S.; Dafalla, H.; Marimuthu, P.R.; Akhtar, S.; Al Hamad, M. Cisplatin delivery, anticancer and antibacterial properties of Fe/SBA-16/ZIF-8 nanocomposite. *RSC Adv.* **2019**, *9*, 42395–42408. [CrossRef]
18. Jose, S.; Senthilkumar, M.; Elayaraja, K.; Haris, M.; George, A.; Raj, A.D.; Sundaram, S.J.; Bashir, A.K.; Maaza, M.; Kaviyarasu, K. Preparation and characterization of Fe doped n-hydroxyapatite for biomedical application. *Surf. Interfaces* **2021**, *25*, 101185. [CrossRef]
19. Janarthanan, G.; Noh, I. Recent trends in metal ion based hydrogel biomaterials for tissue engineering and other biomedical applications. *J. Mater. Sci. Technol.* **2021**, *63*, 35–53. [CrossRef]

20. Jermy, B.R.; Kim, S.Y.; Bineesh, K.V.; Selvaraj, M.; Park, D.W. Easy route for the synthesis of Fe-SBA-16 at weak acidity and its catalytic activity in the oxidation of cyclohexene. *Microporous Mesoporous Mater.* **2009**, *121*, 103–113. [CrossRef]
21. Andreani, T.; de Souza, A.L.R.; Kiill, C.P.; Lorenzon, E.N.; Fangueiro, J.F.; Calpena, A.C.; Chaud, M.V.; Garcia, M.L.; Gremiao, M.P.D.; Silva, A.M.; et al. Preparation and characterization of PEG-coated silica nanoparticles for oral insulin delivery. *Int. J. Pharm.* **2014**, *473*, 627–635. [CrossRef]
22. Nakai, M.; Miyake, K.; Inoue, R.; Ono, K.; Al Jabri, H.; Hirota, Y.; Uchida, Y.; Miyamoto, M.; Nishiyama, N. Synthesis of high silica* BEA type ferrisilicate (Fe-Beta) by dry gel conversion method using dealuminated zeolites and its catalytic performance on acetone to olefins (ATO) reaction. *Microporous Mesoporous Mater.* **2019**, *273*, 189–195. [CrossRef]
23. Bhaumik, A.; Samanta, S.; Mal, N.K. Iron oxide nanoparticles stabilized inside highly ordered mesoporous silica. *Pramana* **2005**, *65*, 855–862. [CrossRef]
24. Hinds, K.D.; Kim, S.W. Effects of PEG conjugation on Insulin properties. *Adv. Drug Deliv. Rev.* **2002**, *54*, 505–530. [CrossRef] [PubMed]
25. Yu, Q.; Feng, Y.; Tang, X.; Yi, H.; Zhao, S.; Gao, F.; Zhou, Y.; Zhang, Y.; Zhuang, R. A novel ferrisilicate MEL zeolite with bi-functional adsorption/catalytic oxidation properties for non-methane hydrocarbon removal from cooking oil fumes. *Microporous Mesoporous Mater.* **2020**, *309*, 110509. [CrossRef]
26. Massaro, M.; Cavallaro, G.; Colletti, C.G.; D'Azzo, G.; Guernelli, S.; Lazzara, G.; Pieraccini, S.; Riela, S. Halloysite nanotubes for efficient loading, stabilization and controlled release of insulin. *J. Colloid Interface Sci.* **2018**, *524*, 156–164. [CrossRef]
27. Kim, M.K.; Ki, D.H.; Na, Y.G.; Lee, H.S.; Baek, J.S.; Lee, J.Y.; Lee, H.K.; Cho, C.W. Optimization of Mesoporous Silica Nanoparticles through Statistical Design of Experiment and the Application for the Anticancer Drug. *Pharmaceutics* **2021**, *13*, 184. [CrossRef]
28. Huang, Q.; Yu, H.; Wang, L.; Shen, D.; Chen, X.; Wang, N. Synthesis and testing of polymer grafted mesoporous silica as glucose-responsive insulin release drug delivery systems. *Eur. Polym. J.* **2021**, *157*, 110651. [CrossRef]
29. Gaballa, H.; Theato, P. Glucose-responsive polymeric micelles via boronic acid–diol complexation for insulin delivery at neutral pH. *Biomacromolecules* **2019**, *20*, 871–881. [CrossRef]
30. Luo, L.; Song, R.; Chen, J.; Zhou, B.; Mao, X.; Tang, S. Fluorophenylboronic acid substituted chitosan for insulin loading and release. *React. Funct. Polym.* **2020**, *146*, 104435. [CrossRef]
31. Zhang, P.; Ma, Q.; He, D.; Liu, G.; Tang, D.; Liu, L.; Wu, J. A new Glucose-Responsive delivery system based on Sulfonamide-phenylboronic acid for subcutaneous insulin injection. *Eur. Polym. J.* **2021**, *157*, 110648. [CrossRef]
32. Lin, K.; Yi, J.; Mao, X.; Wu, H.; Zhang, L.M.; Yang, L. Glucose-sensitive hydrogels from covalently modified carboxylated pullulan and concanavalin A for smart controlled release of insulin. *React. Funct. Polym.* **2019**, *139*, 112–119. [CrossRef]
33. Yin, R.; Bai, M.; He, J.; Nie, J.; Zhang, W. Concanavalin A-sugar affinity based system: Binding interactions, principle of glucose-responsiveness, and modulated insulin release for diabetes care. *Int. J. Biol. Macromol.* **2019**, *124*, 724–732. [CrossRef]
34. Xu, M.; Huang, J.; Jiang, S.; He, J.; Wang, Z.; Qin, H.; Guan, Y.Q. Glucose sensitive konjac glucomannan/concanavalin A nanoparticles as oral insulin delivery system. *Int. J. Biol. Macromol.* **2022**, *202*, 296–308. [CrossRef]
35. Chai, Z.; Dong, H.; Sun, X.; Fan, Y.; Wang, Y.; Huang, F. Development of glucose oxidase-immobilized alginate nanoparticles for enhanced glucose-triggered insulin delivery in diabetic mice. *Int. J. Biol. Macromol.* **2020**, *159*, 640–647. [CrossRef]
36. Wu, J.Z.; Williams, G.R.; Li, H.Y.; Wang, D.; Wu, H.; Li, S.D.; Zhu, L.M. Glucose-and temperature-sensitive nanoparticles for insulin delivery. *Int. J. Nanomed.* **2017**, *12*, 4037. [CrossRef]
37. Jana, B.A.; Shinde, U.; Wadhwani, A. Synthetic enzyme-based nanoparticles act as smart catalyst for glucose responsive release of insulin. *J. Biotechnol.* **2020**, *324*, 1–6. [CrossRef]
38. Ma, R.; Yang, H.; Li, Z.; Liu, G.; Sun, X.; Liu, X.; An, Y.; Shi, L. Phenylboronic acid-based complex micelles with enhanced glucose-responsiveness at physiological pH by complexation with glycopolymer. *Biomacromolecules* **2012**, *13*, 3409–3417. [CrossRef]

pharmaceutics

Review

Target-Specific Delivery and Bioavailability of Pharmaceuticals via Janus and Dendrimer Particles

Jaison Jeevanandam [1], Kei Xian Tan [2], João Rodrigues [1,*,†] and Michael K. Danquah [3,*,†]

[1] CQM—Centro de Química da Madeira, MMRG, Universidade da Madeira, Campus da Penteada, 9020-105 Funchal, Portugal; jaison.jeevanandam@staff.uma.pt
[2] GenScript Biotech (Singapore) Pte. Ltd., 164, Kallang Way, Solaris@Kallang 164, Singapore 349248, Singapore; kxtan5@gmail.com
[3] Department of Chemical Engineering, University of Tennessee, Chattanooga, TN 37403-2598, USA
* Correspondence: joaoc@staff.uma.pt (J.R.); michael-danquah@utc.edu (M.K.D.)
† These authors contributed equally to this work.

Abstract: Nanosized Janus and dendrimer particles have emerged as promising nanocarriers for the target-specific delivery and improved bioavailability of pharmaceuticals. Janus particles, with two distinct regions exhibiting different physical and chemical properties, provide a unique platform for the simultaneous delivery of multiple drugs or tissue-specific targeting. Conversely, dendrimers are branched, nanoscale polymers with well-defined surface functionalities that can be designed for improved drug targeting and release. Both Janus particles and dendrimers have demonstrated their potential to improve the solubility and stability of poorly water-soluble drugs, increase the intracellular uptake of drugs, and reduce their toxicity by controlling the release rate. The surface functionalities of these nanocarriers can be tailored to specific targets, such as overexpressed receptors on cancer cells, leading to enhanced drug efficacy The design of these nanocarriers can be optimized by tuning the size, shape, and surface functionalities, among other parameters. The incorporation of Janus and dendrimer particles into composite materials to create hybrid systems for enhancing drug delivery, leveraging the unique properties and functionalities of both materials, can offer promising outcomes. Nanosized Janus and dendrimer particles hold great promise for the delivery and improved bioavailability of pharmaceuticals. Further research is required to optimize these nanocarriers and bring them to the clinical setting to treat various diseases. This article discusses various nanosized Janus and dendrimer particles for target-specific delivery and bioavailability of pharmaceuticals. In addition, the development of Janus-dendrimer hybrid nanoparticles to address some limitations of standalone nanosized Janus and dendrimer particles is discussed.

Keywords: dendrimers; Janus nanoparticles; drug delivery; biocompatibility; nanoformulation; pharmaceuticals

Citation: Jeevanandam, J.; Tan, K.X.; Rodrigues, J.; Danquah, M.K. Target-Specific Delivery and Bioavailability of Pharmaceuticals via Janus and Dendrimer Particles. *Pharmaceutics* **2023**, *15*, 1614. https://doi.org/10.3390/pharmaceutics15061614

Academic Editor: Barbara Stella

Received: 9 May 2023
Revised: 25 May 2023
Accepted: 26 May 2023
Published: 29 May 2023

1. Introduction

Nanoparticles have gained considerable attention among researchers as a potential drug delivery system due to their unique properties, such as their high surface-to-volume ratio and surface charge-dependent behavior, compared to their bulk counterparts [1–3]. The properties of nanoparticles depend on their size and shape, which can be tailored by selecting an appropriate synthesis approach [4]. Dendrimers, micelles, liposomes, and biopolymers are the most commonly used drug-delivery nanoparticles [5]. Micelles are colloidal suspensions formed by the dispersion of amphiphilic lipid molecules in a liquid and have a hydrophilic head and a hydrophobic tail [6]. Micelles as a drug delivery system have advantages such as improved solubility of highly lipophilic drugs, controlled drug release, the ability to adjust their physiochemical properties, and protection of the drug from environmental factors. However, they have limitations such as low drug-loading capacity, high dependence on critical micelle concentration, and limited applicability to only lipophilic drugs [7]. Liposomes are small artificial spherical vesicles formed using natural, nontoxic phospholipids and cholesterol and have benefits such as biocompatibility and hydrophilic/hydrophobic characteristics [8]. However, liposomes as a drug delivery system face limitations such as high production cost, limited shelf life, vulnerability to oxidation and hydrolysis of phospholipids in certain conditions, instability, fusion, and potential release of encapsulated drugs [9]. Biopolymers (polymers synthesized or extracted from biological source) have also been used for drug formulation, but they often lack solubility or have pH-dependent solubility, which limits their use [10,11].

Dendrimers are synthetic, tree-like hyperbranched polymers with a high number of functional groups and an open molecular structure. They are designed as artificial macromolecules with void spaces for drug storage and targeted release [12–14]. However, dendrimers have limitations such as high non-specific toxicity, drawbacks during scale-up experiments, and low hydro-solubility [15]. Despite these limitations, they have potential as nanoparticles for drug delivery. Janus nanoparticles are a recent addition to the range of nanoparticles, featuring the integration of two or more chemically distinct components into a single structure. They possess unique properties based on their synthesis approaches and the materials infused into the Janus structure [16–19]. However, the complex synthesis process and toxicity due to chemicals involved in the synthesis approach are limitations of Janus nanoparticles [17,20]. The incorporation of Janus and dendrimer into a composite material has been proposed to enhance drug delivery ability and reduce limitations [21,22]. This article discusses various nanosized Janus and dendrimer particles for target-specific delivery and bioavailability of pharmaceuticals. Additionally, the emergence of Janus-dendrimer nanoparticles to overcome the limitations of standalone nanosized Janus and dendrimer particles is discussed.

2. Overview of Nanosized Janus and Dendrimer Particles

2.1. Janus Nanoparticles

Janus nanoparticles were first discovered by Pierre-Gilles de Gennes, the Nobel Laureate who pioneered fabricating microparticles 'Janus grains' with an apolar and polar side [23]. The word 'Janus' comes from the two-faced Roman God of gates, which defines Janus nanoparticles as anisotropic particles that possess two different compartments with varying functionalities, material compositions, morphology, size, shape, and biochemical properties. Janus nanoparticles are originally from polymeride but can be subcategorized as organic/polymeric, inorganic, or hybrid of organic and inorganic Janus particles [24–27]. In addition to the typical spherical shape, Janus nanoparticles can be fabricated into different conformations, which include rod [28], dumbbell [29], platelet [30], and snowman [31,32].

Due to their asymmetric faces, Janus nanoparticles can improve the stability of different phases [33]. This has then broadened their biomedical and clinical applications from emulsion stabilizer, bio-sensing, bio-catalysis, molecular imaging, and diagnostic tools to pharmaceutical targeted drug delivery systems [34], offering significant benefits over the conventional mono-functional particles. This is highly ascribed to the tunable properties of Janus nanoparticles whereby their different surfaces or compartments can be modified with individual functionality. This includes hybrid particles with one amphiphilic surface and another stimuli-responsive surface [19]; Janus nanoparticles made of organic and inorganic compartments [35]; or biocompatible particles [36] for targeted medical treatments.

This enables Janus nanoparticles to be utilized as delivery carriers to carry different drug molecules with the combination of various functionalities. Otherwise, as a delivery system, one hemisphere can load medical drug molecules while another side acts as a targeting element with high specificity toward targeted cells. Janus nanoparticles have practical medical and environmental applications, such as detecting water contaminants and environmental pollutants and serving as superior candidates for cancer theranostics due to their high loading capacity and tunable properties. Janus nanoparticles made of silver/chitosan have also been reported to exhibit high antimicrobial effects against bacteria such as *Escherichia coli*, *Salmonella choleraesuis*, *Bacillus subtilis*, *Staphylococcus aureus*, indicating their potential applications in food sector [37]. Interestingly, there are Janus nanoparticles used to detect DNA and metals for monitoring applications. A streptavidin-modified retroreflective Janus particle can selectively sense the presence of mercury ions with up to 0.027 nM detection limit [38] whilst a hybrid of gold-silver nanorod and polyaniline has also been developed as a Janus nanoparticle, serving as a surface-enhanced Raman scattering sensor for the detection of mercury [39]. In addition, gold-silver Janus nanoparticles have been exploited as aptasensor to detect toxins such as Ochratoxin A quantitatively, which can be widely used in real systems, including red wine monitoring [40]. The above examples highlight the vast potential of Janus nanoparticles for a broad range of applications, offering numerous benefits to various industries.

2.2. Dendrimers

The Greek phrase 'dendron', which means trees or branches, is the source for the word 'dendrimer'. Dendrimers are symmetrical, generation-dependent spherical polymers consisting of a core and dendrons (branches), possessing a hyperbranched, three-dimensional structure [13]. In 1941, Paul John Flory and colleagues (Nobel Prize in Chemistry 1974) introduced the theory of highly branched polymers [41,42], which can be synthesized through polycondensation of a monomer with one or more functional groups, avoiding the gelation process [43]. However, it was not stable and are without a cavity. Later, Vogtle and his team (1978) reported the formation of the first non-skid chain-like and cascade-like molecules with the topology of the molecular cavity, which is considered the earliest dendritic polymer form. The term "hyperbranched polymer" was first coined by Kim and Webster in 1988 in reference to the synthesis of soluble hyperbranched polyphenylene. This term was later used to describe the structure of dendrimers [44]. However, these particular types of polymers attract the academy's attention only with the work of Tomalia et al. (1985) [45] and Newkome et al. (1985) [46]. Further, Tomalia not only coined the term "dendrimer" as made a drastic breakthrough in dendrimers field by forming in a controlled manner using divergent synthesis, poly(amidoamine) (PAMAM) dendrimers with a hollow core in the center and outward branches of tendrils [47]. Currently, there are about 100 dendrimer families, which include beyond poly(amidoamine) (PAMAM) dendrimers, among others, polypropyleneimine (PPI), polyester-, polyamide-, phosphorus, and polyether-based dendrimers [12].

Dendrimers' molecular mass and size are specifically controlled during the polymerization process, which is not possible during linear polymer formation [48]. The unique molecular architecture of dendrimers results in improved physical and chemical properties compared to traditional linear polymers [49]. In general, dendrimers have a tightly packed spherical structure with excellent rheological properties and low viscosity than linear polymers [50,51]. It's worth mentioning that the intrinsic viscosity of a dendrimer reaches its peak at the fourth generation as its molecular mass increases [51,52]. The high solubility, miscibility, and reactivity of dendrimers can be attributed to the multiple chain-ends present in their structure [53]. Similarly, the solubility of the dendrimers depends on their surface group, where dendrimers with hydrophilic and hydrophobic terminations are soluble in both polar and nonpolar solvents, respectively [54]. Furthermore, the spherical shape and presence of internal cavities in dendrimers make them ideal for encapsulating desired molecules or drugs within the macromolecules [55]. These novel polymers are further sub-classified into cationic, neutral and anionic dendrimers, based on their surface charge [56]. It is worth noting that cationic dendrimers are cytotoxic and hemolytic, whereas dendrimers with carboxylate surfaces that are anionic are considered nontoxic for a broad range of concentrations [57,58]. However, the properties of dendrimers are significantly influenced by factors such as pH, solvent, precursor salt, and concentration [59]. Moreover, preparation of dendrimer in the nano-regime will further enhance their properties, due to their exceptional high surface-to-volume ratio and unique structure [60,61]. Figure 1 shows the structural aspects of Janus and dendrimer particles.

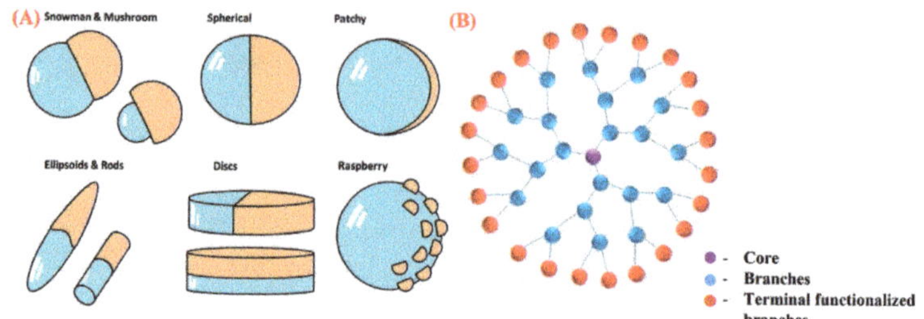

Figure 1. The general structure of (**A**) Janus, Adapted with permission from Honciuc Ref. [62]. Copyright 2019 Springer and (**B**) dendrimer particles, Adapted with permission from Araujo et al. Ref. [63]. Copyright 2018 MDPI.

3. Rational Design and Synthesis Approaches

3.1. Janus Nanoparticles

Janus nanoparticles are known as a new generation of smart building blocks for material evolution. They are innovative materials designed to have two or more physical and biochemical anisotropy as well as assembled layers to serve multiple functionalities. Their unique characteristics and structural designs make them superior to conventional particles. Their superior characteristics are such as the high interfacial activity; engineering of their interfacial activity via external stimulus (i.e., pH, temperature, or ionic strength); recyclable for sustainable usage attributed to the possibility of stimuli-induced separation; high surface-to-volume ratio; manipulation of mass transport across different interfacial layer; and their multi-functionalities in one Janus nanoparticle [64]. This allows them to be flexible as either a macro-phase separation or stable emulsion. Furthermore, many materials especially biodegradable or natural materials such as chitosan, alginates, and cellulose, have been widely used for the Janus nanoparticles fabrication [37,65–69]. Janus nanoparticles that respond to changes in pH have been reported for use in bio-imaging and anti-tumor drug delivery. These nanoparticles enable real-time monitoring and tracking of drug release [70,71].

Various synthetic techniques are available and optimized to fabricate Janus nanoparticles with the desired shape, size, and physical or biochemical functionality on their opposite side. Three widely recognized conservative fabrication methods are masking, phase separation, and self-assembly. Masking is the simplest technique used to fabricate Janus nanoparticles by covering up one side using either solid or liquid blocking surfaces while exposing the other side of the particles for chemical modifications. For instance, inorganic Janus particles were produced via the partial coating of calcium carbonate particles with platinum to create non-Brownian movement in the acidic environment of the tumor site [72]. This approach allows scientists to manipulate the chemical and structural features of Janus nanoparticles, resulting in exceptional design and functionality. However, the masking technique is only appropriate for small-scale laboratory production, making it useful for initial investigation and basic testing. The phase separation technique works by mixing two or more incompatible polymers or components before separating them into different compartments of a single Janus nanoparticle. The pH of the aqueous phase, the makeup of the dispersed phase, spreading coefficients, and interfacial tension can often impact phase separation and particle shape [73]. Self-assembly from di-block copolymer is considered one of the easy and simple methods to synthesize different polymeric Janus nanoparticles by adjusting the cross-linking degree and molecular weights or ratio of the blocks used [34].

Pickering emulsion on the other hand utilizes the interface of two immiscible fluids for both particle adsorption and unique surface modifications to synthesize the anisotropic particles. This technique enables the production of emulsions such as silica-based amphiphilic Janus particles with colloidal stability that lasts for more than 3 weeks [74]. Nonetheless, this technique comes with limitations including the tedious multiple steps that require strict monitoring; limited encapsulation and controlled release capacity; unsuitability for biodegradable polymeric Janus particles due to the high temperature and specific solvents used to eliminate the masking layer [75].

Methods such as microfluidic assembly and a 'lab-on-a-chip' system have also been exploited to fabricate Janus-like dimer capsules. This technique comprises a series of steps such as fluidic assembly, cross-linkage, and droplet formation that occur on a microfluidic chip without manual control. Lu and team [67] have designed Janus-like dimers whereby the functionality and structure of each lobe have been precisely controlled to resume bowling pin and snowman structures. This method allows the formation of Janus-like dimer capsules by having two individual biopolymer mixtures with distinct compositions from two different channels meet at the junction and fuse into one stable formation with distinct compositions via controlled cross-linking and coalescence. There is also a study reported on the use of a microfluidic-mediated fabrication method to produce Janus nanoparticles in a microchannel whereby two adjacent fluid streams with a laminar flow are separated into discrete droplets via the immiscible phase and finally photo-polymerized into Janus nanoparticles [76]. However, this method comes with challenges such as low output efficiency, precise control, and limited polymer choices. Furthermore, there is a fluidic nanoprecipitation system developed to fabricate Janus nanoparticles made of PLGA. The results showed that these polymeric Janus nanoparticles effectively encapsulated two distinct drugs, Paclitaxel (a hydrophobic drug) and Doxorubicin Hydrochloride (a hydrophilic drug), resulting in varied drug release behavior [77]. This is a single-step method that utilizes dual inlets to deliver each half of the Janus compartment into the precipitation flow.

Another team of scientists [75] devised a straightforward, one-step solvent emulsion method for creating hybrid Janus particles. This approach enables the targeted encapsulation of both diagnostic and therapeutic agents into separate compartments of the Janus complex through drug-polymer interactions and the use of specific biopolymer ratios, as illustrated in Figure 2. This technique is capable to offer a more cost-effective, scalable, and seamless approach to synthesizing dual drug-encapsulated theranostic Janus particles with controlled and different release kinetics. With this method, both the immiscible biodegradable and non-biodegradable polymers are emulsified and segregated into two different compartments due to the interfacial tensions. This one synthesis setup technique could replace the single drug-encapsulated particle due to its simplicity and scalability.

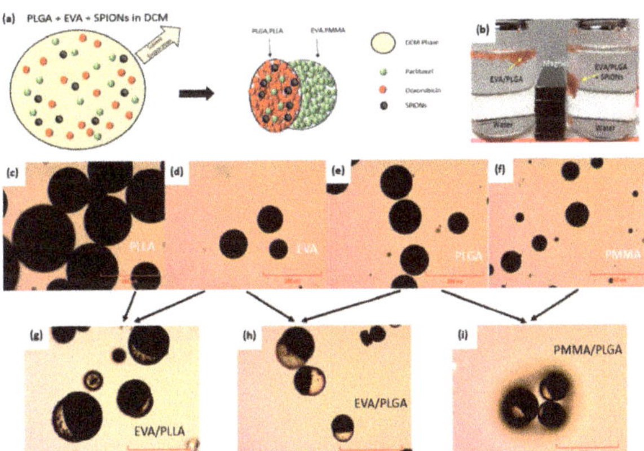

Figure 2. A simple and single-step solvent emulsion technique used to fabricate polymeric Janus particles for theranostic applications. (**a**) Theranostic polymeric Janus particles made of both biodegradable and non-biodegradable polymers, (**b**) magnetic response of drug and SPIONs-encapsulated EVA/PLGA Janus particles, (**c–f**) a comparison of SPION encapsulation in different polymers i.e., PLLA, EVA, PLGA, and PMMA, (**g–i**) the formation of SPIONs-encapsulated Janus particles: EVA/PLLA, EVA/PLGA, and PMMA/PLGA. Reproduced with permission from Lim et al. Ref. [75]. Copyright 2019 Wiley.

3.2. Dendrimers

Generally, dendrimers are synthesized using conventional methods, such as cascade reactions, and divergent as well as convergent approaches [78]. The cascade strategy involves the assembly of three sub-nano/picoscopic-sized repeated units via reaction with a divalent terminus to produce a branched cell with a branch juncture. The generation zero (0) of the dendrimer is formed by attaching the first branch to the reference point for yielding a labeled branch with a valency of four. Later, the first and second generation of dendrimers was formed by stepwise attachment of repeat units according to their incipient valency [79]. Even though the cascade strategy is initially used, convergent and divergent approaches are commonly utilized for the fabrication of dendrimers as shown in Figure 3. In a divergent approach, the dendrimers are synthesized from the core to the branches through repeated coupling and activation steps in an outward direction [45]. Hence, the cascade strategy is also considered a type of divergent approach for dendrimer formation [80]. Recently, Rauch et al. (2020) utilized an iterative divergent approach for the preparation of conjugated starburst borane dendrimers. In this study, the dendrimer was prepared in three steps: the functionalization of iridium-catalyzed carbon-hydrogen borylation and activation of fluorine-generated boronate ester with potassium bi-fluoride and expansion of trifluoroborate salts with aryl Grignard reagent reaction [81]. In a convergent approach, the branches (dendrons) of the dendrimer were prepared initially, which is finally coupled to the moiety in the core after the activation of their focal point [82]. It can be noted that the growth of the dendron is straightforwardly monitored via a convergent method, compared to a divergent approach [83]. Bondareva et al. (2020) prepared sulfonimide-based dendrimers and dendrons with the help of a convergent approach. In this study, both convergent and divergent methods were used to create sulfonimide-based dendrons and dendrimers with chlorosulfonic groups as the central focus. Suprisingly, the results indicated that the convergent method of creating dendrons reached its practical limit in the third generation, making a divergent approach necessary for producing sulfonimide-based dendrimers with more generations [84].

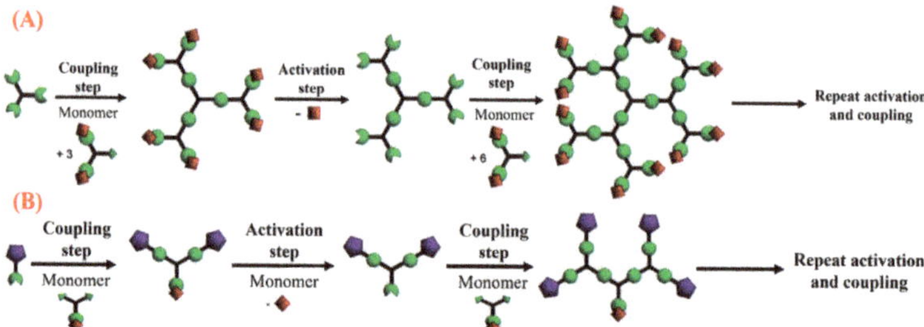

Figure 3. Schematic representation of (**A**) Divergent and (**B**) convergent approach, Concept adapted from Grayson et al. Ref. [85]. Copyright 2001 American Chemical Society.

Apart from the conventional divergent and convergent approach, revised traditional methods, such as the hyper monomer method, double-stage convergent growth, and double exponential growth were also used for the synthesis of dendrimers [86]. Monomers with higher functional group numbers were employed in the hyper monomer approach for dendrimer synthesis in fewer steps, compared to the traditional AB_2 monomer [87]. Balaji and Lewis (2009) synthesized a novel aliphatic polyamide dendrimer by AB_2 hyper monomer strategy and double exponential growth. The study showed that the yield of the dendrimer is about ~93%, which is reduced to about ~89% after 30 min of the purification process [88]. In a final step, low-generation dendrimers and dendrons are linked together through parallel synthesis in a double-stage convergent growth approach [89]. Recently, Agrahari et al. (2020) demonstrated the synthesis of novel glycol-dendrimers and dendrons coated with galactose was achieved through an efficient click approach and double-stage convergent method. The study emphasized the synthesis of a novel galactose coated 9-

peripheral glycodentromer of the zeroth generation and a galactose-coated 27-peripheral glycodentromer of the first generation [90]. Further, low-generation dendrons that were fully protected or deactivated were created, and then actively selected at their periphery or central focus, and joined to produce higher generations of fully protected dendrons through a double exponential growth method [91]. Hartwig et al. (2010) prepared A novel polyglutamate dendrimer of the fourth generation was produced through an iterative binomial exponential growth method. The research demonstrated that this synthesis technique can aid in incorporating either D-alt-L or all-(L) stereochemistry in the peptide backbone and is useful for post-functionalization of the dendritic core and periphery [92].

Additionally, orthogonal, chemoselective reactions (click chemistry), one-pot synthesis, and heterofunctional fabrication approaches were introduced for dendrimer formation. The click approach makes use of Diels-Alder cycloaddition and thiol-ene coupling, which are highly selective chemical reactions, and offer a reliable platform for the synthesis of complex macrostructures [93]. Ma et al. (2009) fabricated a novel polyester dendrimer via a sequential click coupling of asymmetrical monomers. In this study, mechanistic or kinetic chemoselectivity was combined with click reactions between the monomers for efficient dendrimer synthesis via simple sticking of generation by generation together [94]. When different monomers, such as Ab_x and CD_y, instead of AB_n monomer in divergent growth approach for dendrimer formation is termed as orthogonal growth method [95]. Liu et al. (2022) fabricated scaffold-modifiable dendrons via an orthogonal protection strategy. The study revealed that this approach can yield fourth-generation dendrons within 2 days without any significant defects in their structure [96]. Kothari et al. (2013) demonstrated an efficient and accelerated preparation of multifunctional dendrimers via nucleophilic substitution and orthogonal thiol-ene reactions. In this study, multifunctional dendrimers synthesized via an orthogonal approach did not possess the ability to carry out protection-deprotection steps to apply for complex organic molecule construction [97]. Moreover, the heterofunctional method is employed to produce dendrimers with a precise and large number of unique functional groups while maintaining structural integrity within the framework [98]. Goodwin and colleagues synthesized a heterobifunctional, biodegradable dendrimer through the creation of a symmetric aliphatic ester dendrimer derived from 2, 2-bis(hydroxymethyl) propanoic acid. The dendrimer featured a cyclic carbonate periphery and functional amine at the opening of the carbonate moieties. This approach was deemed advantageous for the formation of a dendrimer with eight alkynes and eight protected aldehydes on its periphery, as the resulting dendrimer did not require purification through chromatography [99]. All the above-mentioned methods require purification approach to yield pure dendrimer, either one way or the other. Hence, one pot synthesis has been introduced for the formation of dendrimers, which does not require any purification steps [100]. Recently, Yan et al. (2022) utilized one pot periodate oxidation approach for the preparation of dialdehyde cellulose and are modified with polyamidoamine (PAMAM) dendrimer of zero generation via hyperbranched crosslinking synthesis to yield cellulose-based dialdehyde polymers. The study showed that the dendrimer was grafted onto the backbone of cellulose with a crystallinity of 46.5%, uniform long-strip structure and excellent particle size distribution [101].

In recent times, several novel synthesis approaches were introduced for the preparation of dendrimers. Mahdavijalal et al. (2023) prepared a PAMAM dendrimer that is anchored to tungsten disulfide (WS_2) nano-sheets via a grafting approach. In this study, the stimulus-responsive polymer was prepared by suspending WS_2, 2, 2′ azobisisobutyronitrile (AIBN) as the radical initiator and poly (N-vinyl caprolactam) (PNVCL) in ethanol under a nitrogen atmosphere, stirring was conducted at 65 °C in a paraffin bath that was temperature-regulated, lasting for 7 h. The resultant surface-modified nano-sheets were utilized for the fabrication of three generations of PAMAM dendrimers via grafting, precipitation, and centrifugation technique. The electron micrograph results revealed that before the surface modification, the nano-sheets were 20–50 nm in size and had an inconsistent thickness However, after the surface modification using PAMAM dendrimers, the size of the nano-sheets increased to 80–90 nm, with a more uniform thickness [102]. Further, Sohail et al. (2020) synthesized PAMAM dendrimers via a divergent approach with monomer coupling and monomer end-group transformation for the creation of reactive surface functionality as well as coupling of a novel monomer via solid-phase peptides or oligonucleotide synthesis. In this study, needle-like and spherical structures were produced during the synthesis of PAMAM dendrimers with ester-terminated half-generation and amino-terminated full-generation, respectively with ~827 nm of hydrodynamic size [103]. Furthermore, Bafrooee et al. (2020) utilized the chemical co-precipitation method for carboxyl-terminated hyperbranched PAMAM dendrimer formation, that is grafted with superparamagnetic iron oxide nanoparticles in a core-shell structure. The study revealed that the synthesis process resulted in the formation of a dendrimer with an average of 5.5 generations and an average pore diameter of 11.83 nm. The size of the dendrimer was observed to be in the range of 20–75 nm, with a spherical shape [104].

4. Bioavailability of Janus Nanoparticles

Green chemistry aims to limit or reduce the use and/or creation of hazardous materials starting from the initial design process to production and finally to the application of a product. This promotes prevention over remediation for material sustainability and safety. The production of Janus nanoparticles is via green chemistry. This is because Janus nanoparticles are mostly made of biogenic materials, including both natural and synthetic biopolymers that are biodegradable and biocompatible. This leads to the development of bioavailable, sustainable, and eco-friendly materials, especially for biomedical applications, which can eventually help to address the toxicity issues of some available materials. Organic and soft materials are used to fabricate most biocompatible and biogenic Janus particles using e.g., single/double emulsion, microfluidic, co-jetting, solvent evaporation, polymerization [64]. In addition, many biocompatible synthetic biomaterials are used for Janus nanoparticles synthesis which includes the FDA-approved poly(lactic-co-glycolic acid) (PLGA), polycaprolactone (PCL), and poly(lactic acid) (PLA) for their pharmaceutical delivery applications. They are often been utilized for time-programmed single/dual drug delivery and release at dual-site with simultaneous/stimuli-dependent release of two incompatible drugs due to their difference in biodegradability [73].

Some studies have reported the use of Janus nanoparticles with high bio-distribution for intercellular transportation and cellular uptake. For instance, Shao and their team have developed rod-like magnetic nanosized silica mesoporous particles via the modified sol-gel approach. The results indicated that a more targeted drug delivery efficacy has been achieved with higher retention and accumulation at the tumor sites via varying endocytic pathways, leading to higher intracellular internalization and bioavailability [105]. Another study by Shao, et al. [106] also demonstrated the bioavailability of doxorubicin-loaded Janus nanocomposites by concentrating the accumulation of the drug at the targeted site of the liver tumor with high tumor cell endocytosis whilst offering 'zero' release of doxorubicin to the other surrounding normal cells and blood circulatory system.

Many studies have investigated the cytotoxicity of Janus nanoparticles, indicating their excellent biocompatibility, especially for clinical applications. Cao and team have demonstrated the cytotoxicity of dual-drug loaded Janus nanoparticles in HeLa cells and MDA-MB-231 cancer cell line via the colorimetric cell viability assay [71]. Results revealed more than 90% of cell viability even with such a high concentration of Janus nanoparticles ~500 μg/mL. Moreover, it showed a lower IC_{50} as compared to free drugs. Furthermore, Janus nanoparticles are capable to offer a lowered systematic toxicity with significant tumor growth inhibition as compared to other core-shell nanoparticles [107]. Based on the cytotoxic Sulforhodamine B assessment performed under both normoxic and hypoxic environments, Janus nanoparticles can selectively eliminate the hypoxic liver cancer cells with dose-dependent cytotoxicity without inhibiting the normal surrounding cells for better and safer therapeutic outcomes. Besides, modifying the nanoparticles with silica could be crucial in enhancing their biocompatibility, specifically in the case of Janus nanoparticles.

Additionally, Zhao, et al. [108] have invented Janus hollow spheres with dose-dependent cytotoxicity. According to their research, the safety level of doxorubicin-loaded Janus hollow spheres was found to be favorable in the U87MG cell line when tested using the CCK-8 assay at both pH 6.4 and 7.4 with particle concentrations as high as 200 µg/mL. The results of both the cell viability tests and the pH-dependent toxicity evaluations indicate that Janus hollow spheres have the potential to be a secure and effective vehicle for targeted drug delivery. Another study has described the cytotoxicity evaluation of Janus nanoparticles after penetrating the blood-brain-barrier via the 3-(4,5-Dimethylthiazol-2-yl)-2,5-diphenyltetrazolium-bromide (MTT) assay using the rat C6 glioma cell line, at different time intervals and pH. The study revealed that the optimal conditions for maintaining cell viability above 50% were a minimum pH of 6.2 and a maximum incubation time of 24 h. Also, the results indicated that fewer Janus nanoparticles were present in non-targeted normal cells, likely due to the extended blood circulation time of the Janus nanoparticles, as well as the strong covalent bond between the encapsulated doxorubicin and its Janus compartment. As a result, doxorubicin was only released when triggered by the acidic condition of cancerous cells via breakage of imine bond, indicating the safe usage of Janus nanoparticles for targeted drug therapies.

5. Bioavailability of Dendrimers

It can be noted that dendrimers are biocompatible and less toxic to be used in biomedical applications, similar to Janus particles. Rehman et al. (2021) prepared a novel resorcin arene-based dendrimer vesicles of nano-size based benzyloxy macrocycle. In this study, 4-benzyloxy benzaldehyde and resorcinol were dissolved in acetic acid under constant stirring conditions. Later, their pH was altered using sulphuric acid under reflux conditions for 24 h at 50 °C. A brown precipitate was formed as a resultant sample, which was filtered, washed with water, and dried, where the recrystallization of brown precipitate in tetrahydrofuran or methanol mixture (2:8, *v/v*) with 88% of yield. This study demonstrated that the synthesized dendrimer was highly effective for encapsulating quercetin, which was found to exhibit low toxicity in the NIH-3 T3 cell lines. The lower toxicity of the dendrimer-encapsulated quercetin towards cells is identified to be due to the existence of four hydrophobic benzyl groups with appropriate lipophilicity of benzyloxy macrocycle and their slightly amphiphilic nature with structural saturation [109]. Further, Giorgadze et al. (2020) synthesized PAMAM dendrimers of four generations to be useful for the encapsulation of silver atoms. The results showed that the dendrimers were stable at ~5 nm in size with strong visible light absorption. The study emphasized that the silver atoms in dendrimers can able to overcome the cell nucleus and membrane of 8–9 nm of pore size, which can subsequently reduce their cytotoxicity and improve their biocompatibility [110]. Furthermore, Bhatt et al. (2019) developed a dendrimer via alpha-tocopheryl succinate (alpha-TOS) conjugation and coating polyethylene glycol (PEG) on the surface of PAMAM dendrimer (four generations) as shown in Figure 4. The resultant dendrimer was utilized to improve intracellular paclitaxel delivery, which is a chemotherapeutic drug that is poorly soluble in water. The hemolysis assay revealed that the resultant dendrimer was nontoxic to red blood cells with high biocompatibility. Later, the cellular uptake assay conducted on B16F10 monolayer cells and MDA MB231 3D spheroids showed that the conjugation of alpha-TOS significantly improved the time-dependent uptake of the nanoscale dendrimer [111].

Noorin et al. (2021) created a novel second-generation linear globular dendrimer made of Gadoterate Meglumine and of nanoscale size.The resultant dendrimer was identified to be almost nontoxic towards immortalized human embryonic HEK 293 kidney cells and human glioblastoma cell culture [112]. Similarly, Ahmed et al. (2021) prepared a novel surface modified four generation PAMAM dendrimer via 4-nitrophenyl chloroformate as an activator and 2 kDa of PEG with spherical or semi-spherical morphology for the controlled delivery of anti-tuberculosis rifampicin drug. The findings indicated that the surface of the dendrimer was coated with PEG at a level of 38–100%, leading to a slower release rate of rifampicin compared to the non-PEGylated formulation and the unencapsulated drug. The study found that the fully functionalized dendrimer exhibited reduced toxicity towards raw 264.7 cell lines [113]. Likewise, Alfei et al. (2020) created a new biodegradable fifth-generation dendrimer based on polyester and featuring a free carboxylic group. The dendrimer was synthesized using dichloromethane, N, N'-dicyclohexylcarbodiimide (DCC), 4-(dimethylamino) pyridinium 4-toluene sulfonate and was protected with an acetonide group. Later, the dendrimer was used to encapsulate Etoposide (ETO), a substance derived from podophyllotoxin, the main chemical component found in the herbaceous plants named *Podophyllum hexandrum* and *P. peltatum*. The studies revealed that the dendrimer with ETO was 70 nm in size with nano-spherical morphology, 37% of drug loading, 53% of entrapment efficiency, and possess enhanced solubility in biocompatible solvents, such as ethanol and water. The cytotoxicity studies on neuroblastoma cells emphasized that the dendrimer and ETO possess cytotoxic and pro-oxidant properties and their bioactivity was synergistically improved after encapsulation due to the slow release of ETA [114].

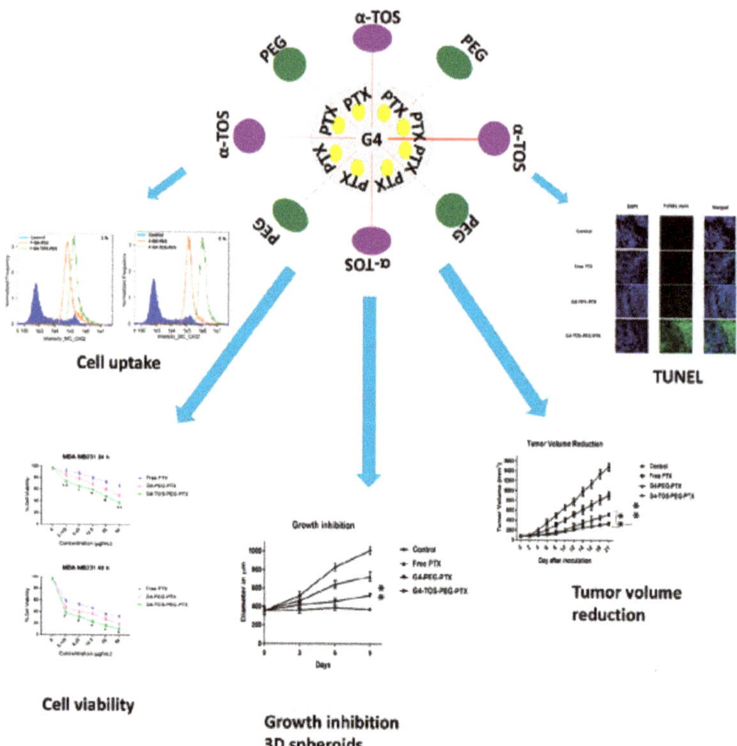

Figure 4. Schematic representation of dendrimers with PEG and alpha-TOS along with their cell uptake, cell viability, growth inhibitions of 3D spheroids, tumor volume reduction, and terminal deoxynucleotidyl transferase biotin-deoxyuridine triphosphate (UTP) nick end labeling (TUNEL) results. Reproduced with permission from Bhatt et al. Ref. [111]. Copyright 2019 ACS.

6. Janus Nanoparticles for Target-Specific Delivery Applications

The advancement in nanotechnology has allowed Janus nanoparticles to serve as nanomedicines to regulate both the bio-distribution and tumor accumulation of administered medical drugs for better efficacy. Janus metallic mesoporous silica nanoparticles (JMMSNs) have been developed to be cancer-targeting for cancer theranostics with multifunctions such as magnetic resonance imaging as well as drug release and therapy [115]. This is attributed to JMMSNs that possess anisotropic compositions with stimuli-responsive properties to provide both tumor diagnosis and therapy synergistically [115]. JMMSNs are capable to load both diagnostic and therapeutic agents in separate compartments and release them spatiotemporally in response to different stimuli for the cancer-targeted delivery. Janus magnetic nanoparticles possess magnetic properties that make them suitable for both enhancing MRI imaging as a contrast agent and delivering drugs when used in conjunction with hyperthermia [116]. Hence, Janus magnetic nanoparticles are widely recognized as safe and effective carriers for biological drugs such as genes, proteins, and other biological molecules due to their controlled and sustained drug release profile and their ability to efficiently load drugs through the use of an external magnetic field. Moreover, the size of nanoparticles allows them to penetrate targeted cells as well as the blood-brain-barrier whilst maintaining the retention effect [117]. Janus nanoparticles have been revealed as a highly promising delivery vehicle for pharmaceuticals due to their superior properties, and they can be decorated with various targeting, diagnostic, and therapeutic biological molecules further to advance the use of nanomedicine with multiple functions. Table 1 shows a summary of reported works that exploited Janus nanoparticles for target-specific delivery and therapy.

Table 1. A summary of reported studies on the targeted delivery applications of Janus nanoparticles.

Janus Nanoparticles	Fabrication Technique	Target-Specific Delivery Applications	References
Anti-epithelial cell adhesion molecule (EpCAM)-coupled JMMSNs	Modified sol-gel method	• To detect the circulating tumor cells and MCF-7 breast cancer cells in both blood samples and spiked cells with the use of magnetic-fluorescent nanoprobes. • Rod-shaped EpCAM-coupled JMMSNs are capable of improving both the capture and binding efficiency of circulating tumor cells that have escaped from the site of the tumor. • Anti-EpCAM serves as a targeting element whilst fluorescent probe loading allows bioimaging of the tumor cells.	[118]
Doxorubicin-loaded Magnetic Fe_3O_4- mesoporous SiO_2 Janus nanoparticles	High-temperature hydrolysis reaction and modified sol-gel process	• It is a multifunctional Janus nanocomposite developed to target and treat liver cancer by inhibiting tumor cell growth. • The tumor-targeted drug delivery is accomplished through magnetic targeting and controlled drug release, resulting in improved permeability and retention. • This helps to enhance the chemotherapeutic effects whilst minimizing the side effects via the magnetic-guided accumulation at the targeted cancer cells.	[106,119]
Stimuli-responsive Doxorubicin and 6-mercaptopurine dual-drug loaded Janus nanoparticles	Sol-gel method and modified Pickering emulsion technique	• This Janus nanoparticle design enables real-time monitoring of dual-drug controlled drug release within living cells through the use of surface-enhanced Raman scattering and Fluorescence resonance energy transfer. • The release of both drugs is triggered by the overexpressed glutathione and acidic environment of tumor sites, leading to better anti-cancer therapeutic effects as compared to single-drug loaded Janus nanoparticles.	[71]
pH-responsive Tirapazamine drug-loaded Janus nanocomposites	Sol-gel method	• To treat liver cancer specifically targeting hypoxic over normoxic liver cancer cells, in response to acid stimuli. • It works as a hypoxia-directed chemotherapy to enhance the synergistic radio-chemo-photothermal therapy that suppresses the growth of hypoxic tumor cells via enhanced folic acid-mediated endocytosis.	[107]
Bi-layered, pH-responsive PCL- poly (diethlyaminoethyl-methacrylate) (PDEAEMA) polymeric Janus hollow spheres	Dual-step polymerization using silica particles: surface-initiated ring-opening polymerization and atom transfer radical polymerization techniques	• A pH-dependent controlled release of encapsulated Doxorubicin at the targeted tumor site can be achieved. • The result reveals a higher intracellular uptake and drug release rate of these doxorubicin-loaded PCL-PDEAEMA Janus particles at the pH of the tumor environment as compared to the physiological pH of other normal cells.	[108]
Doxorubicin-loaded folic acid-coupled Janus nanoparticles	Pickering emulsion technique	• A potential SPION-based pH-sensitive targeted multifunctional Janus nanoparticles have demonstrated the capability to pass through the blood-brain barrier due to their surfactant properties for targeting and treating brain cancerous cells. • Results show that Janus nanoparticles can be delivered specifically to the glioma tumor site with increased tumor accumulation using folic acid as the targeting element. In addition, the release of doxorubicin is highly dependent on the acidic condition of tumor cells with improved cytotoxicity.	[117]

7. Dendrimers for Target-Specific Delivery Applications

Recently, numerous dendrimers were identified to be beneficial as potential targeted drug delivery systems for specific biomedical applications. Swanson et al. (2022) synthesized a novel fifth-generation PAMAM dendrimer, which was surface-modified and functionalized with folic acid. The remaining primary amines in the terminal dendrimer part were conjugated with the bifunctional 1, 4, 6, 10-tetraazacyclododecane-1, 4, 7, 10-tetraacetic acid, alpha-[5-isothiocyanato-2-methoxyphenyl]-, hydrochloride [9Cl] (NCS-DOTA) chelator, where stable gadolinium complexes were loaded into the dendrimer. The resultant dendrimer nanoparticle was about ~5 nm in size with the ability to be a potential target-specific magnetic resonance imaging (MRI) contrast agent with high affinity towards folate receptors in the liver and kidney for identifying human epithelial cancer cells in a murine model [120]. Similarly, Mbatha et al. (2019) prepared PAMAM dendrimer modified with folic acid and functionalized with nanosized gold particles for the efficient and targeted transport of exogenous small interfering ribose nucleic acid (siRNA). The study emphasized that the resultant dendrimer nanosized particles were spherical shaped with 65–128 nm of size and zeta potential above 25 mV, which indicates their high colloidal stability. Additionally, the dendrimer-based nanocomplex was less toxic to human embryo kidney (HEK-293) and HeLa-Tat-*Luc* cells (about 90% of cell viability). The study emphasized that the nanocomplex helped in the 75% increase in the induced transgene-silencing of siRNA and decrease the presence of excess folic acid for effective inhibition of hepatocellular carcinoma (HEPG2) and colon carcinoma (Caco-2) cells [121]. Likewise, Umeda et al. (2010) demonstrated the fabrication of polyethylene glycol (PEG) attached to a PAMAM dendrimer of four generations for the encapsulation of nanosized gold particles as shown in Figure 5. The results revealed that the nanosized gold particles were 2–3 nm in size with spherical morphology, which is engulfed by a single dendrimer particle, while the PEGylated dendrimer encapsulated gold nanoparticles were 15 nm in size. The study revealed that the dendrimers with gold nanoparticles possess enhanced toxicity against HeLa cells under the irradiation of visible light via photoinduced heat generation capability [122].

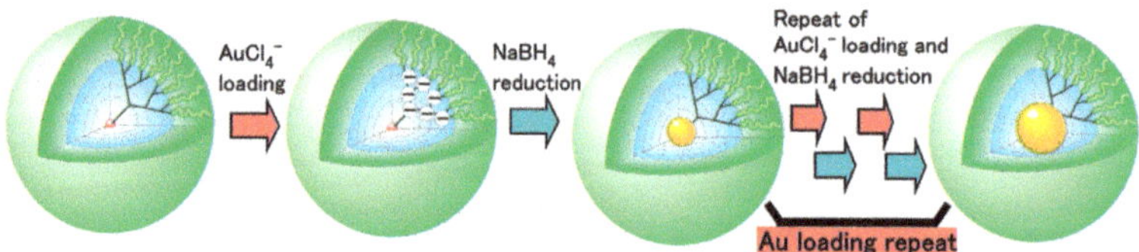

Figure 5. Growth of gold nanoparticles inside the PEGylated PAMAM dendrimers. Reproduced with permission from BUmeda et al. Ref. [122]. Copyright 2010 ACS.

Xiong et al. (2019) exhibited the formation of carboxybetaine acrylamide zwitterion functionalized fifth-generation PAMAM dendrimer and morphine, which is an agent to target lysosome and to entrap nanosized gold particles. The study showed that the nanosized gold particles entrapped by dendrimer are beneficial for the delivery of the H1C1 gene, enhanced by serum. The results showed that the dendrimer-entrapped nanosized gold particles were of size ~1.5 nm, where the vector system helped to carry the H1C1 protein for effective cancer cell migration and metastasis inhibition [123]. Further, Sharma et al. (2021) prepared novel hydroxyl terminated PAMAM dendrimers, that are modified with sugar moieties, such as alpha-D-mannose, beta-D-glucose, or beta-D-galactose via click chemistry approach. The resultant dendrimers were ~4 nm of average particle size with nearly neutral zeta potential and are utilized to target tumor-associated macrophages and microglia. The study showed that the glucose-modified dendrimer has increased brain penetration and cellular internalization, compared to other sugar moieties. It has been proposed that these dendrimers can be potential delivery vehicles for the treatment of glioblastoma and other types of cancers [124]. Furthermore, Zhang et al. (2020) utilized in situ growth approach for the preparation of novel PAMAM dendrimers (0.5, 1.5, 2, and 2.5 generation) that are grafted on the persistent luminescence nanoparticles (PLNPs) surface. The aptamer AS1411 was coupled with the nanoparticle-dendrimer to specifically bind with the over-expressed nucleolin on the tumor cell membrane, thus increasing the intracellular accumulation of the nanoparticle. The PLNPs were spherical and 15.2 nm of average particle size. In addition, an anticancer drug named doxorubicin was also loaded in the dendrimer-nanoparticle-based delivery system via a pH-sensitive hydrazine, which was identified to be released in the targeted site of the intracellular acid environment. The new drug delivery system using dendrimers was found to have the capability of inducing apoptosis in HeLa tumor cells and inhibiting tumor growth [125]. The results of these studies suggest that dendrimers, particularly PAMAM dendrimers, have the potential to serve as effective drug delivery systems for treating specific diseases. Apart from drug delivery, dendrimers are also utilized as a potential gene delivery system. Ebrahimian et al. (2022) developed a novel lipo-polymeric PAMAM dendrimer-liposome, that are functionalized with transactivator of transcription (TAT) peptide and hyaluronic acid for targeted gene delivery system development. The study showed that the lipo-polymer possesses no significant toxicity with enhanced transfection efficiency in murine colon carcinoma cell line (C26), which will be beneficial for gene delivery applications [126]. Additionally, the incorporation or encapsulation of metals and metallocomplexes within the dendritic scaffold (metallodendrimers) holds the potential to yield novel metallodrugs or offer a new approach for the in situ delivery of metallodrugs. In this area, several works with in vitro and in vivo results were published by Rodrigues and coworkers, involving ruthenium compounds (an alternative metal for platinum-based anticancer resistant drugs) (Figure 6) [127,128], as well as platinum and platinum derivatives as anticancer drugs [128,129]. In any case, the research of the dendrimers is not exclusively restricted to the preparation of drug delivery systems. For instance, Rodrigues and coworkers developed new anionic poly (alkylideneamine) dendrimers until generation 3 with carboxylate and sulfonate terminal groups. These dendrimers have shown in vivo, to be very effective as microbicide against HIV-1 infection [130]. Thus, novel dendrimer structures were developed by the team, where their ability for targeted and controlled drug delivery is expected to be improved in the future.

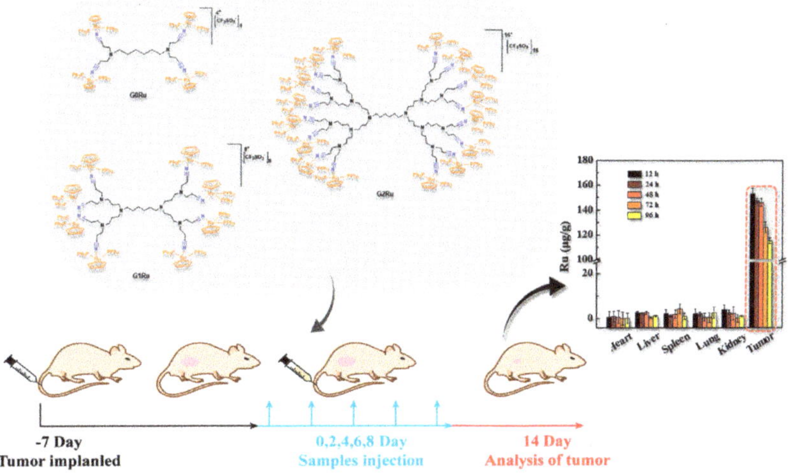

Figure 6. In vivo studies in an MCF-7 xenograft mouse model of low-generation (0–2) nitrile poly(alkylideneamine)-based ruthenium dendrimers. Reproduced with permission from Maciel et al. [127] Copyright 2022 Royal Society of Chemistry.

8. Janus-Dendrimer Particles in Target-Specific Delivery Applications

The enhanced targeted drug delivery ability of Janus structures and dendrimer nanoparticles has led to the emergence of novel Janus-dendrimer nanoparticles to improve their controlled delivery. Further, Janus-dendrimer possess exclusive properties and structures in combinations with different end-group types and distinct surfaces, which makes them a better candidate to form unique drug complexes and conjugates [131]. These Janus-dendrimer particles were fabricated based on chemo-selective coupling, heterogenous double exponential growth, and mixed modular approach to be useful for stochastic and multiple drug conjugation-based combination therapy, solubility enhancement, antioxidant lyophilization, targeted delivery, and as fluorescent labels. In addition, the self-assembled Janus-dendrimer particles were identified to be beneficial for the vesicular delivery system, especially for pH-responsive delivery, targeted delivery, spatiotemporal delivery, and site-specific drug delivery [132]. Pan et al. (2012) created a new Janus-dendrimer particle that was amphiphilic and had peripheral groups made up of acidic amino acids and naproxen molecules, designed for efficient drug delivery to bones. The second-generation dendrimers with >95% of binding rates towards hydroxyapatite in bones, 28-fold enhanced the solubility of naproxen for about 5.37 mg/mL of concentration, compared to the standalone drug. Moreover, the study also revealed that the Janus-dendrimers do not possess any significant cytotoxicity toward HEK293 cells [133]. Further, Iguarbe et al. (2019) created an effective liquid crystal Janus-dendrimer particle made up of mesogenic blocks based on two third-generation Percec-type dendrons with terminal dodecyloxy alkyl chains and one or two carbazole units serving as the electrically active component. The study reported that the carbazole dendrimers were prepared via electrodeposition to form semi-globular particles with electro-polymerizable units. Also, the resultant particles were able to retain the rigid or flexible characteristics of the linker, which eventually influences the size of the particle [134]. Furthermore, a link was established between hydrophobic paclitaxel and the Janus PEGylated peptide dendrimer by Li et al. (2017), through the use of an enzyme-sensitive glycylphenylalanylleucylglycine tetrapeptide as a connector, using an efficient click reaction as shown in Figure 7. The resultant Janus-dendrimer particle possesses the ability to encapsulate 21% of paclitaxel with an average hydrodynamic size of ~69 nm, a narrow polydispersity index of 0.23, the zeta potential of −16.9 mV and spherical morphology. The study showed that the Janus-dendrimer nanoparticle release paclitaxel via enzyme responsive feature and is identified to be highly cytotoxic towards 4T1 (murine breast) cancer cells without any toxicity against normal cells, such as 3T3 murine fibroblast and C2C12 murine myoblast cell lines [135].

Recently, Falanga et al. (2021) created a new type of Janus-like dendrimer that incorporates peptides derived from the glycoproteins (gH and gB) of Herpes Simplex Virus Type 1 (HSV1), aimed at inhibiting viral infection. This was achieved through the combination of copper-catalyzed bio-orthogonal 1,3-dipolar azide/alkyne cycloaddition and photoinitiated thiol-ene coupling, producing both monofunctional and bifunctional peptidodendrimer conjugates. The study revealed that the peptides released by the formulation possess enhanced antiviral activity by inhibiting the DNA replication of HSV1, compared to conventional antiviral drugs, such as foscarnet, acyclovir, and cifofovir [136]. Similarly, Najafi et al. (2020) prepared a novel poly (propyleneimine) (PPI) dendrimer of the fifth generation with a core of cystamine and a hydrophobic surface. Later, the structure scission approach was used to convert disulfide bonds to thiol group and hydrophilic PAMAM dendrons were formed with amine end groups. The study demonstrated that the Janus-like dendrimer, with an average hydrodynamic size of 4.2–28.2 nm, has the ability to enhance the solubility of hydrophobic drugs such as dexamethasone and tetracycline [137]. Likewise, Zhang et al. (2022) designed a hydrophobic multifunctional sequence-defined ionizable amphiphilic Janus-dendrimer region via dissimilar alkyl lengths. The research found that the Janus-dendrimer particles greatly enhanced the activity of the hydrophobic 3, 5-, 3, 4-, and 3,4, 5-substituted phenolic acids they encapsulated, by up to 90.2 times [138].

All these studies showed that the Janus-dendrimer nanoparticles possess enhanced drug delivery capacity, compared to standalone Janus particles and dendrimers.

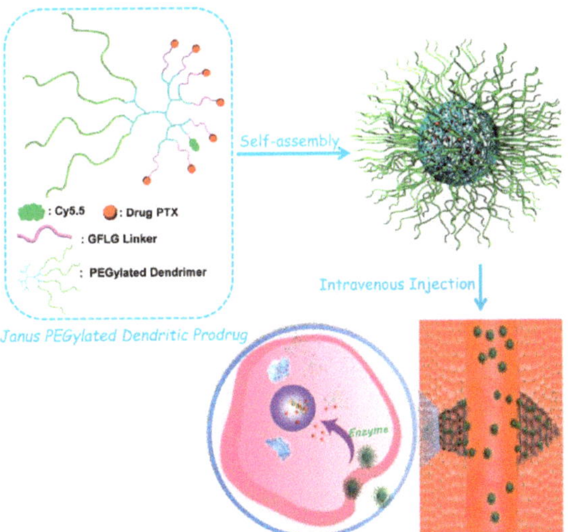

Figure 7. Schematic representation of PEGylated Janus-dendrimer and their self-assembly for intravenous injection in murine models for enhanced stability and anticancer efficacy. Reproduced with permission from Li et al. Ref. [135]. Copyright 2017 ACS.

9. Limitations and Future Perspective

Dendrimers, Janus particles, and Janus-dendrimer nanoparticles have been identified as potential candidates for improved drug delivery applications. However, there remain several limitations that must be overcome before they can be utilized in large-scale commercial applications. The synthesis of Janus particles is one of the major limitations, as it is a multi-step process that can be tedious and challenging to achieve sub-micrometer particle size [139]. The various synthesis approaches, including masking, phase separation, microfluidics, E-jetting, emulsion, and self-assembly, each have their own drawbacks which hinder their scalability. Additionally, general limitations of nanoparticles such as polydispersity, size-dependent toxicity, and the use of hazardous chemicals in synthesis are also applicable to Janus nanoparticles [140]. Dendrimers also face several limitations, particularly at generations higher than 3–4, including high cost of synthesis, a lack of understanding of their effects on biochemical pathways and toxicity, and difficulties in engineering multifunctional dendrimers, especially the tediousness in purification and the yields of the end product. The toxicity of dendrimers has been reported in various studies and can result from improper processing, leading to undesirable side effects, low tolerability, and inefficient drug delivery [141–143]. To address these limitations, the use of lower dendrimers generations with less toxic terminal groups and the use of computational software and machine learning approaches have been employed to optimize Janus, dendrimer, and Janus-dendrimer synthesis parameters. These computational approaches have the potential to reduce the number of experiments and the cost of production, while providing greater insight into the interaction of nanoparticles with biological environments. However, these computational approaches are still in the early stages of research, are highly dependent of the data availability, particularly with regards to the synthesis of nanoparticles and evaluation of their properties and interactions in biological environments [144,145]. In conclusion, the future of Janus and dendrimer nanoparticles for improved drug delivery will likely benefit from a combination of enhanced electron microscopes, synthesis techniques, computational approaches, and machine learning techniques. This will enable the synthesis of novel and highly efficient nanoparticles with improved bioavailability for drug delivery applications.

10. Conclusions

Recently, there has been an increasing interest in developing a novel drug delivery system that can enhance targeted and controlled drug delivery while reducing toxicity and increasing biocompatibility. Janus and dendrimer particles have been introduced as potential drug delivery systems due to their unique properties compared to conventional delivery systems. These particles have the ability to encapsulate and release drugs in a controlled manner, acts as a drug per se, which can improve the efficacy of drugs and reduce their side effects. However, there are limitations to the use of these materials in commercial pharmaceutics, including lack of scalability, high production cost, and the use of toxic chemicals in synthesis. The emergence of Janus-dendrimer particles holds promise for overcoming these limitations. The combination of Janus and dendrimer particles can result in enhanced drug delivery and improved biocompatibility. However, it is important to improve the stability and toxicity of these particles in the future. This can be achieved through simulation, modeling-based computational and machine-learning approaches, as well as by optimizing the synthesis process and reducing the use of toxic chemicals. With these advancements, Janus-dendrimer particles could become a key tool in advancing drug delivery technology and improving patient outcomes. Their utility extends to various applications within the healthcare field and beyond, opening up new possibilities for innovation.

Author Contributions: Writing—original draft preparation, J.J. and K.X.T.; writing—review and editing, J.R. and M.K.D.; supervision, J.R. and M.K.D. All authors have read and agreed to the published version of the manuscript.

Funding: This research received no external funding.

Institutional Review Board Statement: Not applicable.

Informed Consent Statement: Not applicable.

Data Availability Statement: Data sharing not applicable. No new data were created or analyzed in this study. Data sharing is not applicable to this article.

Acknowledgments: The authors (J.J. and J.R.) acknowledge FCT-Fundação para a Ciência e a Tecnologia (Base Fund UIDB/00674/2020 and Programmatic Fund UIDP/00674/2020, Portuguese Government Funds). All the authors thank their respective department for the support during preparation of this article.

Conflicts of Interest: The authors declare no conflict of interest.

References

1. Mitchell, M.J.; Billingsley, M.M.; Haley, R.M.; Wechsler, M.E.; Peppas, N.A.; Langer, R. Engineering precision nanoparticles for drug delivery. *Nat. Rev. Drug Discov.* **2021**, *20*, 101–124. [CrossRef] [PubMed]
2. Yusuf, A.; Almotairy, A.R.Z.; Henidi, H.; Alshehri, O.Y.; Aldughaim, M.S. Nanoparticles as Drug Delivery Systems: A Review of the Implication of Nanoparticles' Physicochemical Properties on Responses in Biological Systems. *Polymers* **2023**, *15*, 1596. [CrossRef] [PubMed]
3. Chandrakala, V.; Aruna, V.; Angajala, G. Review on metal nanoparticles as nanocarriers: Current challenges and perspectives in drug delivery systems. *Emergent Mater.* **2022**, *5*, 1593–1615. [CrossRef]
4. Heuer-Jungemann, A.; Feliu, N.; Bakaimi, I.; Hamaly, M.; Alkilany, A.; Chakraborty, I.; Masood, A.; Casula, M.F.; Kostopoulou, A.; Oh, E. The role of ligands in the chemical synthesis and applications of inorganic nanoparticles. *Chem. Rev.* **2019**, *119*, 4819–4880. [CrossRef] [PubMed]
5. Gorantla, S.; Wadhwa, G.; Jain, S.; Sankar, S.; Nuwal, K.; Mahmood, A.; Dubey, S.K.; Taliyan, R.; Kesharwani, P.; Singhvi, G. Recent advances in nanocarriers for nutrient delivery. *Drug Deliv. Transl. Res.* **2021**, *12*, 2359–2384. [CrossRef] [PubMed]
6. Ghosh, S.; Ray, A.; Pramanik, N. Self-assembly of surfactants: An overview on general aspects of amphiphiles. *Biophys. Chem.* **2020**, *265*, 106429. [CrossRef]
7. Alven, S.; Aderibigbe, B.A. The therapeutic efficacy of dendrimer and micelle formulations for breast cancer treatment. *Pharmaceutics* **2020**, *12*, 1212. [CrossRef]

8. Dymek, M.; Sikora, E. Liposomes as biocompatible and smart delivery systems—The current state. *Adv. Colloid Interface Sci.* **2022**, *309*, 102757. [CrossRef]

9. Guimarães, D.; Cavaco-Paulo, A.; Nogueira, E. Design of liposomes as drug delivery system for therapeutic applications. *Int. J. Pharm.* **2021**, *601*, 120571. [CrossRef]

10. García-González, C.A.; Sosnik, A.; Kalmár, J.; De Marco, I.; Erkey, C.; Concheiro, A.; Alvarez-Lorenzo, C. Aerogels in drug delivery: From design to application. *J. Control. Release* **2021**, *332*, 40–63. [CrossRef]

11. Koyyada, A.; Orsu, P. Natural gum polysaccharides as efficient tissue engineering and drug delivery biopolymers. *J. Drug Deliv. Sci. Technol.* **2021**, *63*, 102431. [CrossRef]

12. Wang, J.; Li, B.; Qiu, L.; Qiao, X.; Yang, H. Dendrimer-based drug delivery systems: History, challenges, and latest developments. *J. Biol. Eng.* **2022**, *16*, 18. [CrossRef] [PubMed]

13. Tomás, H.; Rodrigues, J. Chapter 2—Dendrimers and dendrimer-based nano-objects for oncology applications. In *New Trends in Smart Nanostructured Biomaterials in Health Sciences*; Gonçalves, G., Marques, P., Mano, J., Eds.; Elsevier: Amsterdam, The Netherlands, 2023; pp. 41–78.

14. Mignani, S.; Rodrigues, J.; Tomas, H.; Zablocka, M.; Shi, X.; Caminade, A.-M.; Majoral, J.-P. Dendrimers in combination with natural products and analogues as anti-cancer agents. *Chem. Soc. Rev.* **2018**, *47*, 514–532. [CrossRef] [PubMed]

15. Chis, A.A.; Dobrea, C.; Morgovan, C.; Arseniu, A.M.; Rus, L.L.; Butuca, A.; Juncan, A.M.; Totan, M.; Vonica-Tincu, A.L.; Cormos, G. Applications and limitations of dendrimers in biomedicine. *Molecules* **2020**, *25*, 3982. [CrossRef]

16. Percec, V.; Wilson, D.A.; Leowanawat, P.; Wilson, C.J.; Hughes, A.D.; Kaucher, M.S.; Hammer, D.A.; Levine, D.H.; Kim, A.J.; Bates, F.S.; et al. Self-Assembly of Janus Dendrimers into Uniform Dendrimersomes and Other Complex Architectures. *Science* **2010**, *328*, 1009–1014. [CrossRef] [PubMed]

17. Najafi, F.; Salami-Kalajahi, M.; Roghani-Mamaqani, H. Janus-type dendrimers: Synthesis, properties, and applications. *J. Mol. Liq.* **2022**, *347*, 118396. [CrossRef]

18. Duan, Y.; Zhao, X.; Sun, M.; Hao, H. Research advances in the synthesis, application, assembly, and calculation of Janus materials. *Ind. Eng. Chem. Res.* **2021**, *60*, 1071–1095. [CrossRef]

19. Kirillova, A.; Marschelke, C.; Synytska, A. Hybrid Janus particles: Challenges and opportunities for the design of active functional interfaces and surfaces. *ACS Appl. Mater. Interfaces* **2019**, *11*, 9643–9671. [CrossRef]

20. Rosati, M.; Acocella, A.; Pizzi, A.; Turtù, G.; Neri, G.; Demitri, N.; Nonappa; Raffaini, G.; Donnio, B.; Zerbetto, F.; et al. Janus-Type Dendrimers Based on Highly Branched Fluorinated Chains with Tunable Self-Assembly and 19F Nuclear Magnetic Resonance Properties. *Macromolecules* **2022**, *55*, 2486–2496. [CrossRef]

21. Mollazadeh, S.; Sahebkar, A.; Shahlaei, M.; Moradi, S. Nano drug delivery systems: Molecular dynamic simulation. *J. Mol. Liq.* **2021**, *332*, 115823. [CrossRef]

22. Căta, A.; Ienașcu, I.M.; Ștefănuț, M.N.; Roșu, D.; Pop, O.-R. Properties and Bioapplications of Amphiphilic Janus Dendrimers: A Review. *Pharmaceutics* **2023**, *15*, 589. [CrossRef]

23. de Gennes, P.-G. Soft Matter (Nobel Lecture). *Angew. Chem. Int. Ed. Engl.* **1992**, *31*, 842–845. [CrossRef]

24. Song, Y.; Chen, S. Janus Nanoparticles: Preparation, Characterization, and Applications. *Chem. Asian J.* **2014**, *9*, 418–430. [CrossRef] [PubMed]

25. Zhang, X.; Fu, Q.; Duan, H.; Song, J.; Yang, H. Janus Nanoparticles: From Fabrication to (Bio)Applications. *ACS Nano* **2021**, *15*, 6147–6191. [CrossRef]

26. Lattuada, M.; Hatton, T.A. Synthesis, properties and applications of Janus nanoparticles. *Nano Today* **2011**, *6*, 286–308. [CrossRef]

27. Zhang, Y.; Huang, K.; Lin, J.; Huang, P. Janus nanoparticles in cancer diagnosis, therapy and theranostics. *Biomater. Sci.* **2019**, *7*, 1262–1275. [CrossRef] [PubMed]

28. Yan, J.; Chaudhary, K.; Chul Bae, S.; Lewis, J.A.; Granick, S. Colloidal ribbons and rings from Janus magnetic rods. *Nat. Commun.* **2013**, *4*, 1516. [CrossRef] [PubMed]

29. Wang, C.; Xu, C.; Zeng, H.; Sun, S. Recent Progress in Syntheses and Applications of Dumbbell-like Nanoparticles. *Adv. Mater.* **2009**, *21*, 3045–3052. [CrossRef] [PubMed]

30. Link, J.R.; Sailor, M.J. Smart dust: Self-assembling, self-orienting photonic crystals of porous Si. *Proc. Natl. Acad. Sci. USA* **2003**, *100*, 10607–10610. [CrossRef]

31. Zhao, R.; Yu, X.; Sun, D.; Huang, L.; Liang, F.; Liu, Z. Functional Janus Particles Modified with Ionic Liquids for Dye Degradation. *ACS Appl. Nano Mater.* **2019**, *2*, 2127–2132. [CrossRef]

32. Mou, F.; Chen, C.; Guan, J.; Chen, D.-R.; Jing, H. Oppositely charged twin-head electrospray: A general strategy for building Janus particles with controlled structures. *Nanoscale* **2013**, *5*, 2055–2064. [CrossRef] [PubMed]

33. Bradley, L.C.; Chen, W.-H.; Stebe, K.J.; Lee, D. Janus and patchy colloids at fluid interfaces. *Curr. Opin. Colloid Interface Sci.* **2017**, *30*, 25–33. [CrossRef]

34. Su, H.; Hurd Price, C.A.; Jing, L.; Tian, Q.; Liu, J.; Qian, K. Janus particles: Design, preparation, and biomedical applications. *Mater. Today Bio.* **2019**, *4*, 100033. [CrossRef] [PubMed]

35. Li, X.; Chen, L.; Cui, D.; Jiang, W.; Han, L.; Niu, N. Preparation and application of Janus nanoparticles: Recent development and prospects. *Coord. Chem. Rev.* **2022**, *454*, 214318. [CrossRef]

36. Kim, M.; Jeon, K.; Kim, W.H.; Lee, J.W.; Hwang, Y.-H.; Lee, H. Biocompatible amphiphilic Janus nanoparticles with enhanced interfacial properties for colloidal surfactants. *J. Colloid Interface Sci.* **2022**, *616*, 488–498. [CrossRef]

37. Jia, R.; Jiang, H.; Jin, M.; Wang, X.; Huang, J. Silver/chitosan-based Janus particles: Synthesis, characterization, and assessment of antimicrobial activity in vivo and vitro. *Food Res. Int.* **2015**, *78*, 433–441. [CrossRef]
38. Chun, H.J.; Kim, S.; Han, Y.D.; Kim, D.W.; Kim, K.R.; Kim, H.-S.; Kim, J.-H.; Yoon, H.C. Water-soluble mercury ion sensing based on the thymine-Hg^{2+}-thymine base pair using retroreflective Janus particle as an optical signaling probe. *Biosens. Bioelectron.* **2018**, *104*, 138–144. [CrossRef]
39. Wang, Y.; Shang, M.; Wang, Y.; Xu, Z. Droplet-based microfluidic synthesis of (Au nanorod@Ag)–polyaniline Janus nanoparticles and their application as a surface-enhanced Raman scattering nanosensor for mercury detection. *Anal. Methods* **2019**, *11*, 3966–3973. [CrossRef]
40. Zheng, F.; Ke, W.; Shi, L.; Liu, H.; Zhao, Y. Plasmonic Au–Ag Janus Nanoparticle Engineered Ratiometric Surface-Enhanced Raman Scattering Aptasensor for Ochratoxin A Detection. *Anal. Chem.* **2019**, *91*, 11812–11820. [CrossRef]
41. Flory, P.J. Molecular Size Distribution in Three Dimensional Polymers. I. Gelation1. *J. Am. Chem. Soc.* **1941**, *63*, 3083–3090. [CrossRef]
42. Flory, P.J. Molecular Size Distribution in Three Dimensional Polymers. II. Trifunctional Branching Units. *J. Am. Chem. Soc.* **1941**, *63*, 3091–3096. [CrossRef]
43. Flory, P.J. Molecular size distribution in three dimensional polymers. VI. Branched polymers containing A—R—B_{f-1} type units. *J. Am. Chem. Soc.* **1952**, *74*, 2718–2723. [CrossRef]
44. Kim, Y.H.; Webster, O.W. Hyperbranched polyphenylenes. *Macromolecules* **1992**, *25*, 5561–5572. [CrossRef]
45. Tomalia, D.A.; Baker, H.; Dewald, J.; Hall, M.; Kallos, G.; Martin, S.; Roeck, J.; Ryder, J.; Smith, P. A new class of polymers: Starburst-dendritic macromolecules. *Polym. J.* **1985**, *17*, 117–132. [CrossRef]
46. Newkome, G.R.; Yao, Z.; Baker, G.R.; Gupta, V.K. Micelles. Part 1. Cascade molecules: A new approach to micelles. A [27]-arborol. *J. Org. Chem.* **1985**, *50*, 2003–2004. [CrossRef]
47. Augustus, E.N.; Allen, E.T.; Nimibofa, A.; Donbebe, W. A review of synthesis, characterization and applications of functionalized dendrimers. *Am. J. Polym. Sci.* **2017**, *7*, 8–14.
48. Nikzamir, M.; Hanifehpour, Y.; Akbarzadeh, A.; Panahi, Y. Applications of dendrimers in nanomedicine and drug delivery: A review. *J. Inorg. Organomet. Polym. Mater.* **2021**, *31*, 2246–2261. [CrossRef]
49. Akki, R.; Ramya, M.G.; Sadhika, C.; Spandana, D. A novel approach in drug delivery system using dendrimers. *Pharm. Innov. J.* **2019**, *8*, 166–174.
50. Frechet, J.M.J. Functional polymers and dendrimers: Reactivity, molecular architecture, and interfacial energy. *Science* **1994**, *263*, 1710–1715. [CrossRef]
51. Prakash, P.; Kunjal, K.K.; Shabaraya, A. Dendrimer architecture: A comprehensive review. *World J. Pharm. Res.* **2021**, *10*, 638–659.
52. England, R.M.; Sonzini, S.; Buttar, D.; Treacher, K.E.; Ashford, M.B. Investigating the properties of l-lysine dendrimers through physico-chemical characterisation techniques and atomistic molecular dynamics simulations. *Polym. Chem.* **2022**, *13*, 2626–2636. [CrossRef]
53. Munavalli, B.B.; Naik, S.R.; Torvi, A.I.; Kariduraganavar, M.Y. Dendrimers. In *Functional Polymers*; Springer: Berlin/Heidelberg, Germany, 2019; pp. 289–345.
54. Mittal, P.; Saharan, A.; Verma, R.; Altalbawy, F.; Alfaidi, M.A.; Batiha, G.E.-S.; Akter, W.; Gautam, R.K.; Uddin, M.; Rahman, M. Dendrimers: A new race of pharmaceutical nanocarriers. *BioMed Res. Int.* **2021**, *2021*, 8844030. [CrossRef] [PubMed]
55. Smith, R.J.; Gorman, C.; Menegatti, S. Synthesis, structure, and function of internally functionalized dendrimers. *J. Polym. Sci.* **2021**, *59*, 1028. [CrossRef]
56. Abasian, P.; Ghanavati, S.; Rahebi, S.; Nouri Khorasani, S.; Khalili, S. Polymeric nanocarriers in targeted drug delivery systems: A review. *Polym. Adv. Technol.* **2020**, *31*, 2939–2954. [CrossRef]
57. Klajnert, B.; Bryszewska, M. Dendrimers: Properties and applications. *Acta Biochim. Pol.* **2001**, *48*, 199–208. [CrossRef]
58. Janaszewska, A.; Lazniewska, J.; Trzepiński, P.; Marcinkowska, M.; Klajnert-Maculewicz, B. Cytotoxicity of dendrimers. *Biomolecules* **2019**, *9*, 330. [CrossRef]
59. Gupta, A.; Dubey, S.; Mishra, M. Unique structures, properties and applications of dendrimers. *J. Drug Deliv. Ther.* **2018**, *8*, 328–339. [CrossRef]
60. Mignani, S.; Shi, X.; Rodrigues, J.; Tomas, H.; Karpus, A.; Majoral, J.-P. First-in-class and best-in-class dendrimer nanoplatforms from concept to clinic: Lessons learned moving forward. *Eur. J. Med. Chem.* **2021**, *219*, 113456. [CrossRef]
61. Mignani, S.; Rodrigues, J.; Tomas, H.; Roy, R.; Shi, X.; Majoral, J.-P. Bench-to-bedside translation of dendrimers: Reality or utopia? A concise analysis. *Adv. Drug Deliv. Rev.* **2018**, *136*, 73–81. [CrossRef]
62. Honciuc, A. Amphiphilic Janus Particles at Interfaces. In *Flowing Matter*; Toschi, F., Sega, M., Eds.; Springer: Berlin/Heidelberg, Germany, 2019; pp. 95–136.
63. Araújo, R.V.; Santos, S.D.; Ferreira, E.I.; Giarolla, J. New Advances in General Biomedical Applications of PAMAM Dendrimers. *Molecules* **2018**, *23*, 2849. [CrossRef]
64. Marschelke, C.; Fery, A.; Synytska, A. Janus particles: From concepts to environmentally friendly materials and sustainable applications. *Colloid Polym. Sci.* **2020**, *298*, 841–865. [CrossRef]
65. Sun, X.T.; Zhang, Y.; Zheng, D.H.; Yue, S.; Yang, C.G.; Xu, Z.R. Multitarget sensing of glucose and cholesterol based on Janus hydrogel microparticles. *Biosens. Bioelectron.* **2017**, *92*, 81–86. [CrossRef]

66. Marquis, M.; Renard, D.; Cathala, B. Microfluidic Generation and Selective Degradation of Biopolymer-Based Janus Microbeads. *Biomacromolecules* **2012**, *13*, 1197–1203. [CrossRef] [PubMed]
67. Lu, A.X.; Jiang, K.; DeVoe, D.L.; Raghavan, S.R. Microfluidic Assembly of Janus-Like Dimer Capsules. *Langmuir* **2013**, *29*, 13624–13629. [CrossRef]
68. Khoee, S.; Bakvand, P.M. Synthesis of dual-responsive Janus nanovehicle via PNIPAm modified SPIONs deposition on crosslinked chitosan microparticles and decrosslinking process in the core. *Eur. Polym. J.* **2019**, *114*, 411–425. [CrossRef]
69. Sheikhi, A.; van de Ven, T.G.M. Colloidal aspects of Janus-like hairy cellulose nanocrystalloids. *Curr. Opin. Colloid Interface Sci.* **2017**, *29*, 21–31. [CrossRef]
70. Shao, D.; Zhang, X.; Liu, W.; Zhang, F.; Zheng, X.; Qiao, P.; Li, J.; Dong, W.-F.; Chen, L. Janus Silver-Mesoporous Silica Nanocarriers for SERS Traceable and pH-Sensitive Drug Delivery in Cancer Therapy. *ACS Appl. Mater. Interfaces* **2016**, *8*, 4303–4308. [CrossRef]
71. Cao, H.; Yang, Y.; Chen, X.; Shao, Z. Intelligent Janus nanoparticles for intracellular real-time monitoring of dual drug release. *Nanoscale* **2016**, *8*, 6754–6760. [CrossRef]
72. Guix, M.; Meyer, A.K.; Koch, B.; Schmidt, O.G. Carbonate-based Janus micromotors moving in ultra-light acidic environment generated by HeLa cells in situ. *Sci. Rep.* **2016**, *6*, 21701. [CrossRef]
73. Ekanem, E.E.; Zhang, Z.; Vladisavljević, G.T. Facile Production of Biodegradable Bipolymer Patchy and Patchy Janus Particles with Controlled Morphology by Microfluidic Routes. *Langmuir* **2017**, *33*, 8476–8482. [CrossRef]
74. Jiang, S.; Chen, Q.; Tripathy, M.; Luijten, E.; Schweizer, K.S.; Granick, S. Janus Particle Synthesis and Assembly. *Adv. Mater.* **2010**, *22*, 1060–1071. [CrossRef]
75. Lim, Y.G.J.; Poh, K.C.W.; Loo, S.C.J. Hybrid Janus Microparticles Achieving Selective Encapsulation for Theranostic Applications via a Facile Solvent Emulsion Method. *Macromol. Rapid Commun.* **2019**, *40*, 1800801. [CrossRef] [PubMed]
76. Nie, Z.; Li, W.; Seo, M.; Xu, S.; Kumacheva, E. Janus and ternary particles generated by microfluidic synthesis: Design, synthesis, and self-assembly. *J. Am. Chem. Soc.* **2006**, *128*, 9408–9412. [CrossRef] [PubMed]
77. Xie, H.; She, Z.-G.; Wang, S.; Sharma, G.; Smith, J.W. One-Step Fabrication of Polymeric Janus Nanoparticles for Drug Delivery. *Langmuir* **2012**, *28*, 4459–4463. [CrossRef]
78. Abbasi, E.; Aval, S.F.; Akbarzadeh, A.; Milani, M.; Nasrabadi, H.T.; Joo, S.W.; Hanifehpour, Y.; Nejati-Koshki, K.; Pashaei-Asl, R. Dendrimers: Synthesis, applications, and properties. *Nanoscale Res. Lett.* **2014**, *9*, 247. [CrossRef]
79. Tomalia, D.A. Starburst/Cascade Dendrimers: Fundamental building blocks for a new nanoscopic chemistry set. *Adv. Mater.* **1994**, *6*, 529–539. [CrossRef]
80. Newkome, G.R.; Weis, C.D.; Childs, B.J. Syntheses of 1→ 3 branched isocyanate monomers for dendritic construction. *Des. Monomers Polym.* **1998**, *1*, 3–14. [CrossRef]
81. Rauch, F.; Endres, P.; Friedrich, A.; Sieh, D.; Hähnel, M.; Krummenacher, I.; Braunschweig, H.; Finze, M.; Ji, L.; Marder, T.B. An Iterative Divergent Approach to Conjugated Starburst Borane Dendrimers. *Chem. A Eur. J.* **2020**, *26*, 12951–12963. [CrossRef]
82. Pittelkow, M.; Christensen, J.B. Convergent synthesis of internally branched PAMAM dendrimers. *Org. Lett.* **2005**, *7*, 1295–1298. [CrossRef]
83. Hawker, C.J.; Frechet, J.M.J. Preparation of polymers with controlled molecular architecture. A new convergent approach to dendritic macromolecules. *J. Am. Chem. Soc.* **1990**, *112*, 7638–7647. [CrossRef]
84. Bondareva, J.; Kolotylo, M.; Rozhkov, V.; Burilov, V.; Lukin, O. A convergent approach to sulfonimide-based dendrimers and dendrons. *Tetrahedron Lett.* **2020**, *61*, 152011. [CrossRef]
85. Grayson, S.M.; Fréchet, J.M.J. Convergent Dendrons and Dendrimers: from Synthesis to Applications. *Chem. Rev.* **2001**, *101*, 3819–3868. [CrossRef]
86. Walter, M.V.; Malkoch, M. Simplifying the synthesis of dendrimers: Accelerated approaches. *Chem. Soc. Rev.* **2012**, *41*, 4593–4609. [CrossRef] [PubMed]
87. Wooley, K.L.; Hawker, C.J.; Fréchet, J.M.J. A "Branched-Monomer Approach" for the Rapid Synthesis of Dendimers. *Angew. Chem. Int. Ed. Engl.* **1994**, *33*, 82–85. [CrossRef]
88. Balaji, B.S.; Lewis, M.R. Double exponential growth of aliphatic polyamide dendrimersvia AB2 hypermonomer strategy. *Chem. Commun.* **2009**, 4593–4595. [CrossRef]
89. Miller, T.M.; Neenan, T.X.; Zayas, R.; Bair, H.E. Synthesis and characterization of a series of monodisperse, 1, 3, 5-phenylene-based hydrocarbon dendrimers including C276H186 and their fluorinated analogs. *J. Am. Chem. Soc.* **1992**, *114*, 1018–1025. [CrossRef]
90. Agrahari, A.K.; Singh, A.S.; Mukherjee, R.; Tiwari, V.K. An expeditious click approach towards the synthesis of galactose coated novel glyco-dendrimers and dentromers utilizing a double stage convergent method. *RSC Adv.* **2020**, *10*, 31553–31562. [CrossRef] [PubMed]
91. Malkoch, M.; Malmström, E.; Hult, A. Rapid and efficient synthesis of aliphatic ester dendrons and dendrimers. *Macromolecules* **2002**, *35*, 8307–8314. [CrossRef]
92. Hartwig, S.; Nguyen, M.M.; Hecht, S. Exponential growth of functional poly(glutamic acid)dendrimers with variable stereochemistry. *Polym. Chem.* **2010**, *1*, 69–71. [CrossRef]
93. Morgenroth, F.; Reuther, E.; Müllen, K. Polyphenylene dendrimers: From three-dimensional to two-dimensional structures. *Angew. Chem. Int. Ed. Engl.* **1997**, *36*, 631–634. [CrossRef]
94. Ma, X.; Tang, J.; Shen, Y.; Fan, M.; Tang, H.; Radosz, M. Facile Synthesis of Polyester Dendrimers from Sequential Click Coupling of Asymmetrical Monomers. *J. Am. Chem. Soc.* **2009**, *131*, 14795–14803. [CrossRef] [PubMed]

95. Maraval, V.; Laurent, R.; Marchand, P.; Caminade, A.-M.; Majoral, J.-P. Accelerated methods of synthesis of phosphorus-containing dendrimers. *J. Organomet. Chem.* **2005**, *690*, 2458–2471. [CrossRef]

96. Liu, C.; Wang, R.; Sun, Y.; Yin, C.; Gu, Z.; Wu, W.; Jiang, X. An Orthogonal Protection Strategy for Synthesizing Scaffold-Modifiable Dendrons and Their Application in Drug Delivery. *ACS Cent. Sci.* **2022**, *8*, 258–267. [CrossRef]

97. Kottari, N.; Chabre, Y.M.; Shiao, T.C.; Rej, R.; Roy, R. Efficient and accelerated growth of multifunctional dendrimers using orthogonal thiol–ene and SN2 reactions. *Chem. Commun.* **2014**, *50*, 1983–1985. [CrossRef] [PubMed]

98. Wooley, K.L.; Hawker, C.J.; Fréchet, J.M.J.; Wudl, F.; Srdanov, G.; Shi, S.; Li, C.; Kao, M. Fullerene-bound dendrimers: Soluble, isolated carbon clusters. *J. Am. Chem. Soc.* **1993**, *115*, 9836–9837. [CrossRef]

99. Goodwin, A.P.; Lam, S.S.; Fréchet, J.M.J. Rapid, Efficient Synthesis of Heterobifunctional Biodegradable Dendrimers. *J. Am. Chem. Soc.* **2007**, *129*, 6994–6995. [CrossRef]

100. Lundberg, P.; Hawker, C.J.; Hult, A.; Malkoch, M. Click assisted one-pot multi-step reactions in polymer science: Accelerated synthetic protocols. *Macromol. Rapid Commun.* **2008**, *29*, 998–1015. [CrossRef]

101. Yan, L.; Dang, X.; Yang, M.; Zhang, M.; Rui, L.; Han, W.; Li, Y. One-pot synthesis of PAMAM-grafted hyperbranched cellulose towards enhanced thermal stability and antibacterial activity. *Process Biochem.* **2022**, *121*, 78–86. [CrossRef]

102. Mahdavijalal, M.; Panahi, H.A.; Moniri, E. Synthesis of PAMAM dendrimers anchored to WS2 nano-sheets for controlled delivery of docetaxel: Design, characterization and in vitro drug release. *J. Drug Deliv. Sci. Technol.* **2023**, *79*, 104066. [CrossRef]

103. Sohail, I.; Bhatti, I.A.; Ashar, A.; Sarim, F.M.; Mohsin, M.; Naveed, R.; Yasir, M.; Iqbal, M.; Nazir, A. Polyamidoamine (PAMAM) dendrimers synthesis, characterization and adsorptive removal of nickel ions from aqueous solution. *J. Mater. Res. Technol.* **2020**, *9*, 498–506. [CrossRef]

104. Tabatabaiee Bafrooee, A.A.; Ahmad Panahi, H.; Moniri, E.; Miralinaghi, M.; Hasani, A.H. Removal of Hg^{2+} by carboxyl-terminated hyperbranched poly(amidoamine) dendrimers grafted superparamagnetic nanoparticles as an efficient adsorbent. *Environ. Sci. Pollut. Res.* **2020**, *27*, 9547–9567. [CrossRef]

105. Shao, D.; Lu, M.M.; Zhao, Y.W.; Zhang, F.; Tan, Y.F.; Zheng, X.; Pan, Y.; Xiao, X.A.; Wang, Z.; Dong, W.F.; et al. The shape effect of magnetic mesoporous silica nanoparticles on endocytosis, biocompatibility and biodistribution. *Acta Biomater.* **2017**, *49*, 531–540. [CrossRef] [PubMed]

106. Shao, D.; Li, J.; Zheng, X.; Pan, Y.; Wang, Z.; Zhang, M.; Chen, Q.-X.; Dong, W.-F.; Chen, L. Janus "nano-bullets" for magnetic targeting liver cancer chemotherapy. *Biomaterials* **2016**, *100*, 118–133. [CrossRef]

107. Wang, Z.; Chang, Z.-M.; Shao, D.; Zhang, F.; Chen, F.; Li, L.; Ge, M.-F.; Hu, R.; Zheng, X.; Wang, Y.; et al. Janus Gold Triangle-Mesoporous Silica Nanoplatforms for Hypoxia-Activated Radio-Chemo-Photothermal Therapy of Liver Cancer. *ACS Appl. Mater. Interfaces* **2019**, *11*, 34755–34765. [CrossRef] [PubMed]

108. Zhao, Z.; Zhu, F.; Qu, X.; Wu, Q.; Wang, Q.; Zhang, G.; Liang, F. pH-Responsive polymeric Janus containers for controlled drug delivery. *Polym. Chem.* **2015**, *6*, 4144–4153. [CrossRef]

109. Rehman, K.; Ali, I.; El-Haj, B.M.; Kanwal, T.; Maharjan, R.; Saifullah, S.; Imran, M.; Shafiullah, S.; Simjee, S.U.; Shah, M.R. Synthesis of novel biocompatible resorcinarene based nanosized dendrimer-vesicles for enhanced anti-bacterial potential of quercetin. *J. Mol. Liq.* **2021**, *341*, 116921. [CrossRef]

110. Giorgadze, T.G.; Khutsishvili, I.G.; Melikishvili, Z.G.; Bregadze, V.G. Silver atoms encapsulated in G4 PAMAM (polyamidoamine) dendrimers as a model for their use in nanomedicine for phototherapy. *Eur. Chem. Bull.* **2020**, *9*, 22–27. [CrossRef]

111. Bhatt, H.; Kiran Rompicharla, S.V.; Ghosh, B.; Biswas, S. α-Tocopherol Succinate-Anchored PEGylated Poly(amidoamine) Dendrimer for the Delivery of Paclitaxel: Assessment of in Vitro and in Vivo Therapeutic Efficacy. *Mol. Pharm.* **2019**, *16*, 1541–1554. [CrossRef]

112. Yousefiyeh, N.; Arab, A.; Salehian, E.; Alikhani, R.; Samimi, S.; Assadi, A.; Ardestani, M.S. Gadoterate meglumine-anionic linear globular dendrimer second generation: A novel nano sized theranostic contrast agent. *Nanomed. J.* **2021**, *8*, 298–306.

113. Ahmed, R.; Aucamp, M.; Ebrahim, N.; Samsodien, H. Supramolecular assembly of rifampicin and PEGylated PAMAM dendrimer as a novel conjugate for tuberculosis. *J. Drug Deliv. Sci. Technol.* **2021**, *66*, 102773. [CrossRef]

114. Alfei, S.; Marengo, B.; Domenicotti, C. Polyester-Based Dendrimer Nanoparticles Combined with Etoposide Have an Improved Cytotoxic and Pro-Oxidant Effect on Human Neuroblastoma Cells. *Antioxidants* **2020**, *9*, 50. [CrossRef] [PubMed]

115. Shao, D.; Wang, Z.; Chang, Z.; Chen, L.; Dong, W.-F.; Leong, K.W. Janus metallic mesoporous silica nanoparticles: Unique structures for cancer theranostics. *Curr. Opin. Biomed. Eng.* **2021**, *19*, 100294. [CrossRef]

116. Zhang, L.; Dong, W.-F.; Sun, H.-B. Multifunctional superparamagnetic iron oxide nanoparticles: Design, synthesis and biomedical photonic applications. *Nanoscale* **2013**, *5*, 7664–7684. [CrossRef] [PubMed]

117. Shaghaghi, B.; Khoee, S.; Bonakdar, S. Preparation of multifunctional Janus nanoparticles on the basis of SPIONs as targeted drug delivery system. *Int. J. Pharm.* **2019**, *559*, 1–12. [CrossRef]

118. Chang, Z.-M.; Wang, Z.; Shao, D.; Yue, J.; Xing, H.; Li, L.; Ge, M.; Li, M.; Yan, H.; Hu, H.; et al. Shape Engineering Boosts Magnetic Mesoporous Silica Nanoparticle-Based Isolation and Detection of Circulating Tumor Cells. *ACS Appl. Mater. Interfaces* **2018**, *10*, 10656–10663. [CrossRef]

119. Zhang, L.; Zhang, F.; Dong, W.-F.; Song, J.-F.; Huo, Q.-S.; Sun, H.-B. Magnetic-mesoporous Janus nanoparticles. *Chem. Commun.* **2011**, *47*, 1225–1227. [CrossRef] [PubMed]

120. Swanson, S.D.; Kukowska-Latallo, J.F.; Patri, A.K.; Chen, C.; Ge, S.; Cao, Z.; Kotlyar, A.; East, A.T.; Baker, J.R. Targeted gadolinium-loaded dendrimer nanoparticles for tumor-specific magnetic resonance contrast enhancement. *Int. J. Nanomed.* **2008**, *3*, 201–210. [CrossRef]

121. Mbatha, L.S.; Maiyo, F.C.; Singh, M. Dendrimer functionalized folate-targeted gold nanoparticles for luciferase gene silencing: A proof of principle study. *Acta Pharm.* **2019**, *69*, 49–61. [CrossRef]

122. Umeda, Y.; Kojima, C.; Harada, A.; Horinaka, H.; Kono, K. PEG-Attached PAMAM Dendrimers Encapsulating Gold Nanoparticles: Growing Gold Nanoparticles in the Dendrimers for Improvement of Their Photothermal Properties. *Bioconjugate Chem.* **2010**, *21*, 1559–1564. [CrossRef]

123. Xiong, Z.; Alves, C.S.; Wang, J.; Li, A.; Liu, J.; Shen, M.; Rodrigues, J.; Tomás, H.; Shi, X. Zwitterion-functionalized dendrimer-entrapped gold nanoparticles for serum-enhanced gene delivery to inhibit cancer cell metastasis. *Acta Biomater.* **2019**, *99*, 320–329. [CrossRef]

124. Sharma, R.; Liaw, K.; Sharma, A.; Jimenez, A.; Chang, M.; Salazar, S.; Amlani, I.; Kannan, S.; Kannan, R.M. Glycosylation of PAMAM dendrimers significantly improves tumor macrophage targeting and specificity in glioblastoma. *J. Control. Release* **2021**, *337*, 179–192. [CrossRef] [PubMed]

125. Zhang, H.-J.; Zhao, X.; Chen, L.-J.; Yang, C.-X.; Yan, X.-P. Dendrimer grafted persistent luminescent nanoplatform for aptamer guided tumor imaging and acid-responsive drug delivery. *Talanta* **2020**, *219*, 121209. [CrossRef] [PubMed]

126. Ebrahimian, M.; Hashemi, M.; Farzadnia, M.; Zarei-Ghanavati, S.; Malaekeh-Nikouei, B. Development of targeted gene delivery system based on liposome and PAMAM dendrimer functionalized with hyaluronic acid and TAT peptide: In vitro and in vivo studies. *Biotechnol. Prog.* **2022**, *38*, e3278. [CrossRef] [PubMed]

127. Maciel, D.; Nunes, N.; Santos, F.; Fan, Y.; Li, G.; Shen, M.; Tomás, H.; Shi, X.; Rodrigues, J. New insights into ruthenium(ii) metallodendrimers as anticancer drug nanocarriers: From synthesis to preclinic behaviour. *J. Mater. Chem. B* **2022**, *10*, 8945–8959. [CrossRef] [PubMed]

128. Gouveia, M.; Figueira, J.; Jardim, M.G.; Castro, R.; Tomás, H.; Rissanen, K.; Rodrigues, J. Poly(alkylidenimine) Dendrimers Functionalized with the Organometallic Moiety [Ru(η5-C5H5)(PPh3)2]+ as Promising Drugs Against Cisplatin-Resistant Cancer Cells and Human Mesenchymal Stem Cells. *Molecules* **2018**, *23*, 1471. [CrossRef]

129. Camacho, C.; Tomás, H.; Rodrigues, J. Use of Half-Generation PAMAM Dendrimers (G0.5–G3.5) with Carboxylate End-Groups to Improve the DACHPtCl2 and 5-FU Efficacy as Anticancer Drugs. *Molecules* **2021**, *26*, 2924. [CrossRef]

130. Maciel, D.; Guerrero-Beltrán, C.; Ceña-Diez, R.; Tomás, H.; Muñoz-Fernández, M.Á.; Rodrigues, J. New anionic poly (alkylideneamine) dendrimers as microbicide agents against HIV-1 infection. *Nanoscale* **2019**, *11*, 9679–9690. [CrossRef]

131. Sathe, R.Y.; Bharatam, P.V. Drug-dendrimer complexes and conjugates: Detailed furtherance through theory and experiments. *Adv. Colloid Interface Sci.* **2022**, *303*, 102639. [CrossRef]

132. Sikwal, D.R.; Kalhapure, R.S.; Govender, T. An emerging class of amphiphilic dendrimers for pharmaceutical and biomedical applications: Janus amphiphilic dendrimers. *Eur. J. Pharm. Sci.* **2017**, *97*, 113–134. [CrossRef]

133. Pan, J.; Wen, M.; Yin, D.; Jiang, B.; He, D.; Guo, L. Design and synthesis of novel amphiphilic Janus dendrimers for bone-targeted drug delivery. *Tetrahedron* **2012**, *68*, 2943–2949. [CrossRef]

134. Iguarbe, V.; Barberá, J.; Serrano, J.L. Functional Janus dendrimers containing carbazole with liquid crystalline, optical and electrochemical properties. *Liq. Cryst.* **2020**, *47*, 301–308. [CrossRef]

135. Li, N.; Cai, H.; Jiang, L.; Hu, J.; Bains, A.; Hu, J.; Gong, Q.; Luo, K.; Gu, Z. Enzyme-Sensitive and Amphiphilic PEGylated Dendrimer-Paclitaxel Prodrug-Based Nanoparticles for Enhanced Stability and Anticancer Efficacy. *ACS Appl. Mater. Interfaces* **2017**, *9*, 6865–6877. [CrossRef] [PubMed]

136. Falanga, A.; Del Genio, V.; Kaufman, E.A.; Zannella, C.; Franci, G.; Weck, M.; Galdiero, S. Engineering of Janus-Like Dendrimers with Peptides Derived from Glycoproteins of Herpes Simplex Virus Type 1: Toward a Versatile and Novel Antiviral Platform. *Int. J. Mol. Sci.* **2021**, *22*, 6488. [CrossRef] [PubMed]

137. Najafi, F.; Salami-Kalajahi, M.; Roghani-Mamaqani, H. Synthesis of amphiphilic Janus dendrimer and its application in improvement of hydrophobic drugs solubility in aqueous media. *Eur. Polym. J.* **2020**, *134*, 109804. [CrossRef]

138. Zhang, D.; Atochina-Vasserman, E.N.; Lu, J.; Maurya, D.S.; Xiao, Q.; Liu, M.; Adamson, J.; Ona, N.; Reagan, E.K.; Ni, H.; et al. The Unexpected Importance of the Primary Structure of the Hydrophobic Part of One-Component Ionizable Amphiphilic Janus Dendrimers in Targeted mRNA Delivery Activity. *J. Am. Chem. Soc.* **2022**, *144*, 4746–4753. [CrossRef]

139. Safaie, N.; Ferrier, R.C. Janus nanoparticle synthesis: Overview, recent developments, and applications. *J. Appl. Phys.* **2020**, *127*, 170902. [CrossRef]

140. Tohidi, Z.; Teimouri, A.; Jafari, A.; Gharibshahi, R.; Omidkhah, M.R. Application of Janus nanoparticles in enhanced oil recovery processes: Current status and future opportunities. *J. Pet. Sci. Eng.* **2022**, *208*, 109602. [CrossRef]

141. Mignani, S.; Shi, X.; Rodrigues, J.; Roy, R.; Muñoz-Fernández, Á.; Ceña, V.; Majoral, J.-P. Dendrimers toward Translational Nanotherapeutics: Concise Key Step Analysis. *Bioconjugate Chem.* **2020**, *31*, 2060–2071. [CrossRef]

142. Madaan, K.; Kumar, S.; Poonia, N.; Lather, V.; Pandita, D. Dendrimers in drug delivery and targeting: Drug-dendrimer interactions and toxicity issues. *J. Pharm. Bioallied Sci.* **2014**, *6*, 139.

143. Hossen, S.; Hossain, M.K.; Basher, M.K.; Mia, M.N.H.; Rahman, M.T.; Uddin, M.J. Smart nanocarrier-based drug delivery systems for cancer therapy and toxicity studies: A review. *J. Adv. Res.* **2019**, *15*, 1–18. [CrossRef]

144. Wazir, M.B.; Daud, M.; Ali, F.; Al-Harthi, M.A. Dendrimer assisted dye-removal: A critical review of adsorption and catalytic degradation for wastewater treatment. *J. Mol. Liq.* **2020**, *315*, 113775. [CrossRef]
145. Grammatikopoulos, P.; Sowwan, M.; Kioseoglou, J. Computational modeling of nanoparticle coalescence. *Adv. Theory Simul.* **2019**, *2*, 1900013. [CrossRef]

 *pharmaceutics*

Article

Catheters with Dual-Antimicrobial Properties by Gamma Radiation-Induced Grafting

Lorena Duarte-Peña [1],*, Héctor Magaña [2] and Emilio Bucio [1],*

[1] Departamento de Química de Radiaciones y Radioquímica, Instituto de Ciencias Nucleares, Universidad Nacional Autónoma de México, Circuito Exterior, Ciudad Universitaria, Ciudad de Mexico 04510, Mexico

[2] Facultad de Ciencias Químicas e Ingeniería, Universidad Autónoma de Baja California, Calzada Universidad 14418, Parque Industrial Internacional Tijuana, Tijuana 22390, Mexico

* Correspondence: lorena.duarte@correo.nucleares.unam.mx (L.D.-P.); ebucio@nucleares.unam.mx (E.B.)

Abstract: Dual antimicrobial materials that have a combination of antimicrobial and antifouling properties were developed. They were developed through modification using gamma radiation of poly (vinyl chloride) (PVC) catheters with 4-vinyl pyridine (4VP) and subsequent functionalization with 1,3-propane sultone (PS). These materials were characterized by infrared spectroscopy, thermogravimetric analysis, swelling tests, and contact angle to determine their surface characteristics. In addition, the capacity of the materials to deliver ciprofloxacin, inhibit bacterial growth, decrease bacterial and protein adhesion, and stimulate cell growth were evaluated. These materials have potential applications in the manufacturing of medical devices with antimicrobial properties, which can reinforce prophylactic potential or even help treat infections, through localized delivery systems for antibiotics.

Keywords: antimicrobial; antifouling; drug delivery; pH sensitivity; zwitterionic polymers; gamma radiation

1. Introduction

The biocontamination of both urinary and central line catheters is one of the principal causes of nosocomial infections, mainly in patients who are in intensive care units [1,2]. One of the reasons for this is that these devices are made of polymeric materials that have a tendency towards microorganism contamination in biological environments. According to surveys carried out in different countries, it is estimated that one in seven hospitalizations presents an incidence of nosocomial infection, of which approximately 25% are associated with the use of medical devices [3–5]. The National Healthcare Safety Network (NHSN) reports eight types of microorganisms that cause the most nosocomial infections, among which, the following are prominent: *Staphylococcus aureus*, *Escherichia coli*, and coagulase-negative *staphylococci* [6].

Therefore, searching for materials that are resistant to bacterial contamination is relevant to the medical field [7–9]. A material can present resistance to contamination by microorganisms through two mechanisms [10]. The first mechanism consists of the incorporation of active agents into the material; it can be in its internal structure as groups of quaternary amines [11,12] or stored to be released at a site of interest. Among the most widely used active agents for release are antibiotics and silver or zinc metallic nanoparticles with antibacterial properties [13–15]. The second mechanism is based on the generation of materials whose surface prevents the adhesion of the microorganism and its proliferation. These materials generally owe their antifouling capacity to the formation of superficial hydration layers stabilized by van der Waals interactions or electrostatic interactions, as in the case of zwitterionic polymers. The development of materials with dual antimicrobial capacity, that is, materials capable of both preventing adhesion and releasing an active agent, is a challenge for materials science [16,17].

Citation: Duarte-Peña, L.; Magaña, H.; Bucio, E. Catheters with Dual-Antimicrobial Properties by Gamma Radiation-Induced Grafting. *Pharmaceutics* **2023**, *15*, 960. https://doi.org/10.3390/pharmaceutics15030960

Academic Editor: Ana Isabel Fernandes

Received: 14 February 2023
Revised: 10 March 2023
Accepted: 14 March 2023
Published: 16 March 2023

Modified systems for the release of active agents are a point of interest in biomedicine because they can provide the optimal amount of drug at the right time and place, giving rise to a continuous release of therapeutic dose without reaching maximum levels, thus avoiding side effects caused by large drug discharges and concentrating the drug in the affected area [18,19]. Within these systems are smart polymers, that is, polymers that respond to external stimuli, such as temperature, pH, ionic strength, or light, by changing their structure, which allows more control over the load and release of active agents depending on the environmental conditions [20]. The poly 4-vinylpyridine (P4VP) is pH-sensitive polymer that undergoes protonation at a pH below its pK_a. This leads to the formation of cations that repel each other thus increasing the distance between the chains of the material and changing its structure.

Zwitterionic polymers have a high antifouling capacity because their ion distribution allows them to create an electrostatically stabilized surface hydration layer, which significantly reduces bacterial adhesion to these surfaces, in addition to being highly hydrophilic systems [21–23]. However, the synthesis of these materials is limited by the low solubility of the polymer in the solvents commonly used for polymerization. Due to this, alternative techniques are used to obtain these materials, such as the functionalization of ionic polymers with an ion of opposite charge, to form the zwitterion in situ [24]. Hydrogels developed with this type of polymer have shown relevant antifouling properties and good biocompatibility [25–28].

This work presents the development of PVC catheters modified with a 4-vinylpyrinidine and zwitterionic polymer to provide their surface with antifouling capacity and pH sensitivity. This material constitutes a dual antimicrobial system that has the ability to load and release ciprofloxacin. This system allows the localized release of the antibiotic because the drug is stored in the device, which can help improve drug efficiency. The modification was carried out by graft polymerization of 4VP using gamma radiation as the initiator and subsequently a zwitterion was formed by the functionalizing of the grafted 4VP with 1,3-propane sultone (PS). The synthesized materials were characterized to determine their antimicrobial capacity. Materials with dual antimicrobial capacity have potential applications in the manufacturing of medical devices that can reinforce their prophylactic potential or even help treat infections. In this case, modified catheters represent an alternative device which can reduce the nosocomial infections associated with traditional catheter use.

2. Materials and Methods

2.1. Materials

PVC catheters (outer diameter 3 mm and thickness 0.5 mm) were from Biçakcilar (Istanbul, Turkey). 4VP (95%), 1,3-propane sultone (PS), and dimethylformamide anhydrous were purchased from Aldrich Chemical, Saint Louis, MO, USA. 4VP was purified by vacuum distillation to remove the inhibitor. Chloride (NaCl), potassium chloride (KCl), sodium phosphate dibasic (NaH_2PO_4), and potassium phosphate monobasic ($KHPO_4$) were also purchased from Aldrich Chemical, Saint Louis, MO, USA; these materials were used as received. Ciprofloxacin was from Sigma Aldrich. Distillate water was used for all the assays. Software DDSolver from Excel was used for modeling drug delivery. The gamma-ray source was a ^{60}Co Gammabeam 651-PT of Nordion International Inc from Ottawa, ON, Canada proportioned by the Nuclear Science Institute at Universidad Nacional Autónoma de México (ICN-UNAM).

2.2. Synthesis of PVC-g-4VP

The 4VP graft on PVC was performed using the direct irradiation method, following the parameters used in previous studies to obtain graft percentages of 12 and 23%. A sample of PVC approximately 6 cm in length was placed in a glass ampoule, a solution of 4VP in H_2O/MeOH was added, and oxygen was removed by air displacement with argon bubbling for 15 min. The sealed ampoule was kept at 5 °C for 4 h and irradiated using gamma radiation. The grafted catheters were removed and cleaned with methanol. Finally, the samples were dried for 12 h at 30 °C in a vacuum oven, and the percentage of grafting was calculated by the difference in weight using Equation (1), where W_f is the weight of the grafted sample (g) and W_i is the weight of the sample without modification (g).

$$\text{Grafting (\%)} = (W_f - W_i) \times 100/W_i \tag{1}$$

2.3. Formation of PVC-g-4VP/4VPPS Graft by Functionalization

A dry and weighed sample of PVC-g-4VP was placed in a glass ampoule and left under vacuum for 20 min. Then, a solution of PS in dimethylformamide anhydrous was added, the ampoule was sealed, and the solution was heated for a certain period of time. Finally, the modified material was removed, washed with methanol and water for 12 h, and dried at 30 °C under a vacuum for 12 h. PS reacts quickly with water, hydrolyzing to hydroxysulfonic acid, so the reaction must be carried out under anhydrous conditions. The reaction yield was calculated using Equation (2), where M_f is the final weight of the material, M_i is the initial weight of the material, and 4VP (%) is the percentage of 4VP grafting in the initial material.

$$\text{Reaction yield (\%)} = (M_f - M_i) \times 8607.7)/(M_i \times 4VP(\%)) \tag{2}$$

The effect of the different reaction conditions was studied, varying the temperature (50, 60, and 70 °C), the reaction time (30, 45, 60, and 75 min), and the concentration of PS (0.35, 0.5, 0.65, 0.8, 0.8, and 1 M).

2.4. Infrared Spectroscopy and Thermal Analysis

Infrared spectroscopy was performed on a Perkin Elmer Spectrum 100 spectrophotometer (Perkin Elmer Cetus Instruments, Norwalk, CT, USA) with 16 scans, in the ATR module, in the range of 4000 to 650 cm^{-1}. On the other hand, the thermal behavior was monitored by TGA under a nitrogen atmosphere from 30 to 700 °C at a heating rate of 10 °C/min using a TGA Q50 (TA Instruments, New Castle, DE, USA).

2.5. Swelling and Contact Angle

For the swelling tests, a dry sample was weighed and placed in a glass with distilled water at 25 °C. Once removed from the beaker, excess solvent was removed from the sample and it was weighed every 5 min for the first 15 min and then at 0.5, 1, 2, 4, 6, and 12 h. The swelling percentage was determined using Equation (3), where W_2 is the weight of the swollen sample and W_1 is the dry sample weight.

$$\text{Swelling (\%)} = (W_2 - W_1) \times 100/W_1 \tag{3}$$

The contact angle provided information on the degree of wettability; this determination was measured using a DSA 100 Krüss GmbH, German goniometer from Hamburg, using the sessile drop method with water. The samples were split, flattened, using glass plates, and dried at 40 °C in a vacuum oven for 4 h. For the determination, a drop of distilled water was deposited on the flat surface, and the angle formed between the surface and the liquid was measured. All of the measurements were carried out six times.

2.6. pH-Responsiveness

To determine the pH response of the samples, phosphate buffer solutions of pH 2, 3, 4, 5, 6, 8, 10, and 12 were prepared. A dry sample was weighed, and the solution with pH 2 was added, maintaining a controlled temperature at 25 °C for 2 h. Later the sample was removed, and the swelling percentage was calculated. The same procedure was used with the other solutions.

2.7. Load and Realese of Ciprofloxacin

2.7.1. Ciprofloxacin Load

Samples around 100 mg were placed in vials with 3 mL of 0.012 μg/mL ciprofloxacin aqueous solution, at 25 °C, for 30 h. The loading time was determined by measuring the absorbance at different time intervals to assess the loading progress (4, 6, 8, 24, and 30 h). The amount of drug loaded was determined by measuring the difference in absorbance between a 0.012 μg/g ciprofloxacin solution without material and the solution with the material at each time interval at a wavelength of 266 nm, using a calibration curve. A SPECORD 200 PLUS brand spectrophotometer from Analytikjena (Germany) was used for the test. After that time, the samples were extracted and gently washed with distilled water. The calibration curve is shown in Supplementary Materials, in Figure S1.

2.7.2. Ciprofloxacin Release

The samples loaded with ciprofloxacin were deposited in vials containing 3 mL of phosphate buffer solution at pH 7.4 and 37 °C with constant stirring (130 rpm). The cumulative release was monitored by taking measurements at 0.25, 0.5, 1, 2, 4, 6, 8, 24, 30, and 48 h at 266 nm using a UV–Vis spectrophotometer. The calibration curve is shown in Supplementary Materials, Figure S2. The release profiles were analyzed by the software DDSolver, and the detailed results are presented in Supplementary Materials.

2.8. Protein Adsorption Test

Approximately 80 mg of a sample were placed in PBS buffer solution at 37 °C for 2 h, then the sample was extracted and incubated in bovine serum albumin (BSA) protein solution at a 30 mg/mL concentration in PBS at 37 °C for 2 h. After this time, the materials were washed three times with PBS, and the adhered protein was extracted for later quantification.

For the extraction, 600 μL of 1% wt. SDS solution were added to the sample, previously incubated in BSA, and proceeded to be shaken at 130 rpm for 20 min and sonicated for 10 min, and finally vortexed for 30 s; this process was repeated 3 times. Finally, the protein concentration was quantified using the bicinchoninic acid method: 1 mL of the extraction solution was placed in a glass vial, 2 mL of the working solution (Table S1) was added, the mixture was gently shaken and heated at 60 °C for 30 min. After this time, the sample was cooled to room temperature, and the absorbance was measured using UV–Vis spectroscopy at 556 nm. Quantification was performed using a calibration curve (Supplementary Materials, Figure S3).

2.9. Cell Viability Assay

Cell viability was studied by observing the growth of BALB/3T3 murine embryonic fibroblasts (ATCC CCL-163, Manassas, VA, USA) in the presence of the modified materials. The assays were performed in 96-well plates seeded with 4×10^4 cells/mL, using Dulbecco's modified Eagle's medium (DMEM) with FBS (fetal bovine serum, 10% v/v), penicillin-streptomycin (1% w/v) and gentamicin (10 μg/mL), for 24 h, under standard culture conditions (a humidified atmosphere of 5% CO_2, at 37 °C). Samples of approximately six μg were added to the cell culture and incubated for 24 h. After this, the samples and the medium were removed, and the MIT kit was used to quantify them, measuring the absorbance at 620 nm (Multiskan FC, Thermo Scientific). Cell viability was obtained by comparing cell growth in the presence of samples with that obtained by cultures without them. The assay was performed in triplicate.

2.10. Bacterial Inhibition Test

A solution of *Escherichia coli* (ATCC25922) of 0.5 MF (1.5×10^8 cfu/mL) was prepared in peptone water (pH: 7.2) and 2 mL of the solution was placed in a test tube, which had been previously sterilized (121 °C, 15 min and pressure of 1.6 kg/cm^2). The material to be analyzed was placed in the medium and incubated at 37 °C for 24 h. The material was removed from the culture medium, and the optical density was measured at 600 nm by spectroscopy to quantify the inhibition of growth by comparing the difference between it and a solution whose bacterial growth was not affected by any external material [29]; the experiments were carried out in triplicate.

2.11. Bacterial Adhetion Test

Samples of approximately 30 mg were incubated in 0.5 MF *E. coli* solution for 24 h. Once removed and gently washed with sterile water to remove non-adhering bacteria, the samples were placed in a glass tube with 1 mL of sterile water and taken to the vortex for 3 min [30,31]. After vortexing, a 0.1 mL aliquot was taken and diluted to 1 mL. An amount of 0.1 mL of this solution was removed and seeded using the inverted plate technique. The solution was left to incubate for 24 h, and the former colonies were counted [32].

3. Results

3.1. Synthesis of PVC-g-4VP/4VPPS

Once the 12 and 23% 4VP grafts were obtained using the conditions reported in Duarte-Peña [33], we proceeded to functionalize them to form the PVC-*g*-4VP/4VPPS. One of the ways to obtain zwitterionic polymers is via the formation of the zwitterion after polymerization, that is, cationic or anionic polymers are functionalized with oppositely charged groups. This was the method used in the case of PVC-*g*-4VP/4VPPS materials, where 4VP was initially grafted onto the PVC matrix and later functionalized with PS to form the zwitterionic polymer 4-vinylpyridine propylsulfobetaine (4VPPS), resulting in a graft composed of 4VP/4VPPS. The process for functionalization consisted of a ring-opening alkylation reaction (Figure 1) [25].

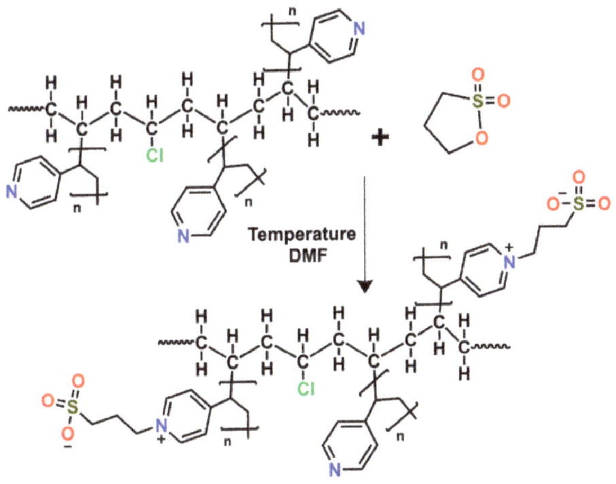

Figure 1. Reaction scheme for PVC-g-4VP/4VPSS.

Three factors were analyzed on the samples with 12 and 23% of 4VP to obtain the best reaction conditions: PS concentration, reaction time, and temperature. For this process, the reaction time was limited by the stability of the polymeric matrix in DMF, so the maximum reaction time used was 75 min. Figure 2A,B shows the behavior of the reaction yield when increasing the concentration of PS for samples with 12 and 23% of 4VP, respectively. In both cases, the reaction yield increased with a higher concentration; samples with 12% of 4VP required 0.35 M PS to reach 80 % of reaction yield, whereas the samples with 23% of 4VP reached these yields at a concentration of 0.8 and 1 M under the conditions used. Figure 2C,D show that the reaction yield increases with increasing reaction time, reaching yield values of 80 to 95% between 60 and 75 min for samples with 23% of 4VP and close to 100% at 20 min for 12% of 4VP.

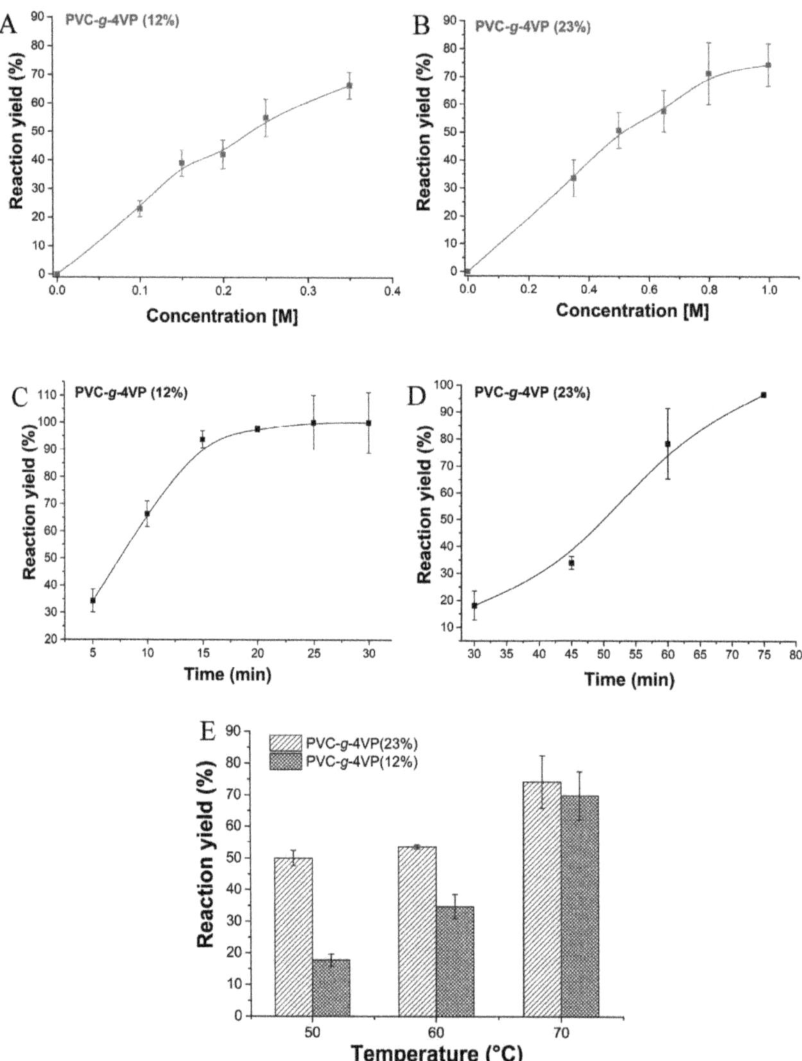

Figure 2. Effect of the reaction conditions in the yield percentage. (**A**) PS concentration effect for PVC-g-4VP(12%), conditions: 70 °C and 10 min, (**B**) PS concentration effect for PVC-g-4VP(23%), conditions: 70 °C and 75 min; (**C**) Time reaction effect for PVC-g-4VP(12%), conditions: 70 °C and 0.35 M PS; (**D**) Time reaction effect for PVC-g-4VP(23%), conditions: 70 °C and 0.8 M PS; and (**E**) Temperature effect, conditions for PVC-g-4VP(12%): 0.35 M PS and 10 min, and conditions for PVC-g-4VP(23%): 0.8 M PS and 75 min. Reported: mean ± standard error of the mean, n = 3.

On the other hand, the temperature played an important role during the reaction because this is an endothermic process that requires energy to overcome the activation barrier and shift the equilibrium towards the formation of the product. Three temperatures were tested: 50, 60, and 70 °C. Figure 2E shows the yields as a function of temperature for the two samples, the best yield for both was found at 70 °C.

Table 1 shows four graft ranges that were obtained from this synthesis. The material with grafts of 13% 4VP and 32% 4VPPS was discarded for characterization because there was a significant deformation of the catheter.

Table 1. Materials synthesized.

Samples
-◄- PVC-*g*-4VP$_{4\%}$/4VPPS$_{32\%}$
-▲- PVC-*g*-4VP$_{10\%}$/4VPPS$_{13\%}$
-●- PVC-*g*-4VP$_{16\%}$/4VPPS$_{22\%}$
PVC-*g*-4VP$_{13\%}$/4VPPS$_{32\%}$

3.2. Infrared Spectroscopy and Themal Analysis

The PVC-*g*-4VP/4VPPS materials were characterized by infrared spectroscopy to determine the presence of the different functional groups, as shown in Figure 3. The control PVC catheter spectrum presented bands at 2992 cm^{-1} of C-H stretching of -CH$_2$- groups, 1267 cm^{-1} of C-H flexions; additionally, bands at 1745 and 1459 cm^{-1} were presented, which are characteristics of the plasticizer. The PVC-*g*-4VP spectrum, apart from the previous bands, presented C=C stretching signals at 1597 and 1557 cm^{-1}, and at 830 cm^{-1} of the C-H bends outside the plane of the aromatic amine [34,35]. Finally, the PVC-*g*-4VP/4VPPS spectra, together with the characteristic bands of the aromatic amine, show bands at 1039 and 1265 cm^{-1} corresponding to the symmetric and asymmetric stretching of the sulfonate group (SO$_3^-$), and the band 1680 cm^{-1} of the C=N stretches of the pyridine, corroborating the formation of the zwitterion 4VPPS [36].

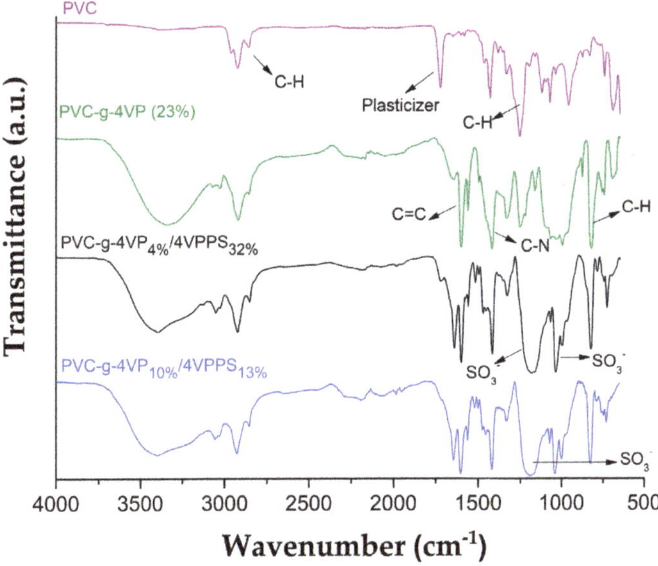

Figure 3. Infrared spectra of the neat and grafted PVC catheters using ATR module.

TGA thermal analysis showed that the catheter PVC-*g*-4VP/4VPPS has greater thermal stability than PVC and PVC-*g*-4VP because it lost 10% weight at a temperature 18 °C higher (Table 2). PVC and PVC-*g*-4VP presented three decomposition temperatures: the first was attributed to the decomposition of the plasticizer, the second and more intense was produced by the dehydrochlorination of PVC, and the last was attributed to the total chain decomposition; the materials modified with 4VP presented an increase of around

20 degrees at all decomposition temperatures. Finally, the PVC-*g*-4VP/4VPPS catheters showed four decomposition temperatures: the first was at 272 °C due to the loss of the plasticizer, the second was at 333 °C due to the dehydrochlorination of PVC [37], the third was at 364 °C, which is consistent with the decomposition of the quaternary amine [38], and the last was at 453 °C, which corresponds to the carbon chain decomposition. Figure 4 shows the thermograms of each sample.

Table 2. Thermogravimetric analysis.

Sample	10% Weight Loss (°C)	Decomposition Temperatures (°C)
PVC	236	243
		287
		457
PVC-*g*-4VP	240	261
		301
		470
PVC-*g*-4VP/4VPPS	256	272
		333
		364
		453

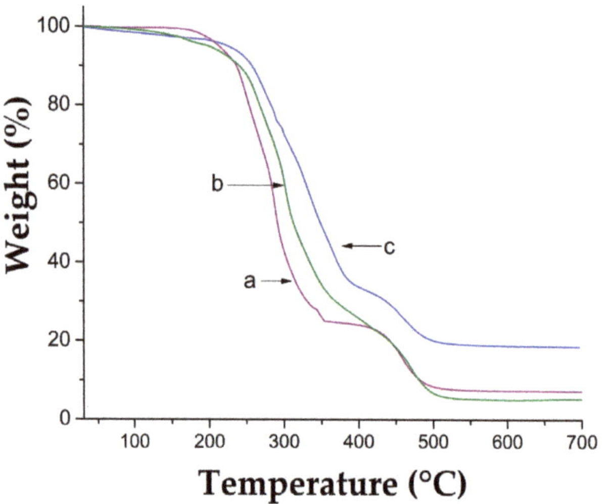

Figure 4. Thermograms: (**a**) PVC, (**b**) PVC-g-4VP, and (**c**) PVC-g-4VP/4VPPS.

3.3. Swelling and Contact Angle

Figure 5A shows the swelling curves in water at 25 °C; PVC has a hydrophobic character for what it did not swell. Samples modified only with 4VP showed their maximum swell at 2 h. On the other hand, the catheters with 4VP/4VPPS reached their maximum swelling at 30 min and presented maximum percentages three times those reached by the samples without functionalization. Figure 5B shows the contact angles for the different materials; the PVC surface showed a contact angle of approximately 100°, which corresponds to its hydrophobic character; the modified materials, on the other hand, presented contact angles of less than 90°, indicating that the surface acquired hydrophilicity. The surfaces modified with 4VP/4VPPS initially had higher contact angles than those modified only with 4VP; however, after 10 min of water-surface interaction, the contact angles decreased to similar values.

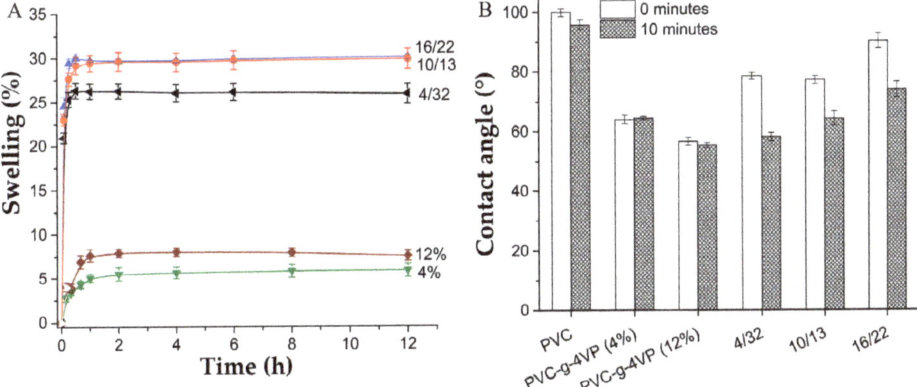

Figure 5. (**A**) Swelling curves in water at 25 °C, reported: mean ± standard error of the mean, $n = 3$. and (**B**) Water contact angle, reported: mean ± standard error of the mean, $n = 6$. PVC-*g*-4VP$_{4\%}$/4VPSS$_{32\%}$ (4/32), PVC-*g*-4VP$_{10\%}$/4VPSS$_{13\%}$ (10/13), and PVC-*g*-4VP$_{16\%}$/4VPSS$_{22\%}$ (16/22).

3.4. pH-Sensitivity

4VP is a pH-sensitive polymer because it has an amino group in its structure that presents an acid-base balance (Figure 6A), with a pK$_a$ around 5.4. At pHs below its pK$_a$, the polymer is in its ionic form, so the chains suffer repulsion between them and swelling is greater. Figure 6B shows the behavior of the swelling as a function of the pH for the pristine PVC, PVC-*g*-4VP, and PVC-*g*-4VP/4VPPS. PVC did not present a pH response, whereas all the materials modified with 4VP showed this capability, with a critical pH in the range of 6.3 to 7.

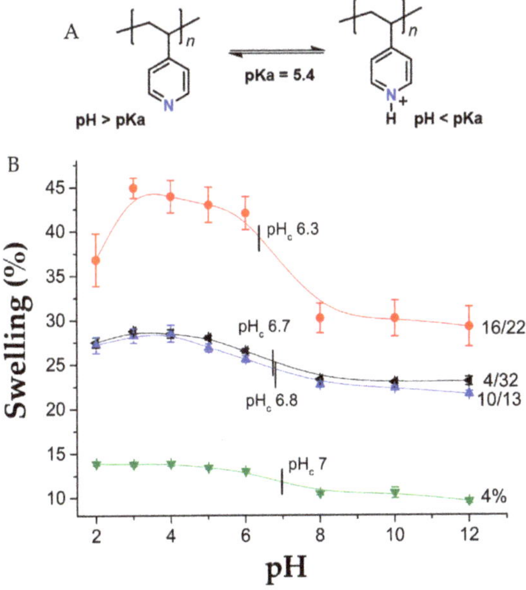

Figure 6. (**A**) Acid-base balance of 4VP and (**B**) pH-sensitivity, reported: mean ± standard error of the mean, $n = 3$.

3.5. Protein Adsortion Test

The presence of the 4VPPS zwitterionic polymer decreased the percentage of protein adsorbed on the surface. Figure 7 shows the results; the material modified with 10% 4VP and 13% 4VPPS was the one that showed the highest antifouling capacity, reducing BSA adsorption by 74% compared to the unmodified PVC surface. This shows that the functionalization of the grafted 4VP was successful and the 4VPPS formed expresses its antifouling characteristics.

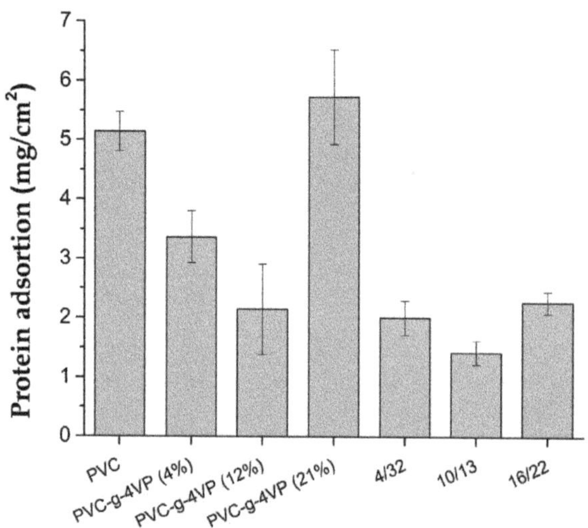

Figure 7. Protein adsorption on the polymer surface of different composition (incubation: 30% BSA solution in PBS, 37 °C, 2 h). Reported: mean ± standard error of the mean, *n* = 3.

3.6. Load and Release of Ciprofloxacin

Ciprofloxacin loading and release assays were performed on PVC, PVC-*g*-4VP(4%), PVC-*g*-4VP$_{4\%}$/4VPPS$_{32\%}$, PVC-*g*-4VP$_{10\%}$/4VPPS$_{13\%}$, and PVC-*g*-4VP$_{16\%}$/4VPSS$_{22\%}$ samples. PVC and PVC-*g*-4VP(4%) samples did not show ciprofloxacin loading capacity, contrary to the samples modified with the zwitterionic polymer. Figure 8A shows the ciprofloxacin load for the three zwitterionic graft samples; in all the cases, the load reaches maximum levels after 30 h of interaction, but the amount of ciprofloxacin loaded by the PVC-*g*-4VP$_{10\%}$/4VPSS$_{13\%}$ and PVC-*g*-4VP$_{16\%}$/4VPSS$_{22\%}$ catheters was higher. Figure S4 shows the load profiles. However, release profiles (Figure 8B), showed that the sample PVC-*g*-4VP$_{10\%}$/4VPPS$_{13\%}$ had a final release of approximately 40 µg/mL, and the samples PVC-*g*-4VP$_{4\%}$/4VPPS$_{32\%}$ and PVC-*g*-4VP$_{16\%}$/4VPPS$_{22\%}$ released around 20 µg/mL; in all cases, the release was gradual until 30 h.

The release profiles were processed using DDSolver Excel software to determine the fitting model (details of the fit are specified in Supplementary Materials, Table S2). In all cases, following the values of the Akaike Information Criterion (AIC), the Model Selection Criterion (MSC), and the adjustment of squares (r^2), the model that presented the highest affinity with the behavior of the materials was the Peppas–Sahlin model [39].

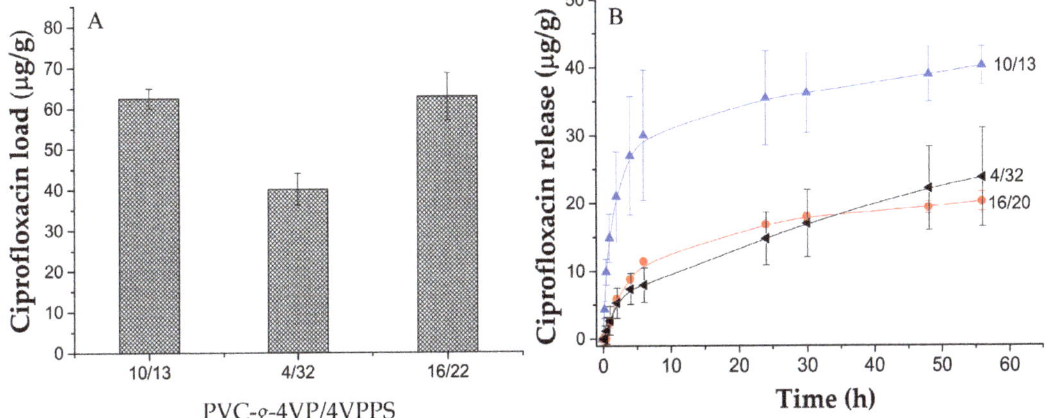

Figure 8. (**A**) Ciprofloxacin load (condition: 0.012 µg/mL, 25 °C por 30 h) and (**B**) Ciprofloxacin release profiles (condition: PBS a 37 °C). Reported: mean ± standard error of the mean, $n = 3$.

The Peppas–Sahlin model describes systems that present the contribution of two mechanisms (diffusional and relaxational) in the release process. This model is described by Equation (4), where M_∞ is the amount of drug in an equilibrium state, M_t is the amount of drug released in the determined time, k_1 represents the Fickian diffusion contribution, k_2 the polymer chains relaxation contribution, and m is the Fickian diffusion exponent [40].

$$M_t/M_\infty = k_1 t^m + k_2 t^{2m} \tag{4}$$

Table 3 shows the values of the Peppas–Sahlin model parameters for the different materials; all materials presented values of k_1 that predominate over k_2, indicating that the principal release mechanism is diffusion. However, for the material, the contribution of the chain relaxation mechanism is higher, possibly due to the greater amount of the zwitterionic polymer. The PVC-*g*-4VP$_{10\%}$/4VPPS$_{13\%}$ catheter released the highest amount of ciprofloxacin, with a release rate of approximately 60%.

Table 3. Peppas–Sahlin model parameters.

Parameter	PVC-*g*-4VP$_{4\%}$/4VPPS$_{32\%}$	PVC-*g*-4VP$_{10\%}$/4VPPS$_{13\%}$	PVC-*g*-4VP$_{16\%}$/4VPPS$_{22\%}$
k_1	13.22	14.48	13.63
k_2	0.523	−0.678	−0.657
m	0.573	0.377	0.601

The PVC-*g*-4VP$_{10\%}$/4VPPS$_{13\%}$ materials presented the best properties and fast synthesis conditions (0.35 M, 70 °C, 5 min, and 12% 4VP), which is why they were chosen to carry out cell viability, antimicrobial activity, and antifouling capacity tests.

3.7. Cell Viability

Figure 9 shows the results of cell viability using BALB/3T3 murine embryonic fibroblasts; it was observed that the control PVC increased cell growth, in the same way as the material modified with 4% 4VP. The material grafted with 4VP/4VPPS showed a decrease in cell growth of approximately 5%; when the material was loaded with ciprofloxacin, the cell growth decreased by 18%; however, the material can be considered non-cytotoxic according to ISO-10993-5-2009 [41].

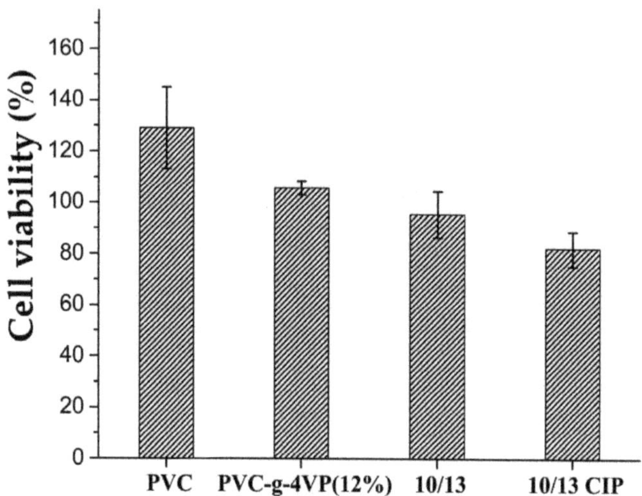

Figure 9. Cell viability result using BALB/3T3 murine embryonic fibroblasts and MIT assay. Reported: mean ± standard error of the mean, *n* = 3.

3.8. Antimicrobial Activity and Antifouling Capability

Figure 10A presents the antibacterial capacity of the material based on its ability to inhibit the growth of *E. coli*. It was observed that PVC, PVC-*g*-4VP(4%), and PVC-*g*-4VP$_{10\%}$/4VPPS$_{13\%}$ do not have antimicrobial activity, on the contrary, PVC-*g*-4VP$_{10\%}$/4VPPS$_{13\%}$ loaded with ciprofloxacin achieved 82% inhibition at 24 h. On the other hand, Figure 10B presents the antifouling capacity of the material determined by the number of colony-forming units of *E. coli* bacteria, which adhered to the material when it was in direct contact with the *E. coli* solution of 1×10^8 cfu for 24 h. It is observed that the material with binary graft presents a 42% decrease in bacterial adhesion in the control PVC and this effect is enhanced when loading ciprofloxacin, reaching a 55% decrease.

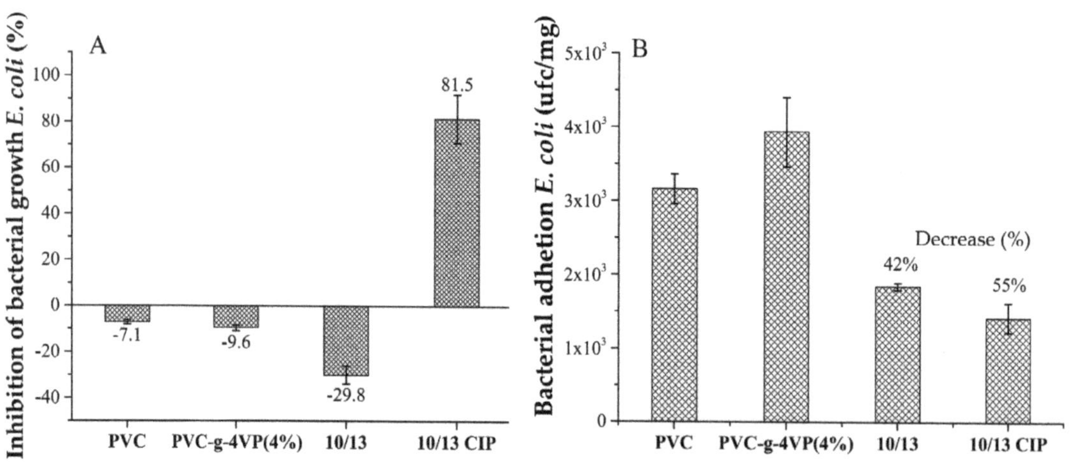

Figure 10. (**A**) Inhibition of bacterial growth for *E. coli* and (**B**) Bacterial adhesion of *E. coli*. Reported: mean ± standard error of the mean, *n* = 3.

4. Discussion

Ionic polymer functionalization using oppositely charged ions is one of the ways to synthesize zwitterionic polymers. In this case, the functionalization of 4VP grafted onto PVC catheters made it possible to obtain the characteristics of the 4VPPS zwitterion while maintaining the pH-sensitivity of the 4VP system. The 4VP is a cationic polymer that reacts with PS to produce 4VPPS, by a ring-opening alkylation reaction (Figure 1). This form of zwitterion synthesis resulted in different percentages of grafting, which was impossible with a direct grafting method, that is, directly grafting the 4VPPS monomer. The difficulty of working with the 4VPPS monomer was due to its insolubility in most solvents; a 1 M saline solution is required to solubilize it [42,43]. The formation of grafts using ionizing radiation represents an advantageous method for medical device modification because the obtained materials do not present contamination by agents used for the polymerization [44,45].

The poly (4VPPS) is an aromatic zwitterionic polymer that has demonstrated antifouling capabilities in biomedicine, electronics [46], and water treatment. Venault, et. al. synthesized 4VPPS hydrogels to be used as medical device coatings. They demonstrated that 4VPPS coatings present a bio-antifouling capability similar to that of sulfobetaine methacrylate (SBMA), one of the most common zwitterionic polymers, employed to produce antifouling surfaces by coating, grafting, or copolymerization [47–50]. Additionally, 4VPPS coatings were more resistant to temperature than SBMA coatings because they retained their bio-antifouling properties after being heated at 121 °C for 1 h (sterilization conditions), representing an advantage in the biomedical field, where most of the devices require sterilization [25]. On the other hand, Gui, et. al. copolymerized acrylamide with 4VPPS to produce a zwitterionic poly-acrylamide that showed thermic stability and good properties as a flocculant of cationic and anionic inorganic contaminants due to its electrostatic characteristics [51].

Materials grafting with 4VPPS showed the ability to load and release ciprofloxacin due to the hydrophilicity and pH-sensitivity of the system. In addition, the modification provided the material with the antifouling property of zwitterionic polymers, allowing a significant decrease in both the adhesion of the BSA protein and *E. coli*. A synergistic effect was observed when loading the material with ciprofloxacin, increasing its capacity to prevent the adhesion of *E. coli* by 13%, for the PVC-*g*-4VP$_{10\%}$/4VPPS$_{13\%}$ catheter. The materials were also shown to be non-toxic since they did not significantly affect cell viability. The modification of the catheters allows the creation of antimicrobial systems with the capacity to release an antibiotic at the site of interest, making it a localized drug delivery system, increasing its effectiveness and decreasing the side effects.

5. Conclusions

The functionalization of the 4VP grafted onto PVC catheters was successful and obtained materials with different percentages of 4VP and 4VPPS in their composition. Although all the modified catheters showed hydrophilicity and pH sensitivity, the PVC-*g*-4VP$_{10\%}$/4VPPS$_{13\%}$ catheter was the one that showed the best capacities for loading and releasing ciprofloxacin, achieving an inhibition of *E. coli* growth of approximately 82% in 24 h. In addition, due to the presence of zwitterion, this material showed antifouling characteristics, which decreased the adhesion of BSA by 74% and the adhesion of *E. coli* by 55%. This material is a pH-sensitive system with dual antimicrobial capacity and is therefore a promising alternative for developing devices with less tendency towards biocontamination.

Supplementary Materials: The following supporting information can be downloaded at: https://www.mdpi.com/article/10.3390/pharmaceutics15030960/s1, Figure S1: Calibration curves to quantify load of ciprofloxacin; Figure S2. Calibration curves to quantify release of ciprofloxacin; Figure S3. Calibration curve to BSA quantification (Abs 556 nm); Figure S4. Profiles of ciprofloxacin load; Table S1: Working solution preparation; Table S2. Parameters to select the drug release model.

Author Contributions: Conceptualization, L.D-P. and E.B.; methodology, L.D.-P. and H.M.; formal analysis, L.D.-P.; investigation, L.D.-P.; resources, E.B. and H.M.; writing—original draft preparation, L.D.-P.; writing—review and editing, E.B. and H.M.; visualization, L.D.-P.; supervision, E.B.; project administration, E.B.; and funding acquisition, E.B. All authors have read and agreed to the published version of the manuscript.

Funding: This work was supported by the Dirección General de Asuntos del Personal Académico (DGA-PA), Universidad Nacional Autónoma de México under Grant IN204223.

Institutional Review Board Statement: Not applicable.

Informed Consent Statement: Not applicable.

Data Availability Statement: Not applicable.

Acknowledgments: L. Duarte Peña (887494) acknowledges CONACyT for the doctoral scholarship. This work was supported by DGAPA-UNAM [Grant IN202320] (Mexico). Benjamin Leal and Martín Cruz from ICN-UNAM are acknowledged for their technical assistance.

Conflicts of Interest: The authors declare no conflict of interest.

References

1. Khan, H.A.; Baig, F.K.; Mehboob, R. Nosocomial infections: Epidemiology, prevention, control and surveillance. *Asian Pac. J. Trop. Biomed.* **2017**, *7*, 478–482. [CrossRef]
2. Jenkins, D.R. Nosocomial infections and infection control. *Medicine* **2017**, *45*, 629–633. [CrossRef]
3. Rosenthal, V.D.; Maki, D.G.; Graves, N. The International Nosocomial Infection Control Consortium (INICC): Goals and objectives, description of surveillance methods, and operational activities. *Am. J. Infect. Control* **2008**, *36*, e1–e12. [CrossRef]
4. Aitken, C.; Jeffries, D.J. Nosocomial spread of viral disease. *Clin. Microbiol. Rev.* **2001**, *14*, 528–546. [CrossRef]
5. Fätkenheuer, G.; Cornely, O.; Seifert, H. Clinical management of catheter-related infections. *Clin. Microbiol. Infect.* **2002**, *8*, 545–550. [CrossRef] [PubMed]
6. Pujol, M.; Limón, E. General epidemiology of nosocomial infections. Surveillance systems and programs. *Enferm. Infecc. Microbiol. Clin.* **2013**, *31*, 108–113. [CrossRef] [PubMed]
7. D'Agata, R.; Bellassai, N.; Jungbluth, V.; Spoto, G. Recent Advances in Antifouling Materials for Surface Plasmon Resonance Biosensing in Clinical Diagnostics and Food Safety. *Polymers* **2021**, *13*, 1929. [CrossRef] [PubMed]
8. Wang, Z.; Scheres, L.; Xia, H.; Zuilhof, H. Developments and Challenges in Self-Healing Antifouling Materials. *Adv. Funct. Mater.* **2020**, *30*, 1908098. [CrossRef]
9. Choudhury, R.R.; Gohil, J.M.; Mohanty, S.; Nayak, S.K. Antifouling, fouling release and antimicrobial materials for surface modification of reverse osmosis and nanofiltration membranes. *J. Mater. Chem. A* **2018**, *6*, 313–333. [CrossRef]
10. Zander, Z.K.; Becker, M.L. Antimicrobial and antifouling strategies for polymeric medical devices. *ACS Macro Lett.* **2018**, *7*, 16–25. [CrossRef]
11. Liu, Y.; Ren, X.; Huang, T.S. Antimicrobial cotton containing N-halamine and quaternary ammonium groups by grafting copolymerization. *Appl. Surf. Sci.* **2014**, *296*, 231–236. [CrossRef]
12. Massi, L.; Guittard, F.; Levy, R.; Géribaldi, S. Enhanced activity of fluorinated quaternary ammonium surfactants against Pseudomonas aeruginosa. *Eur. J. Med. Chem.* **2009**, *44*, 1615–1622. [CrossRef]
13. Maleki Dizaj, S.; Lotfipour, F.; Barzegar-Jalali, M.; Hossein Zarrintan, M.; Adibkia, K. Antimicrobial activity of the metals and metal oxide nanoparticles. *Mater. Sci. Eng. C* **2014**, *44*, 278–284. [CrossRef]
14. Affes, S.; Maalej, H.; Aranaz, I.; Kchaou, H.; Acosta, N.; Heras, Á.; Nasri, M. Controlled size green synthesis of bioactive silver nanoparticles assisted by chitosan and its derivatives and their application in biofilm preparation. *Carbohydr. Polym.* **2020**, *236*, 116063. [CrossRef] [PubMed]
15. Rajivgandhi, G.; Maruthupandy, M.; Muneeswaran, T.; Anand, M.; Quero, F.; Manoharan, N.; Li, W.J. Biosynthesized silver nanoparticles for inhibition of antibacterial resistance and biofilm formation of methicillin-resistant coagulase negative Staphylococci. *Bioorg. Chem.* **2019**, *89*, 103008. [CrossRef] [PubMed]
16. Zhao, J.; Ma, L.; Millians, W.; Wu, T.; Ming, W. Dual-Functional Antifogging/Antimicrobial Polymer Coating. *ACS Appl. Mater. Interfaces* **2016**, *8*, 8737–8742. [CrossRef] [PubMed]
17. Qiu, H.; Si, Z.; Luo, Y.; Feng, P.; Wu, X.; Hou, W.; Zhu, Y.; Chan-Park, M.B.; Xu, L.; Huang, D. The Mechanisms and the Applications of Antibacterial Polymers in Surface Modification on Medical Devices. *Front. Bioeng. Biotechnol.* **2020**, *8*, 910. [CrossRef]
18. Do Nascimento Marques, N.; Da Silva Maia, A.M.; De Carvalho Balaban, R. Development of dual-sensitive smart polymers by grafting chitosan with poly (N-isopropylacrylamide): An overview. *Polimeros* **2015**, *25*, 237–246. [CrossRef]
19. Edis, Z.; Bloukh, S.H.; Ibrahim, M.R.; Sara, H.A. "Smart" Antimicrobial Nanocomplexes with Potential to Decrease Surgical Site Infections (SSI). *Pharmaceutics* **2020**, *12*, 361. [CrossRef]

20. Rizwan, M.; Yahya, R.; Hassan, A.; Yar, M.; Azzahari, A.D.; Selvanathan, V.; Sonsudin, F.; Abouloula, C.N. pH sensitive hydrogels in drug delivery: Brief history, properties, swelling, and release mechanism, material selection and applications. *Polymers* **2017**, *9*, 137. [CrossRef]

21. Blackman, L.D.; Gunatillake, P.A.; Cass, P.; Locock, K.E.S. An introduction to zwitterionic polymer behavior and applications in solution and at surfaces. *Chem. Soc. Rev.* **2019**, *48*, 757–770. [CrossRef]

22. Sin, M.C.; Chen, S.H.; Chang, Y. Hemocompatibility of zwitterionic interfaces and membranes. *Polym. J.* **2014**, *46*, 436–443. [CrossRef]

23. Ladd, J.; Zhang, Z.; Chen, S.; Hower, J.C.; Jiang, S. Zwitterionic Polymers Exhibiting High Resistance to Nonspecific Protein Adsorption from Human Serum and Plasma. *Biomacromolecules* **2008**, *9*, 1357–1361. [CrossRef]

24. Laschewsky, A. Structures and synthesis of zwitterionic polymers. *Polymers* **2014**, *6*, 1544–1601. [CrossRef]

25. Venault, A.; Lai, M.-W.; Jhong, J.-F.; Yeh, C.-C.; Yeh, L.-C.; Chang, Y. Superior Bioinert Capability of Zwitterionic Poly(4-vinylpyridine propylsulfobetaine) Withstanding Clinical Sterilization for Extended Medical Applications. *ACS Appl. Mater. Interfaces* **2018**, *10*, 17771–17783. [CrossRef] [PubMed]

26. Yang, W.; Sundaram, H.S.; Ella, J.-R.; He, N.; Jiang, S. Low-fouling electrospun PLLA films modified with zwitterionic poly(sulfobetaine methacrylate)-catechol conjugates. *Acta Biomater.* **2016**, *40*, 92–99. [CrossRef] [PubMed]

27. Kao, C.W.; Cheng, P.H.; Wu, P.T.; Wang, S.W.; Chen, I.C.; Cheng, N.C.; Yang, K.C.; Yu, J. Zwitterionic poly(sulfobetaine methacrylate) hydrogels incorporated with angiogenic peptides promote differentiation of human adipose-derived stem cells. *RSC Adv.* **2017**, *7*, 51343–51351. [CrossRef]

28. Harijan, M.; Singh, M. Zwitterionic polymers in drug delivery: A review. *J. Mol. Recognit.* **2022**, *35*, e2944. [CrossRef] [PubMed]

29. Stevenson, K.; McVey, A.F.; Clark, I.B.N.; Swain, P.S.; Pilizota, T. General calibration of microbial growth in microplate readers. *Sci. Rep.* **2016**, *6*, 38828. [CrossRef]

30. Raad, I.I.; Sabbagh, M.F.; Rand, K.H.; Sherertz, R.J. Quantitative tip culture methods and the diagnosis of central venous catheter-related infections. *Diagn. Microbiol. Infect. Dis.* **1992**, *15*, 13–20. [CrossRef]

31. Aldea Mansilla, C.; Martínez-Alarcón, J.; Gracia Ahufinger, I.; Guembe Ramírez, M. Microbiological diagnosis of catheter-related infections. *Enferm. Infecc. Microbiol. Clin.* **2019**, *37*, 668–672. [CrossRef]

32. Brugger, S.D.; Baumberger, C.; Jost, M.; Jenni, W.; Brugger, U.; Mühlemann, K. Automated counting of bacterial colony forming units on agar plates. *PLoS ONE* **2012**, *7*, e33695. [CrossRef]

33. Duarte-Peña, L.; López-Saucedo, F.; Concheiro, A.; Alvarez-Lorenzo, C.; Bucio, E. Modification of indwelling PVC catheters by ionizing radiation with temperature- and pH-responsive polymers for antibiotic delivery. *Radiat. Phys. Chem.* **2022**, *193*, 110005. [CrossRef]

34. Xue, Y.; Xiao, H. Antibacterial/Antiviral Property and Mechanism of Dual-Functional Quaternized Pyridinium-type Copolymer. *Polymers* **2015**, *7*, 2290–2303. [CrossRef]

35. Wu, C.; Zheng, J.; Hu, J. Novel antifouling polysulfone matrix membrane modified with zwitterionic polymer. *J. Saudi Chem. Soc.* **2021**, *25*, 101281. [CrossRef]

36. Shafi, H.Z.; Wang, M.; Gleason, K.K.; Khan, Z. Synthesis of surface-anchored stable zwitterionic films for inhibition of biofouling. *Mater. Chem. Phys.* **2020**, *239*, 121971. [CrossRef]

37. Lucio, D.S.V.; Rivera-Armenta, J.L.; Rivas-Orta, V.; Díaz-Zavala, N.P.; Páramo-García, U.; Rivas, N.V.G.; Cinco, M.Y.C. Manufacturing of composites from chicken feathers and polyvinyl chloride (PVC). In *Handbook of Composites from Renewable Materials*; John Wiley & Sons: New York, NY, USA, 2017; pp. 159–174. [CrossRef]

38. Ibrahim, G.P.S.; Isloor, A.M.; Inamuddin; Asiri, A.M.; Ismail, N.; Ismail, A.F.; Ashraf, G.M. Novel, one-step synthesis of zwitterionic polymer nanoparticles via distillation-precipitation polymerization and its application for dye removal membrane. *Sci. Rep.* **2017**, *7*, 15889. [CrossRef]

39. Zhang, Y.; Huo, M.; Zhou, J.; Zou, A.; Li, W.; Yao, C.; Xie, S. DDSolver: An Add-In Program for Modeling and Comparison of Drug Dissolution Profiles. *AAPS J.* **2010**, *12*, 263–271. [CrossRef]

40. Bruschi, M.L. (Ed.) 5—Mathematical models of drug release. In *Strategies to Modify the Drug Release from Pharmaceutical Systems*; Woodhead Publishing: Cambridge, UK, 2015; pp. 63–86. ISBN 978-0-08-100092-2.

41. International Organization for Standardization (ISO). *ISO 10993-5:2009 Biological Evaluation of Medical Devices—Part 5: Tests for In Vitro Cytotoxicity*, 3rd ed.; International Organization for Standardization: Geneva, Switzerland, 2009.

42. Lowe, A.B.; McCormick, C.L. Synthesis and Solution Properties of Zwitterionic Polymers. *Chem. Rev.* **2002**, *102*, 4177–4190. [CrossRef]

43. Brown, M.U.; Seong, H.-G.; Margossian, K.O.; Bishop, L.; Russell, T.P.; Muthukumar, M.; Emrick, T. Zwitterionic Ammonium Sulfonate Polymers: Synthesis and Properties in Fluids. *Macromol. Rapid Commun.* **2022**, *43*, 2100678. [CrossRef] [PubMed]

44. Meléndez-Ortiz, H.I.; Varca, G.H.C.; Lugão, A.B.; Bucio, E. Smart polymers and coatings obtained by ionizing radiation: Synthesis and biomedical applications. *Open J. Polym. Chem.* **2015**, *5*, 17–33. [CrossRef]

45. Ashfaq, A.; Clochard, M.-C.; Coqueret, X.; Dispenza, C.; Driscoll, M.S.; Ulański, P.; Al-Sheikhly, M. Polymerization reactions and modifications of polymers by ionizing radiation. *Polymers* **2020**, *12*, 2877. [CrossRef] [PubMed]

46. Jin, S.; Chen, P.-Y.; Qiu, Y.; Zhang, Z.; Hong, S.; Joo, Y.L.; Yang, R.; Archer, L.A. Zwitterionic Polymer Gradient Interphases for Reversible Zinc Electrochemistry in Aqueous Alkaline Electrolytes. *J. Am. Chem. Soc.* **2022**, *144*, 19344–19352. [CrossRef] [PubMed]

47. Pimenta, A.F.R.; Vieira, A.P.; Colaço, R.; Saramago, B.; Gil, M.H.; Coimbra, P.; Alves, P.; Bozukova, D.; Correia, T.R.; Correia, I.J.; et al. Controlled release of moxifloxacin from intraocular lenses modified by Ar plasma-assisted grafting with AMPS or SBMA: An in vitro study. *Colloids Surf. B Biointerfaces* **2017**, *156*, 95–103. [CrossRef]
48. Han, L.; Tan, Y.Z.; Xu, C.; Xiao, T.; Trinh, T.A.; Chew, J.W. Zwitterionic grafting of sulfobetaine methacrylate (SBMA) on hydrophobic PVDF membranes for enhanced anti-fouling and anti-wetting in the membrane distillation of oil emulsions. *J. Memb. Sci.* **2019**, *588*, 117196. [CrossRef]
49. Yang, Y.F.; Li, Y.; Li, Q.L.; Wan, L.S.; Xu, Z.K. Surface hydrophilization of microporous polypropylene membrane by grafting zwitterionic polymer for anti-biofouling. *J. Memb. Sci.* **2010**, *362*, 255–264. [CrossRef]
50. Yang, B.; Wang, C.; Zhang, Y.; Ye, L.; Qian, Y.; Shu, Y.; Wang, J.; Li, J.; Yao, F. A thermoresponsive poly(N-vinylcaprolactam-co-sulfobetaine methacrylate) zwitterionic hydrogel exhibiting switchable anti-biofouling and cytocompatibility. *Polym. Chem.* **2015**, *6*, 3431–3442. [CrossRef]
51. Gui, Z.; Qian, J.; An, Q.; Xu, H.; Zhao, Q. Synthesis, characterization and flocculation performance of zwitterionic copolymer of acrylamide and 4-vinylpyridine propylsulfobetaine. *Eur. Polym. J.* **2009**, *45*, 1403–1411. [CrossRef]

 *pharmaceutics*

Article

Evaluation of the In Vitro Antimicrobial Efficacy against *Staphylococcus aureus* and *epidermidis* of a Novel 3D-Printed Degradable Drug Delivery System Based on Polycaprolactone/Chitosan/Vancomycin—Preclinical Study

Iván López-González [1], Ana Belén Hernández-Heredia [2], María Isabel Rodríguez-López [2], David Auñón-Calles [2], Mohamed Boudifa [3], José Antonio Gabaldón [2,*] and Luis Meseguer-Olmo [1,*]

[1] Tissue Regeneration and Repair Group: Orthobiology, Biomaterials and Tissue Engineering, UCAM—Universidad Católica de Murcia, Campus de los Jerónimos 135, Guadalupe, 30107 Murcia, Spain; ilopez27@ucam.edu

[2] Molecular Recognition and Encapsulation Research Group (REM), Health Sciences Department, UCAM—Universidad Católica de Murcia, Campus de los Jerónimos 135, Guadalupe, 30107 Murcia, Spain; abhernandez@ucam.edu (A.B.H.-H.); mirodriguez@ucam.edu (M.I.R.-L.); daunon@ucam.edu (D.A.-C.)

[3] CRITT—Matériaux Innovation, 9 Rue Claude Chrétien, Campus Sup Ardenne, 08000 Charleville-Mézières, France; m.boudifa@critt-mi.com

* Correspondence: jagabaldon@ucam.edu (J.A.G.); lmeseguer@ucam.edu (L.M.-O.); Tel.: +34-968-278800 (L.M.-O.)

Citation: López-González, I.; Hernández-Heredia, A.B.; Rodríguez-López, M.I.; Auñón-Calles, D.; Boudifa, M.; Gabaldón, J.A.; Meseguer-Olmo, L. Evaluation of the In Vitro Antimicrobial Efficacy against *Staphylococcus aureus* and *epidermidis* of a Novel 3D-Printed Degradable Drug Delivery System Based on Polycaprolactone/Chitosan/Vancomycin—Preclinical Study. *Pharmaceutics* 2023, 15, 1763. https://doi.org/10.3390/pharmaceutics15061763

Academic Editor: Ana Isabel Fernandes

Received: 6 May 2023
Revised: 10 June 2023
Accepted: 14 June 2023
Published: 18 June 2023

Abstract: Acute and chronic bone infections, especially those caused by methicillin-resistant *Staphylococcus aureus* (MRSA), remains a major complication and therapeutic challenge. It is documented that local administration of vancomycin offers better results than the usual routes of administration (e.g., intravenous) when ischemic areas are present. In this work, we evaluate the antimicrobial efficacy against *S. aureus* and *S. epidermidis* of a novel hybrid 3D-printed scaffold based on polycaprolactone (PCL) and a chitosan (CS) hydrogel loaded with different vancomycin (Van) concentrations (1, 5, 10, 20%). Two cold plasma treatments were used to improve the adhesion of CS hydrogels to the PCL scaffolds by decreasing PCL hydrophobicity. Vancomycin release was measured by means of HPLC, and the biological response of *ah*-BM-MSCs growing in the presence of the scaffolds was evaluated in terms of cytotoxicity, proliferation, and osteogenic differentiation. The PCL/CS/Van scaffolds tested were found to be biocompatible, bioactive, and bactericide, as demonstrated by no cytotoxicity (LDH activity) or functional alteration (ALP activity, alizarin red staining) of the cultured cells and by bacterial inhibition. Our results suggest that the scaffolds developed would be excellent candidates for use in a wide range of biomedical fields such as drug delivery systems or tissue engineering applications.

Keywords: 3D printing; hybrid scaffold; polycaprolactone; chitosan; vancomycin; mesenchymal stem cells; tissue engineering; drug delivery systems (DDSs); osteomyelitis

1. Introduction

Bone infections, such as osteomyelitis, remain a major clinical challenge in the field of bone surgery due to their serious rate of mortality and morbidity [1]. The most common causative species of surgical site infections and medical device-associated infections are the opportunistic Gram-positive staphylococci ($\simeq$75% of cases), particularly *Staphylococcus aureus* and *Staphylococcus epidermidis* [2,3]. In terms of pathogenesis, osteomyelitis is complex and varied, with bacteria reaching the bone in two ways: (i) endogenously via blood or originating from another nearby or distant source of infection (hematogenous osteomyelitis); and (ii) exogenously through direct inoculation, contamination of an open trauma, or postsurgical procedures [4].

The traditional treatment for chronic osteomyelitis includes extensive resection of the infected tissue, repair of the bone defect, and intravenous and local administration of antibiotics over long periods of time [5,6] (Figure 1A). In more specific cases such as infections associated with the implantation of prosthetic materials (hip, knee, and shoulder prosthesis) or osteosynthesis materials for fracture stabilization (plates, screws, pins, etc.), as a first step, treatment requires the simultaneous removal of the implant and intravenous and local antibiotherapy maintenance until normalization of biochemical parameters, as well as the performance of osteoarticular reconstruction surgery as required by the case [5].

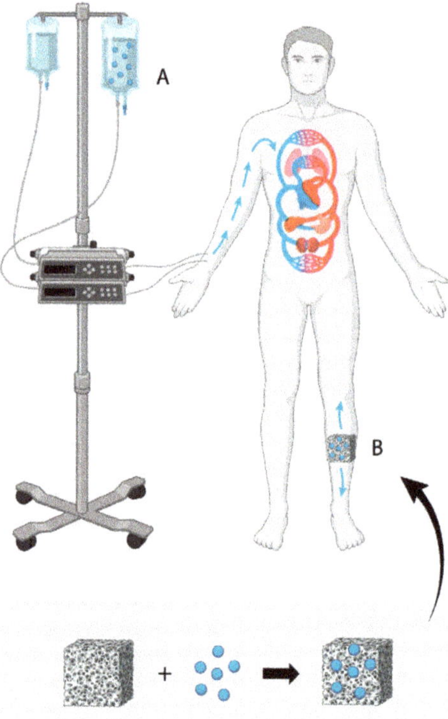

Figure 1. Schematic representation of current antibiotic treatment strategies used to control bone infections (blue spheres: antibiotics; (**A**): intravenous route; (**B**): implantation of a local drug delivery system (DDS) fabricated by loading a scaffold with antibiotics).

Another strategy for the local treatment of chronic bone infections has been based on the administration of antibiotics by means of implantation of drug delivery systems (DDSs) at the site of infection (Figure 1B). One of the most used DDSs for the treatment of osteomyelitis have been antibiotic-loaded poly(methyl methacrylate) (PMMA) beads [7,8]. PMMA, also referred to as (acrylic) bone cement, is a non-resorbable biomaterial that works by slowly releasing antibiotics (generally gentamicin and vancomycin) over time, which can help to eradicate the bacteria causing the infection [9,10]. Despite having been used for decades in clinical practice, PMMA beads or blocks are far from being an ideal antibiotic carrier. The non-degradability of this biomaterial requires a second surgery to remove the beads 2 or 3 weeks after implantation [11].

Vancomycin (Van) is the most commonly used antibiotic in treatment of infections in arthroplasty surgery and chronic osteomyelitis of any etiology [10]. This glycopeptide antibiotic acts primarily as a cell wall synthesis inhibitor in susceptible organisms. It binds rapidly and irreversibly to the cell wall of susceptible bacteria, inhibiting the synthesis of peptidoglycan, which forms the structure of the cell wall [12,13].

Polycaprolactone (PCL) is a synthetic polymer that has been commonly used in 3D-printing applications as a scaffolding component for bone and cartilage reconstruction [14–16]. Among its main advantages, we find its biocompatibility, biodegradability, good mechanical properties, and its low melting temperature ($\simeq$60 °C), which makes it more versatile than other synthetic polymers used for 3D printing applications [17]. However, the 3D printing temperatures used to fabricate well-defined scaffolds with controlled architecture are around 120–160 °C, which makes it impossible to combine them with cells, growth factors, or other bioactive molecules during the printing process [18,19]. On the other hand, PCL lacks natural motifs that provide specific binding sites for cells that facilitate tissue integration [20]. Because of this, different strategies have been developed to overcome PCL native hydrophobicity, such as surface modification using NaOH treatment or its combination with other synthetic or naturally derived materials (hydrogels) to create hybrid scaffolds with enhanced properties [21–24]. Synthetic and naturally derived hydrogels have been widely used for different biomedical applications due to their ability to encapsulate cells, drugs, growth factors, or other bioactive molecules [25,26]. Naturally derived hydrogels principally include collagen, chitosan, alginate, silk fibroin, gelatine, cellulose, hyaluronic acid, and hydrogels derived from decellularized tissues [27].

Chitosan (CS) is a naturally derived semicrystalline polymer that is obtained by partial deacetylation of chitin under alkaline conditions [28]. It is one of the materials most widely used to prepare hydrogels due to its excellent biocompatibility, nontoxicity, and biodegradability [29,30]. However, CS hydrogels present several problems due to their unstable structures with large-sized and poorly defined pores, weak formability, and low mechanical properties, limiting their further utilization for in vivo studies [31,32]. For this reason, they have been combined with curable polymers such as polycaprolactone (PCL) and polylactic acid (PLA), which provide the scaffolds basic mechanical support [33,34].

The fabrication of hybrid materials that combine natural and synthetic polymers is a promising approach for creating new scaffolds that combine the intrinsic advantages of both materials and meet several requirements, such as being biodegradable, bioactive, having good mechanical properties, and being easy to fabricate [35]. In this work, we focus on the fabrication of a novel hybrid 3D-printed scaffold based on polycaprolactone and a chitosan hydrogel loaded with different vancomycin concentrations (1%, 5%, 10%, and 20%) as a DDS to evaluate its antimicrobial efficacy against *S. aureus* and *S. epidermidis*. In addition, we propose a novel method for improving the adhesion of hydrophobic polymers (PCL) to hydrogels using two different cold plasma treatments. The obtained scaffolds combine the natural biocompatibility, biodegradability, and antibacterial properties of CS with the excellent mechanical properties of PCL. The morphological characteristics of the scaffolds were studied by means of optical and scanning electron microscopy (SEM), showing that the CS/Van hydrogel successfully coated the PCL matrix homogeneously after the plasma treatments. The antibacterial efficacy of the scaffolds was tested against *Staphylococcus aureus* and *Staphylococcus epidermidis*, and vancomycin release was studied at different time periods by means of high-performance liquid chromatography (HPCL). Finally, we evaluated the possible systemic adverse effects of scaffolds at the cellular level by analyzing the viability, proliferation, and differentiation of a population of adult human bone marrow-derived mesenchymal stem cells (*ah*-BM-MSC).

2. Materials and Methods

2.1. Materials

Polycaprolactone (PCL) 3D printing filament (molecular weight: 50,000 g/mol) was purchased from 3D4makers (Haarlem, The Netherlands). Chitosan (#448869, 75–85% deacetylated, low molecular weight), acetic acid, and NaOH were obtained from Sigma-Aldrich (Saint Louis, MO, USA) and used as received. Vancomycin hydrochloride was purchased from Lab. Reig Jofre S.A. (Barcelona, Spain).

Staphylococcus aureus (CECT 239) and *Staphylococcus epidermidis* (CECT 231) strains were purchased from the Spanish Type Culture Collection (CECT) (Valencia, Spain). Tryptic Soy Broth (TSB), Tryptic Soy Agar (TSA), and buffered peptone water were obtained from Scharlau (Barcelona, Spain).

2.2. Methods

2.2.1. Fabrication of Porous Polycaprolactone/Chitosan Scaffolds Loaded with Vancomycin Design and Fabrication of 3D-Printed PCL Scaffolds with Controlled Porosity Using the Fused Deposition Modelling (FDM) Method

The scaffolds were designed by using REGEMAT 3D Designer software (v1.4.4) and 3D-printed using REGEMAT 3D Bio V1® bioprinter (REG4Life, REGEMAT 3D, Granada, Spain) equipped with a glass bed and a thermoplastic extruder with a 0.4 mm diameter nozzle. The scaffolds were designed with the following parameters: scaffold size 1.50 × 20 × 20 mm (height, width, length), pore size 200 μm, layer height 0.25 mm, perimeters 0, solid bottom/top layers 0, infill pattern triangular; and manufactured using a medical-grade PCL filament printed at 160 °C with an infill speed of 11 mm/s. As described in a previous work, an 8 mm biopsy punch was used to prepare defined and reproducible disk-shaped scaffolds of 8 × 1.50 mm (diameter, height) from each printed scaffold (Figure 2A) [18]. In addition, nonporous (solid) PCL scaffolds of 12 × 12 × 1.50 mm (width, length, height) were printed in order to avoid porosity interfering with some of the experimental results (Figure 2B). Porous disk-shaped scaffolds were characterized and used for all biological and microbiological experiments and solid scaffolds were only used to evaluate the effect of cold plasma treatment on the adhesion of CS/Van hydrogel to the PCL matrix (see Section 2.3.2), as it was noticed that the porosity interfered in the experimental results.

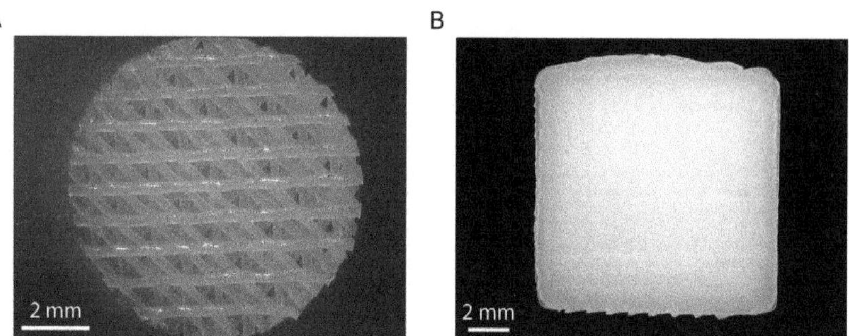

Figure 2. Micrographs of (**A**) porous and (**B**) solid PCL scaffolds printed using REGEMAT 3D Bio V1® bioprinter.

Vancomycin-Loaded Chitosan Hydrogel Preparation

The chitosan hydrogels were prepared by dissolving low molecular weight chitosan at a concentration of 4% (w/v) in demineralized water containing 1.5% (v/v) acetic acid. The solutions were mechanically stirred for 24 h (250 rpm) and vancomycin was added to the chitosan hydrogels in a content of 1%, 5%, 10%, and 20% (w/w of chitosan). After 24 h, the solutions were centrifuged (3000 rpm, 1 h) to remove air bubbles.

Hybrid Scaffold Preparation

The 3D-printed PCL scaffolds were dip-coated in the chitosan hydrogel and left to dry overnight. Then, the scaffolds were neutralized in a 1 M NaOH solution for 2 h and rinsed three times with distilled water to remove residual acids. The whole process was performed in a laminar flow cabinet to avoid sample contamination.

Preliminary tests showed that the chitosan coating remained firmly adhered to the porous PCL scaffolds, as it was trapped between the 3D-printed strands. However, that was not the case for solid PCL scaffolds, which repelled the chitosan coating as seen in

Figure 3C. This could be due to the fact that PCL is a hydrophobic polymer, which makes it difficult to adhere to hydrogels with more than 90% water content.

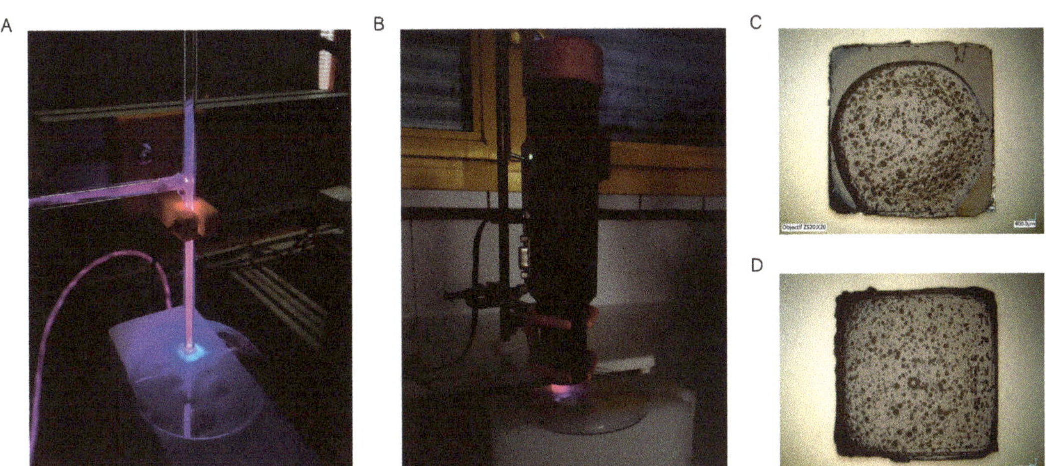

Figure 3. Plasma treatments performed on the PCL scaffolds. (**A**) Cold atmospheric plasma jet; (**B**) homemade torch with a piezoelectric plasma generator. Optical microscopy images of (**C**) Solid PCL scaffold (untreated) coated using CS hydrogel (scale bar 400 μm); (**D**) solid PCL scaffold coated using CS hydrogel after being exposed to cold plasma for 30 s (scale bar 400 μm).

Plasma treatments have recently been used to lower the hydrophobicity of polymer surfaces by forming reactive oxygen species (ROS) [36]. These newly formed reactive species generate hydrophilic groups (hydroxyl or carboxylic) on the polymer surface that can interact with the hydrogels [37]. Even if plasma treatment was not necessary for porous scaffolds, two different plasma technologies were tested on the surfaces of solid PCL scaffolds: (i) a cold atmospheric pressure plasma jet developed in GREMI-CNRS consisting of a plasma gun that is a dielectric barrier discharge (DBD) plasma and helium fed and powered using a microsecond voltage pulse [38] (Figure 3A); and (ii) a homemade torch with a piezoelectric plasma generator (CeraPlas®) developed by TDK Electronics GmbH & Co [39] (Figure 3B). This technology is based on a piezo ceramic component and a driving circuit that allows for generating cold atmospheric plasma. After being exposed for 30 s to plasma treatments with 5 mm gap, the PCL scaffolds were dip-coated in the chitosan hydrogels and left to dry overnight. As can be seen in Figure 3D, the CS hydrogel was not repelled by the scaffold, showing a homogeneous distribution.

In order to determine the amount of chitosan adhered to the scaffolds, the weight of the PCL scaffolds was compared to the weight of PCL/CS/Van scaffolds after the coating process. After weighing the scaffolds (n = 10) in an analytical balance, an increment from 33.34 ± 0.34 mg (native PCL) to 41.16 ± 1.01 mg (PCL/CS/Van scaffolds) was observed. No significant differences were obtained between the weight of PCL/CS scaffolds at different vancomycin concentrations.

2.3. Characterization of the 3D-Printed Scaffolds

2.3.1. Morphological Characterization of the 3D-Printed Scaffolds

The morphological characterization of the 3D-printed scaffolds was carried out using a digital camera coupled to a stereomicroscope. The surface morphology of the scaffolds was examined using scanning electron microscopy (SEM; model JEOL-6100, JEOL Ltd., Tokyo, Japan). Prior to visualization, the scaffolds were mounted on aluminum stubs with conductive paint and sputter-coated with platinum (10 mA, 120 s). SEM micrographs were obtained at magnifications of ×20 and ×45 under an accelerating voltage of 15–20 kV and a working distance of 4–5 mm.

2.3.2. Wettability Assay

Static water contact angle (WCA) measurements were performed on a drop shape analyzer (DSA100S KRÜSS GmbH) to assess the effect of cold plasma treatment on PCL hydrophobicity. The water droplet volume was set to 1 µL and the deposit speed at 2.67 µL/s. Briefly, solid PCL scaffolds (n = 3) were treated using a cold atmospheric plasma jet (Plasma 1) and a homemade torch with a piezoelectric plasma generator (Plasma 2) (Figure 3). Three untreated scaffolds were used as control. Each measurement was performed in triplicate in three different scaffolds for better statistical significance.

2.3.3. In Vitro Degradation Kinetics

The in vitro degradation of the scaffolds was assessed by measuring the weight loss of the constructs at 1, 3, 7, 14, 21, and 28 days. Briefly, PCL/CS/Van scaffolds were weighed (W_0, initial weight) and subsequently immersed in eppendorf tubes containing 1 mL of complete growth medium (DMEM). Then, the scaffolds were incubated at 37 °C in a 5% CO_2 atmosphere at 95% relative humidity. At different time periods, the scaffolds were recovered from medium, sightly wiped using filter paper, and reweighed (W_d, final weight). The weight loss (WL) was calculated by applying the following equation [40]:

$$WL\% = [(W_0 - W_d)/W_0] \times 100 \tag{1}$$

where W_0 and W_d indicate the weight of the scaffold before and after the scheduled immersion time, respectively.

2.3.4. Water Absorption Assay

The swelling behavior of PCL/CS/Van scaffolds was tested in PBS solutions of pH 7.4. Three PCL/CS/Van scaffolds of known weight were put into stainless steel baskets in order to prevent scaffolds from floating. At different timepoints (0.5, 1, 3, 6, 24, 48 h), the scaffolds were collected and gently wiped using filter paper before being weighed. Water absorption percentage was calculated using the following equation [41]:

$$Water\ absorption\ (\%) = \frac{w_t - w_0}{w_0} \times 100 \tag{2}$$

where w_t is the weight of the scaffolds at time t, and w_0 is the initial weight of the dry scaffolds.

2.3.5. Vancomycin Release Quantification

Vancomycin-loaded chitosan scaffolds were immersed into 1.5 mL eppendorf tubes containing 1 mL phosphate-buffered saline (PBS, pH 7.4) medium at 37 °C in darkness. At predetermined timepoints (1 h, 3 h, 6 h, 24 h, 48 h, 72 h, 7 d, and 14 d), the medium was collected, and fresh PBS was refilled accordingly. The vancomycin concentration was measured using high-performance liquid chromatography (HPLC) (Agilent 1200 series, Agilent Technologies, Waldbronn, Germany) with diode array (DAD) and fluorescence (FLD) detector, equipped with a Vydac 218TP C18 column (250 × 4.6 mm, 5 µm, Avantor®, Radnor Township, PA, USA). The analysis was performed under isocratic conditions (0.8 mL/min) using 89% ammonium phosphate 0.5 M and 11% acetonitrile as mobile phase. Prior to the measurements, a calibration curve of vancomycin in PBS was determined by measuring absorbance values of stock solutions of vancomycin from 100 to 0.5 µg/mL at 282 nm (linear range 12.4 to 0.5 µg/mL; $R^2 > 0.998$).

2.4. Evaluation of Antimicrobial Properties

2.4.1. Bacterial Culture

Strains of *Staphylococcus aureus* (CECT 239) and *Staphylococcus epidermidis* (CECT 231) were obtained from the Spanish Type Culture Collection (CECT) (Valencia, Spain). To activate the *S. aureus* and *S. epidermidis* strains, they were grown in TSB medium and incubated aerobically at 35 °C for 24 h. The bacterial cultures were stored in TSA medium at 4 °C for over three months. For daily use, a working culture was prepared by transferring a single colony from TSA to 10 mL of TSB, followed by incubation at 35 °C for 24 h.

2.4.2. Agar Diffusion Method

In order to assess the antimicrobial properties of the chitosan coating loaded with vancomycin, experiments were conducted to measure the inhibition zones. Initially, a suspension of each microorganism (*S. aureus* and *S. epidermidis*) was prepared with a concentration of 5.0 log10 colony-forming units/mL (CFU/mL) in TSB (2X) medium. Subsequently, 100 µL aliquots of the bacterial suspension were added to 6.0 mm diameter well plates containing TSA agar.

Following a brief drying period, the samples were brought into contact with the PCL/Chitosan/Van scaffolds and incubated overnight at 37 °C. Finally, the inhibition halos were measured using a ruler at 24 and 48 h. All the experiments were replicated three times.

2.5. Biological Assays

2.5.1. Isolation, Purification, Characterization, and Culture of Adult Human Bone Marrow-Derived Mesenchymal Stem Cells (*ah*-BM-MSCs)

Multipotent *ah*-BM-MSCs were isolated from bone marrow as previously described (additional information related to the applied methodology, cell isolation, culture, and expansion can be found in previous publications [42,43]). Briefly, bone marrow was collected from the iliac crest of three volunteer patients by direct puncture. Then, bone marrow aspirate was transferred into transfer bags containing heparin. The mononuclear cell fraction was obtained using Ficoll density gradient media and a cell-washing closed automated SEPAX^{TM} S-100 system (Biosafe, Eysines, Switzerland). After assessing cell viability by using trypan blue staining (#T8154; Sigma-Aldrich, Saint Louis, MO, USA), cells were plated out at a density of 3.75×10^5 mL in 75 cm^2 culture flasks (Sarstedt, Nümbrecht, Germany) with 10 mL of basal culture growth medium (GM). The growth medium used was Dulbecco's Minimum Essential Media (DMEM) (#31885-023, Gibco, Bleiswijk, The Netherlands) complemented with 10% (*v/v*) inactivated fetal bovine serum (FBS) (#F7524, Sigma-Aldrich, Saint Louis, MO, USA) and routine antibiotics such as Penicillin and streptomycin (100 U mL^{-1} penicillin and 100 µg mL^{-1} streptomycin) (#P4333, Sigma-Aldrich, Saint Louis, MO, USA) before incubating cells under standard conditions of normoxia (21% O$_2$) at 37 °C and 5% CO$_2$. *ah*-BM-MSCs in passages 3–5 were used for in vitro experiments.

All biological experiments were in full compliance with the established regulations and the experimental protocol was reviewed and approved by the Institutional Ethics Committee of UCAM—Universidad Católica de Murcia (authorization n° CE051904) UCAM Ethics Committee (CE n° 052114/05.28.2022).

2.5.2. Immunophenotypic Profiles of *ah*-BM-MSC Cultures

The isolated *ah*-BM-MSCs were characterized by means of flow cytometry (Beckman Dickinson & Co., Franklin Lakes, NJ, USA; Navios Software v1.2) for mesenchymal (CD90, CD73, 105) and hematopoietic (CD34, CD45) markers as previously described [42,44,45]. Before performing the in vitro assays, single-cell suspensions obtained using culture trypsinization were labeled with fluorochrome-conjugated antibodies: CD73-PE, CD90-APC, CD105-FITC, CD34-APC, and CD45-FITC (Human MSC Phenotyping Cocktail,

Miltenyi Biotec, Bergisch Gladbach, Germany) in order to verify the purity of *ah*-BM-MSCs populations.

2.5.3. Cell Seeding Methods

For cell viability, proliferation, and osteogenic differentiation assays, *ah*-BM-MSCs were seeded at the bottom of 24 WP at a density of 10×10^3 cells cm^{-2}, and PCL/CS/Van scaffolds were placed in 0.4 µm pore culture well inserts (Falcon®) to be in indirect contact with the culture. Cells seeded onto tissue culture-treated polystyrene wells (TCPS; Sigma-Aldrich, Corning, NY, USA) (without materials) were taken as positive control. The culture media were changed three times a week for all the experiments performed.

2.5.4. Cell Viability and Proliferation Assays
In Vitro Cytotoxicity Assay

The cell viability of *ah*-BM-MSCs was evaluated using LDH cytotoxicity detection kit (#11644793001, Roche Diagnostics, Roche. Applied Science, Mannheim, Germany) after placing PCL/CS/Van scaffolds in indirect contact with cells for 24 and 72 h. Before performing the assay, it is important to determine the optimal cell concentration for the cell line used, as different cell types may contain different amounts of LDH. In general, we choose the concentration at which the difference between the low and high control is at a maximum. For our culture of *ah*-BM-MSCs, the optimal cell concentration to be used for the subsequent experiment was found to be 3×10^4 cells/well for 24 WP (Figure 4).

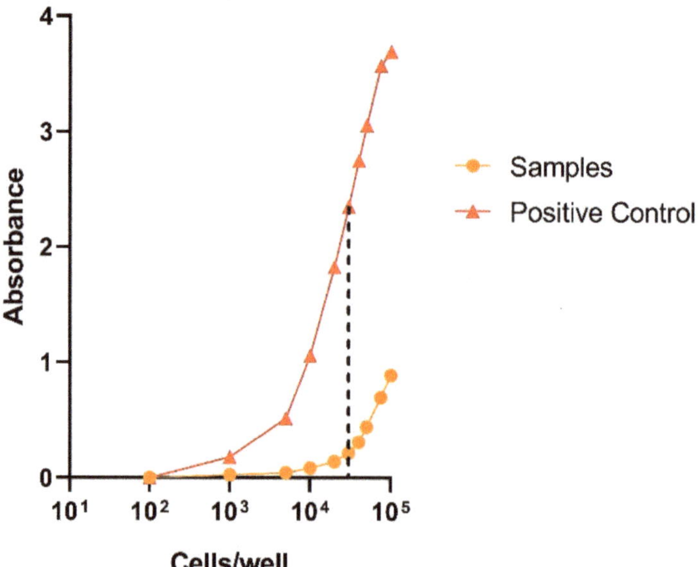

Figure 4. Determination of the optimal target cell concentration for *ah*-BM-MSCs. *ah*-BM-MSCs were titrated in 24 WP at densities of 100, 1000, 5000, 10,000, 20,000, 30,000, 40,000, 50,000, 75,000, and 100,000 cells/well. Culture medium was added for the determination of the spontaneous LDH activity (samples), and culture medium containing 1% Triton X-100 was added to determine the maximum LDH release (positive control).

Twenty-four hours after seeding *ah*-BM-MSCs on 24 WP, the culture medium was replaced using fresh assay medium containing 1% FBS to remove LDH activity released from the cells during the incubation step. Then, the PCL/CS/Van scaffolds were placed in indirect contact with cells using 0.4 µm pore culture well inserts and LDH activity was assessed at 24 and 72 h. Cells growing on tissue culture polystyrene (TCPS) plates without PCL/CS/Van scaffolds were used as low control (spontaneous LDH release) and cells cultured using assay medium containing 1% Triton X-100 solution were used as high control (maximum LDH release). At different timepoints, 100 μL aliquots of each well were transferred to an optically clear flat-bottomed 96 WP followed by the addition of 100 μL of the reaction mixture. After 30 min incubation at RT, the absorbance was read directly using a Spectramax iD3 plate reader (Molecular Devices, San Jose, CA, USA) at 490 nm.

Cellular Metabolic Activity Assay

The cellular metabolic activity of *ah*-BM-MSCs grown in the presence of PCL/CS scaffolds loaded with different vancomycin concentrations was evaluated using the AlamarBlue® assay (#DAL1100, Invitrogen, Carlsbad, CA, USA) on days 1, 3, 7, 14 after seeding. Cells growing onto TCPS (without PCL/CS/Van scaffolds) were used as control. Briefly, at different time periods, the inserts containing PCL/CS/Van scaffolds were temporarily removed (prior to the incubation step), and fresh medium (1 mL) containing 10% (v/v) AlamarBlue® reagent was added to each well. Then, the plate was incubated for 4 h in darkness (wrapped with aluminum foil) at 37 °C in a 5% CO_2 atmosphere with 95% of relative humidity. Finally, 150 μL aliquots of each well were transferred to a black-walled 96 WP and fluorescence was assessed at excitation and emission wavelengths of 530 and 590 nm (respectively) using a Spectramax iD3 plate reader (Molecular Devices, USA).

2.5.5. Osteoblastic Differentiation Assays
Alkaline Phosphatase (ALP) Activity

The ALP activity of *ah*-BM-MSCs was assessed at 7 and 14 days using an Alkaline Phosphatase Detection Kit (#SCR004, Merck Millipore, Billerica, MA, USA) after seeding the cells in the presence of PCL, PCL/CS, and PCL/CS/10%Van scaffolds. At each time period, cells were detached using trypsin, and an aliquot of 1.5×10^4 cells (per sample) was treated by adding 100 μL p-nitrophenylphosphate (p-NPP). After 20 min incubation at RT, the catalytic reaction (p-NNP → phosphate + p-nitrophenol) affords a yellow colored byproduct that is proportional to the amount of ALP present within the reaction. The absorbance was measured using a Spectramax iD3 plate reader (Molecular Devices, USA) at a wavelength of 405 nm.

In Vitro Mineralization Assay

In vitro mineralization was evaluated using an Osteogenesis Assay Kit (#ECM815; Millipore, Billerica, MA, USA) 21 days after seeding *ah*-BM-MSCs in the presence (indirect contact) of PCL, PCL/CS, and PCL/CS/10%Van scaffolds. Briefly, the *ah*-BM-MSCs were fixed using 10% formaldehyde, stained using alizarin red stain solution, and visualized using an optical microscope (Axio Vert.A1, ZEISS, Jena, Germany) coupled to a digital camera (Axiocam 305 color, ZEISS, Germany). Then, the samples were treated following the manufacturer's protocol to quantify matrix mineralization and the optical density (OD) was measured using a Spectramax iD3 plate reader (Molecular Devices, USA) at 405 nm.

2.6. Statistics

All data were represented as mean ± SD, n = 3. The statistical significance was determined by a two-way ANOVA using GraphPrism 8.0.1 (GraphPad Software Inc., San Diego, CA, USA) for Windows. Comparison between groups was evaluated using the *t*-test, the significance level being $p < 0.05$.

3. Results and Discussion
3.1. Characterization of the 3D-Printed Scaffolds
3.1.1. Morphological Characterization of the 3D-Printed Scaffolds

Microstructural examination of the scaffolds' morphology using SEM imaging showed differences in the scaffolds' surfaces depending on the vancomycin concentration loaded. Figure 5(A1,B1) shows a micrograph of the native PCL constructs before being coated with chitosan hydrogel. The scaffolds show a parallel distribution of the printed strands in each printed layer, and the triangular printing pattern results in scaffolds with interconnected porosity with an approximate pore size of 200 μm and an overall porosity of 59%, as calculated in a previous publication [18]. As can be seen in Figure 5(A2,B2), the chitosan coating causes a discrete thickening of the 3D-printed strands as well as at the points of contact between the strands as a result of the thin coating layer provided by CS. However, this finding intensifies progressively and proportionally when the chitosan hydrogel is loaded with increasing vancomycin concentrations (Figure 5(A3–A6,B3–B6)). The latter is observed in the micrographs corresponding to PCL/CS/20%Van scaffolds, in which almost the entire surface of the PCL scaffold was covered, making it difficult to distinguish its pores (Figure 5(A6,B6)).

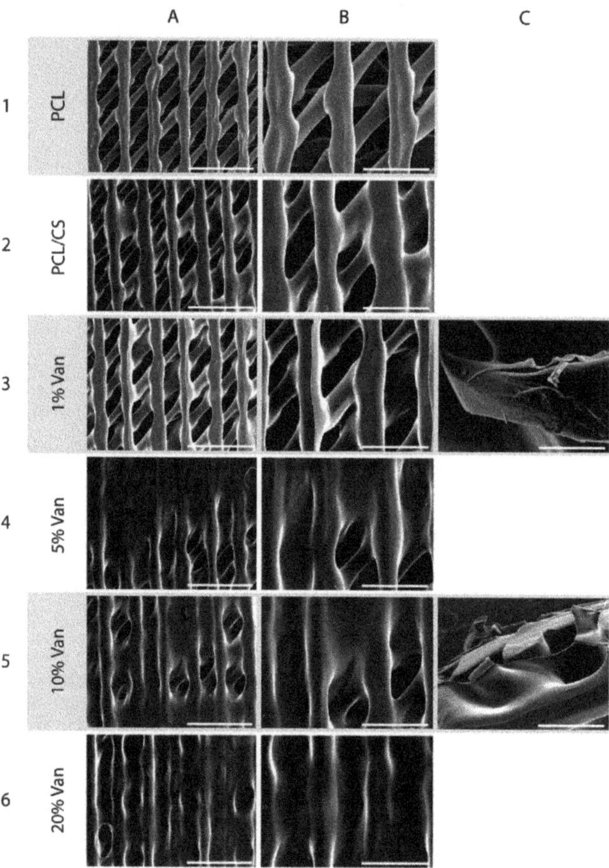

Figure 5. SEM micrographs of the (**1**) PCL, (**2**) PCL/CS, and (**3–6**) PCL/CS/Van scaffolds loaded with 1%, 5%, 10%, and 20% *w/t* vancomycin, respectively, at different magnifications: (**A1–A6**) 22×, scale bar 2 mm; (**B1–B6**) 40×, scale bar 1 mm; (**C3**) scale bar 200 μm; (**C5**) scale bar 700 μm.

The functional characteristics of the construct developed in this study agree with the premises established by different authors for the design and fabrication of 3D scaffolds to be used for tissue engineering applications [46,47]. In fact, we have taken into account that an ideal construct or scaffold, in general, should have sufficient porosity to accommodate both differentiated and undifferentiated cells of different lineages. In addition, it was combined with a natural-derived polymer (chitosan) capable of acting as a carrier or DDS for the controlled release of drugs or biomolecules. Likewise, these reticular pore systems should have open and interconnected porosity to facilitate the diffusion of nutrients, metabolites, and cells from the external environment to the inner parts of the scaffold [20].

3.1.2. Wettability Assay

The wettability of solid 3D-printed PCL scaffolds was evaluated by measuring the static water contact angle of 1 µL water droplets before and after the plasma treatments. As shown in Figure 6, the static WCA of the untreated PCL scaffolds was $79.93 \pm 2.36°$, demonstrating hydrophobic behavior without any surface modification. However, for plasma-activated scaffolds, we observed water contact angles of $55.40 \pm 3.16°$ and $59.62 \pm 4.22°$ for Plasma 1 and Plasma 2, respectively, showing a decrease of around 20° that attests to a decrease in surface hydrophobicity.

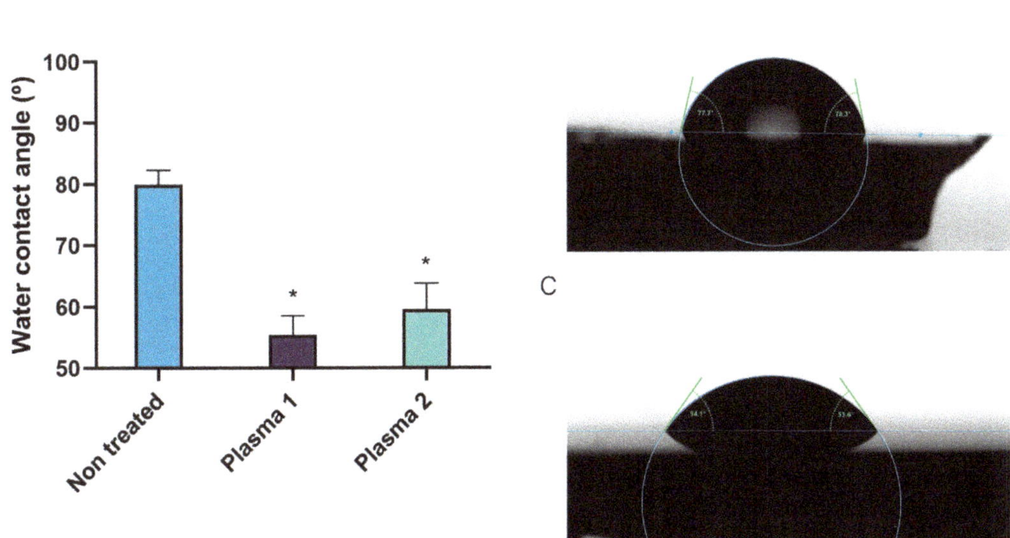

Figure 6. (**A**) Water contact angle on 3D-printed solid (nonporous) PCL scaffolds before and after plasma treatments (Plasma 1: cold atmospheric pressure plasma jet; Plasma 2: homemade torch with piezoelectric plasma generator). Water droplets with a volume of 1 µL deposited on (**B**) nontreated PCL scaffolds and (**C**) PCL scaffolds treated using Plasma 1. Data represent mean $\pm$ SD. * Significant differences with the water contact angle of nontreated PCL scaffold.

A change in WCA is a clear indicator of a polymer surface chemistry modification [48]. The results obtained are in total accordance with the ones obtained by Carette et al., who studied the effects of atmospheric plasma on the surface of a polylactic acid (PLA) scaffold in order to improve the adhesion of a 1% chitosan solution, obtaining a decrease of about 30° in WCA (from 85° to 55°) after plasma activation [36].

3.1.3. In Vitro Degradation Kinetics

The in vitro degradation assay carried out using PCL/CS/Van scaffolds did not result in any changes in the scaffold weights ($W_0 = W_d$) after 28 days of immersion in culture medium at 37 °C in a 5% CO_2 atmosphere at 95% relative humidity. It is well known that CS, as well as other polysaccharides, is degraded in the body by depolymerization, oxidation, and hydrolysis (enzymatic or nonenzymatic) [49]. Lim et al. studied the degradation of CS porous beads with different degrees of acetylation (DA 10–50%) in different enzyme solutions (lysozyme, NAGase, and a mixture of both) [50]. As main findings, they determined that (i) CS beads did not degrade in the NAGase solution, (ii) CS beads with high DA degraded faster than those with low DA in lysozyme and enzymatic mixture solutions, and (iii) CS beads with DA values around 30 to 50% rapidly degraded into monomers in the enzymatic mixture in less than 30 to 60 days.

3.1.4. Water Absorption Assay

The water absorption of the chitosan coating of PCL/CS and PCL/CS/Van scaffolds was tested in PBS solutions. As can be seen in Figure 7, the total weight of the scaffolds increased between 34.61 ± 1.13% (PCL/CS) and 42.14 ± 1.46% (PCL/CS/20%Van) after 30 min of immersion in PBS, and remained more or less stable for at least for 48 h. The minimum absorption values were shown by the PCL/CS scaffolds (34.61 ± 1.13%), while the maximum ones were reached by the PCL/CS/20%Van scaffolds 45.32 ± 8.41%. The latter can be justified by the results obtained in the scaffolds characterization section (SEM imaging), which showed that increasing concentrations of vancomycin caused a thickening of the 3D-printed strands, having more chitosan surface to absorb PBS.

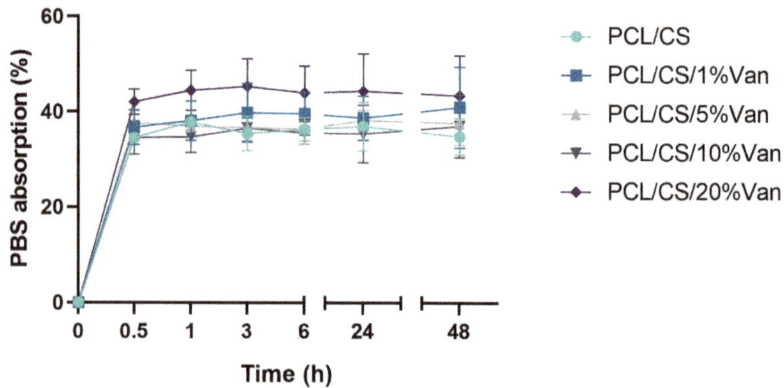

Figure 7. Water absorption as a function of time of PCL/CS and PCL/CS/Van scaffolds (1%, 5%, 10%, and 20%) in contact with PBS medium (pH = 7.4).

3.1.5. Vancomycin Release Quantification

The vancomycin release from the PCL/CS scaffolds loaded with 1%, 5%, 10%, and 20% vancomycin followed first-order release kinetics (Figure 8). Fast vancomycin release was observed during the first hour, followed by slower release during the next hours for all the scaffolds studied. As can be seen in the graph, vancomycin release finishes after 24 h and the cumulative drug release profiles seem to reach a plateau. These release profiles could be explained by the presence of two different drug fractions: (i) drug weakly bound or deposited on the CS hydrogel surface, and (ii) drug included in the CS network by chemical interactions. It is worth noting that the scaffolds loaded with 10% and 20% Van had similar drug release profiles, indicating that the solubility limit of the drug in PBS is probably between both concentrations.

These kinetics profiles are in total accordance with the results obtained by López-Iglesias et al., who studied vancomycin release from porous chitosan aerogels obtaining similar drug release profiles [51]. On the other hand, Kausar et al. developed chitosan films loaded with vancomycin (10 and 20% with respect to dry polymer mass), obtaining 100% dissolution of free vancomycin in 7 h with more than 50% of drug dissolved within the first hour [52].

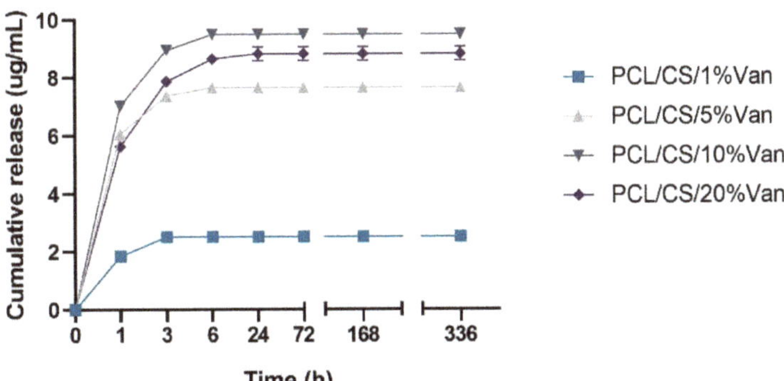

Figure 8. In vitro cumulative drug release profiles of PCL/CS scaffolds loaded with 1%, 5%, 10%, and 20% vancomycin for 14 days (336 h).

3.2. Evaluation of Antimicrobial Properties

PCL/Chitosan scaffolds loaded with different vancomycin concentrations have been shown to exhibit antimicrobial activity against both *S. epidermidis* and *S. aureus*. This is because, in all cases, the minimum inhibitory concentration for vancomycin was exceeded (2 µg/mL for *S. aureus* and 2–4 µg/mL for *S. epidermidis*) [53,54]. The inhibition halos were measured at 24 and 48 h, with no significant differences being observed between them. This may be due to the fact that no more vancomycin is released between 6 and 24 h (Figure 9).

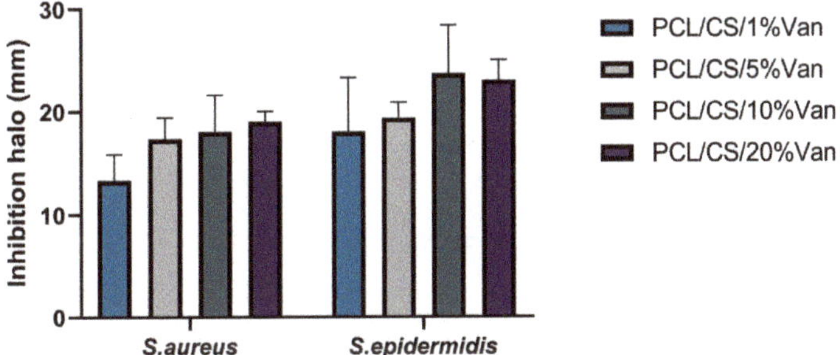

Figure 9. Inhibition halos (mm) produced by PCL/CS/Van scaffolds loaded with 1%, 5%, 10%, and 20% *w/t* vancomycin for *S. aureus* and *S. epidermidis*. Data represent mean ± SD. No significant differences were found between the scaffolds for both microorganisms studied.

As seen in Figure 9, a halo of inhibition is shown in all cases. In the case of *S. aureus*, the highest inhibition halo is for PCL/CS/20%Van, but there are no significant differences with the PCL/CS/10%Van and PCL/CS/5%Van scaffolds. This is because the vancomycin concentrations in the media are similar to each other, 9.5 mg/mL and 8.82 mg/mL (Figure 9), respectively. In the case of *S. epidermidis*, the highest inhibition halo was obtained for the PCL/CS/10%Van scaffolds. If the PCL/CS/10%Van and PCL/CS/20%Van scaffolds are compared, there are no significant differences between them because the concentrations of vancomycin in the media are similar (Figure 9).

If we compare the inhibition halos of *S. epidermidis* and *S. aureus*, it is observed that although both bacteria are Gram-positive; vancomycin is more effective against *S. epidermidis*. No antimicrobial effect of PCL/CS was observed, this may be due to the low solubility and the limited amount of positive charges of the polysaccharide at the pH (7.4 ± 0.1) of the culture medium [55,56].

3.3. Biological Assays

3.3.1. Cell Viability and Proliferation Assays

In Vitro Cytotoxicity Assay

Cytotoxicity was assessed using an LDH activity assay after placing the PCL/CS/Van scaffolds in indirect contact with *ah*-BM-MSCs for 24 and 72 h (Figure 10). The results of LDH release demonstrated that none of the tested scaffolds were found to be cytotoxic. However, the maximum LDH release values were achieved by PCL, PCL/CS, and PCL/CS/1%Van scaffolds, showing cytotoxicity values of $11.4 \pm 6.4\%$, $8.3 \pm 4.5\%$, and $10.0 \pm 8.0\%$ at 24 h, and $11.1 \pm 4.9\%$, $6.9 \pm 5.0\%$, and $11.7 \pm 6.4\%$ at 72 h, respectively, with respect to low and high control. On another hand, PCL/CS scaffolds loaded with 5, 10, and 20% vancomycin reached cytotoxicity values of $5.7 \pm 5.5\%$, $6.1 \pm 4.7\%$, and $4.9 \pm 5.5\%$ at 24 h, and $4.4 \pm 3.4\%$, $4.8 \pm 4.0\%$, and $4.7 \pm 2.9\%$ at 72 h, respectively. No statistically significant differences were observed between the tested samples.

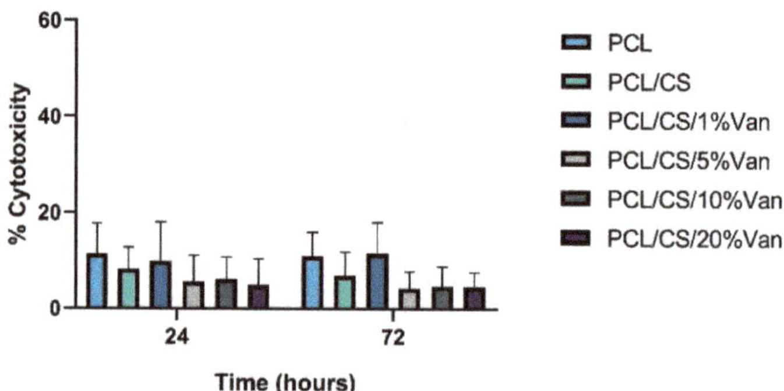

Figure 10. Cytotoxicity results using the LDH assay after placing PCL, PCL/CS, and PCL/CS/Van scaffolds in indirect contact with *ah*-BM-MSCs for 24 and 72 h. The mean cytotoxicity percentage was calculated and normalized with respect to spontaneous LDH release (low control) and maximum LDH release (high control). Bars represent standard deviations of the mean. No significant differences were observed between the tested samples.

Both polymers used in this study (PCL and CS) have been commonly used as DDSs or scaffolds components for tissue engineering applications due to their biocompatibility, biodegradability, and ease of modification [57,58]. Cytocompatibility of CS has already been studied, with results showing that it does not have an acute cytotoxic effect [59]. Our results are in line with previous publications that demonstrate the good biocompatibility of PCL and CS-based structures [33,60].

Cellular Metabolic Activity Assay

The cellular metabolic activity of *ah*-BM-MSCs growing in the presence of PCL, PCL/CS, and PCL/CS/Van scaffolds was evaluated using the AlamarBlue® assay on days 1, 3, 7, and 14 after seeding (Figure 11). As can be seen in the graph, the metabolic activity of *ah*-BM-MSCs increased gradually from days 1 to 14 for all the scaffolds studied except for PCL/CS/20%Van, in which a slight decrease was observed after day 7. No significant differences were observed at early periods (days 1 and 3), where all the scaffolds showed a similar proliferation rate. On the other hand, significant differences were noted between PCL, PCL/CS scaffolds, and the control on day 7, and between PCL, PCL/CS/20%Van, and the control on day 14. It is worth noting that the PCL/CS/10%Van scaffolds showed higher metabolic activity with respect to the scaffolds studied.

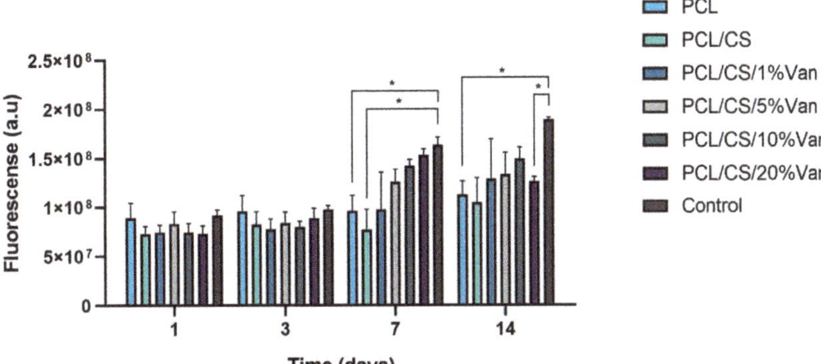

Figure 11. Cellular metabolic activity of *ah*-BM-MSCs using the AlamarBlue® assay at different time periods. The metabolic activity of cells seeded on plastic (TCPs) was used as positive control. Bars represent standard deviations of the mean. * Significant differences between the bracketed groups at the same time period.

3.3.2. Osteoblastic Differentiation Assays

Osteogenic differentiation was assessed on PCL, PCL/CS, and PCL/CS/10%Van scaffolds, as it provided the best results in the cytotoxicity and proliferation assays, as well as in the microbiological assays, showing a similar bactericidal effect to the scaffold loaded with 20% vancomycin.

Alkaline Phosphatase (ALP) Activity

The Alkaline Phosphatase activity of *ah*-BM-MSCs cultured in the presence of PCL, PCL/CS, and PCL/CS/10%Van scaffolds is shown in Figure 12. From days 7 to 14, using growth medium (GM), no significant differences in ALP activity were observed between the different time periods or within the scaffolds studied. On the other hand, when using osteogenic medium (OM), the ALP activity significantly increased at 14 days, reaching the maximum values in the presence of PCL/CS/10%Van. In fact, significant differences were observed between the scaffold loaded with 10% vancomycin and the other experimental groups, indicating that PCL/CS/10%Van seems to increase the ALP activity of the cultured cells.

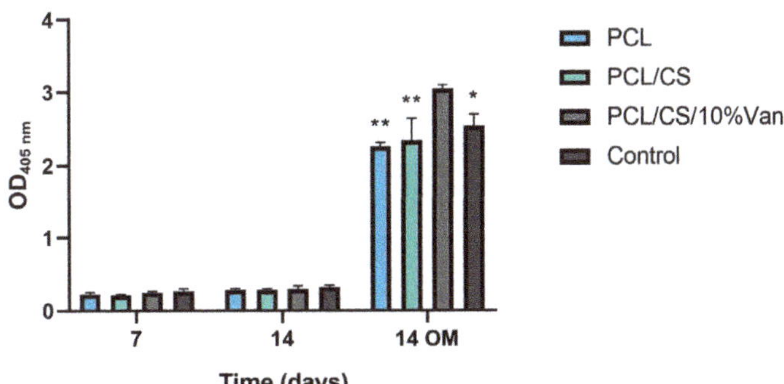

Figure 12. Alkaline phosphatase activity of *ah*-BM-MSCs after being cultured for 7 and 14 days in the presence of PCL, PCL/CS, and PCL/CS/10%Van scaffolds using growth medium (GM) and osteogenic medium (OM). Cells seeded on plastic (TCPs) were taken as positive control. Results are shown as a function of optical density (OD_{405nm}) units. Data represent mean ± SD. Significant differences were found between PCL/CS/10%Van and the other samples at 14 days using OM; * $p < 0.001$ and ** $p < 0.0001$.

However, the PCL/CS scaffold did not show an increase in ALP activity, so this finding cannot be attributed to the presence of the chitosan coating. Other authors have followed novel strategies to increase the osteogenic activity of chitosan, consisting of its combination with hydroxyapatite (HA) via the fabrication of carboxymethyl chitosan–HA nanofibers [61] or HA/resveratrol/chitosan composite microspheres [62]. This kind of HA composite can provide necessary calcium sources that can consequently induce cell differentiation, deposition of extracellular matrix, and mineralization [63].

In Vitro Mineralization Assay

The osteogenic differentiation of *ah*-BM-MSCs was determined by quantifying the presence of calcium deposits after placing the PCL/CS/Van scaffolds in indirect contact with *ah*-BM-MSCs for 21 days. As shown in Figure 13A, no major visual differences were observed between the samples, although the control appears to have a slightly higher reddish coloration. Quantitative examination (Figure 13B) showed that the alizarin red staining concentration was higher in the control than in the cells exposed to the scaffolds, even if no significant differences were observed between the samples. This could be due to a cellular adaptation phase to varying environmental changes caused by the presence of the scaffolds, which could have slightly delayed or affected mineralization.

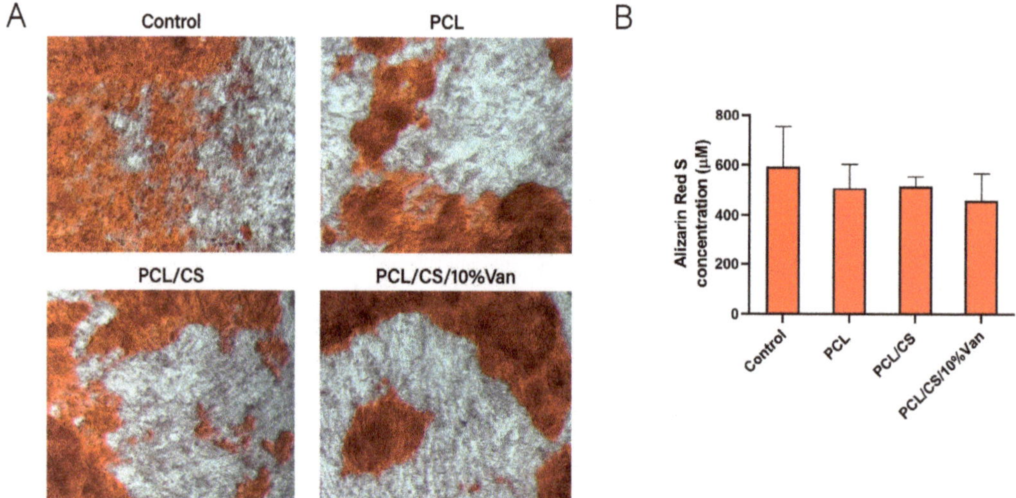

Figure 13. Alizarin red staining of *ah*-BM-MSCs after being cultured for 21 days in the presence of PCL, PCL/CS, and PCL/CS/10%Van scaffolds. (**A**) Alizarin red staining showing mineralization (original magnification ×10); (**B**) quantitative determination of alizarin red staining. Cells seeded on plastic (TCPs) were taken as positive control. Quantitative data are presented as mean ± SD. No significant differences were observed between groups.

As demonstrated by both osteoblastic differentiation assays, the PCL, PCL/CS, and PCL/CS/10%Van scaffolds did not affect *ah*-BM-MSCs differentiation capacity, indicating that both the scaffold components and the vancomycin dose used did not cause adverse effects or alterations at the cellular level.

4. Conclusions

Biocompatible 3D-printed hybrid PCL/CS/Van scaffolds were developed as drug delivery systems with antimicrobial efficacy against *S. aureus* and *S. epidermidis*. Two cold plasma treatments were presented as novel methods to improve the adhesion of hydrophobic polymers to hydrogels, resulting in a decrease in PCL water contact angle of

20°. All the scaffolds studied (1%, 5%, 10%, and 20%Van) had antimicrobial activity against both *S. aureus* and *S. epidermidis*, showing higher inhibition halos for PCL/CS/10%Van and PCL/CS/20%Van scaffolds. In vitro assays demonstrated that the PCL/CS/Van scaffolds were found to be biocompatible and bioactive as demonstrated by no cytotoxicity or functional alteration of the cultured *ah*-BM-MSCs. In conclusion, the developed DDS could be useful in achieving a controlled and effective local release of vancomycin against *S. aureus* and *S. epidermidis*, considered as the main causes of bone infections, but further preclinical studies in vivo using animal models are needed to confirm these results.

Author Contributions: Conceptualization, I.L.-G. and L.M.-O.; methodology, I.L.-G., A.B.H.-H., M.B., J.A.G. and L.M.-O.; software, I.L.-G., M.B. and D.A.-C.; investigation and validation, I.L.-G., M.I.R.-L., M.B., J.A.G. and L.M.-O.; formal analysis, I.L.-G., A.B.H.-H. and L.M.-O.; data curation, I.L.-G., D.A.-C., M.I.R.-L. and L.M.-O.; writing—original draft preparation, I.L.-G., D.A.-C., M.I.R.-L., M.B. and L.M.-O.; writing—review and editing, I.L.-G., A.B.H.-H., D.A.-C., M.I.R.-L., J.A.G. and L.M.-O.; visualization, I.L.-G. and L.M.-O.; supervision, L.M.-O.; project administration, I.L.-G. All authors have read and agreed to the published version of the manuscript.

Funding: This research received no external funding.

Institutional Review Board Statement: This study was conducted according to the guidelines of the Declaration of Helsinki and approved by the Institutional Ethics Committee of UCAM-Universidad Católica de Murcia (authorization no. CE051904) UCAM ethics committee (CE n° 052114/05.28.2022).

Informed Consent Statement: Informed consent was obtained from all subjects involved in the study.

Data Availability Statement: The data presented in this study are disclosed in the main text.

Acknowledgments: The authors of this study would like to thank the Centre Régional d'Innovation et de Transfert de Technologie—Matériaux Innovation (CRITT-MI) for support in this study.

Conflicts of Interest: The authors declare no conflict of interest.

References

1. Wassif, R.K.; Elkayal, M.; Shamma, R.N.; Elkheshen, S.A. Recent advances in the local antibiotics delivery systems for management of osteomyelitis. *Drug Deliv.* **2021**, *28*, 2392–2414. [CrossRef] [PubMed]
2. Walter, G.; Kemmerer, M.; Kappler, C.; Hoffmann, R. Treatment algorithms for chronic osteomyelitis. *Dtsch. Arztebl. Int.* **2012**, *109*, 257–264. [CrossRef]
3. Kavanagh, N.; Ryan, E.J.; Widaa, A.; Sexton, G.; Fennell, J.; O'Rourke, S.; Cahill, K.C.; Kearney, C.J.; O'Brien, F.J.; Kerrigan, S.W. Staphylococcal Osteomyelitis: Disease Progression, Treatment Challenges, and Future Directions. *Clin. Microbiol. Rev.* **2018**, *31*, e00084-17. [CrossRef] [PubMed]
4. Seebach, E.; Kubatzky, K.F. Chronic Implant-Related Bone Infections-Can Immune Modulation be a Therapeutic Strategy? *Front. Immunol.* **2019**, *10*, 1724. [CrossRef] [PubMed]
5. Masters, E.A.; Trombetta, R.P.; de Mesy Bentley, K.L.; Boyce, B.F.; Gill, A.L.; Gill, S.R.; Nishitani, K.; Ishikawa, M.; Morita, Y.; Ito, H.; et al. Evolving concepts in bone infection: Redefining "biofilm", "acute vs. chronic osteomyelitis", "the immune proteome" and "local antibiotic therapy". *Bone Res.* **2019**, *7*, 20. [CrossRef]
6. Pincher, B.; Fenton, C.; Jeyapalan, R.; Barlow, G.; Sharma, H.K. A systematic review of the single-stage treatment of chronic osteomyelitis. *J. Orthop. Surg. Res.* **2019**, *14*, 393. [CrossRef]
7. Walenkamp, G.H.; Vree, T.B.; van Rens, T.J. Gentamicin-PMMA beads. Pharmacokinetic and nephrotoxicological study. *Clin. Orthop. Relat. Res.* **1986**, *205*, 171–183. [CrossRef]
8. Walenkamp, G.H.; Kleijn, L.L.; de Leeuw, M. Osteomyelitis treated with gentamicin-PMMA beads: 100 patients followed for 1–12 years. *Acta. Orthop. Scand.* **1998**, *69*, 518–522. [CrossRef]
9. Patel, K.H.; Bhat, S.N.; Mamatha, H. Outcome analysis of antibiotic-loaded poly methyl methacrylate (PMMA) beads in musculoskeletal infections. *J. Taibah. Univ. Med. Sci.* **2021**, *16*, 177–183. [CrossRef]
10. van Vugt, T.A.G.; Arts, J.J.; Geurts, J.A.P. Antibiotic-Loaded Polymethylmethacrylate Beads and Spacers in Treatment of Orthopedic Infections and the Role of Biofilm Formation. *Front. Microbiol.* **2019**, *10*, 1626. [CrossRef]
11. Makarov, C.; Cohen, V.; Raz-Pasteur, A.; Gotman, I. In vitro elution of vancomycin from biodegradable osteoconductive calcium phosphate-polycaprolactone composite beads for treatment of osteomyelitis. *Eur. J. Pharm. Sci.* **2014**, *62*, 49–56. [CrossRef] [PubMed]
12. Gallarate, M.; Chirio, D.; Chindamo, G.; Peira, E.; Sapino, S. Osteomyelitis: Focus on Conventional Treatments and Innovative Drug Delivery Systems. *Curr. Drug Deliv.* **2021**, *18*, 532–545. [CrossRef] [PubMed]

13. Wilhelm, M.P. Vancomycin. *Mayo. Clin. Proc.* **1991**, *66*, 1165–1170. [CrossRef] [PubMed]

14. Chung, J.H.Y.; Kade, J.C.; Jeiranikhameneh, A.; Ruberu, K.; Mukherjee, P.; Yue, Z.; Wallace, G.G. 3D hybrid printing platform for auricular cartilage reconstruction. *Biomed. Phys. Eng. Express* **2020**, *6*, 035003. [CrossRef]

15. Nyberg, E.; Rindone, A.; Dorafshar, A.; Grayson, W.L. Comparison of 3D-Printed Poly-varepsilon-Caprolactone Scaffolds Functionalized with Tricalcium Phosphate, Hydroxyapatite, Bio-Oss, or Decellularized Bone Matrix. *Tissue Eng. Part A* **2017**, *23*, 503–514. [CrossRef] [PubMed]

16. Zhang, Z.Z.; Wang, S.J.; Zhang, J.Y.; Jiang, W.B.; Huang, A.B.; Qi, Y.S.; Ding, J.X.; Chen, X.S.; Jiang, D.; Yu, J.K. 3D-Printed Poly(epsilon-caprolactone) Scaffold Augmented with Mesenchymal Stem Cells for Total Meniscal Substitution: A 12- and 24-Week Animal Study in a Rabbit Model. *Am. J. Sports Med.* **2017**, *45*, 1497–1511. [CrossRef]

17. Siddiqui, N.; Asawa, S.; Birru, B.; Baadhe, R.; Rao, S. PCL-Based Composite Scaffold Matrices for Tissue Engineering Applications. *Mol. Biotechnol.* **2018**, *60*, 506–532. [CrossRef]

18. Lopez-Gonzalez, I.; Zamora-Ledezma, C.; Sanchez-Lorencio, M.I.; Tristante Barrenechea, E.; Gabaldon-Hernandez, J.A.; Meseguer-Olmo, L. Modifications in Gene Expression in the Process of Osteoblastic Differentiation of Multipotent Bone Marrow-Derived Human Mesenchymal Stem Cells Induced by a Novel Osteoinductive Porous Medical-Grade 3D-Printed Poly(epsilon-caprolactone)/beta-tricalcium Phosphate Composite. *Int. J. Mol. Sci.* **2021**, *22*, 1216. [CrossRef]

19. Ragaert, K.; De Somer, F.; Van de Velde, S.; Degrieck, J.; Cardon, L. Methods for Improved Flexural Mechanical Properties of 3D-Plotted PCL-Based Scaffolds for Heart Valve Tissue Engineering. *Stroj. Vestn.-J. Mech. Eng.* **2013**, *59*, 669–676. [CrossRef]

20. Atala, A.; Yoo, J.J. *Essentials of 3D Biofabrication and Translation*; Elsevier: Amsterdam, The Netherlands, 2015.

21. Yeo, A.; Wong, W.J.; Khoo, H.H.; Teoh, S.H. Surface modification of PCL-TCP scaffolds improve interfacial mechanical interlock and enhance early bone formation: An in vitro and in vivo characterization. *J. Biomed. Mater. Res. A* **2010**, *92*, 311–321. [CrossRef]

22. Mirtaghi, S.M.; Hassannia, H.; Mahdavi, M.; Hosseini-Khah, Z.; Mellati, A.; Enderami, S.E. A novel hybrid polymer of PCL/fish gelatin nanofibrous scaffold improves proliferation and differentiation of Wharton's jelly-derived mesenchymal cells into islet-like cells. *Artif. Organs* **2022**, *46*, 1491–1503. [CrossRef]

23. Merk, M.; Chirikian, O.; Adlhart, C. 3D PCL/Gelatin/Genipin Nanofiber Sponge as Scaffold for Regenerative Medicine. *Materials* **2021**, *14*, 2006. [CrossRef]

24. Jang, C.H.; Kim, M.S.; Cho, Y.B.; Jang, Y.S.; Kim, G.H. Mastoid obliteration using 3D PCL scaffold in combination with alginate and rhBMP-2. *Int. J. Biol. Macromol.* **2013**, *62*, 614–622. [CrossRef] [PubMed]

25. Do, N.H.N.; Truong, Q.T.; Le, P.K.; Ha, A.C. Recent developments in chitosan hydrogels carrying natural bioactive compounds. *Carbohydr. Polym.* **2022**, *294*, 119726. [CrossRef]

26. Dang, P.A.; Palomino-Durand, C.; Elsafi Mabrouk, M.; Marquaille, P.; Odier, C.; Norvez, S.; Pauthe, E.; Corte, L. Rational formulation design of injectable thermosensitive chitosan-based hydrogels for cell encapsulation and delivery. *Carbohydr. Polym.* **2022**, *277*, 118836. [CrossRef]

27. Catoira, M.C.; Fusaro, L.; Di Francesco, D.; Ramella, M.; Boccafoschi, F. Overview of natural hydrogels for regenerative medicine applications. *J. Mater. Sci. Mater. Med.* **2019**, *30*, 115. [CrossRef]

28. Hamedi, H.; Moradi, S.; Hudson, S.M.; Tonelli, A.E. Chitosan based hydrogels and their applications for drug delivery in wound dressings: A review. *Carbohydr. Polym.* **2018**, *199*, 445–460. [CrossRef]

29. Nolan, K.; Millet, Y.; Ricordi, C.; Stabler, C.L. Tissue engineering and biomaterials in regenerative medicine. *Cell Transplant.* **2008**, *17*, 241–243. [CrossRef] [PubMed]

30. Domalik-Pyzik, P.; Chłopek, J.; Pielichowska, K. Chitosan-Based Hydrogels: Preparation, Properties, and Applications. In *Cellulose-Based Superabsorbent Hydrogels*; Polymers and Polymeric Composites: A Reference Series; Springer: Berlin/Heidelberg, Germany, 2018; pp. 1–29.

31. Seo, J.W.; Shin, S.R.; Lee, M.Y.; Cha, J.M.; Min, K.H.; Lee, S.C.; Shin, S.Y.; Bae, H. Injectable hydrogel derived from chitosan with tunable mechanical properties via hybrid-crosslinking system. *Carbohydr. Polym.* **2021**, *251*, 117036. [CrossRef] [PubMed]

32. Hoffman, A.S. Hydrogels for biomedical applications. *Adv. Drug Deliv. Rev.* **2012**, *64*, 18–23. [CrossRef]

33. Li, P.; Fu, L.; Liao, Z.; Peng, Y.; Ning, C.; Gao, C.; Zhang, D.; Sui, X.; Lin, Y.; Liu, S.; et al. Chitosan hydrogel/3D-printed poly(epsilon-caprolactone) hybrid scaffold containing synovial mesenchymal stem cells for cartilage regeneration based on tetrahedral framework nucleic acid recruitment. *Biomaterials* **2021**, *278*, 121131. [CrossRef]

34. Osman, M.A.; Virgilio, N.; Rouabhia, M.; Mighri, F. Development and Characterization of Functional Polylactic Acid/Chitosan Porous Scaffolds for Bone Tissue Engineering. *Polymers* **2022**, *14*, 5079. [CrossRef]

35. Zhang, L.; Yang, G.; Johnson, B.N.; Jia, X. Three-dimensional (3D) printed scaffold and material selection for bone repair. *Acta Biomater.* **2019**, *84*, 16–33. [CrossRef]

36. Carette, X.; Mincheva, R.; Herbin, M.; Noirfalise, X.; Nguyen, T.C.; Leclere, P.; Godfroid, T.; Kerdjoudj, H.; Jolois, O.; Boudhifa, M.; et al. Atmospheric plasma: A simple way of improving the interface between natural polysaccharides and polyesters. *IOP Conf. Ser. Mater. Sci. Eng.* **2021**, *1056*, 012005. [CrossRef]

37. Wang, R.; Shen, Y.; Zhang, C.; Yan, P.; Shao, T. Comparison between helium and argon plasma jets on improving the hydrophilic property of PMMA surface. *Appl. Surf. Sci.* **2016**, *367*, 401–406. [CrossRef]

38. Robert, E.; Barbosa, E.; Dozias, S.; Vandamme, M.; Cachoncinlle, C.; Viladrosa, R.; Pouvesle, J.M. Experimental Study of a Compact Nanosecond Plasma Gun. *Plasma Process. Polym.* **2009**, *6*, 795–798. [CrossRef]

39. EPCOS AG. *CeraPlas® HF Series, Piezoelectric Based Plasma Generator*; Data Sheet; EPCOS AG: Munich, Germany, 2018.
40. Hu, Z.; Das, S.K.; Yan, S.; You, R.; Li, X.; Luo, Z.; Li, M.; Zhang, Q.; Kaplan, D.L. Stability and biodegradation of silk fibroin/hyaluronic acid nerve conduits. *Compos. Part B Eng.* **2020**, *200*, 108222. [CrossRef]
41. Jannah Mohd Sebri, N.; Anuar Mat Amin, K. Composite Film of Chitosan Loaded Norfloxacin with Improved Flexibility and Antibacterial Activity for Wound Dressing Application. *Orient. J. Chem.* **2017**, *33*, 628–636. [CrossRef]
42. De Aza, P.N.; Garcia-Bernal, D.; Cragnolini, F.; Velasquez, P.; Meseguer-Olmo, L. The effects of Ca_2SiO_4-$Ca_3(PO_4)_2$ ceramics on adult human mesenchymal stem cell viability, adhesion, proliferation, differentiation and function. *Mater. Sci. Eng. C Mater. Biol. Appl.* **2013**, *33*, 4009–4020. [CrossRef]
43. Rabadan-Ros, R.; Revilla-Nuin, B.; Mazón, P.; Aznar-Cervantes, S.; Ros-Tarraga, P.; De Aza, P.N.; Meseguer-Olmo, L. Impact of a Porous Si-Ca-P Monophasic Ceramic on Variation of Osteogenesis-Related Gene Expression of Adult Human Mesenchymal Stem Cells. *Appl. Sci.* **2018**, *8*, 46. [CrossRef]
44. Dominici, M.; Le Blanc, K.; Mueller, I.; Slaper-Cortenbach, I.; Marini, F.; Krause, D.; Deans, R.; Keating, A.; Prockop, D.; Horwitz, E. Minimal criteria for defining multipotent mesenchymal stromal cells. The International Society for Cellular Therapy position statement. *Cytotherapy* **2006**, *8*, 315–317. [CrossRef]
45. Horwitz, E.M.; Le Blanc, K.; Dominici, M.; Mueller, I.; Slaper-Cortenbach, I.; Marini, F.C.; Deans, R.J.; Krause, D.S.; Keating, A.; International Society for Cellular, T. Clarification of the nomenclature for MSC: The International Society for Cellular Therapy position statement. *Cytotherapy* **2005**, *7*, 393–395. [CrossRef] [PubMed]
46. Liu, X.; Ma, P.X. Polymeric scaffolds for bone tissue engineering. *Ann. Biomed. Eng.* **2004**, *32*, 477–486. [CrossRef] [PubMed]
47. Raeisdasteh Hokmabad, V.; Davaran, S.; Ramazani, A.; Salehi, R. Design and fabrication of porous biodegradable scaffolds: A strategy for tissue engineering. *J. Biomater. Sci. Polym. Ed.* **2017**, *28*, 1797–1825. [CrossRef]
48. Neuendorf, R.E.; Saiz, E.; Tomsia, A.P.; Ritchie, R.O. Adhesion between biodegradable polymers and hydroxyapatite: Relevance to synthetic bone-like materials and tissue engineering scaffolds. *Acta Biomater.* **2008**, *4*, 1288–1296. [CrossRef] [PubMed]
49. Jennings, J.A. Controlling chitosan degradation properties in vitro and in vivo. In *Chitosan Based Biomaterials Volume 1*; Woodhead Publishing: Sawston, UK, 2017; pp. 159–182.
50. Lim, S.M.; Song, D.K.; Oh, S.H.; Lee-Yoon, D.S.; Bae, E.H.; Lee, J.H. In vitro and in vivo degradation behavior of acetylated chitosan porous beads. *J. Biomater. Sci. Polym. Ed.* **2008**, *19*, 453–466. [CrossRef]
51. Lopez-Iglesias, C.; Barros, J.; Ardao, I.; Monteiro, F.J.; Alvarez-Lorenzo, C.; Gomez-Amoza, J.L.; Garcia-Gonzalez, C.A. Vancomycin-loaded chitosan aerogel particles for chronic wound applications. *Carbohydr. Polym.* **2019**, *204*, 223–231. [CrossRef] [PubMed]
52. Kausar, R.; Khan, A.U.; Jamil, B.; Shahzad, Y.; Ul-Haq, I. Development and pharmacological evaluation of vancomycin loaded chitosan films. *Carbohydr. Polym.* **2021**, *256*, 117565. [CrossRef]
53. Hadder, B.; Dexter, F.; Robinson, A.D.M.; Loftus, R.W. Molecular characterisation and epidemiology of transmission of intraoperative Staphylococcus aureus isolates stratified by vancomycin minimum inhibitory concentration (MIC). *Infect. Prev. Pract.* **2022**, *4*, 100249. [CrossRef]
54. Peixoto, P.B.; Massinhani, F.H.; Netto Dos Santos, K.R.; Chamon, R.C.; Silva, R.B.; Lopes Correa, F.E.; Barata Oliveira, C.; Oliveira, A.G. Methicillin-resistant Staphylococcus epidermidis isolates with reduced vancomycin susceptibility from bloodstream infections in a neonatal intensive care unit. *J. Med. Microbiol.* **2020**, *69*, 41–45. [CrossRef]
55. Chang, S.H.; Lin, H.T.; Wu, G.J.; Tsai, G.J. pH Effects on solubility, zeta potential, and correlation between antibacterial activity and molecular weight of chitosan. *Carbohydr. Polym.* **2015**, *134*, 74–81. [CrossRef]
56. Garcia-Gonzalez, C.A.; Barros, J.; Rey-Rico, A.; Redondo, P.; Gomez-Amoza, J.L.; Concheiro, A.; Alvarez-Lorenzo, C.; Monteiro, F.J. Antimicrobial Properties and Osteogenicity of Vancomycin-Loaded Synthetic Scaffolds Obtained by Supercritical Foaming. *ACS Appl. Mater. Interfaces* **2018**, *10*, 3349–3360. [CrossRef] [PubMed]
57. Baranwal, A.; Kumar, A.; Priyadharshini, A.; Oggu, G.S.; Bhatnagar, I.; Srivastava, A.; Chandra, P. Chitosan: An undisputed bio-fabrication material for tissue engineering and bio-sensing applications. *Int. J. Biol. Macromol.* **2018**, *110*, 110–123. [CrossRef] [PubMed]
58. Backes, E.H.; Harb, S.V.; Beatrice, C.A.G.; Shimomura, K.M.B.; Passador, F.R.; Costa, L.C.; Pessan, L.A. Polycaprolactone usage in additive manufacturing strategies for tissue engineering applications: A review. *J. Biomed. Mater. Res. B Appl. Biomater.* **2022**, *110*, 1479–1503. [CrossRef]
59. Ferreira, M.O.G.; Leite, L.L.R.; de Lima, I.S.; Barreto, H.M.; Nunes, L.C.C.; Ribeiro, A.B.; Osajima, J.A.; da Silva Filho, E.C. Chitosan Hydrogel in combination with Nerolidol for healing wounds. *Carbohydr. Polym.* **2016**, *152*, 409–418. [CrossRef]
60. Tardajos, M.G.; Cama, G.; Dash, M.; Misseeuw, L.; Gheysens, T.; Gorzelanny, C.; Coenye, T.; Dubruel, P. Chitosan functionalized poly-epsilon-caprolactone electrospun fibers and 3D printed scaffolds as antibacterial materials for tissue engineering applications. *Carbohydr. Polym.* **2018**, *191*, 127–135. [CrossRef]
61. Zhao, X.; Zhou, L.; Li, Q.; Zou, Q.; Du, C. Biomimetic mineralization of carboxymethyl chitosan nanofibers with improved osteogenic activity in vitro and in vivo. *Carbohydr. Polym.* **2018**, *195*, 225–234. [CrossRef]

62. Li, L.; Yu, M.; Li, Y.; Li, Q.; Yang, H.; Zheng, M.; Han, Y.; Lu, D.; Lu, S.; Gui, L. Synergistic anti-inflammatory and osteogenic n-HA/resveratrol/chitosan composite microspheres for osteoporotic bone regeneration. *Bioact. Mater.* **2021**, *6*, 1255–1266. [CrossRef]

63. Cai, B.; Zou, Q.; Zuo, Y.; Li, L.; Yang, B.; Li, Y. Fabrication and cell viability of injectable n-HA/chitosan composite microspheres for bone tissue engineering. *RSC Adv.* **2016**, *6*, 85735–85744. [CrossRef]

Article

Molecular Interactions between APIs and Enteric Polymeric Excipients in Solid Dispersion: Insights from Molecular Simulations and Experiments

Krishna M. Gupta [1,*], Xavier Chin [1] and Parijat Kanaujia [1,2,*]

[1] Institute of Sustainability for Chemicals, Energy and Environment (ISCE2), Agency for Science, Technology and Research (A*STAR), 1 Pesek Road, Jurong Island, Singapore 627833, Singapore
[2] Department of Pharmacy, National University of Singapore, 18 Science Drive 4, Singapore 117559, Singapore
* Correspondence: krishna_mohan_gupta@isce2.a-star.edu.sg (K.M.G.); parijat_kanaujia@isce2.a-star.edu.sg (P.K.)

Abstract: Solid dispersion of poorly soluble APIs is known to be a promising strategy to improve dissolution and oral bioavailability. To facilitate the development and commercialization of a successful solid dispersion formulation, understanding of intermolecular interactions between APIs and polymeric carriers is essential. In this work, first, we assessed the molecular interactions between various delayed-release APIs and polymeric excipients using molecular dynamics (MD) simulations, and then we formulated API solid dispersions using a hot melt extrusion (HME) technique. To assess the potential API–polymer pairs, three quantities were evaluated: (a) interaction energy between API and polymer [electrostatic (E_{coul}), Lenard-Jones (E_{LJ}), and total (E_{total})], (b) energy ratio (API–polymer/API–API), and (c) hydrogen bonding between API and polymer. The E_{total} quantities corresponding to the best pairs: NPX-Eudragit L100, NaDLO–HPMC(P), DMF–HPMC(AS) and OPZ–HPMC(AS) were −143.38, −348.04, −110.42, and −269.43 kJ/mol, respectively. Using a HME experimental technique, few API–polymer pairs were successfully extruded. These extruded solid forms did not release APIs in a simulated gastric fluid (SGF) pH 1.2 environment but released them in a simulated intestinal fluid (SIF) pH 6.8 environment. The study demonstrates the compatibility between APIs and excipients, and finally suggests a potential polymeric excipient for each delayed-release API, which could facilitate the development of the solid dispersion of poorly soluble APIs for dissolution and bioavailability enhancement.

Keywords: molecular dynamics simulation; interaction energy; hydrogen bonding; solid dispersion; hot melt extrusion; amorphous formulation

Citation: Gupta, K.M.; Chin, X.; Kanaujia, P. Molecular Interactions between APIs and Enteric Polymeric Excipients in Solid Dispersion: Insights from Molecular Simulations and Experiments. *Pharmaceutics* **2023**, *15*, 1164. https://doi.org/10.3390/pharmaceutics15041164

Academic Editor: Ana Isabel Fernandes

Received: 24 February 2023
Revised: 30 March 2023
Accepted: 4 April 2023
Published: 6 April 2023

1. Introduction

One of the most important challenges in the delivery of active pharmaceutical ingredients (APIs) faced by formulation scientists is how to counter the low aqueous solubility of APIs [1–3]. Amorphous solid dispersion formulations of APIs with excipients is considered to be a promising strategy to improve the oral bioavailability of poorly soluble APIs [4]. Consequently, solid dispersions of APIs in excipients (water-soluble polymeric carriers), which enhance aqueous solubility over the crystalline counterpart, have been widely applied in pharmaceutical formulations. Molecules in amorphous solids exist at higher energy than those in a crystalline state; therefore, the energy penalty required to dissociate these molecules is lower, resulting in a higher solubility than a crystalline form [5].

To develop the amorphous solid dispersion of APIs, a hot melt extrusion (HME) technique has emerged as a potent processing technology and several commercial HME products are available on the market or are under late-stage development [6,7]. In HME, a mixture of drug, excipient and plasticizer, if required, is heated at high temperatures (below the melting point of API) and intensively mixed using a twin-screw extruder to yield a homogeneous product. On the other hand, when using HME, an API is dispersed into a polymer matrix to produce solid dispersions with improved bioavailability of poorly soluble drugs [8,9]. The other method of solid dispersion preparation is the common solvent method, in which both drug and carrier are dissolved in a common solvent and then the solvent is evaporated by spray drying [10] or freeze drying to obtain the solid dispersion product [11]. Compared to the traditional milling or solvent-based methods, HME technology has received significant attention for solid dispersion formulations owing to various advantages such as fewer processing steps, decreased processing time, continuous operation, solvent-free operation, superior mixing capabilities, and potential for automation [12,13]. Products developed using a hot melt extrusion process have been approved by regulatory agencies worldwide for human use [6]. Some of the marketed products include Kaletra®, Norvir®, FenoglideTM, Verapamil and Posaconazole.

A major challenge for the development of amorphous solid dispersion formulations occurs due to the amorphous nature of APIs as they are thermodynamically unstable and tend to recrystallize [14]. To mitigate this issue, polymeric additives are usually added as excipients in the formulation; these additives prevent the recrystallization process and hence improve physical stability [15]. A wide range of polymeric excipients have been commercially utilized including, but not limited to, hydroxypropyl methylcellulose (HPMC), hydroxypropyl methylcellulose phthalate [HPMC(P)], hydroxypropyl methylcellulose acetate succinate [HPMC(AS)], hydroxypropyl cellulose (HPC), polyvinylpyrrolidone (PVP), polyvinylpyrrolidone-vinyl acetate copolymer (PVP-VA), and polyethylene glycol (PEG) [16]. Among the employed excipients, cellulosic polymers are known to be superior in the inhibition of API crystallization [17,18]. Owing to three substitution positions on each D-glucose monomeric unit of cellulose polymer, a large degree of freedom exists regarding the design of new candidates in terms of both the substitution patterns and the degree of substitutions. Nonetheless, this presents a technological challenge in designing a polymer with optimal properties for a given API [19].

For the successful development of a solid dispersion formulation, understanding of the intermolecular interactions between APIs and polymeric carriers is crucial [20,21]. In particular, molecular interactions between various API molecules and polymeric excipients correlating their compatibility/miscibility are essential for the rational design and screening of formulation systems [19,22–25]. From this perspective, Meng et al. highlighted the role of molecular interactions between a poorly soluble drug (curcumin) and various hydrophilic polymers such as PVP, Eudragit EPO (EPO), HPMC and PEG in the successful formulation of solid dispersion using Fourier transform infrared and Raman spectroscopy [26]. Various experimental methods such as glass transition temperature (Tg), Raman mapping, X-ray diffraction data, solid state nuclear magnetic resonance (NMR) spectroscopy, and atomic force microscopy (AFM) have been used to qualitatively examine drug–polymer miscibility [27]. Recently, Lu et al. have investigated molecular interaction between posaconazole and HPMCAS polymer in amorphous solid dispersions using 19F magic angle spinning (MAS) nuclear magnetic resonance (NMR) techniques [28].

In addition to experimental explorations, molecular dynamics (MD) simulation has been widely employed as a promising tool to determine API-polymer miscibility in solid dispersions as well as formulation design [29–31]. For instance, Yani et al. have performed MD simulations to predict the miscibility of API in various ionic and non-ionic polymeric excipients for solid dispersion systems by evaluating Hansen's solubility parameter, hydrogen-bonding interaction energy and hydrogen-bond lifetime analysis [25]. By combining experimental and MD simulation techniques, Gong and co-workers have investigated the state evolution of norfloxacin (NFX) in solid dispersions with three commonly used excipients, namely, PVP, HPMC, and HPMC(P). It was demonstrated that conversion from an amorphous NFX to a hydrated state is possibly due to the dominating self-protonation of NFX over other interactions (NFX—polymer hydrogen bonding or ionic interaction); however, enhanced NFX—NFX aggregation leads to conversion to an anhydrous crystalline state [32]. Very recently, by integrating experimental and MD simulation approaches,

Kabedev et al. evaluated the underlying mechanism of β-lactoglobulin stability in solid dispersions of indomethacin [33]. Although the pace of growth in the understanding of drug—excipient interaction towards stabilizing amorphous solid dispersions of poorly water-soluble drugs has been rather encouraging, understanding of molecular interactions between delayed-release APIs and polymers governing API formulation and dissolution is largely elusive and hinders the development of solid dispersion in pharmaceutical applications.

Our work involved using both MD simulation and experimental techniques to gain a better understanding of how delayed-release drug APIs interact with enteric polymeric excipients at the molecular level. This knowledge can help speed up the development of solid dispersions for delayed-release applications. It is worthwhile to note that the selected APIs are available in enteric-coated dosage form on the market; therefore, they are the ideal candidates for solid dispersion formulations. Precisely, first the compatibility between APIs and polymeric excipients was investigated by accessing molecular interactions using molecular dynamics (MD) simulations, which were later used to identify the potential API-polymeric pairs for solid dispersions. Then, solid dispersions of the suggested API-excipient pairs were prepared using HME to gauge the feasibility of the pairs, followed by release studies in both simulated gastric fluid (SGF) with a pH of 1.2 and simulated intestinal fluid (SIF) with a pH of 6.8.

2. Materials and Methods

2.1. Simulation Models and Methods

For each API, first, MD simulations were carried out to identify the potential polymeric excipients suitable for pairing. To do so, three commonly used polymeric excipients, namely, hydroxypropyl methylcellulose phthalate [HPMC(P)], hydroxypropyl methylcellulose acetate succinate [HPMC(AS)], and Eudragit L100 were chosen. Figure 1 depicts the molecular structures of drug APIs, namely, naproxen (NPX), diclofenac sodium (NaDLO), dimethyl fumarate (DMF), and omeprazole (OPZ). For each polymer, Figure 2 represents the molecular structure of a polymeric chain with 10 monomer units. Similar to our previous work [34], we first constructed a repeat unit of each polymer, and then created a polymer chain with 10 monomer units using the *Polymer Builder* module in Materials Studio [35]. These polymer chains were geometrically optimized using the *forcite* module in Materials Studio. After optimizing the structures, input parameter files were generated for the MD simulations. The optimized potentials for liquid simulations all-atom (OPLS-AA) force field was used to describe the parameters for both APIs and polymeric structures [36]. Using the MKTOP tool, parameter files were created for most structures, and the TPP-MKTOP tool was used for a few structures [37,38]. The atomic charges were adjusted based on the OPLS-AA force field. The Lennard-Jones (LJ) and Coulombic (Coul.) potentials were used to describe the non-bonded interactions.

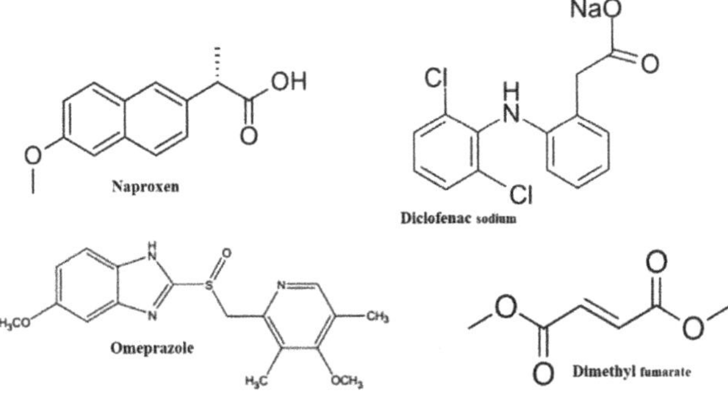

Figure 1. 2D structures of APIs.

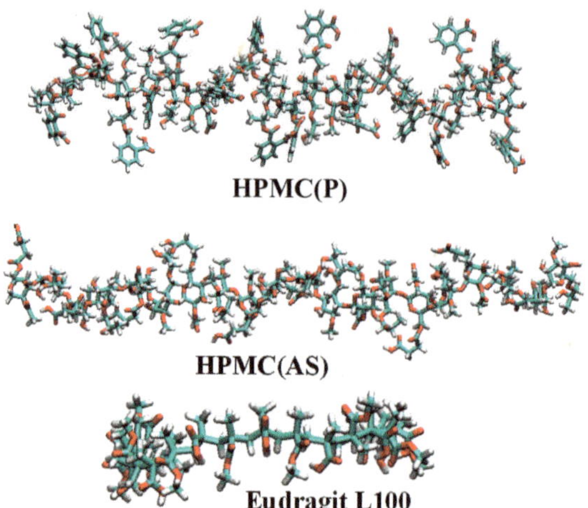

Figure 2. Atomic structures of polymer chains. Color code: O, red; N, blue; C, cyan; H, white.

MD simulations were conducted in four separate sets. In the first set, the crystal structure of each drug API (NPX or NaDLO or DMF or OPZ) was simulated for 5 ns using isothermal and isobaric (NPT) MD simulation. This set of simulations was performed to verify the force field used herein. In the second set, a cubic simulation box (~6 nm each side) was built for each polymer excipient [HPMC(P) or HPMC(AS) or Eudragit L100]. Then, 10 API molecules of each drug were ramdomly added into the simulation boxes, resulting in 12 different simulation systems. The number of polymer chains was added in such a way that all systems maintained 5 wt% API and 95 wt% polymer excipients. For each system, 10 ns isothermal and isochoric (NVT) MD simulations were performed at 300 K, and the last 8 ns trajectory was used for analysis. A representative simulation snapshot for this type of system is shown in Figure 3. The 2nd set was aimed at screening and identifying a suitable polymeric excipient for each drug API.

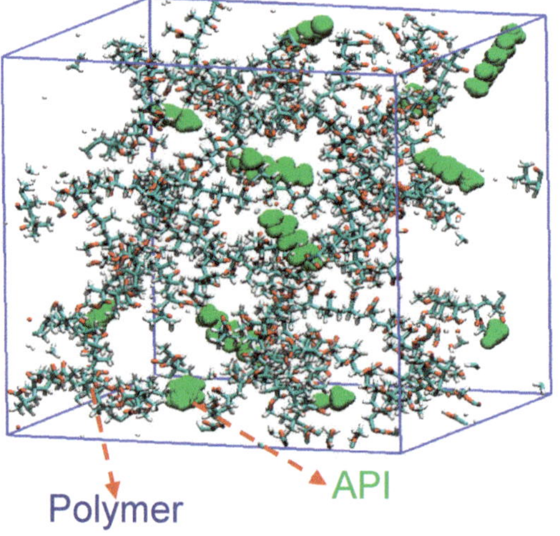

Figure 3. A representative simulation snapshot for the mixture of API molecules and polymeric excipient. The API molecules are shown in green.

Experiments were conducted to assess the feasibility of mixing the API and polymeric excipient in a solid state once appropriate pairs of drug APIs and polymeric excipients were identified through MD simulations. The experimental procedures are elaborated later in this section. Afterwards, simulation systems were built for feasible API and excipient pairs in the third set, with a different API loading of 50 wt%. Similarly to the 2nd set, for this set, each system was simulated for 10 ns NVT MD simulations. The aim of this set was to analyze the effect of loading. In the final set, the simulations were performed at various temperatures.

The GROMACS v5.1.2 package was used to conduct all simulations [39]. The first step in all simulations was energy minimization using the steepest descent with a truncation force of 1000 kJ/mol.nm, followed by MD simulations. The Maxwell–Boltzmann distribution was used to generate initial velocities for the MD simulations. The velocity-rescale thermostat with a relaxation time of 0.1 ps was applied to control the temperature of the simulation systems. The particle-mesh Ewald summation method was adopted to calculate Coul interactions with a grid spacing of 1.2 Å. The LJ interactions were calculated using a cutoff of 14 Å. The leap-frog algorithm with a time step of 2 fs was used to integrate the equations of motion. In all three dimensions, periodic boundary conditions were enforced. Simulation snapshots were created in VMD (version 1.9.3) software using simulation trajectories [40].

2.2. Experimental Description

Following the simulation, API polymer mixtures corresponding to 5 wt% and 50 wt% were prepared by accurately weighing powdered API and polymers in a screw-cap bottle and mixing them homogenously using Powder mixer (Alphie, 2.5L 3D Powder Mixer, Hexagon Products, Por, Guj, India) at 50 RPM for 30 min. NaDLO and OPZ were purchased from Jai Radhe Sales India. DMF and NPX base were supplied by Sigma-Aldrich, St. Louis, MO, USA. HPMC AS (LF grade) and HPMC-P (55 grade) was kindly provided by Shine Etsu Chemical Co. (Tokyo, Japan). Eudragit L100-55 was provided by Evonik Industries (Darmstadt, Germany). High-performance liquid chromatography (HPLC) grade solvents were supplied by Fischer Scientific Pte. Ltd., (Pandan Cres., Singapore) and other reagents were supplied by Sigma, St. Louis, MO, USA and used as supplied. The solubilities, melting points, and degradation temperatures of the API and polymers are shown in Table S1. The API polymer physical mixture was fed manually to a preheated co-rotating twin-screw hot melt extruder (Prism Eurolab 16 Melt extruder from Thermo Scientific, Karlsruhe, Germany) rotating at 100 RPM. The temperature profile and screw speed were set to ensure transport, melting and mixing at the respective zones on the screw. The temperatures used in the different zones are shown below in Table 1.

Table 1. Temperature profile used for HME of various API polymer combinations.

Barrel Zone Temperature (°C)	1	2	3	4	5	6	Rod Die
NaDLO-HPMC(P)	Feeding	110	140	140	145	145	145
DMF-HPMC(AS)	Feeding	75	100	110	120	120	115
NPX-Eudragit L100	Feeding	130	200	215	220	220	215
OPZ-HPMC(AS)	Feeding	75	100	110	120	120	115

For extruding solid dispersions in the form of cylindrical strands, a rod die with a 2 mm orifice was employed. The extruded strands were gathered on a conveyor belt, cooled with air, and then kept in screw-capped glass bottles under low humidity conditions (25% relative humidity) at room temperature. The resulting extrudates were subjected to ball milling using a stainless-steel vessel with a 1.5-cm diameter stainless-steel ball, at a frequency of 30 Hz for 2 min, using the MM 200 Retsch GmbH equipment from

Haan, Germany. The milled extrudates were utilized for analyzing and characterizing the solid dispersions.

2.2.1. Powder X-ray Diffraction Analysis

A Bruker D8 Advance powder X-ray diffractometer, which utilizes Cu Kα radiation (λ = 1.540 60 Å), an acceleration voltage of 35 kV, and a current of 40 mA power, was used to collect Powder X-ray diffraction (PXRD) data. Samples were scanned using a continuous scanning mode in the 2θ range from 5° to 50°, with a scan rate of 5° min^{-1}.

2.2.2. FTIR Spectroscopy

A Frontier FT-IR/NIR Spectrometer (PIKE Technologies I, PerkinElmer) equipped with a mid-infrared (MIR) triglycine sulfate (TGS) detector was used to obtain FTIR transmission spectra of the APIs, physical mixtures, and extruded samples. A small quantity of powdered sample was secured on the sample holder and scanned over a range of 4000–650 cm^{-1} at a scan speed of 0.2 cm^{-1}S^{-1} with a spectral resolution of 4 cm^{-1}.

2.2.3. Dissolution Study

The extruded samples were broken as granules and passed through sieve no. 18 and retained in sieve no. 30 for dissolution studies. USP type 2 dissolution apparatus (Agilent Technologies) with baskets was employed for this study. All the samples were tested using the following sequence:

(1) In Simulated Gastric fluid (SGF) pH 1.2 for 2 h: The samples were placed in a basket and dilution was tested for 2 h in SGF pH 1.2 with sample collections at 15, 30, 60, 90 and 120 min. A 2 mL sample was withdrawn after each time interval and fresh media were replenished. The samples were filtered through a 0.45 μm syringe filter and analyzed by HPLC.

(2) In Simulated Intestinal fluid (SIF) pH 6.5 for 2 h: After 2 h in SGF, the dissolution medium was changed to SIF and the basket was lowered to run the same sample again. Samples were collected at 5, 10, 15, 30, 60, 90 and 120 min and 2 mL of fresh SIF was replenished after each collection. The samples were filtered through a 0.45 μm syringe filter and analyzed by HPLC.

For NaDLO solid dispersion dissolution, granules equivalent to 50 mg of NaDLO were used, whereas for DMF solid dispersion dissolution, granules equivalent to 120 mg DMF were used.

3. Results and Discussion

3.1. Force Field Validation of Drug API

To validate the force field of API molecules, first, the unit cell of each drug crystal was downloaded from the Cambridge Crystallographic Data Centre (CCDC) database. Then, the unit cell of each drug was extended in all directions to ensure that the dimensions in each direction were either close to or higher than 50 Å. To do so, NPX, NaDLO, DMF, and OPZ were extended to 4 × 9 × 7, 5 × 2 × 5, 14 × 10 × 7, and 5 × 5 × 5, respectively. After MD simulation, from the first set of simulations, densities and lattice parameters of all crystals were estimated and compared with the literature, as tabulated in Table 2. For all crystal structures, the simulated densities and lattice parameters show <3% deviation from the experimental values, which reflects the reliability of the force field adopted for the simulation.

Table 2. Simulated densities and lattice parameters of drug crystals. The experimental values are mentioned in ().

API	Temp (K)	Density (g/cc)	Lattice Parameters						Ref.
			a (Å)	b (Å)	c (Å)	α (°)	β (°)	γ (°)	
Naproxen	300 (283–303)	1.27 (1.25)	13.30 (13.38)	5.76 (5.79)	7.87 (7.91)	90.0 (90.0)	93.9 (93.9)	90.0 (90.0)	[41]
Diclofenac sodium	150 (150)	1.48 (1.44)	9.48 (9.55)	39.19 (39.49)	9.77 (9.84)	90.0 (90.0)	90.7 (90.7)	90.0 (90.0)	[42]
Dimethyl fumarate	150 (150)	1.43 (1.43)	3.87 (3.87)	5.64 (5.64)	8.36 (8.36)	100.8 (100.8)	100.3 (100.3)	105.7 (105.7)	[43]
Omeprazole	300 (283–303)	1.31 (1.33)	9.81 (9.70)	10.49 (10.29)	10.45 (10.62)	90.0 (90.4)	111.5 (112.1)	116.5 (115.9)	[44]

3.2. Evaluation of Drug-Polymer Pairs for Solid Dispersion Formulations

To identify a suitable polymeric excipient for a stabilized amorphous API in solid dispersion formulation, solubility measurement is crucial. Typically, the solubility of a molecule in a specific medium can be gauged based on the interaction energy between interacting molecules; in general, the higher the interaction energy (stronger interaction), the higher the solubility [45–47]. Recently, we estimated the interaction energies and energy ratios between actives and lipid excipients to accurately predict the active encapsulation tendency in excipients, while identifying the best active–excipient pairs for nanoparticle formulations [48]. Similarly, to determine the most suitable polymer excipient for each API, the study employed MD simulation trajectory to estimate interaction energies and energy ratios. A negative energy value typically indicates an attractive interaction, with a higher absolute value indicating a stronger interaction and greater compatibility. Conversely, a positive value indicates a repulsive or unfavorable interaction. Figure 4 illustrates the estimated interaction energies, including electrostatic (E_{coul}), Lenard-Jones (E_{LJ}), and total (E_{total}), between drug API molecules and polymer excipients, which were determined in the second set of simulations.

The interaction energies between NPX and polymeric excipients were negative and were found to increase in the order of HPMC(AS) < HPMC(P) < Eudragit L100 (Figure 4a). This indicates that NPX exhibits favourable interactions with all polymeric excipients. Among the tested excipients, Eudragit L100 showed the highest interaction. The highest interaction between NPX and Eudragit L100 might be due to the presence of similar functional groups (-OCH3 and -COOH) in both (see Figures 1 and 2). Owing to the strongest interaction of NPX, Eudragit L100 was expected to show higher solubility. It is worthwhile to note that the E_{total} between NPX and polymeric excipient is dominated by van der Waals interactions (E_{LJ}). Similarly to NPX, NaDLO–excipient interactions were negative (Figure 4b). The E_{coul}, E_{LJ}, and E_{total} increased in the order of Eudragit L100 < HPMC(AS) < HPMC(P). The E_{total} between NaDLO and HPMC(P) was the highest, and thus is expected to be a potential choice. The possible reason for the highest interaction between NaDLO and HPMC(P) is the favourable interaction between hexagonal aromatic rings that are present in both structures (see Figures 1 and 2). Owing to the strongest interaction of NaDLO, HPMC(P) was expected to show higher solubility. Interestingly, in contrast to NPX, the E_{total} between NaDLO and the polymeric excipient was dominated by electrostatic (E_{coul}) interactions due to the ionic nature of NaDLO.

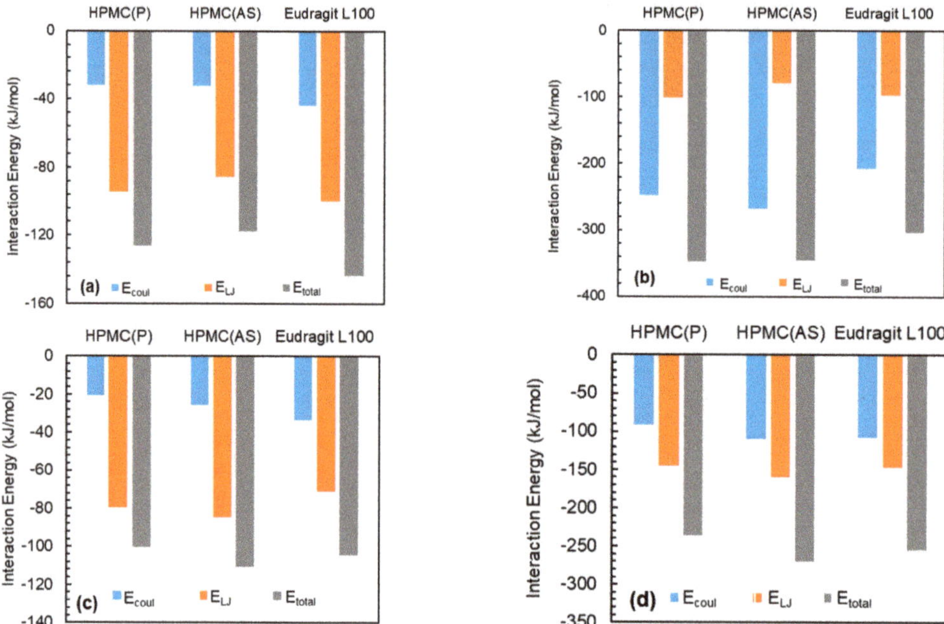

Figure 4. Interaction energies in terms of electrostatic (E_{coul}), Lenard–Jones (E_{LJ}), and total (E_{total}) between API molecules and polymer excipients for (**a**) NPX, (**b**) NaDLO, (**c**) DMF, and (**d**) OPZ.

Similarly to NPX and NaDLO, DMF–excipient and OPZ–excipient interactions were negative (Figure 4c,d), indicating favorable interactions among them. For DMF, the E_{coul}, E_{LJ}, and E_{total} increased in the order of Eudragit L100 < HPMC(P) < HPMC(AS), whereas for OPZ, these quantities increased in the order of HPMC(P) < Eudragit L100 < HPMC(AS). By carefully observing the chemical structures, one can say that, with a linear chain in Eudragit L100, a higher interaction with DMF (also linear chain) is expected. Interestingly, this was not observed here, instead HPMC(AS) showed the highest interaction with both DMF and OPZ. To unveil this behaviour, we further evaluated the intra-atomic Eudragit L100-Eudragit L100 and HPMC(AS)–HPMC(AS) interactions by radial distribution function $g(r)$ as

$$g_{ij}(r) = \frac{N_{ij}(r, r + \Delta r)\, V}{4\pi r^2 \Delta r\, N_i N_j} \tag{1}$$

where r is the distance between atoms i and j, N_i and N_j are the numbers of atoms i and j, $N_{ij}(r, r + \Delta r)$ is the number of atoms j around i within a shell from r to $r + \Delta r$, respectively. Figure 5 shows the $g(r)$ of Eudragit L100 around Eudragit L100 and HPMC(AS) around HPMC(AS) in a mixture of these excipients with DMF and OPZ, based on all atoms. For DMF, as shown in Figure 5a, two prominent peaks are observed at $r \sim 0.96$ and 1.1 Å, indicating strong interaction between Eudragit L100 and Eudragit L100 as well as HPMC(AS) and HPMC(AS). However, Eudragit L100-Eudragit L100 interaction was stronger than HPMC(AS)-HPMC(AS), as reflected by a greater peak height in the former. Thus, owing to weaker intra-atomic HPMC(AS)–HPMC(AS) interaction, DMF is more accessible to HPMC(AS) compared to Eudragit L100 and thus has the highest interaction. Similarly, for OPZ (Figure 5b), two prominent peaks are observed at $r \sim 0.94$ and 1.08 Å, indicating strong interaction between Eudragit L100 and Eudragit L100 as well as HPMC(AS) and HPMC(AS). Due to weaker intra-atomic HPMC(AS)-HPMC(AS) interaction, OPZ is more accessible to HPMC(AS) compared to Eudragit L100, and thus shows the highest interaction. In brief, the potential choice for both DMF and OPZ is

HPMC(AS). Because both DMF and OPZ are neutral molecules, as is NPX, the E_{total} between DMF/OPZ and the polymeric excipient is dominated by the van der Waals interaction rather than the electrostatic interaction.

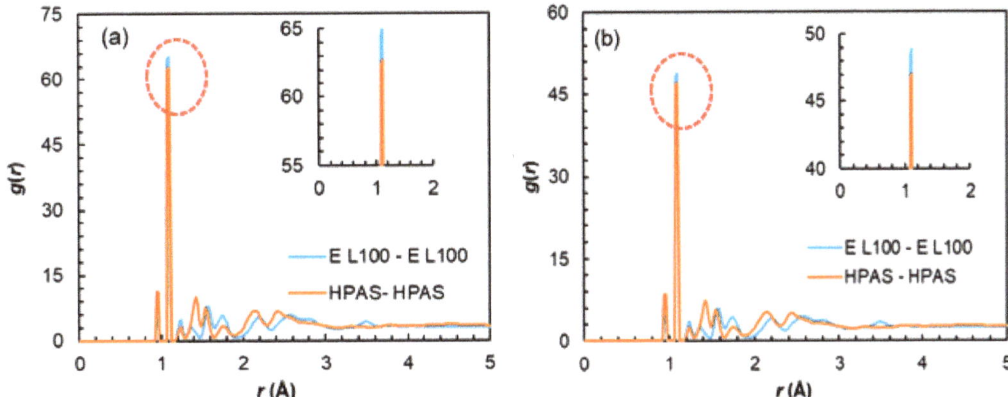

Figure 5. Radial distribution functions, $g(r)$ of Eudragit L100 around Eudragit L100 and HPMC(AS) around HPMC(AS) in a mixture of these excipients with (**a**) DMF and (**b**) OPZ. The E L100 and HPAS in the figure legend indicate Eudragit L100 and HPMC(AS), respectively. The insets provide a zoomed-in view of the red circled portion.

To better understand the relative interactions between drug API and polymer excipients, energy ratios (API–polymer/API–API) were estimated for all 12 systems (Figure 6). The energy ratios lie between 0 and 2.2 depending on the system. For NPX, the energy ratios increased in the order of HPMC(AS) < HPMC(P) < Eudragit L100. For NaDLO, the energy ratios increased in the order of Eudragit L100 < HPMC(AS) < HPMC(P). For DMF, the energy ratios increased in the order of HPMC(P) < Eudragit L100 < HPMC(AS), whereas for OPZ, these quantities increased in the order of HPMC(P) < Eudragit L100 < HPMC(AS). Consistent with the API–polymer interactions prediction, the energy ratios indicate that the best pairs among the examined combinations are NPX–Eudragit L100, NaDLO–HPMC(P), DMF–HPMC(AS), and OPZ–HPMC(AS). Further, we also estimated hydrogen bonds between API and polymer excipients as they play an important role in the solubility analysis [49]. Two geometrical criteria were implemented to calculate the hydrogen bonds: (1) the distance (r) between a donor and an acceptor ≤ 3.5 Å and (2) the angle of hydrogen-donor-acceptor, $\alpha \leq 30°$ [50]. Table S2 represents the number of hydrogen bonds between APIs and polymers per API on a molecular basis. All the APIs were observed to form hydrogen bonds with polymer excipients. Being ionic in nature, NaDLO showed a greater number of hydrogen bonds compared to the other APIs. Among non-ionic APIs, OPZ had a higher number of hydrogen bonds due to the presence of more oxygen atoms. To visualize the interaction patterns between APIs and polymers corresponding to the best pairs, Figure S1 shows MD simulation snapshots in NPX–Eudragit L100, NaDLO–HPMC(P), DMF–HPMC(AS), and OPZ–HPMC(AS). Following this, experiments were performed to analyze the feasibility of the suggested API-excipient pairs by simulations. Particularly, HME experiments were performed to prepare the solid dispersions of the recommended pairs, followed by dissolution experiments in an actual interstitial environment.

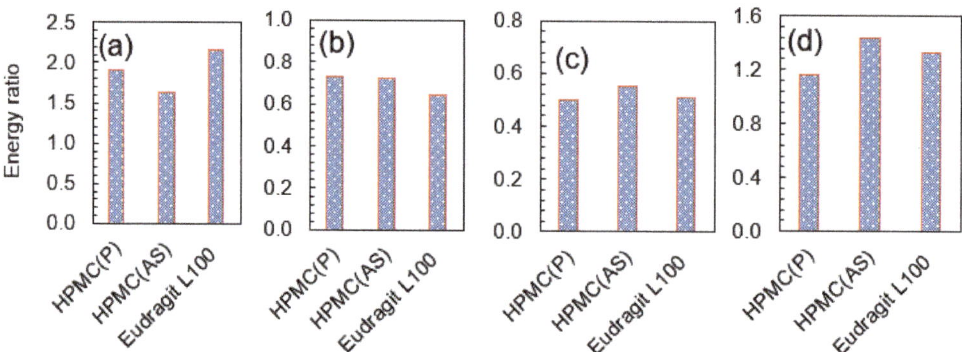

Figure 6. The energy ratios (**a**) NPX–polymer/NPX–NPX, (**b**) NaDLO–polymer/NaDLO–NaDLO, (**c**) DMF–polymer/DMF–DMF, and (**d**) OPZ–polymer/OPZ–OPZ.

3.3. Evaluation of Drug-Polymer Pairs by Experiment

3.3.1. Solid Dispersion by Hot Melt Extruder Experiments

The selected API polymer physical mixtures were melt-extruded under the conditions given in the Table 1. Extrudates of NaDLO and DMF containing 5% w/w API were glassy and transparent, whereas extrudates containing 50% w/w API were opaque. The charring of API was observed during the extrusion of NPX with Eudragit L-100. The degradation temperature of NPX has been reported to be 196 °C [51], which is higher than the extrusion temperature required to extrude Eudragit L100. The NPX was successfully extruded with Eudragit L100 by adding 10% triethyl citrate as a plasticizer to lower the extrusion temperature [52]. The OPZ degraded and formed a black product when extruded with HPMC-AS. Therefore, it was concluded that, among the suggested pairs, NaDLO–HPMC(P) and DMF–HPMC(AS) were successfully prepared. OPZ is an acid liable drug, and it is reported to degrade in the presence of enteric polymer in solution [53] and even in a solid state [54]. Sharma et al. studied the solid–state interactions between OPZ and various enteric polymers at the core–coat interface. In the acidic medium (including in the presence of enteric polymers), OPZ molecules rearrange and form a pyridinium salt, which binds selectively and irreversibly with the proton pump H+/K+-ATPase at the parietal cell secretory membrane [55]. To protect OPZ molecules from the enteric polymer, a sub-coating of neutral polymer such as HPMC and amylopectin was applied, which acts as barrier between OPZ and enteric polymer [56].

3.3.2. PXRD and FTIR

The PXRD of crystalline NaDLO exhibited peaks at 15.23, 19.96, 25.0, 25.90 and 27.16° at 2θ values. After melt extrusion with 50% w/w HPMC(P), the characteristic peaks of crystalline NaDLO were present, but the intensity of the peaks was reduced. The 5% w/w NaDLO solid dispersion with HPMC(P) produced an amorphous halo indicating the solubilization of APIs in polymer melt forming solid solution [57] (Figure 7a). The PXRD of crystalline DMF (Figure 7b) showed characteristic diffraction peaks at 2θ values 11.08°, 17.68°, 22.11°, 24.12°, and 27.5° [58]. The PXRD of 50% w/w DMF solid dispersion with HPMC(AS) exhibited diffraction peak characteristics of crystalline DMF with reduced intensity, indicating incomplete crystalline–amorphous transformation. The PXRD of pure polymers are shown in Figure S2d. Solid dispersion with HPMC (AS) containing 5% w/w DMF produced an amorphous halo, confirming the complete amorphization of DMF. In addition to PXRD, the FTIR spectra of NaDLO exhibited distinctive peaks at 3388.57 cm^{-1} due to NH stretching of the secondary amine, at 1576.82 cm^{-1} owing to C=O stretching of the carboxyl ion and at 747.35 cm^{-1} because of C-Cl stretching (Figure S2). In the FTIR of the melt extruded sample of 5% w/w NaDLO with HPMC(P) containing 5% w/w NaDLO, a peak at 1576 was still present, showing no interaction between APIs and the polymer

(Figure S2a) [59]. The FTIR spectra of crystalline DMF showed 1672 cm^{-1} for the C=C stretching vibration, 1719 and 3430 cm^{-1} for the C=O stretching vibration, 1160 cm^{-1} for the C−O stretching vibration, and 670–890 cm^{-1} and 2850–3080 cm^{-1} for the C−H bending vibration. These peaks were not found in either the 5% *w/w* DMF-HPMC(AS) physical mixture or the melt extruded solid dispersion containing 5% DMF [60] (Figure S2b). For comparison, the FTIR of pure polymers has been shown in Figure S2c.

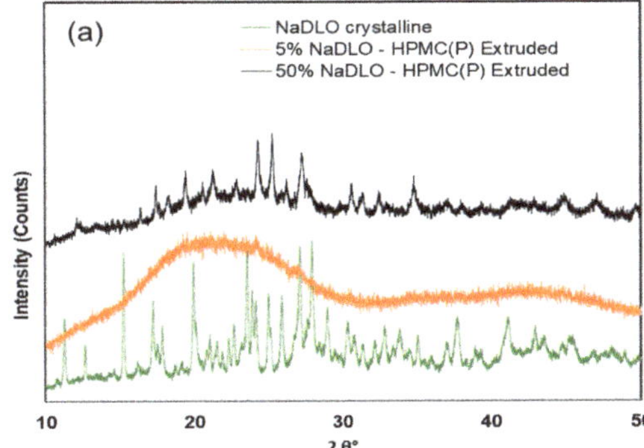

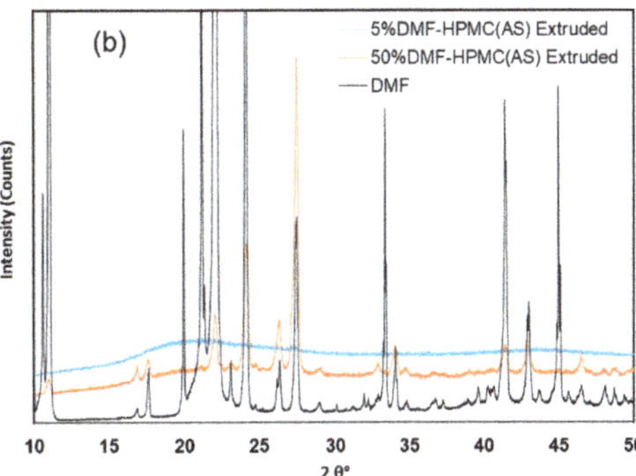

Figure 7. PXRD pattern of (**a**) NaDLO–HPMC(P) extruded, (**b**) DMF–HPMC(AS) extruded at 5% and 50% API loading.

3.3.3. Dissolution Experiments

The dissolutions of the extruded samples were performed according to the protocol for enteric dosage forms. The solid dispersion of 5% *w/w* NaDLO with HPMC(P) showed no release in SGF pH 1.2 for 2 h. When exposed to SIF pH 6.8, HPMC(P) solid dispersion released 10% of the drug. In the case of solid dispersion containing 50% *w/w* NaDLO, 65–75% drug release was observed in 2 h (Figure 8a). At a pH below the pKa of the carboxylic acid group in diclofenac (around 4.0), majority of diclofenac sodium will be in the neutral form. As the pH increases above the pKa, the proportion of ionized diclofenac will increase, with a maximum at around pH 4.5 to 5.5. Above this pH range, the proportion of ionized diclofenac will decrease as the carboxylic acid group becomes fully deprotonated and the molecule becomes negatively charged [61]. This could be a possible explanation as to why solid dispersion of 5% *w/w* NaDLO with HPMC(P) showed no release in SGF pH 1.2 for 2 h. Solid dispersion of 5% *w/w* DMF with HPMC(AS) showed no release in both SGF pH 1.2 and SIF pH 6.8, whereas solid dispersion containing 50% DMF started releasing DMF in SGF pH 1.2 (5% in 2 h) and in SIF pH 6.8, incomplete release (42% in 2 h) was observed (Figure 8b). Although both polymers are soluble in SIF pH 6.8, the negligible dissolution of the drug at 5% *w/w* loading could be due to a strong interaction between the APIs and the polymer.

3.4. Effect of Drug Loading

Analysis from experiments reflects that higher loading is suitable for faster API release. To gain insights into this observation at the atomic level, further MD simulations were performed at higher loadings corresponding to the feasible pairs, i.e., NaDLO–HPMC(P) and DMF–HPMC(AS) obtained by experiments. Figure 9a shows the interaction energy between NaDLO and HPMC(P). As the loading of NaDLO in HPMC(P) increases from 5 wt% to 50 wt%, all interaction energy terms E_{coul}, E_{LJ}, and E_{total} decrease. This indicates that the strength of binding between NaDLO and HPMC(P) is reduced. Due to reduced interaction or weaker binding, NaDLO shows faster release at higher loading, as observed in experiments. Consistent with interaction energies, the energy ratio also indicates weaker relative interaction between NaDLO and HPMC(P) at higher loading, as shown in Figure 9b, which supports faster NaDLO release. It is instructive to observe

NaDLO–NaDLO at both loadings. Figure S3a depicts E_{coul}, E_{LJ}, and E_{total} at both loadings. Similarly to NaDLO–HPMC(P) interactions, all interaction terms in NaDLO–NaDLO decrease as loading increases; however, Na^+-DLO^- interaction increases with loading (Figure S3b). Interestingly, E_{LJ} interactions between Na^+ and DLO^- ions are repulsive, as indicated by positive energy values. Owing to very strong ionic interaction between Na^+ and DLO^-, these ions reside very close to each other and hence LJ interactions are repulsive. Figure 10 shows the interaction energies between DMF and HPMC(AS) as well as the energy ratios at both loadings. With the increase in DMF loading from 5 wt% to 50 wt%, both the DMF–HPMC(AS) interaction and the energy ratio are reduced, indicating weaker binding between DMF and HPMC(AS), hence supporting faster DMF release at higher loading, as observed in experiments. Despite a decrease in DMF–HPMC(AS) interaction, DMF–DMF interaction showed a reverse trend with loading (Figure S4) as seen in the case of Na^+-DLO^- interaction.

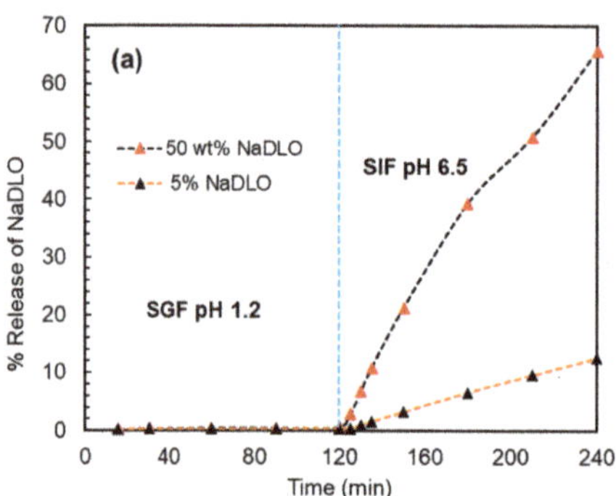

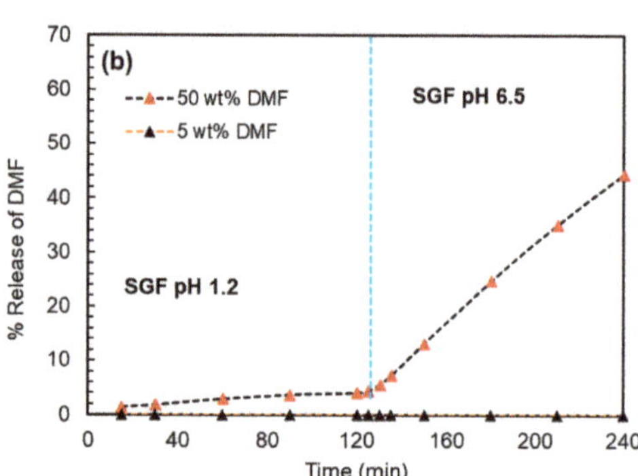

Figure 8. The release of (**a**) NaDLO from extruded samples with HPMC(P), and (**b**) DMF from extruded samples with HPMC(AS) under simulated gastric fluid (SGF) and simulated intestinal fluid (SIF) environments.

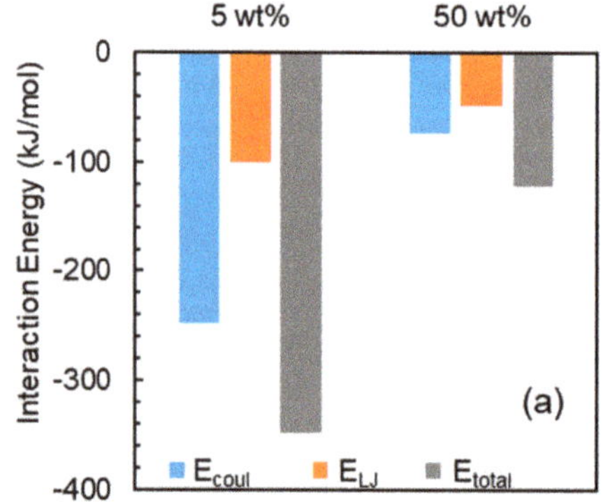

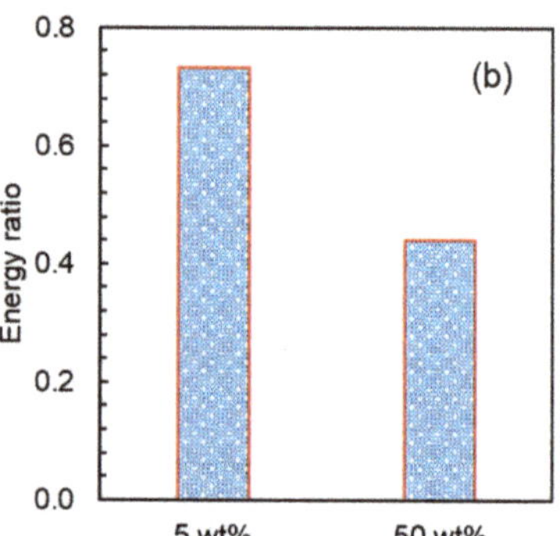

Figure 9. (**a**) Interaction energies between NaDLO and HPMC(P) and (**b**) energy ratio: NaDLO–HPMC(P)/NaDLO–NaDLO at 5 and 50 wt% NaDLO.

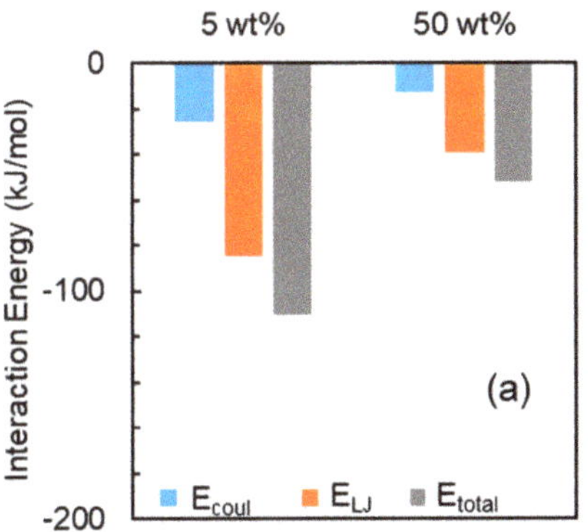
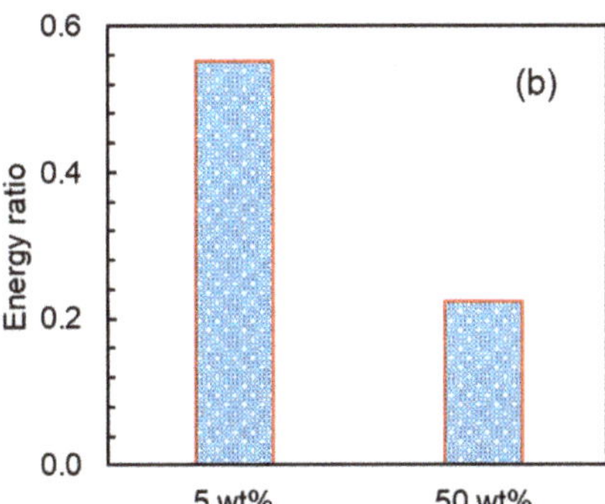

Figure 10. (**a**) Interaction energies between DMF and HPMC(AS) and (**b**) energy ratio: DMF–HPMC(AS)/DMF–DMF at 5 and 50 wt% DMF.

The release phenomena of an API from solid dispersion could efficiently be illustrated by dynamics of the API in solid dispersion. Usually, the dynamics of a molecule in a mixture environment is evaluated by mean-squared displacement (MSD) [62,63]. Not only the dynamics, but also the diffusion modes of a molecule/particle can be characterized by MSD [64]. The MSD is quantified as

$$\mathrm{MSD}(t) = \frac{1}{N}\sum_{i=1}^{N}\left\langle |\mathbf{r}_i(t) - \mathbf{r}_i(0)|^2 \right\rangle \tag{2}$$

where N is the number of active molecules and $\mathbf{r}_i(t)$ is the position of the ith active molecule at time t. To be precise, MSD was computed based on the trajectory from 2 ns to 10 ns. Figure 11 depicts the MSDs of NaDLO and DMF in HPMC(P) and HPMC(AS), respectively, at 5 and 50 wt% loadings. As simulation time proceeds, the MSDs continuously increase. As the loading of NaDLO increases from 5 to 50 wt%, the mobility of the NaDLO is enhanced, as indicated by the increased MSD value at 50 wt% (Figure 11a), which, in turn, indicates faster release of NaDLO from solid dispersion. This supports the experimental observation obtained from the release study (Figure 8a). Similar to the MSD of NaDLO in HPMC(P), the MSD of DMF in HPMC(AS) also depicts that, with an increase in loading, the mobility of the DMF is raised (Figure 11b). This is also consistent with experimental observation (Figure 8b). Overall, upon increasing the loading of API, API–polymer interaction decreases, which results in higher mobility of APIs in solid dispersions, thus boosting the release of APIs in the SIF environment.

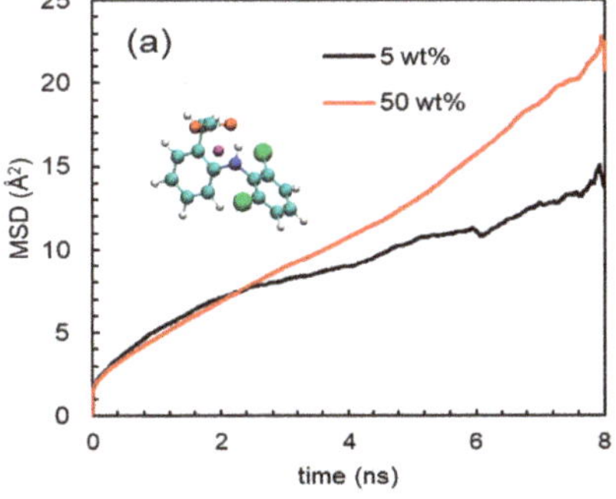
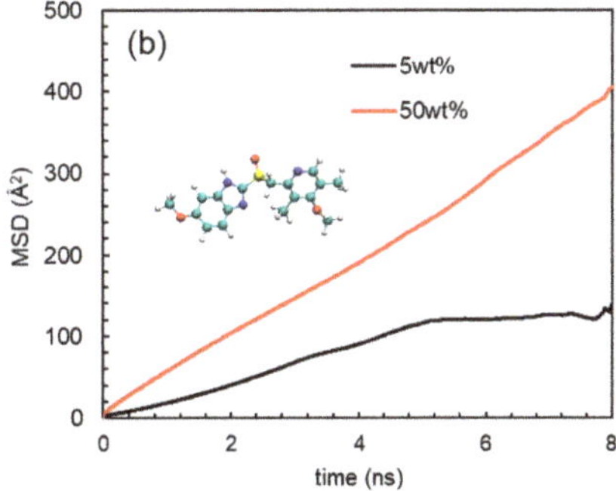

Figure 11. Mean-squared deviations (MSDs) of (**a**) DLO in HPMC(P) and (**b**) DMF in HPMC(AS) at 5 and 50 wt% loadings of each API.

3.5. Effect of Temperature

To elucidate the effect of temperature at the molecular level, a model system, i.e., a mixture of DMF and HPMC(AS) was considered. Particularly, simulations were performed at three different temperatures at 300 K, 373 K and 433 K, respectively. It should be noted that the examined temperatures are below the degradation temperature of DMF. Figure 12a shows the interaction energies between DMF and HPMC(AS) at various temperatures. The hierarchy of all the DMF–HPMC(AS) interaction terms (E_{coul}, E_{LJ}, and E_{total}) is [DMF–HPMC(AS)]$_{433\,K}$ < [DMF-HPMC(AS)]$_{373\,K}$ < [DMF-HPMC(AS)]$_{300\,K}$. This indicates that DMF would be strongly bounded to HPMC(AS) at a lower temperature. As expected, energy ratios indicate the same hierarchy as the interaction trend (Figure 12b). Furthermore, the dynamics of DMF in the mixture of DMF and HPMC(AS) at these temperatures are illustrated to predict the release behaviour. Figure 13 depicts the MSDs of DMF at 300 K, 373 K, and 433 K, particularly for the last 8 ns simulation. As the temperature increases, the mobility of DMF is enhanced, as indicated by significantly increased MSD values at 433 K compared to 300 K. Weaker DMF–HPMC(AS) interaction at a higher temperature, as explained above, results in higher mobility of DMF, which would support the release of DMF from solid dispersion. In brief, a higher temperature would be expected to boost the API release phenomena.

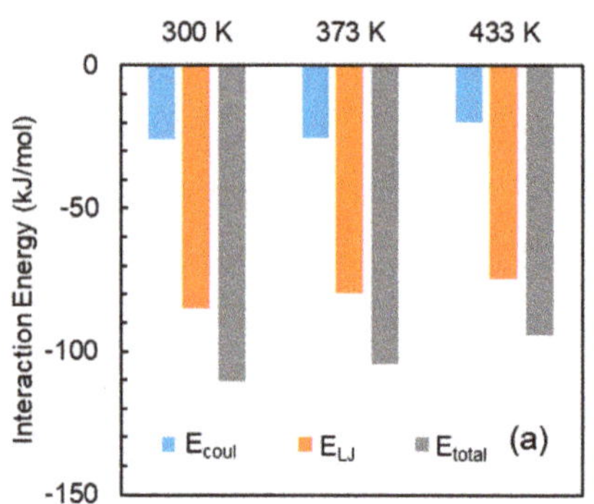
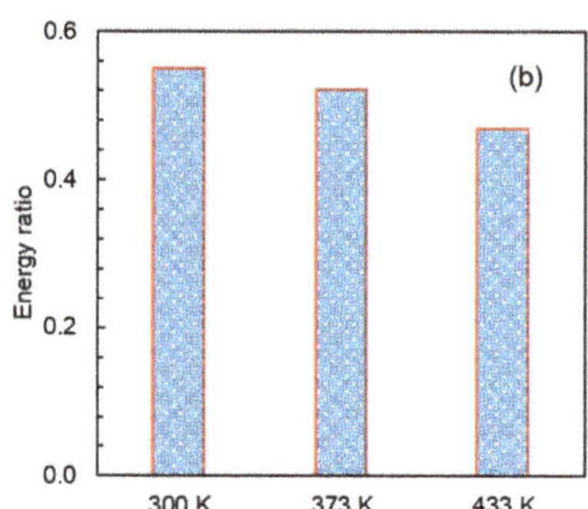

Figure 12. (**a**) Interaction energies between DMF and HPMC(AS), and (**b**) energy ratio: DMF–HPMC(AS)/DMF–DMF at temperatures of 300, 373, and 433 K.

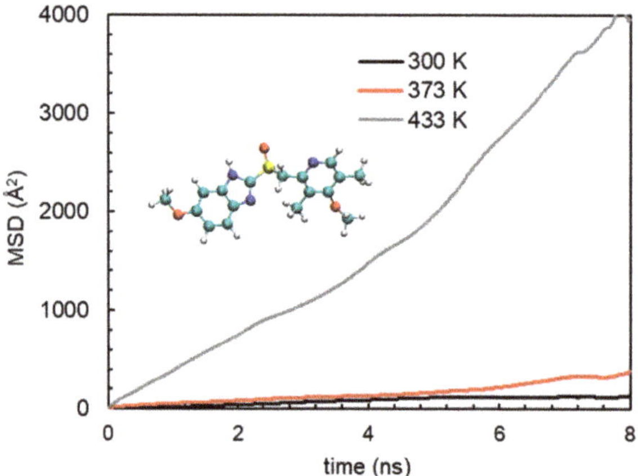

Figure 13. Mean-squared deviations (MSDs) of DMF in HPMC(AS) at temperatures of 300, 373, and 433 K.

4. Conclusions

To investigate molecular interactions in solid dispersions, MD simulations and experimental approaches were adopted. The interaction energies between NPX and polymer in terms of E_{coul}, E_{LJ}, and E_{total} increased in the order of HPMC(AS) < HPMC(P) < Eudragit L100, whereas for NaDLO, these quantities increased in the order of Eudragit L100 < HPMC(AS) < HPMC(P). For DMF and OPZ, the hierarchies of the energy terms are Eudragit L100 < HPMC(P) < HPMC(AS) and HPMC(P) < Eudragit L100 < HPMC(AS). The suggested API–polymer pairs are NPX-Eudragit L100, NaDLO-HPMC(P), DMF-HPMC(AS) and OPZ-HPMC(AS), respectively. The energy ratio trends of API–polymer combinations are consistent with the interaction energy hierarchies. All APIs were able to form hydrogen bonds with polymeric excipients. Being ionic in nature, NaDLO showed a higher number of hydrogen bonds compared to the other APIs. Among non-ionic APIs, OPZ had a higher number of hydrogen bonds. The pairs NaDLO–HPMC(P) and DMF–HPMC(AS) were successfully extruded by HME experiments. The release studies revealed that NaDLO and DMF APIs could be loaded up to 50 wt% in HPMC(P) and HPMC(AS), respectively, as these dispersions are rarely released in SGF but are mostly released in SIF. API–polymer interaction decreases as API loading increases, which leads to a higher mobility of API, thus boosting the release of API. The MSD indicates that with an increase in temperature, the mobility of API increases, resulting in a faster release of API. This study, in suggesting a potential polymeric excipient for delayed-release APIs, provides atomic-level insights into the compatibility between APIs and polymeric carriers, which could accelerate the rational design and development of a solid dispersion system for poorly soluble APIs.

Supplementary Materials: The following supporting information can be downloaded at: https://www.mdpi.com/article/10.3390/pharmaceutics15041164/s1, Table S1. Solubilities, melting points, and degradation temperatures of the API and polymers; Table S2: Number of Hydrogen bonds between polymer excipients and APIs; Figure S1: MD simulation snapshots; Figure S2: FTIR spectra of at 5% *w/w* API loading in (a) NaDLO–HPMC(P) physical mixture and extruded, (b) DMF–HPMC(AS) physical mixture and extruded, (c) FTIR and (d) PXRD of pure polymers; Figure S3: Interaction energies between: NaDLO and NaDLO, and Na^+ and DLO at 5 and 50 wt% NaDLO; Figure S4: Interaction energies between DMF and DMF at 5 and 50 wt% DMF.

Author Contributions: K.M.G.: Conceptualization of the research idea, MD simulation design, data analysis, and writing original draft; X.C.: experimental data collection and analysis and manuscript review; P.K.: Conceptualization of the research idea, experimental design, data analysis, and writing experimental section of the manuscript. All authors have read and agreed to the published version of the manuscript.

Funding: This work was supported by the internal research grant (SC22/20-1A0120-0AAK), provided by the Science and Engineering Research Council of A*STAR (Agency for Science, Technology and Research), Singapore.

Institutional Review Board Statement: Not applicable.

Informed Consent Statement: Not applicable.

Data Availability Statement: Literature data that support our results are properly cited in the respective sections.

Acknowledgments: This work was supported by the A*STAR Computational Resource Centre through the use of its high-performance computing facilities for most of the simulations. We also acknowledge the National Supercomputing Centre of Singapore for providing computational resources for conducting part of the work.

Conflicts of Interest: The authors declare no conflict of interest.

References

1. Di, L.; Kerns, E.; Carter, G. Drug-like property concepts in pharmaceutical design. *Curr. Pharm. Des.* **2009**, *15*, 2184–2194. [CrossRef] [PubMed]
2. Leuner, C.; Dressman, J. Improving drug solubility for oral delivery using solid dispersions. *Eur. J. Pharm. Biopharm.* **2000**, *50*, 47–60. [CrossRef] [PubMed]
3. Williams, H.D.; Trevaskis, N.L.; Charman, S.A.; Shanker, R.M.; Charman, W.N.; Pouton, C.W.; Porter, C.J. Strategies to address low drug solubility in discovery and development. *Pharmacol. Rev.* **2013**, *65*, 315–499. [CrossRef]
4. Qian, F.; Huang, J.; Hussain, M. Drug-polymer solubility and miscibility: Stability consideration and practical challenges in amorphous solid dispersion development. *J. Pharm. Sci.* **2010**, *99*, 2941–2947. [CrossRef]
5. Murdande, S.B.; Pikal, M.J.; Shanker, R.M.; Bogner, R.H. Solubility advantage of amorphous pharmaceuticals: I. A thermodynamic analysis. *J. Pharm. Sci.* **2010**, *99*, 1254–1264. [CrossRef] [PubMed]
6. Shah, S.; Maddineni, S.; Lu, J.; Repka, M.A. Melt extrusion with poorly soluble drugs. *Int. J. Pharm.* **2013**, *453*, 233–252. [CrossRef]
7. Dinunzio, J.; Zhang, F.; Martin, C.; Mcginity, J. *Formulating Poorly Water Soluble Drugs*; Springer: New York, NY, USA, 2012; pp. 311–362.
8. Stanković, M.; Frijlink, H.; Hinrichs, W. Polymeric formulations for drug release prepared by hot melt extrusion: Application and characterization. *Drug Discov. Today* **2015**, *20*, 812–823. [CrossRef]
9. Breitenbach, J. Melt extrusion: From process to drug delivery technology. *Eur. J. Pharm. Biopharm.* **2002**, *54*, 107–117. [CrossRef]
10. Singh, A.; Van den Mooter, G. Spray drying formulation of amorphous solid dispersions. *Adv. Drug Deliv. Rev.* **2016**, *100*, 27–50. [CrossRef]
11. Al-Japairai, K.A.S.; Alkhalidi, H.M.; Mahmood, S.; Almurisi, S.H.; Doolaanea, A.A.; Al-Sindi, T.A.; Chatterjee, B. Lyophilized Amorphous Dispersion of Telmisartan in a Combined Carrier–Alkalizer System: Formulation Development and In Vivo Study. *ACS Omega* **2020**, *5*, 32466–32480. [CrossRef]
12. Repka, M.A.; Battu, S.K.; Upadhye, S.B.; Thumma, S.; Crowley, M.M.; Zhang, F.; Martin, C.; McGinity, J.W. Pharmaceutical applications of hot-melt extrusion: Part II. *Drug Dev. Ind. Pharm.* **2007**, *33*, 1043–1057. [CrossRef]
13. Li, Y.; Pang, H.; Guo, Z.; Lin, L.; Dong, Y.; Li, G.; Lu, M.; Wu, C. Interactions between drugs and polymers influencing hot melt extrusion. *J. Pharm. Pharmacol.* **2013**, *66*, 148–166. [CrossRef]
14. Ahlneck, C.; Zografi, G. The molecular basis of moisture effects on the physical and chemical stability of drugs in the solid state. *Int. J. Pharm.* **1990**, *62*, 87–95. [CrossRef]
15. Huang, C.; Powell, C.T.; Sun, Y.; Cai, T.; Yu, L. Effect of Low-Concentration Polymers on Crystal Growth in Molecular Glasses: A Controlling Role for Polymer Segmental Mobility Relative to Host Dynamics. *J. Phys. Chem. B* **2017**, *121*, 1963–1971. [CrossRef] [PubMed]
16. Vo, C.L.-N.; Park, C.; Lee, B.-J. Current trends and future perspectives of solid dispersions containing poorly water-soluble drugs. *Eur. J. Pharm. Biopharm.* **2013**, *85*, 799–813. [CrossRef] [PubMed]
17. Konno, H.; Handa, T.; Alonzo, D.E.; Taylor, L.S. Effect of polymer type on the dissolution profile of amorphous solid dispersions containing felodipine. *Eur. J. Pharm. Biopharm.* **2008**, *70*, 493–499. [CrossRef] [PubMed]
18. Curatolo, W.; Nightingale, J.; Herbig, S. Utility of Hydroxypropylmethylcellulose Acetate Succinate (HPMCAS) for Initiation and Maintenance of Drug Supersaturation in the GI Milieu. *Pharm. Res.* **2009**, *26*, 1419–1431. [CrossRef] [PubMed]
19. Jha, P.K.; Larson, R.G. Assessing the Efficiency of Polymeric Excipients by Atomistic Molecular Dynamics Simulations. *Mol. Pharm.* **2014**, *11*, 1676–1686. [CrossRef] [PubMed]
20. Prasad, D.; Chauhan, H.; Atef, E. Role of Molecular Interactions for Synergistic Precipitation Inhibition of Poorly Soluble Drug in Supersaturated Drug–Polymer–Polymer Ternary Solution. *Mol. Pharm.* **2016**, *13*, 756–765. [CrossRef] [PubMed]
21. Punčochová, K.; Heng, J.Y.; Beránek, J.; Štěpánek, F. Investigation of drug-polymer interaction in solid dispersions by vapour sorption methods. *Int. J. Pharm.* **2014**, *469*, 159–167. [CrossRef]
22. Zhao, Y.; Inbar, P.; Chokshi, H.P.; Malick, A.W.; Choi, D.S. Prediction of the thermal phase diagram of amorphous solid dispersions by Flory–Huggins theory. *J. Pharm. Sci.* **2011**, *100*, 3196–3207. [CrossRef]
23. Marsac, P.J.; Shamblin, S.; Taylor, L. Theoretical and practical approaches for prediction of drug–polymer miscibility and solubility. *Pharm. Res.* **2006**, *23*, 2417. [CrossRef]
24. Maniruzzaman, M.; Pang, J.; Morgan, D.J.; Douroumis, D. Molecular Modeling as a Predictive Tool for the Development of Solid Dispersions. *Mol. Pharm.* **2015**, *12*, 1040–1049. [CrossRef]
25. Yani, Y.; Kanaujia, P.; Chow, P.S.; Tan, R.B.H. Effect of API-Polymer Miscibility and Interaction on the Stabilization of Amorphous Solid Dispersion: A Molecular Simulation Study. *Ind. Eng. Chem. Res.* **2017**, *56*, 12698–12707. [CrossRef]
26. Meng, F.; Trivino, A.; Prasad, D.; Chauhan, H. Investigation and correlation of drug polymer miscibility and molecular interactions by various approaches for the preparation of amorphous solid dispersions. *Eur. J. Pharm. Sci.* **2015**, *71*, 12–24. [CrossRef]
27. Meng, F.; Dave, V.; Chauhan, H. Qualitative and quantitative methods to determine miscibility in amorphous drug–polymer systems. *Eur. J. Pharm. Sci.* **2015**, *77*, 106–111. [CrossRef]
28. Lu, X.; Li, M.; Huang, C.; Lowinger, M.B.; Xu, W.; Yu, L.; Byrn, S.R.; Templeton, A.C.; Su, Y. Atomic-Level Drug Substance and Polymer Interaction in Posaconazole Amorphous Solid Dispersion from Solid-State NMR. *Mol. Pharm.* **2020**, *17*, 2585–2598. [CrossRef] [PubMed]

29. Gupta, J.; Nunes, C.; Vyas, S.; Jonnalagadda, S. Prediction of Solubility Parameters and Miscibility of Pharmaceutical Compounds by Molecular Dynamics Simulations. *J. Phys. Chem. B* **2011**, *115*, 2014–2023. [CrossRef] [PubMed]

30. Maus, M.; Wagner, K.G.; Kornherr, A.; Zifferer, G. Molecular dynamics simulations for drug dosage form development: Thermal and solubility characteristics for hot-melt extrusion. *Mol. Simul.* **2008**, *34*, 1197–1207. [CrossRef]

31. Walden, D.M.; Bundey, Y.; Jagarapu, A.; Antontsev, V.; Chakravarty, K.; Varshney, J. Molecular Simulation and Statistical Learning Methods toward Predicting Drug–Polymer Amorphous Solid Dispersion Miscibility, Stability, and Formulation Design. *Molecules* **2021**, *26*, 182. [CrossRef]

32. Liu, S.; Jia, L.; Xu, S.; Chen, Y.; Tang, W.; Gong, J. Insight into the State Evolution of Norfloxacin as a Function of Drug Concentration in Norfloxacin-Vinylpyrrolidone/Hydroxypropyl Methylcellulose/Hydroxypropyl Methylcellulose Phthalate Solid Dispersions. *Cryst. Growth Des.* **2019**, *19*, 6239–6251. [CrossRef]

33. Kabedev, A.; Zhuo, X.; Leng, D.; Foderà, V.; Zhao, M.; Larsson, P.; Bergström, C.A.S.; Löbmann, K. Stabilizing Mechanisms of β-Lactoglobulin in Amorphous Solid Dispersions of Indomethacin. *Mol. Pharm.* **2022**, *19*, 3922–3933. [CrossRef] [PubMed]

34. Gupta, K.M.; Liu, J.; Jiang, J. A molecular simulation study for efficient separation of 2,5-furandiyldimethanamine by a microporous polyarylate membrane. *Polymer* **2019**, *175*, 8–14. [CrossRef]

35. Batwa, A.; Norrman, A. Blockchain Technology and Trust in Supply Chain Management: A Literature Review and Research Agenda. *Oper. Supply Chain. Manag. Int. J.* **2021**, *14*, 203–220. [CrossRef]

36. Jorgensen, W.L.; Maxwell, D.; Tirado-Rives, J. Development and Testing of the OPLS All-Atom Force Field on Conformational Energetics and Properties of Organic Liquids. *J. Am. Chem. Soc.* **1996**, *118*, 11225–11236. [CrossRef]

37. Ribeiro, A.A.; Horta, B.; Alencastro, R. MKTOP: A program for automatic construction of molecular topologies. *J. Braz. Chem. Soc.* **2008**, *19*, 1433–1435. [CrossRef]

38. Available online: http://erg.biophys.msu.ru/wordpress/archives/32 (accessed on 23 February 2023).

39. Hess, B.; Kutzner, C.; van der Spoel, D.; Lindahl, E. GROMACS 4: Algorithms for highly efficient, load-balanced, and scalable molecular simulation. *J. Chem. Theory Comput.* **2008**, *4*, 435–447. [CrossRef] [PubMed]

40. Humphrey, W.; Dalke, A.; Schulten, K. VMD: Visual molecular dynamics. *J. Mol. Graph.* **1996**, *14*, 33–38. [CrossRef]

41. Kim, Y.B.; Song, H.; Park, I. Refinement of the structure of naproxen, (+)-6-methoxy-α-methyl-2-naphthaleneacetic acid. *Arch. Pharmacal Res.* **1987**, *10*, 232–238. [CrossRef]

42. Llinàs, A.; Burley, J.C.; Box, K.J.; Glen, R.C.; Goodman, J.M. Diclofenac Solubility: Independent Determination of the Intrinsic Solubility of Three Crystal Forms. *J. Med. Chem.* **2007**, *50*, 979–983. [CrossRef]

43. Kooijman, H.; Sprengers, J.W.; Agerbeek, M.J.; Elsevier, C.J.; Spek, A.L. Di-methyl fumarate. *Acta Crystallogr. Sect. E* **2004**, *60*, o917–o918. [CrossRef]

44. Deng, J.; Chi, Y.; Fu, F.; Cui, X.; Yu, K.; Zhu, J.; Jiang, Y. Resolution of omeprazole by inclusion complexation with a chiral host BINOL. *Tetrahedron Asymmetry* **2000**, *11*, 1729–1732. [CrossRef]

45. Cadden, J.; Gupta, K.M.; Kanaujia, P.; Coles, S.J.; Aitipamula, S. Cocrystal Formulations: Evaluation of the Impact of Excipients on Dissolution by Molecular Simulation and Experimental Approaches. *Cryst. Growth Des.* **2021**, *21*, 1006–1018. [CrossRef]

46. Gupta, K.M.; Jiang, J. Systematic Investigation of Nitrile Based Ionic Liquids for CO_2 Capture: A Combination of Molecular Simulation and ab Initio Calculation. *J. Phys. Chem. C* **2014**, *118*, 3110–3118. [CrossRef]

47. Gupta, K.M. Tetracyanoborate based ionic liquids for CO_2 capture: From ab initio calculations to molecular simulations. *Fluid Phase Equilibria* **2016**, *415*, 34–41. [CrossRef]

48. Gupta, K.M.; Das, S.; Wong, A.B.H.; Chow, P.S. Formulation and Skin Permeation of Active-Loaded Lipid Nanoparticles: Evaluation and Screening by Synergizing Molecular Dynamics Simulations and Experiments. *Langmuir* **2023**, *39*, 308–319. [CrossRef] [PubMed]

49. Li, C.; Wang, J.-X.; Le, Y.; Chen, J.-F. Studies of Bicalutamide–Excipients Interaction by Combination of Molecular Docking and Molecular Dynamics Simulation. *Mol. Pharm.* **2013**, *10*, 2362–2369. [CrossRef] [PubMed]

50. Luzar, A.; Chandler, D. Hydrogen-bond kinetic in liquid water. *Nature* **1996**, *379*, 55–57. [CrossRef]

51. Sovizi, M.R. Thermal behavior of drugs. *J. Therm. Anal. Calorim.* **2010**, *102*, 285–289. [CrossRef]

52. Vynckier, A.K.; De Beer, M.; Monteyne, T.; Voorspoels, J.; De Beer, T.; Remon, J.P.; Vervaet, C. Enteric protection of naproxen in a fixed-dose combination product produced by hot-melt co-extrusion. *Int. J. Pharm.* **2015**, *491*, 243–249. [CrossRef]

53. Riedel, A.; Leopold, C.S. Degradation of Omeprazole Induced by Enteric Polymer Solutions and Aqueous Dispersions: HPLC Investigations. *Drug Dev. Ind. Pharm.* **2005**, *31*, 151–160. [CrossRef]

54. Stroyer, A.; McGinity, J.; Leopold, C. Solid state interactions between the proton pump inhibitor omeprazole and various enteric coating polymers. *J. Pharm. Sci.* **2006**, *95*, 1342–1353. [CrossRef] [PubMed]

55. Sharma, V.D.; Akocak, S.; Ilies, M.A.; Fassihi, R. Solid-State Interactions at the Core-Coat Interface: Physicochemical Characterization of Enteric-Coated Omeprazole Pellets without a Protective Sub-Coat. *AAPS PharmSciTech* **2015**, *16*, 934–943. [CrossRef] [PubMed]

56. Erickson, M.; Josefsson, L. Pharmaceutical Formulation of Omeprazole. U.S. Patent 6090827, 18 May 1998.

57. Shivakumar, H.N.; Desai, B.; Deshmukh, G. Design and optimization of diclofenac sodium controlled release solid dispersions by response surface methodology. *Indian J. Pharm. Sci.* **2008**, *70*, 22–30. [CrossRef]

58. Chen, C.W.; Lee, T. Round Granules of Dimethyl Fumarate by Three-in-One Intensified Process of Reaction, Crystallization, and Spherical Agglomeration in a Common Stirred Tank. *Org. Process Res. Dev.* **2017**, *21*, 1326–1339. [CrossRef]

59. Aielo, P.B.; Borges, F.A.; Romeira, K.M.; Miranda, M.C.R.; Arruda, L.B.d.; Filho, P.N.L.; Drago, B.d.C.; Herculano, R.D. Evaluation of sodium diclofenac release using natural rubber latex as carrier. *Mater. Res.* **2014**, *17*, 146–152. [CrossRef]
60. Sinha, S.; Garg, V.; Sonali; Singh, R.P.; Dutt, R. Chitosan-alginate core-shell-corona shaped nanoparticles of dimethyl fumarate in orodispersible film to improve bioavailability in treatment of multiple sclerosis: Preparation, characterization and biodistribution in rats. *J. Drug Deliv. Sci. Technol.* **2021**, *64*, 102645. [CrossRef]
61. Adeyeye, C.M.; Li, P.-K. Diclofenac Sodium. *Anal. Profiles Drug Subst.* **1990**, *19*, 123–144.
62. Gupta, K.M.; Das, S.; Chow, P.S.; Macbeath, C. Encapsulation of Ferulic Acid in Lipid Nanoparticles as Antioxidant for Skin: Mechanistic Understanding through Experiment and Molecular Simulation. *ACS Appl. Nano Mater.* **2020**, *3*, 5351–5361. [CrossRef]
63. Gupta, K.M.; Yani, Y.; Poornachary, S.K.; Chow, P.S. Atomistic Simulation to Understand Anisotropic Growth Behavior of Naproxen Crystal in the Presence of Polymeric Additives. *Cryst. Growth Des.* **2019**, *19*, 3768–3776. [CrossRef]
64. Metzler, R.; Klafter, J. The random walk's guide to anomalous diffusion: A fractional dynamics approach. *Phys. Rep.* **2000**, *339*, 1–77. [CrossRef]

 *pharmaceutics*

Article

Application of Sucrose Acetate Isobutyrate in Development of Co-Amorphous Formulations of Tacrolimus for Bioavailability Enhancement

Eman M. Mohamed [1,2], Sathish Dharani [1], Mohammad T. H. Nutan [3], Phillip Cook [4], Rajendran Arunagiri [4], Mansoor A. Khan [1] and Ziyaur Rahman [1,*]

[1] Irma Lerma Rangel School of Pharmacy, Texas A&M Health Science Center, Texas A&M University, College Station, TX 77843, USA; eman_nabil24@tamu.edu (E.M.M.)
[2] Department of Pharmaceutics, Faculty of Pharmacy, Beni-Suef University, Beni-Suef 62514, Egypt
[3] Irma Lerma Rangel School of Pharmacy, Texas A&M Health Science Center, Texas A&M University, Kingsville, TX 78363, USA
[4] Eastman Chemical Company, Kingsport, TN 37662, USA
* Correspondence: rahman@tamu.edu; Tel.: +1-979-436-0873; Fax: +1-979-436-0087

Abstract: The focus of the present work was to develop co-amorphous dispersion (CAD) formulations of tacrolimus (TAC) using sucrose acetate isobutyrate as a carrier, evaluate by in vitro and in vivo methods and compare its performance with hydroxypropyl methylcellulose (HPMC) based amorphous solid dispersion (ASD) formulation. CAD and ASD formulations were prepared by solvent evaporation method followed by characterization by Fourier transformed infrared spectroscopy, X-ray powder diffraction (XRPD), differential scanning calorimetry (DSC), dissolution, stability, and pharmacokinetics. XRPD and DSC indicated amorphous phase transformation of the drug in the CAD and ASD formulations, and dissolved more than 85% of the drug in 90 min. No drug crystallization was observed in the thermogram and diffractogram of the formulations after storage at 25 °C/60% RH and 40 °C/75% RH. No significant change in the dissolution profile was observed after and before storage. SAIB-based CAD and HPMC-based ASD formulations were bioequivalent as they met 90% confidence of 90–11.1% for C_{max} and AUC. The CAD and ASD formulations exhibited C_{max} and AUC 1.7–1.8 and 1.5–1.8 folds of tablet formulations containing the drug's crystalline phase. In conclusion, the stability, dissolution, and pharmacokinetic performance of SAIB-based CAD and HPMC-based ASD formulations were similar, and thus clinical performance would be similar.

Keywords: tacrolimus; sucrose acetate isobutyrate; amorphous solid dispersion; dissolution; stability; pharmacokinetics

Citation: Mohamed, E.M.; Dharani, S.; Nutan, M.T.H.; Cook, P.; Arunagiri, R.; Khan, M.A.; Rahman, Z. Application of Sucrose Acetate Isobutyrate in Development of Co-Amorphous Formulations of Tacrolimus for Bioavailability Enhancement. *Pharmaceutics* **2023**, *15*, 1442. https://doi.org/10.3390/pharmaceutics15051442

Academic Editor: Ana Isabel Fernandes

Received: 10 March 2023
Revised: 5 May 2023
Accepted: 6 May 2023
Published: 9 May 2023

1. Introduction

Most drugs fail to reach the market during development due to poor water solubility, even though they have desirable safety and efficacy profiles [1,2]. Several approaches were employed to enhance solubility and dissolution. These include chemical modification, particle size reduction, and formulation development [3,4]. Chemical modification of the drug results in significant changes in safety and efficacy profile that may result in elimination of potential lead compounds [3]. Particle size reduction improves dissolution rate but may not increase extent of crystalline solubility of drugs, even with an increase in surface-to-volume ratio [5,6]. Among the formulation development strategies, co-amorphous dispersion (CAD) and amorphous solid dispersion (ASD) are commonly used approaches for a poorly soluble drug where a crystalline drug is transformed into an amorphous form [7–12]. Dissolution of crystalline drugs includes disruption of the crystalline lattice, solvation/hydration, and breakdown of hydrogen bonds [13]. This approach exploits solubility advantage of an amorphous drug over a crystalline one due to lack of lattice order [14]. However, lack of lattice order increases free energy, making the amorphous form thermodynamically unstable. The amorphous drug reverts to stable crystalline form after high temperature and humidity exposure. The amorphous drug is stabilized to a certain extent against crystallization by adding a high melting/glass transition temperature polymer to form an ASD, which increases the glass transition temperature of the drug. The polymer also provides means of dosage form development and manufacturability [15]. CAD is formed when the polymer is replaced by a small molecule [7,8,11]. Small molecules stabilize the amorphous drugs through intermolecular interactions, e.g., hydrogen bonds, π–π, or even ionic etc. This technique is widely used for solubility, dissolution, and oral bioavailability enhancement of poorly water-soluble drugs [16,17]. The CAD and ASD can be formulated into tablet and or capsule dosage forms. FDA has approved many ASD since 2007 [18].

Tacrolimus (TAC) is a BCS class II drug, meaning solubility and dissolution is the rate-limiting step in its absorption [19]. The ASD of TAC has been reported to increase the oral bioavailability of the drug [9,20]. Immediate release ASD dosage form of TAC has been reported using hydroxypropyl cellulose [21], polyvinyl pyrrolidone, polyethylene glycol, hydroxypropyl methylcellulose (HPMC), [22], Eudragit® [23], HPMC and sodium lauryl sulfate [24]. Similarly, extended-release ASD of TAC has also been reported using HPMC and ethyl cellulose [25,26]. Among all the reported polymers for ASD, HPMC was the most effective in maintaining supersaturation during in vitro and in vivo dissolution, thus, enhancing oral bioavailability [22]. Therefore, FDA approved immediate and extended-release ASD dosage forms of TAC. However, immediate and extended-release ASD formulations of TAC have been recalled due to failure to meet dissolution specification during stability testing [27,28]. This could be related to drug crystallization during stability if the formulation composition is not optimized, the manufacturing method is incorrect, and/or the packaging is defective, etc. Drug crystallization is a long-standing problem in ASD dosage forms [9,20]. This dictates to search for a new polymer or excipient that may inhibit or reduce the drug crystallization while maintaining supersaturation during dissolution and in vivo absorption. In this paper, an attempt was made to develop CAD of TAC using sucrose acetate isobutyrate (SAIB) as a new carrier and characterize for physicochemical, stability, and pharmacokinetic attributes and compared with HPMC-based ASD formulation. SAIB is a pale straw glassy solid at room temperature that liquefies at 60 °C. It is synthesized by controlled esterification of sucrose with acetic anhydride and isobutyric anhydride. The molecular formula and weight of SAIB are $C_{40}H_{62}O_{19}$ and 846.9 g/mL, respectively (Figure 1A). It is insoluble in water and soluble in most organic solvents with a LogP ranging from 3.4 to 7. The calculated hydrogen donor and acceptor counts in the molecule is zero and 19, respectively [29]. The physicochemical properties of TAC are similar to SAIB (Figure 1B). The molecular weight and LogP values of the drug are 804.0 and 2.7, respectively. Unlike SAIB, TAC has eleven hydrogen acceptor and three hydrogen donor counts. Reported solubility of TAC in SAIB was 115 mg/gm [29]. It is possible that both molecules may form CAD by hydrogen and hydrophobic interactions based on structural and physicochemical attributes. SAIB has never been reported in the literature for CAD formulation development.

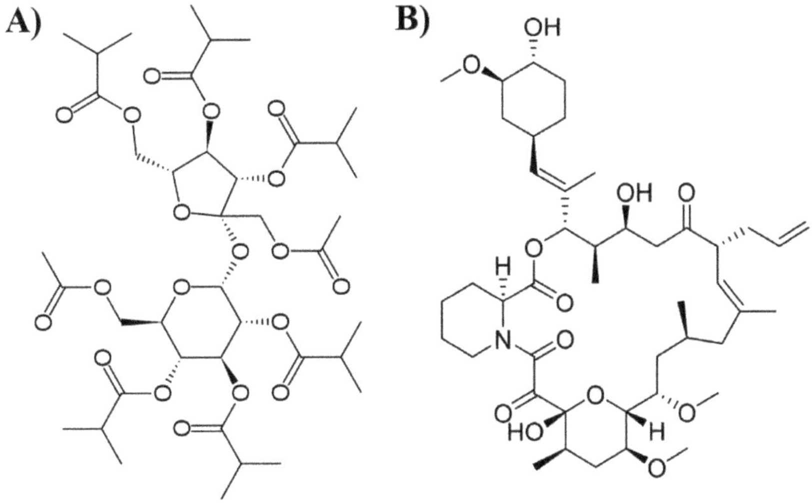

Figure 1. (**A**) Structure of SAIB, (**B**) Structure of Tacrolimus.

2. Materials and Methods

2.1. Materials

TAC monohydrate >98% and hydroxypropyl cellulose (HPC, MW 100,000) were purchased from Sigma-Aldrich (St Louis, MO, USA). Deuterated TAC-^{13}C$_3$D$_2$ (>85%) and Beagle dog plasma were obtained from Toronto Research Chemicals, Ontario, Canada, and BioChemed Services, Winchester, VA, USA, respectively. BioSustane™ SAIB and microcrystalline cellulose (MCC, Avicel® PH-101) were obtained from Eastman Chemical Company (Kingsport, TN, USA) and FMC Corporation (Princeton, NJ, USA), respectively. Hydroxypropyl methylcellulose (HPMC, 50 cps), magnesium stearate (MGS), croscarmellose sodium (CCS), lactose monohydrate, orthophosphoric acid (OPA), methanol, ammonium acetate, zinc sulfate (ZnSO$_4$) and formic acid were purchased from Fisher Scientific (Asheville, NC, USA). Sodium lauryl sulfate (SLS) was purchased from VWR Chemicals, LLC (Fountain Parkway, OH, USA). All reagents were of analytical grade and used as received. In-house water (18 MΩcm, Millipore Milli-Q Gradient A-10 water purification system) was used in the study.

2.2. Preparation of CAD and ASD

The formulations were prepared by solvent evaporation method as per Table 1. Briefly, the drug and the SAIB or HPMC were dissolved/dispersed in ethanol by sonication, followed by the addition SLS with sonication, and MCC and CCS. Solvents were evaporated under the hood at room temperature with stirring to form a solid mass. The solid mass was crushed and passed through #60 sieve. The physical mixtures (PM) of CAD (F16) were prepared by adding the drug to the placebo formulation. Dried powder equivalent to 5 mg TAC either filled into hard gelatin capsules or mixed with MGS and compressed into tablets using a Mini Press-1 (Globe Pharma, New Brunswick, NJ, USA) 10-station tableting machine with 5-mm concave die and punches (Natoli Engineering Company, Saint Charles, MO, USA). A tablet formulation (F18) compositionally identical to F16 or F17 without SAIB or HPMC, and contained crystalline form of the drug was also prepared for pharmacokinetic study. Various solutions/dispersions of TAC and SAIB were also prepared by heating at 100–120 °C. All samples were stored in a desiccator until further analysis.

Table 1. Composition of tacrolimus CAD and ASD formulations.

Formulation	Tacrolimus (mg)	SAIB (mg)	SLS (mg)	MCC (mg)	LMH (mg)	HPMC (mg)	CCS (mg)	Dosage Form
F1	5	5	0	0	0	0	0	Capsule
F2	5	5	0	0	25	0	0	Capsule
F3	5	5	0	0	50	0	0	Capsule
F4	5	5	0	0	100	0	0	Capsule
F5	5	5	0	25	0	0	0	Capsule
F6	5	5	0	50	0	0	0	Capsule
F7	5	5	0	100	0	0	0	Capsule
F8	5	0.5	0	50	0	0	0	Capsule
F9	5	3.75	0	50	0	0	0	Capsule
F10	5	5	0	50	0	0	10	Capsule
F11	5	7.5	0	50	0	0	10	Capsule
F12	5	10	0	50	0	0	10	Capsule
F13	5	7.5	5	50	0	0	10	Capsule
F14	5	7.5	5	50	0	0	15	Capsule
F15	5	7.5	5	50	0	0	20	Capsule
F16	5	7.5	5	50	0	0	15	Tablet
F17	5	0	5	50	0	7.5	15	Tablet
F18	5	0	5	50	0	0	15	Tablet

2.3. Fourier Transform Infrared Spectroscopy

The Fourier-transform infrared (FT-IR) spectra of the samples were collected by a modular NicoletTM iS™ 50 system (Thermo Fisher Scientific, Austin, TX, USA). The spectra were obtained in absorbance mode over a wavelength range of 400–4000 cm^{-1} with a data resolution of 8 cm^{-1} and 100 scans. A small amount of powder was placed on the diamond crystal and pressed with the attached arm to avoid any air entrapment in the sample. OMNIC software, version 9.0 (ThermoFisher Scientific), was used to capture and analyze the spectra.

2.4. X-ray Powder Diffractometry

X-ray powder diffraction (XRPD) of the samples was performed using Bruker D2 Phaser SSD 160 Diffractometer (Bruker AXS, Madison, WI) equipped with LYNXEYE scintillation detector and Cu Kα radiation (λ = 1.54184 Å) at a voltage of 30 KV and a current of 10 mA. Approximately 400 mg sample was evenly filled in a sample holder. The diffraction angle was set as $5 < 2\theta < 15°$ at a rate of 2°/min and 1 s per step with an increment of 0.1778° and rotated at 15 rpm to collect average diffractograms. Data was evaluated using Diffrac. EVA Suite version V4.2.1 and further processed using File Exchange 5.0 (Bruker AXS, Madison, WI, USA).

2.5. Differential Scanning Calorimetry

The thermal behavior of formulation components, PM, CAD, and ASD, was assessed by differential scanning calorimetry (DSC) using Q2000 instruments (TA Instruments Co., New Castle, DE, USA). Approximately 5 mg sample was hermetically sealed in an aluminum pan. The samples were scanned over a temperature range of 10 to 250 °C at a rate of 10 °C/min to cover the melting point of the drug and excipients. Nitrogen gas was purged at a pressure of 20 psi and 50 mL/min flow rate to provide an inert atmosphere during the measurement.

2.6. Dissolution

Dissolution of the formulations was performed using USP dissolution apparatus 2 (Agilent 708-DS, Santa Clara, CA, USA) equipped with an autosampler (Agilent 850-DS Dissolution Sampling Station) in a 900 mL dissolution media (water containing 1 in 20,000 HPC, pH adjusted to 4.5) at 37 ± 0.5 °C and 50 rpm. The samples were withdrawn at 30, 60, 90, and 120 min, and the amount of drug dissolved was determined by the validated HPLC method. Dissolution samples were diluted with SLS solution (1%) in a ratio of 9:1 to ensure drug solubility and to prevent crystallization during analysis. The dissolution experiment was performed in triplicate.

2.7. Stability

Short-term stability of CAD formulation F16 (SAIB) and ASD formulation F17 (HPMC) was performed by packing in an HDPE bottle and storing at 25 °C/60% RH and 40 °C/75% RH for three months and one month, respectively. The samples were examined for physical and chemical changes by dissolution, FTIR, XRPD, and DSC.

2.8. Pharmacokinetics

This study was carried out to compare the pharmacokinetic profiles of HPMC-based ASD formulation (F17), SAIB-based CAD formulation (F16), and a tablet containing the crystalline form of the drug (F18). Four beagle dogs (2 males and 2 females, 10 ± 2 Kg) were used in this study. The study (IACUC 2019-0241) was approved by the Institutional Animal Care and Use Committee (IACUC) of Texas A&M University. The animals were given F16 (SAIB as a carrier), F17 (HPMC as a carrier), or crystalline drug tablets (F18) containing 5 mg TAC with 15 days washout period between the studies. The animals were fasted overnight (midnight to 7 a.m.) before administration of the dose and 2 h post-dosing. The animals had free access to water during fasting and food and water 2 h post-dosing. Lidocaine cream was applied to the catheterization site before catheterization/needle stick and applied at each puncture site. A bitter apple was applied to the catheter to prevent licking and chewing of the catheter, and E-collar was also used to prevent dogs from reaching the catheter. 3 mL of blood was collected at 0, 0.25, 0.5, 1, 2, 3, 4, 6, and 8 h through the cephalic or saphenous catheter. Blood samples at 12, 24, 36, 48, and 72 h were directly withdrawn from the cephalic or saphenous vein. Blood was immediately transferred to a heparinized tube and stored at -80 °C until analysis. The protein precipitation method was used to extract the drug from the sample. Whole blood samples (500 µL) were treated with 0.1 M $ZnSO_4$ (50 µL) to break red blood cells, followed by addition of methanol (900 µL) and internal standard (IS) (100 µL TAC-$^{13}C_3D_2$, 50 ng/mL). The samples were vortexed for 2 min and centrifuged at 13,300 rpm and 4 °C for 15 min. The supernatant was analyzed for TAC and IS by the UPLC-MS method.

2.9. High-Performance Liquid Chromatography

A reported HPLC method was modified and validated for dissolution and assay analysis of the formulations [30]. The HPLC consisted of Agilent 1260 series (Agilent Technologies, Wilmington, DE, USA) equipped with a quaternary pump, online degasser, column heater, autosampler, and UV/Vis detector. Separation of the analyte was achieved on a 4.6 × 150 mm, 3 µm Luna C18 (Phenomenex, Torrance, CA, USA) column and a C18, 4.6 × 2.5 mm (5 µm packing) Luna C18 guard column (Phenomenex, Torrance, CA). The mobile phase was ACN and 0.05 M phosphoric acid (65:35 *v/v*) flowing at 1.0 mL/min. The column and auto-sampler were maintained at 60 °C. A sample volume of 200 µL was injected into the system and detected at 210 nm. Two injections per sample were analyzed to demonstrate reproducibility of the method. Data was collected and analyzed using OpenLab software (Agilent Technologies, Wilmington, DE, USA).

2.10. Ultra-Performance Liquid Chromatography-Mass Spectroscopy

UPLC was performed on a Waters Acquity® UPLC system (Waters Corporation, Milford, MA, USA) equipped with Agilent InfinityLab Poroshell 120 EC-C18 (4.6 × 50 mm, 2.7 μm) and maintained at 60 °C in the column oven. Separation of the analytes was performed using a mobile phase consisting of 5 mM ammonium acetate adjusted to pH 5 with formic acid and 0.1% formic acid (v/v) in methanol (5:95, v/v). The flow rate and run time were 0.5 mL/min, and 3 min, respectively. The retention times of TAC and TAC-$^{13}C_3D_2$ were 1.405 and 1.402 min, respectively, and the peaks were well separated from the baseline. Mass spectrometry parameters were electrospray positive ionization (ESI+) mode with 0.8 KV capillary voltage and 15V collision energy. The TAC and TAC-$^{13}C_3D_2$ molecular masses were detected at 804 and 806 Dalton by the QDa detector. The calibration range was 4-100 ng/mL. Concentrations of 4, 10, 50, and 100 ng/mL TAC were used in precision and accuracy assessment, and met the requirement of ±15% of nominal concentration. The method was validated per FDA bioanalytical method validation guidance [31].

3. Results and Discussion

3.1. Fourier Transformed Infrared Spectroscopy

TAC showed absorption bands due to stretching vibrations of O-H at 3438 cm^{-1}, C=O (ester and ketone) at 1738, 1723, and 1692 cm^{-1}, C=O (keto-amide), and C=C at 1637 cm^{-1}, C–O (ester) at 1172 cm^{-1}, and C–O–C (ether) at 1087 cm^{-1}. SAIB showed a strong absorption band at 1744 cm^{-1}, which is related to its ester carbonyl. Major absorption bands of TAC appeared with reduced intensity in the PM due to dilution with the excipients. Furthermore, TAC absorption bands at 1738 and 1723 cm^{-1} were masked by SAIB at 1644 cm^{-1}. In the case of CAD (F16), many characteristic absorption bands of TAC disappeared, broadened, or shifted to a new wavenumber. The absorption band of TAC at 783 and 1692 cm^{-1} disappeared, which may indicate the phase transformation of the drug. Similarly, F17, an ASD formulation based on HPMC, showed similar vibration bands (Figure 2). Phase transformation of the drug was further supported by XRPD and DSC data.

3.2. X-ray Powder Diffraction

TAC showed characteristics reflection peaks at 8.4, 10.2, 11.1, 11.6, 12.5, 13.6, 14.0, 15.2, 17.0, 18.0, 18.9, and 23.4° 2θ values. SAIB showed halo diffractograms, characteristics of amorphous or glassy materials. The degree of crystalline to amorphous phase ratio varied with TAC to SAIB. The crystalline phase decreased with SAIB content in the mixture. The crystalline phase completely disappeared when TAC: SAIB ratio was 1:1 or higher (Figure 3). Adding the excipients did not change the amorphous nature of the drug in the SAIB. XRPD was collected from 6 to 15° 2θ value as reflection peaks of the drug before 15° were not interfered with by the excipients (lactose, microcrystalline cellulose, etc.). Characteristics reflection peaks of the drug appeared with reduced peak height in the PM due to dilution with the excipients. In the case of F16 (SAIB) and F17 (HPMC) formulations, the drug peaks completely disappeared due to the formation of an amorphous phase (Figure 4).

3.3. Differential Scanning Calorimetry

TAC showed a sharp melting endothermic peak at 131.5 °C that indicated the drug's crystalline nature. The melting peak of the drug concurred with the literature-reported value [32]. SAIB showed no thermal event confirming its glassy nature. The addition of TAC to SAIB causes the transformation of the drug from crystalline to amorphous phase. However, the extent of phase transformation depends on the TAC to SAIB ratio. Height of the drug melting endothermic peak decreased with an increase in SAIB proportion in the formulation. This was due to drug solubilization in the glassy SAIB. The drug melting peak completely disappeared when TAC: SAIB ratio was 1:1 or higher (Figure 5).

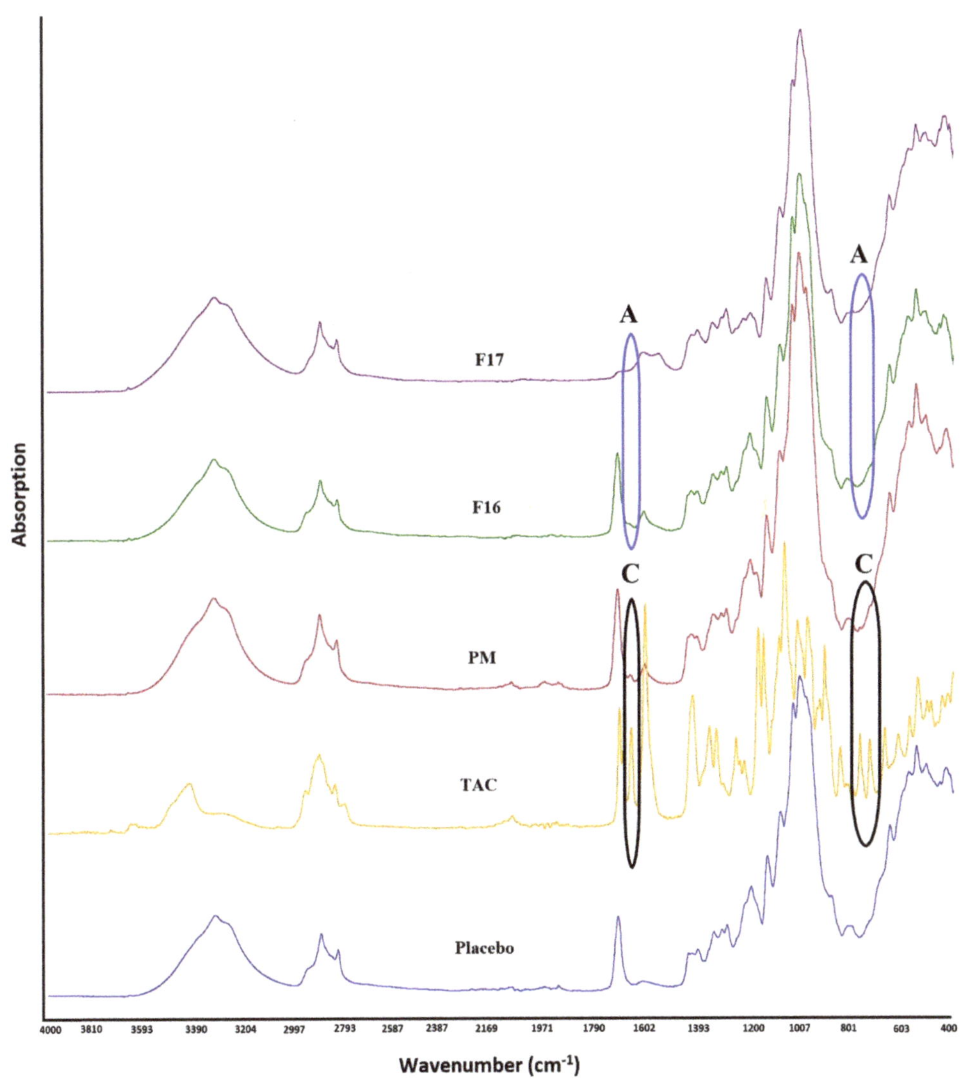

Figure 2. FTIR spectra of tacrolimus, physical mixture (PM), placebo, CAD, and ASD formulations (A: amorphous drug, C: Crystalline drug).

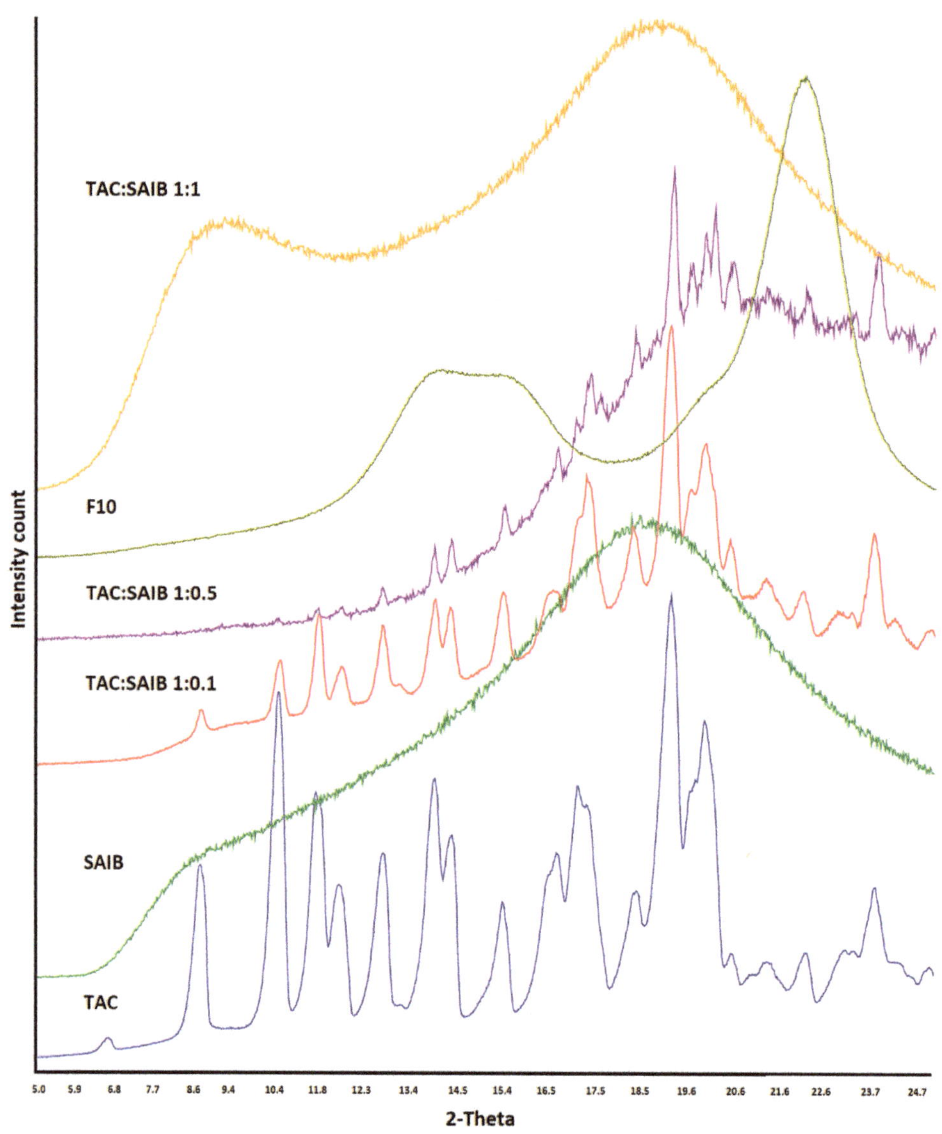

Figure 3. X-ray powder diffractogram of tacrolimus, SAIB, and various solutions/dispersions of drug and SAIB.

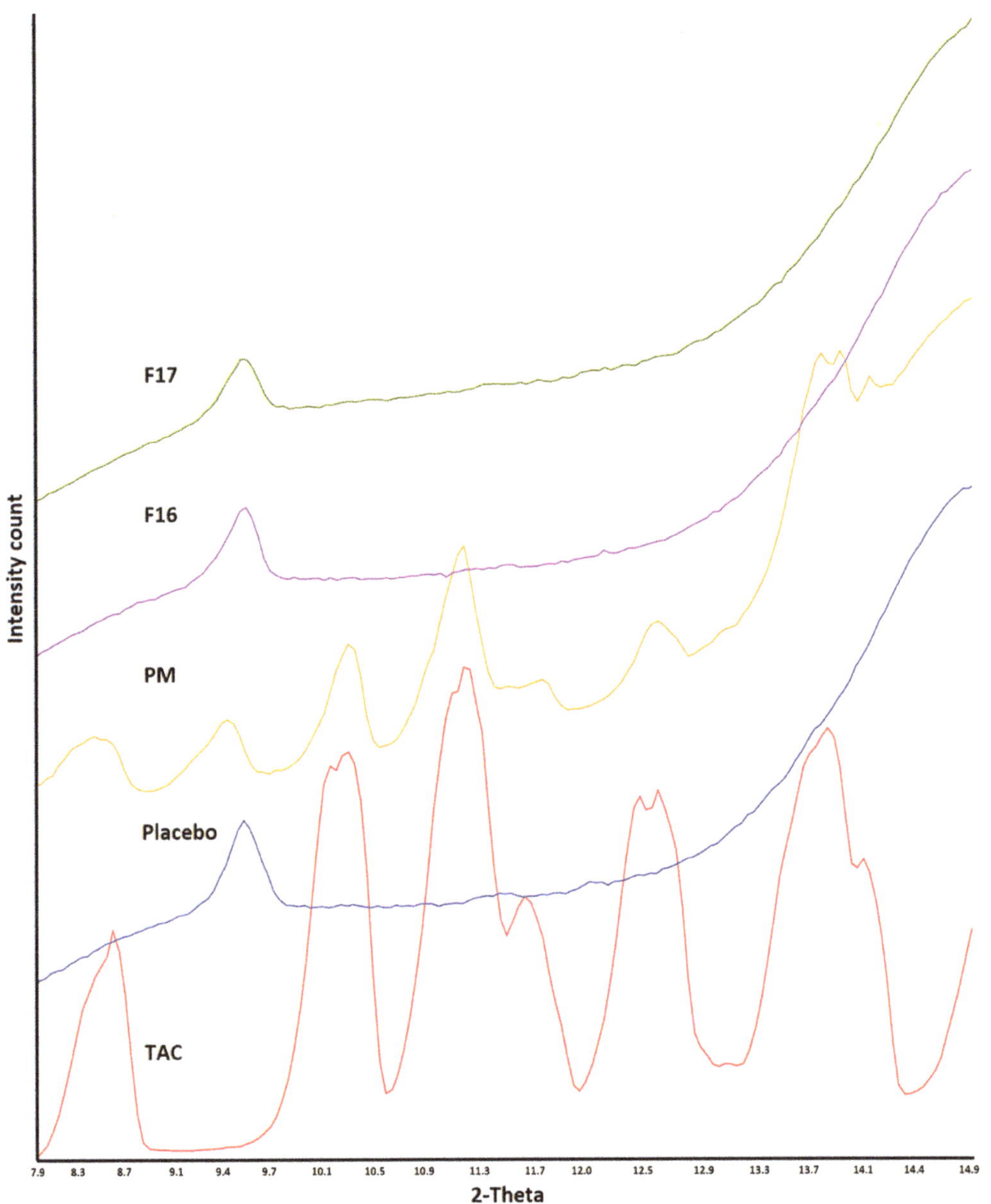

Figure 4. X-ray powder diffractogram, physical mixture, placebo, CAD, and ASD formulations.

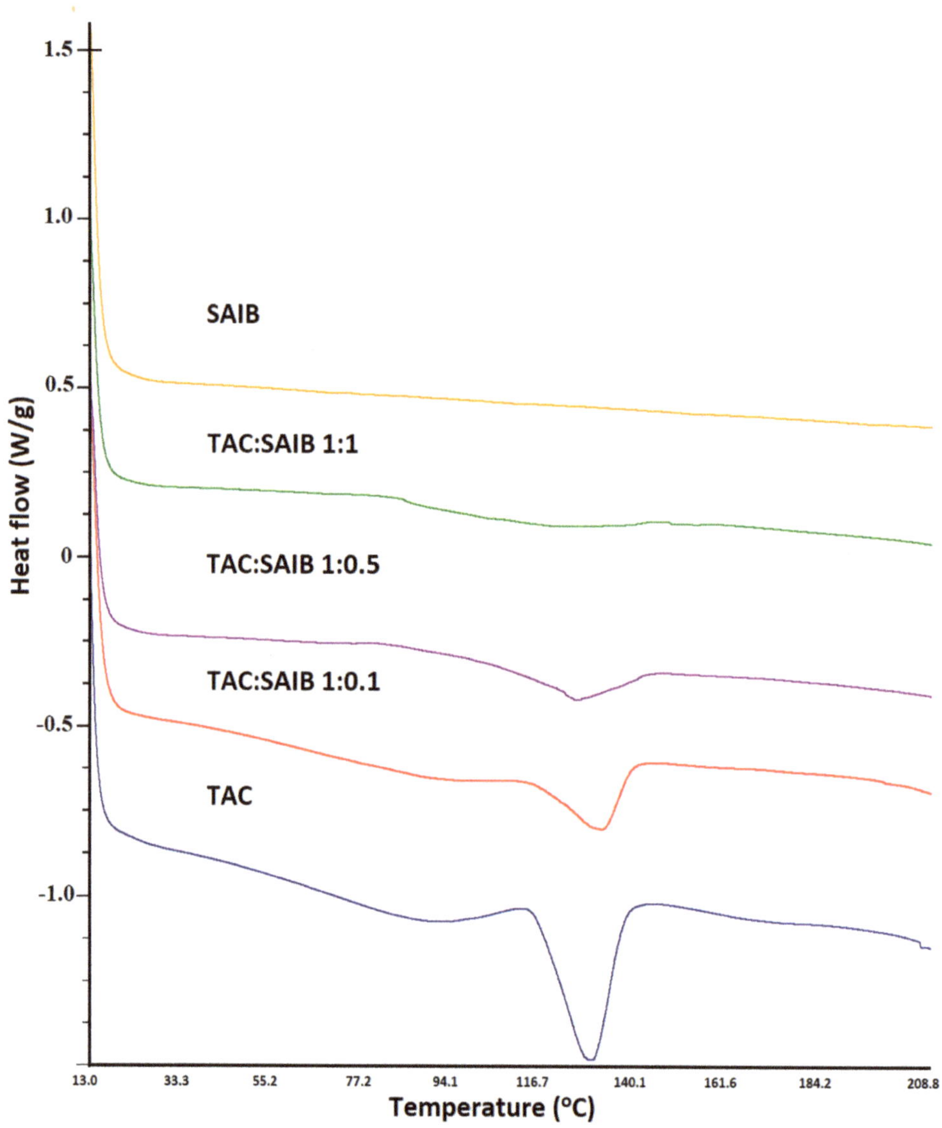

Figure 5. DSC thermograms of tacrolimus, SAIB, and various solutions/dispersions of drug and SAIB.

DSC thermogram of CCS and MCC exhibited shallow peaks in the region 60–130 °C, which could be due to physically adsorbed water. However, SLS showed multiple thermal events that may be due to its complex nature or the presence of numerous polymorphic forms. It is a mixture of sodium alkyl sulfate, mainly lauryl [33]. SLS showed thermal events at 103.9, 111.1, and 200.8 °C, which matched with the literature-reported values [34]. The PM showed additive thermograms with some differences from thermograms of individual components. The drug peak intensity was reduced due to dilution with excipients. Additionally, the drug melting peak was shifted to 134.2 °C due to excipients acting as impurities, thus causing a shift in melting point to a higher temperature. Additionally, the thermal event peak at 200.8 °C became broad and shifted to 180.8 °C. This was possibly due to the melting of the drug and partial dissolution that resulted in a broad peak at lower melting. The CAD and ASD formulations did not show an endothermic peak of the drug as observed in the PM, indicating conversion of the crystalline drug to the amorphous phase in formulations F16 and F17. However, a broad and shallow doublet appeared at 116–124 °C, which could be related to the glass transition temperature of the drug (Figure 6).

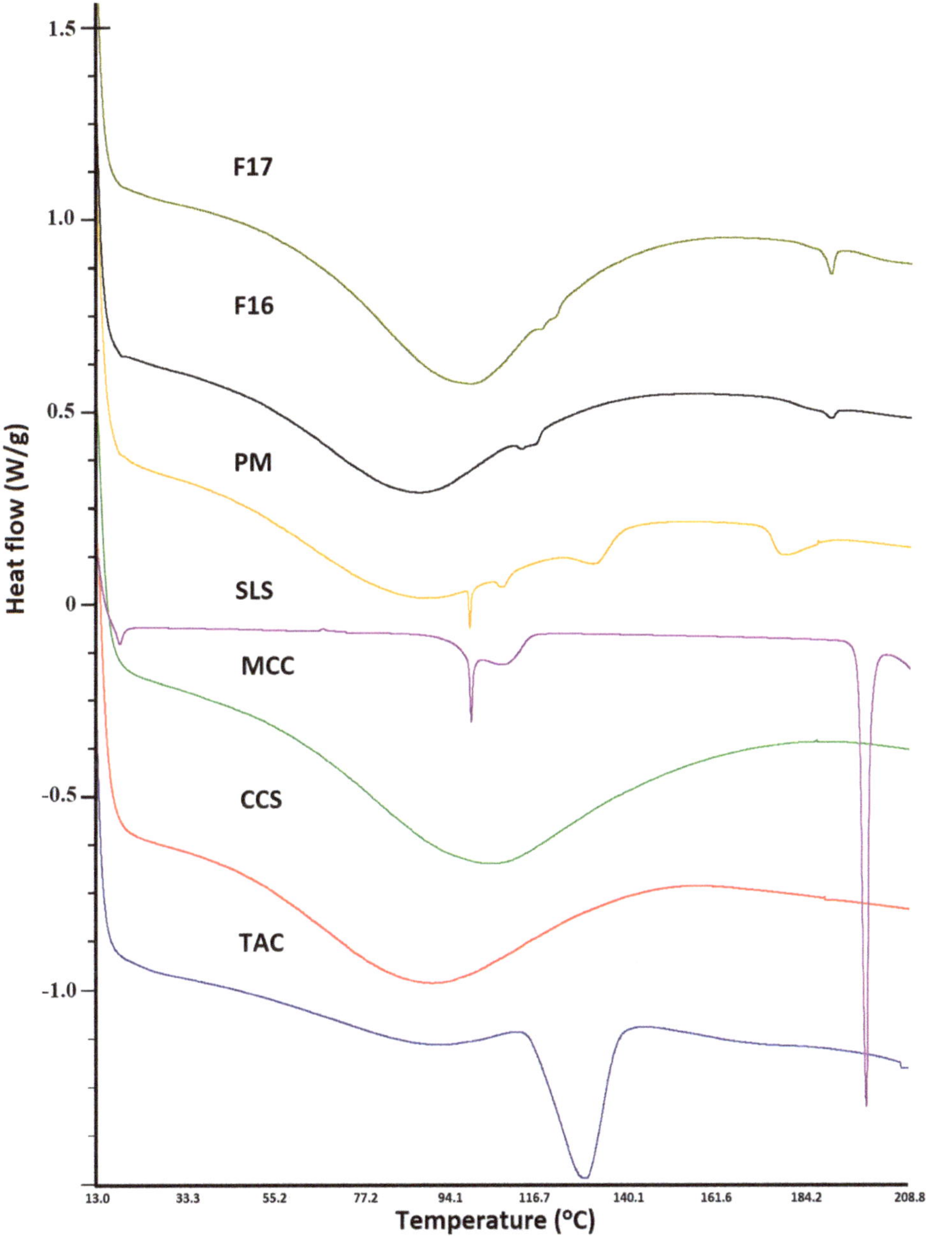

Figure 6. DSC thermograms of tacrolimus, physical mixture, placebo, CAD, and ASD formulations.

3.4. Dissolution

The dissolution specification of immediate release TAC capsule is 85% in 90 min in 900 mL dissolution media (water containing 1 in 20,000 HPC, pH adjusted to 4.5) as per USP [35]. This criterion was used to develop the CAD formulation of TAC. The drug dissolved in 2 h was $5.4 \pm 0.9\%$ from the pure TAC. Various ratios of TAC to SAIB and excipients were explored to increase rate and extent of the dissolution. F1 formulation contained a drug-to-SAIB ratio of 1:1. It showed almost no dissolution due to the hydrophobic nature of the drug and SAIB, and formed a blob in the dissolution medium. To disperse the blob and increase the surface area available for dissolution, LMH or MCC was added to the formulation in various ratios from 1:5 to 1:20 drug to LMH (F2 to F4) or MCC (F5 to F7). Dissolution was 1.9% in 2 h from the F3 formulation with TAC: SAIB: LMH ratio of 1:1:10. Increasing or decreasing the LMH in the formulations did not increase the dissolution. The dissolution was almost negligible when TAC: SAIB: LMH (F2) was 1:1:5. Increasing the LMH proportion in the formulation F4 from 1:10 to 1:20 did not significantly increase dissolution. The dissolution was $3.7 \pm 1.4\%$ in 2 h from F4 formulation, respectively.

On the other hand, addition of MCC in the formulations caused a significant increase in dissolution compared to LMH-based formulations. Dissolution was 20.6 ± 0.9 and $22.4 \pm 2.2\%$ in 2 h from F5 and F6, respectively (Figure 7A). These formulations contained TAC: SAIB:MCC 1:1:5 and 1:1:10, respectively. The dissolution in MCC and LMH-based formulations can be explained by solubility, porosity/pore formation phenomenon, and adsorbed layer thickness. LMH must dissolve first to make pores for the dissolution medium to penetrate through the formulation matrix in order to dissolve the drug. This is not the case for MCC since it is porous, and a dissolution medium can easily penetrate through the matrix [36]. TAC-SAIB solution was physically adsorbed over the surface of LMH or MCC. The thickness of adsorbed TAC-SAIB layer would be thicker in LMH due to its non-porous nature. Thus, less surface area would be available compared to MCC-based formulations. Dissolution is proportional to the surface area as per the Noyes-Whitney equation [37]. Further, an increase in MCC in the formulation did not increase the extent of dissolution but increased the rate. For example, the drug dissolved was $20.4 \pm 1.6\%$ in 2 h in the F7 formulation (Figure 7A). A faster dissolution rate in the F7 formulation with an increase in MCC was due to the thinner adsorb layer of the TAC-SAIB solution. Thus, a higher surface area was available for the drug to dissolve.

To understand the effect of SAIB on the dissolution, TAC: SAIB proportion was decreased from 1:1 to 1:0.1 (F8) and 1:0.75 (F9) while keeping the proportion of TAC: MCC constant (1:10). Compared to F6, rate and extent of the dissolution in F8 and F9 decreased significantly. The dissolution was 4.1 ± 0.2 and $15.3 \pm 1.0\%$ in 2 h from F8 and F9, respectively (Figure 7A). A decrease in dissolution can be explained by crystallinity of the drug, which was supported by diffractograms and thermograms (Figures 3 and 5). These formulations exhibited characteristic reflection peaks of the drug without MCC. Thus, a lower proportion of SAIB was insufficient to keep the drug in its molecular form. Dissolution was further increased by adding CCS outside the CAD formulation, which means CCS was added to the formulation after CAD manufacturing. Addition of CCS resulted in a significant increase in dissolution from 22.4 ± 2.2 (F6) to $50.2 \pm 2.2\%$ (F10) in 2 h (Figure 7B). Proportional ratio of TAC:SAIB:MCC:CCS was 1:1:10:2 in F10. An increase in dissolution in F10 can be explained by the fast dispersion of the capsule compact that resulted in a significant increase in rate and extent of the dissolution. The addition of CCS during the manufacturing step significantly impacted the dissolution. CCS was added during the CAD manufacturing step and, at the same time, increased the proportion of SAIB in the formulation. F11 formulation proportional composition was TAC:SAIB:MCC:CCS 1:1.5:10:2. The dissolution of TAC was increased from 50.2 ± 2.2 (F10) to $72.8 \pm 5.6\%$ (F11) in 2 h. Adding CCS during manufacturing step resulted in a thin coating of the formulation over the insoluble excipients, which increased the surface area available for dissolution. Further, an increase in SAIB proportion relative to the drug from 1:1.5 (F11) to 1:2 (F12) while keeping MCC and CCS constant did not increase dissolution. On the contrary, it decreased the dissolution, which can be explained by the hydrophobic nature of SAIB and the thick coating of the TAC-SAIB solution over the excipients and, thus, decreased available surface area for dissolution (Figure 7B). Surfactants were explored to increase rate and extent of the dissolution in formulation F11. SLS was added during the CAD manufacturing step in the ratio 1:1 ratio with respect to TAC in the formulation F13. The dissolution rate and extent were increased from 72.8 ± 5.6% (F11) to 81.3% in 2 h in F13 (Figure 7B). Surfactants are known to improve the dissolution of hydrophobic compounds by forming micelles structures [38]. Dissolution of 85% in 1.5 h can be achieved by increasing CCS proportion from 1 to 1.5 (F14) and 2 (F15) with respect to the drug. Increasing CCS resulted in 86.2 ± 1.4% dissolution in 1.5 h from the F14 formulation. However, increasing CCS to two proportions with respect to the drug in F15 did not result in a significant change in dissolution and did not achieve ≥85% dissolution in 1.5 h (Figure 7C). Formulation F14 was considered the optimized formulation that met the USP dissolution criterion [35]. Finally, F14 was converted from capsule to tablet dosage form to understand its impact on dissolution. The dissolution was increased from 86.2 ± 1.4 to 91.1 ± 6.6% in 1.5 h by changing the dosage forms (Figure 7C). This behavior can be explained by the tablet's disintegration and the capsule's bursting. The bursting time was about 5-10 min compared to less than 1 min disintegration time of the tablet. To compare with the HPMC formulation, the HPMC-based formulation (F17) was prepared by replacing the SAIB with HPMC and was compositionally identical to the F16 formulation. The dissolution was 89.2 ± 2.1% in 1.5 h from the HPMC-based formulation. The dissolution of tablet formulation with crystalline drug (F18) without SAIB or HPMC was 4.8 ± 1.1% in 2 h (Figure 7C).

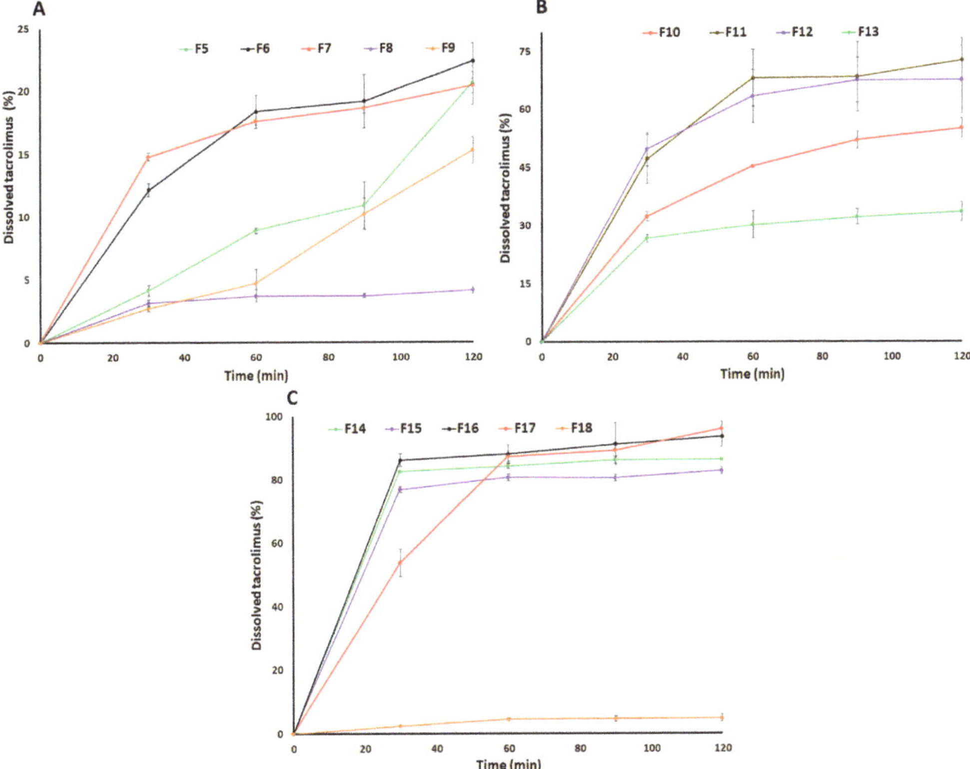

Figure 7. Dissolution profiles of (**A**) F5-F9, (**B**) F10-F13, and (**C**) F14-F18 formulations.

3.5. Stability

The stability conditions exposed samples showed no significant changes in the physical and chemical attributes of the CAD and ASD formulations. FTIR spectrum indicated insignificant changes in the F16 formulation on exposure to high humidity and temperature conditions, which suggested no chemical interactions and maintenance of the amorphous phase of the drug during storage. This was supported by the nonappearance of the absorption bands of TAC at 783 and 1692 cm^{-1}. Similar findings were observed in the F17 formulation, and the spectrum was identical to before exposed sample (Figure 8).

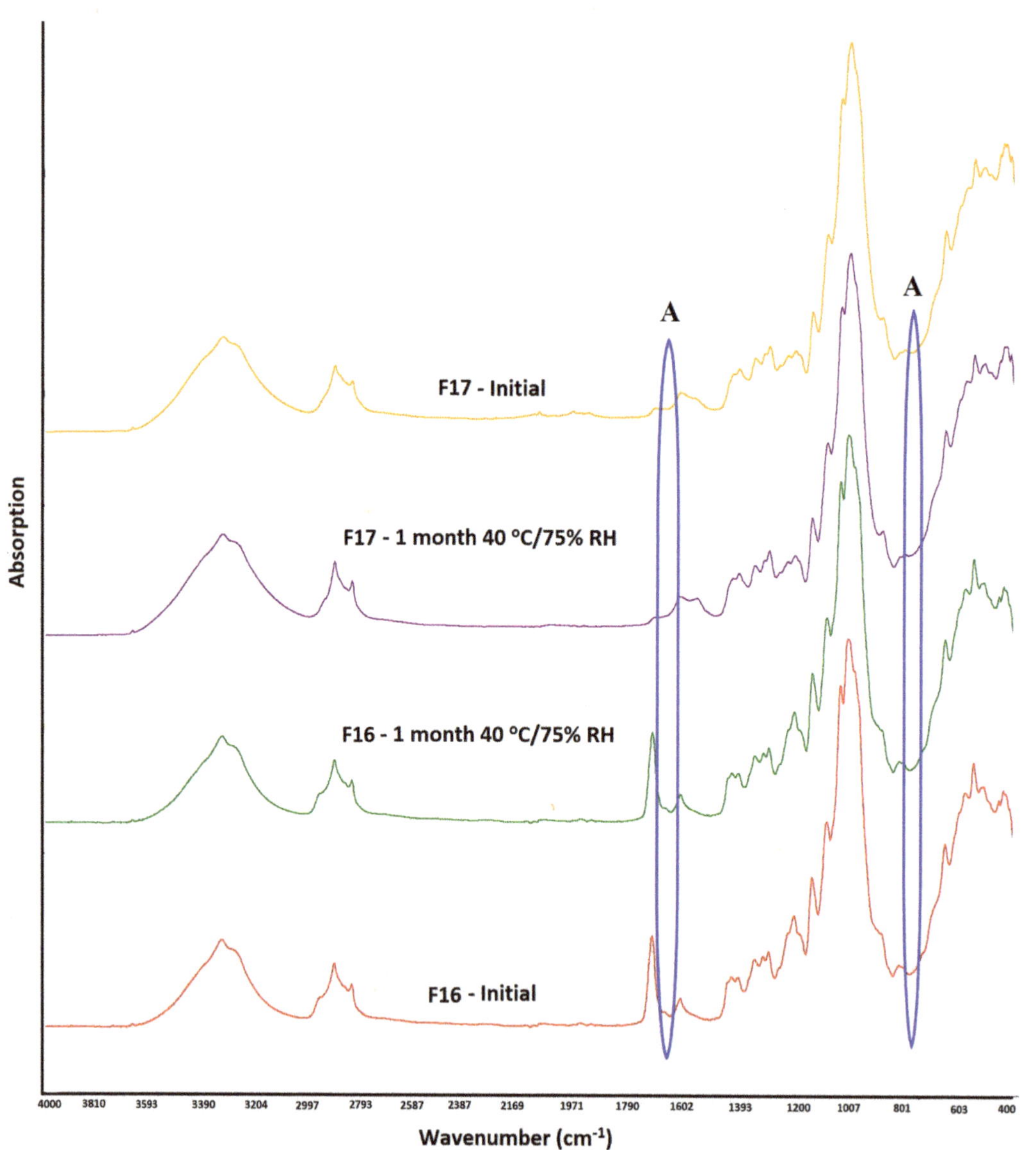

Figure 8. FTIR spectra of CAD and ASD formulations before and after storage at 25 °C/60% RH and 40 °C/75% RH conditions (A: amorphous drug).

This was further supported by DSC and XRPD data. The diffractogram of F16 and F17 formulations did not show appearance of characteristics reflection peak of the drug at 8.4, 10.2, 11.1, 11.6, 12.5, and 13.6° that could have indicated amorphous to crystalline reversion (Figure 9). Similarly, the melting peak of the drug in the region 130–135 °C did not appear in thermograms of the exposed F16 and F17 formulations (Figure 10). The data of orthogonal techniques indicated no transformation of the amorphous drug to its crystalline form on exposure to high humidity and temperature condition.

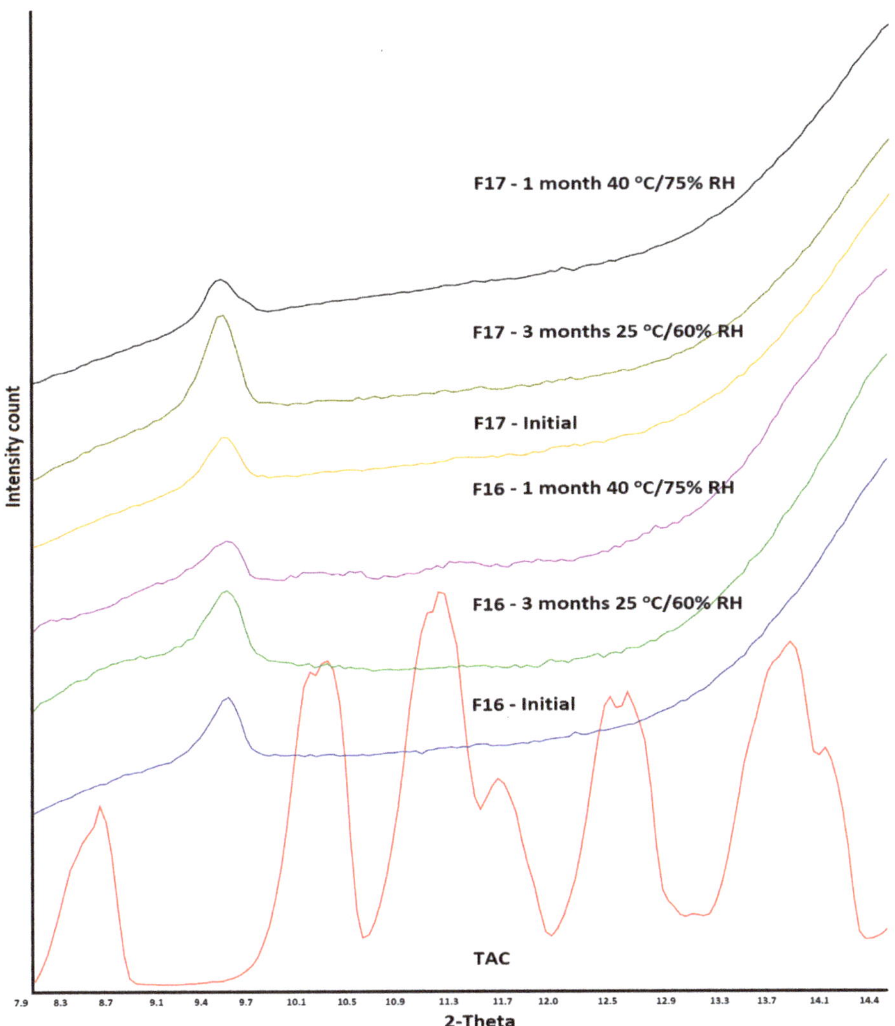

Figure 9. X-ray powder diffractogram of CAD and ASD formulations before and after storage at 25 °C/60% RH and 40 °C/75% RH conditions.

No significant change in dissolution was expected as no reversion of the drug phase was observed. Both formulations (F16 and F17) met the USP limit of 85% in 90 min after exposure to stability conditions (Figure 11). The dissolution changed from 91.1 ± 6.6% to 90.1 ± 1.8 and 98.5 ± 1.9% after exposure of F16 to 25 °C/60% and 40 °C/75% RH, respectively. Similarly, the dissolution changed from 89.2 ± 2.2% to 96.6 ± 0.2 and 95.4 ± 2.7%

after exposure of F17 to stability conditions. An increase in the dissolution rate was observed at 30 min, possibly due to decreased disintegration time after exposure.

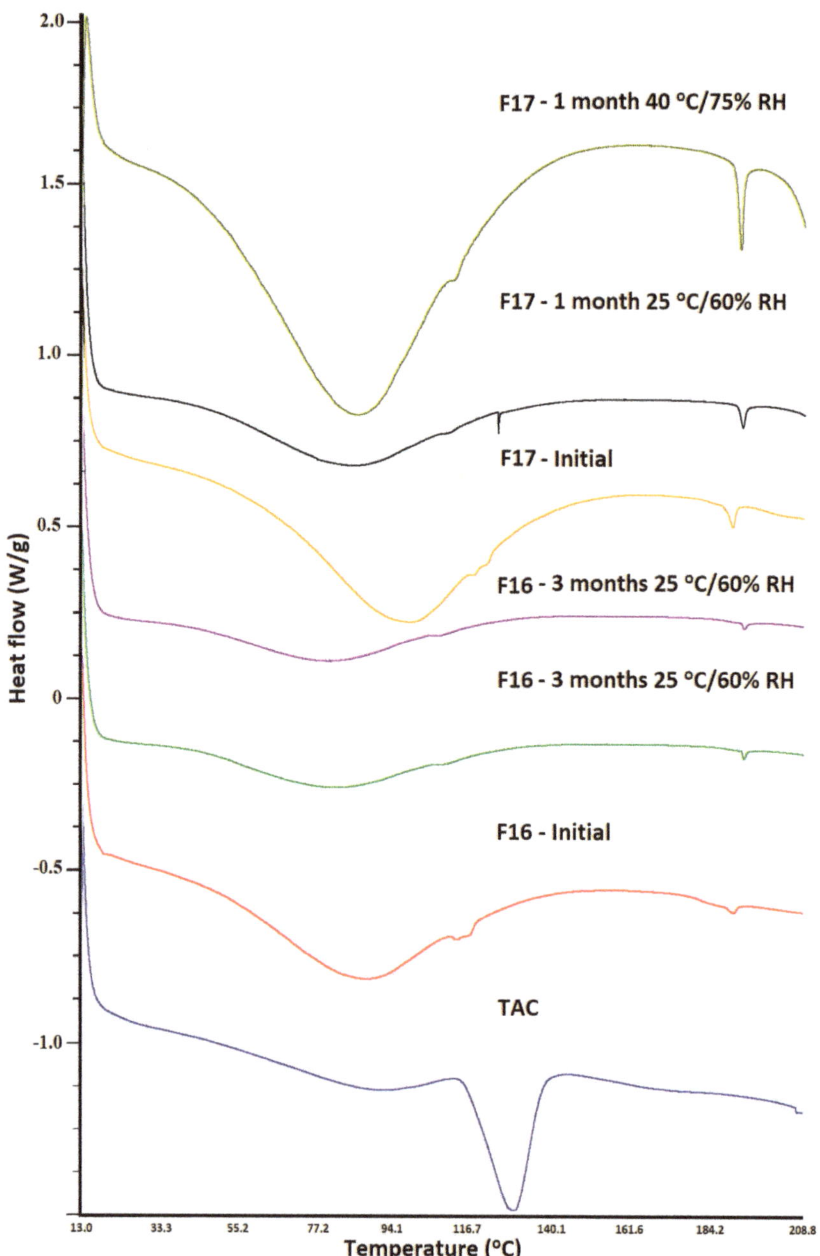

Figure 10. DSC thermograms of CAD and ASD formulations before and after storage at 25 °C/60% RH and 40 °C/75% RH conditions.

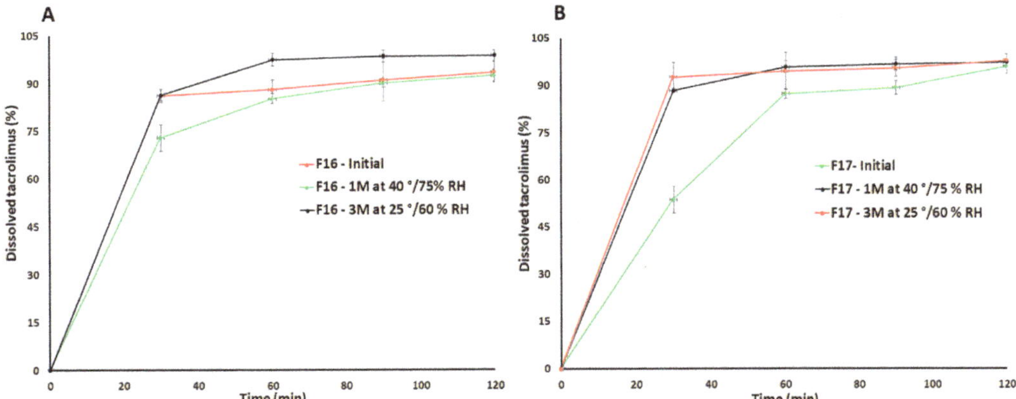

Figure 11. Dissolution profiles of (**A**) F16 and (**B**) F17 formulations before and after storage at 25 °C/60% RH and 40 °C/75% RH conditions.

3.6. Pharmacokinetics

Oral absorption of TAC is incomplete and variable. The absolute bioavailability of TAC is 17 ± 10% in adult kidney transplant patients, 22 ± 6% in adult liver transplant patients, 23 ± 9% in adult heart transplant patients, and 18 ± 5% in healthy volunteers [39]. Comparative bioequivalence was performed among F16 (SAIB-based CAD), F17 (HPMC-based ASD), and Control formulation (F18 formulation contained crystalline drug without SAIB or HPMC). The pharmacokinetic profiles of F16 and F17 formulations were similar and almost superimposable but different from F18 (Figure 12). The drug was detectable up to 72 h post-dosing but below the quantification limit of 4 ng/mL after 12, 8, and 6 h from F16, F17, and F18, respectively.

The pharmacokinetic parameters of F16 and F17 were also very similar. Maximum plasma concentration (C_{max}) and $AUC_{0-\infty}$ (area under the plasma concentration curve) of F16 and F17 were 38.0 ± 11.4 and 35.9 ± 5.9 ng/mL, and 270.3 ± 222.2 and 224.7 ± 55.3 ng/mL.h, respectively. The time to achieve (T_{max}) C_{max} was one hour for both formulations. No statistical differences were observed in the pharmacokinetic parameters of both formulations ($p < 0.05$). On the other hand, the pharmacokinetic profile of F18 was completely different from CAD and ASD formulations. Time to achieve C_{max} was longer in PM compared to CAD and ASD formulations, and was reached in 2 h. Furthermore, the AUC of F18 was lower than F16 and F17 formulations which also concurred with dissolution data. Longer T_{max} and lower C_{max} and $AUC_{0-\infty}$ value in F18 were due to crystalline drug that accounted for lower in vitro and in vivo dissolution compared to CAD and ASD formulations. CAD and ASD formulation can achieve supersaturation due to the amorphous nature of the drug [5,40,41]. T_{max}, C_{max}, and $AUC_{0-\infty}$ of F18 were two h, 20.6 ± 0.6 ng/mL, and 142.9 ± 10.8 ng/mL.h, respectively. The formulations (F16 and F17) exhibited C_{max} and AUC 1.7–1.8 and 1.5–1.8 folds of the F18 formulation, respectively. Higher values of pharmacokinetic parameters in CAD and ASD formulations were due to faster and higher dissolution. Dissolution is the rate-determining step in the absorption and bioavailability of BCS class II drugs such as TAC [9,42]. Furthermore, statistically significant differences ($p < 0.05$) were observed between the pharmacokinetic parameters of CAD and ASD (F16 and F17) and F18 formulations. Therefore, more drug is expected to be bioavailable from CAD and ASD formulations compared to F18.

Furthermore, FDA standards were used to determine bioequivalence among F16, F17, and F18 formulations. The two formulations were considered bioequivalent when the confidence interval (CI) at the alpha level of 90% is 80–125% of the geometric mean of the log-transformed ratio of test/reference or reference/test of C_{max} and AUC for regular drug [43]. However, the CI criterion is tightened for narrow therapeutic index (NTI) drugs to ensure clinical similarity when formulations are switched. FDA-recommended CI criterion is 90–11.1% for NTI [44,45]. The ratio of C_{max} of F16/F17 and F17/F18 ranged from 99.7 to 100.3%. Similarly, the AUC of F16/F17 and F17/F18 ranged from 98.0 to 102.0%. Therefore, the F16 and F17 formulations can be considered bioequivalent as they met the CI criterion for both pharmacokinetic parameters. Thus, the clinical performance of F16 would be similar to F17, and these formulations may be interchangeable.

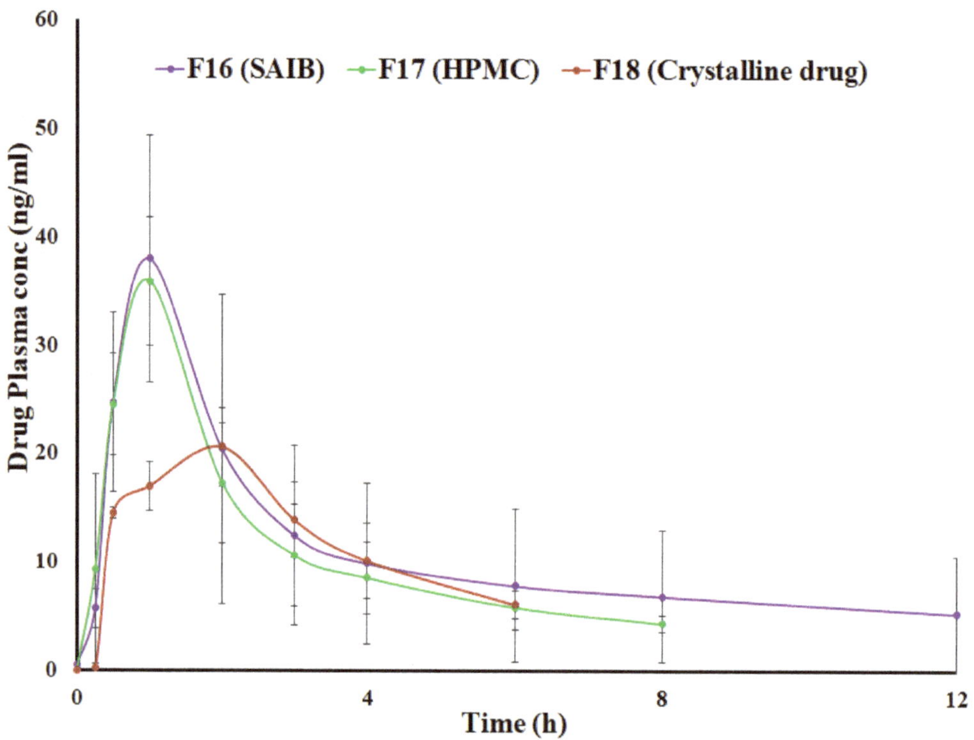

Figure 12. Comparative pharmacokinetic profiles of F16 (SAIB), F17 (HPMC), and F18 formulations.

Bioequivalence of the F16 or F17 formulation was also compared with the F18 formulation. The ratio of C_{max} and AUC of F16 /F18 and F18/F16 ranged from 84.3 to 118.7 and 90.2 to 110.9%, respectively. Both C_{max} and AUC criteria should be met to be considered bioequivalent. Thus, F16 and F18 were not bioequivalent. F17 and F18 were not bioequivalent and interchangeable as they did not meet the CI criteria of pharmacokinetic parameters. The ratio of C_{max} and AUC of F17/F18 and F18/F17 ranged from 84.6 to 118.3 and 90.0 to 108.7%, respectively.

4. Conclusions

CAD and ASD formulations offer a tremendous opportunity to convert the poorly insoluble drug into a soluble, dissolvable, and absorbable molecule, provided it maintains its amorphous phase at various stability conditions. The prepared SAIB-based CAD formulation showed that the drug was in the amorphous phase as indicated by XRPD and DSC data. The dissolution can be modulated by SAIB proportion and other excipients to meet regulatory or pharmacopeial requirements. Short-term stability at room temperature and accelerated conditions indicated that the drug's amorphous phase was maintained and met dissolution specifications. The dissolution and stability of the SAIB-based CAD formulation were similar to the HPMC-based ASD formulation. Therefore, the pharmacokinetic and clinical performance of SAIB-based CAD would be similar to HPMC-based ASD formulations. However, long-term stability and human study data must correlate with the current findings.

Author Contributions: Investigation, conceptualization, writing—review, and editing, visualization—E.M.M.; investigation, conceptualization—S.D.; conceptualization, review and editing—P.C. and R.A.; writing—review, and editing—M.T.H.N.; conceptualization, writing—review and editing, M.A.K.; conceptualization, formal analysis, writing—review and editing, supervision, project administration—Z.R. All authors have read and agreed to the published version of the manuscript.

Funding: This research was funded by Eastman Chemicals Inc. research agreement #M1901663.

Informed Consent Statement: Not Applicable.

Acknowledgments: Authors would like to thank Phillip Cook, Technical Associate, Eastman Chemical Company, for his helpful advice on various technical issues and comments.

Conflicts of Interest: The authors declare no conflict of interest.

References

1. Coty, J.B.; Martin, C.; Telò, I.; Spitzer, D. Use of Spray Flash Evaporation (SFE) technology to improve dissolution of poorly soluble drugs: Case study on furosemide nanocrystals. *Int. J. Pharm.* **2020**, *589*, 119827. [CrossRef]
2. Al-Ani, A.J.; Sugden, P.; Wilson, C.C.; Castro-Dominguez, B. Elusive Seed Formation via Electrical Confinement: Control of a Novel Cocrystal in Cooling Crystallization. *Cryst. Growth Des.* **2021**, *21*, 3310–3315. [CrossRef]
3. Jornada, D.H.; dos Santos Fernandes, G.F.; Chiba, D.E.; de Melo, T.R.; dos Santos, J.L.; Chung, M.C. The Prodrug Approach: A Successful Tool for Improving Drug Solubility. *Molecules* **2015**, *21*, 42. [CrossRef]
4. Sigfridsson, K.; Lundqvist, A.J.; Strimfors, M. Particle size reduction for improvement of oral absorption of the poorly soluble drug UG558 in rats during early development. *Drug Dev. Ind. Pharm.* **2009**, *35*, 1479–1486. [CrossRef]
5. He, Y.; Ho, C. Amorphous Solid Dispersions: Utilization and Challenges in Drug Discovery and Development. *J. Pharm. Sci.* **2015**, *104*, 3237–3258. [CrossRef]
6. Savjani, K.T.; Gajjar, A.K.; Savjani, J.K. Drug solubility: Importance and enhancement techniques. *ISRN Pharm.* **2012**, *2012*, 195727. [CrossRef] [PubMed]
7. Kasten, G.; Grohganz, H.; Rades, T.; Löbmann, K. Development of a screening method for co-amorphous formulations of drugs and amino acids. *Eur. J. Pharm. Sci.* **2016**, *95*, 28–35. [CrossRef]
8. Korhonen, O.; Pajula, K.; Laitinen, R. Rational excipient selection for co-amorphous formulations. *Expert Opin. Drug Deliv.* **2017**, *14*, 551–569. [CrossRef] [PubMed]
9. Rahman, Z.; Mohammad, A.; Akhtar, S.; Siddiqui, A.; Korang-Yeboah, M.; Khan, M.A. Chemometric Model Development and Comparison of Raman and (13)C Solid-State Nuclear Magnetic Resonance-Chemometric Methods for Quantification of Crystalline/Amorphous Warfarin Sodium Fraction in the Formulations. *J. Pharm. Sci.* **2015**, *104*, 2550–2558. [CrossRef] [PubMed]
10. Salem, H.F.; Abdelhaleem Ali, A.M.; Maher, E.M. Formulation and in-vitro evaluation of fast dissolving tablets containing a poorly soluble antipsychotic drug. *Int. J. Drug Deliv.* **2015**, *7*, 113–125.
11. Ueda, H.; Muranushi, N.; Sakuma, S.; Ida, Y.; Endoh, T.; Kadota, K.; Tozuka, Y. A Strategy for Co-former Selection to Design Stable Co-amorphous Formations Based on Physicochemical Properties of Non-steroidal Inflammatory Drugs. *Pharm. Res.* **2016**, *33*, 1018–1029. [CrossRef] [PubMed]
12. Zhang, M.; Suo, Z.; Peng, X.; Gan, N.; Zhao, L.; Tang, P.; Wei, X.; Li, H. Microcrystalline cellulose as an effective crystal growth inhibitor for the ternary Ibrutinib formulation. *Carbohydr. Polym.* **2020**, *229*, 115476. [CrossRef] [PubMed]
13. Van den Mooter, G. The use of amorphous solid dispersions: A formulation strategy to overcome poor solubility and dissolution rate. *Drug Discov. Today Technol.* **2012**, *9*, e71–e174. [CrossRef]
14. Hamed, R.; Mohamed, E.M.; Sediri, K.; Khan, M.A.; Rahman, Z. Development of stable amorphous solid dispersion and quantification of crystalline fraction of lopinavir by spectroscopic-chemometric methods. *Int. J. Pharm.* **2021**, *602*, 120657. [CrossRef] [PubMed]
15. Zidan, A.S.; Rahman, Z.; Sayeed, V.; Raw, A.; Yu, L.; Khan, M.A. Crystallinity evaluation of tacrolimus solid dispersions by chemometric analysis. *Int. J. Pharm.* **2012**, *423*, 341–350. [CrossRef]
16. Maher, E.M.; Ali, A.M.; Salem, H.F.; Abdelrahman, A.A. In vitro/in vivo evaluation of an optimized fast dissolving oral film containing olanzapine co-amorphous dispersion with selected carboxylic acids. *Drug Deliv.* **2016**, *23*, 3088–3100. [CrossRef]
17. Wang, R.; Han, J.; Jiang, A.; Huang, R.; Fu, T.; Wang, L.; Zheng, Q.; Li, W.; Li, J. Involvement of metabolism-permeability in enhancing the oral bioavailability of curcumin in excipient-free solid dispersions co-formed with piperine. *Int. J. Pharm.* **2019**, *561*, 9–18. [CrossRef]

18. Anane-Adjei, A.B.; Jacobs, E.; Nash, S.C.; Askin, S.; Soundararajan, R.; Kyobula, M.; Booth, J.; Campbell, A. Amorphous solid dispersions: Utilization and challenges in preclinical drug development within AstraZeneca. *Int. J. Pharm.* **2022**, *614*, 121387. [CrossRef]

19. Kang, J.H.; Chon, J.; Kim, Y.I.; Lee, H.J.; Oh, D.W.; Lee, H.G.; Han, C.S.; Kim, D.W.; Park, C.W. Preparation and evaluation of tacrolimus-loaded thermosensitive solid lipid nanoparticles for improved dermal distribution. *Int. J. Nanomed.* **2019**, *14*, 5381–5396. [CrossRef]

20. Rahman, Z.; Siddiqui, A.; Bykadi, S.; Khan, M.A. Determination of tacrolimus crystalline fraction in the commercial immediate release amorphous solid dispersion products by a standardized X-ray powder diffraction method with chemometrics. *Int. J. Pharm.* **2014**, *475*, 462–470. [CrossRef]

21. Ponnammal, P.; Kanaujia, P.; Yani, Y.; Ng, W.K.; Tan, R.B.H. Orally Disintegrating Tablets Containing Melt Extruded Amorphous Solid Dispersion of Tacrolimus for Dissolution Enhancement. *Pharmaceutics* **2018**, *10*, 35. [CrossRef]

22. Yamashita, K.; Nakate, T.; Okimoto, K.; Ohike, A.; Tokunaga, Y.; Ibuki, R.; Higaki, K.; Kimura, T. Establishment of new preparation method for solid dispersion formulation of tacrolimus. *Int. J. Pharm.* **2003**, *267*, 79–91. [CrossRef] [PubMed]

23. Yoshida, T.; Kurimoto, I.; Yoshihara, K.; Umejima, H.; Ito, N.; Watanabe, S.; Sako, K.; Kikuchi, A. Effect of aminoalkyl methacrylate copolymer E/HCl on in vivo absorption of poorly water-soluble drug. *Drug Dev. Ind. Pharm.* **2013**, *39*, 1698–1705. [CrossRef]

24. Jung, H.J.; Ahn, H.I.; Park, J.Y.; Ho, M.J.; Lee, D.R.; Cho, H.R.; Park, J.S.; Choi, Y.S.; Kang, M.J. Improved oral absorption of tacrolimus by a solid dispersion with hypromellose and sodium lauryl sulfate. *Int. J. Biol. Macromol.* **2016**, *83*, 282–287. [CrossRef]

25. Cho, J.H.; Kim, Y.I.; Kim, D.W.; Yousaf, A.M.; Kim, J.O.; Woo, J.S.; Yong, C.S.; Choi, H.G. Development of novel fast-dissolving tacrolimus solid dispersion-loaded prolonged release tablet. *Eur. J. Pharm. Sci.* **2014**, *54*, 1–7. [CrossRef]

26. Tsunashima, D.; Yamashita, K.; Ogawara, K.; Sako, K.; Higaki, K. Preparation of extended release solid dispersion formulations of tacrolimus using ethylcellulose and hydroxypropylmethylcellulose by solvent evaporation method. *J. Pharm. Pharmacol.* **2016**, *68*, 316–323. [CrossRef]

27. HMP Global Learning Network—Immunosuppressant Recalled, 3 April 2020. Available online: https://www.hmpgloballearningnetwork.com/site/pln/content/immunosuppressant-recalled-0 (accessed on 1 February 2023).

28. Fierce Pharma. European Medicines Agency Aggress to Precautionary Recall of Advagraf 0.5 mg Capsule Batches. 2011. Available online: https://www.fiercepharma.com/pharma/european-medicines-agency-agrees-to-precautionary-recall-of-advagraf-0-5-mg-capsule-batches (accessed on 1 February 2023).

29. Dharani, S.; Sediri, K.; Cook, P.; Arunagiri, R.; Khan, M.A.; Rahman, Z. Preparation and Characterization of Stable Amorphous Glassy Solution of BCS II and IV Drugs. *AAPS PharmSciTech* **2022**, *23*, 35. [CrossRef] [PubMed]

30. Binkhathlan, Z.; Badran, M.M.; Alomrani, A.; Aljuffali, I.A.; Alghonaim, M.; Al-Muhsen, S.; Halwani, R.; Alshamsan, A. Reutilization of Tacrolimus Extracted from Expired Prograf® Capsules: Physical, Chemical, and Pharmacological Assessment. *AAPS PharmSciTech* **2016**, *17*, 978–987. [CrossRef]

31. FDA. Bioanalytical Method Validation Guidance. 2018. Available online: https://www.fda.gov/regulatory-information/search-fda-guidance-documents/bioanalytical-method-validation-guidance-industry (accessed on 2 February 2023).

32. Wu, X.; Hayes, D., Jr.; Zwischenberger, J.B.; Kuhn, R.J.; Mansour, H.M. Design and physicochemical characterization of advanced spray-dried tacrolimus multifunctional particles for inhalation. *Drug Des. Dev. Ther.* **2013**, *7*, 59–72. [CrossRef]

33. PubChem. Sodium Dodecyl Sulfate. Available online: https://pubchem.ncbi.nlm.nih.gov/compound/Sodium-dodecyl-sulfate (accessed on 2 February 2023).

34. Freire, F.; Aragão, C.; de Lima e Moura, T.; Raffin, F. Compatibility study between chlorpropamide and excipients in their physical mixtures. *J. Therm. Anal. Calorim.* **2009**, *97*, 355–357. [CrossRef]

35. USP43-NF38. Tacrolimus Capsules. 2022; p. 4196. Available online: https://online.uspnf.com/uspnf/document/1_GUID-3E902 2BA-83DC-47EC-80DC-45D54BE1C4B1_3_en-US?source=Search%20Results&highlight=tacrolimus (accessed on 4 February 2023).

36. Westermarck, S.; Juppo, A.M.; Kervinen, L.; Yliruusi, J. Microcrystalline cellulose and its microstructure in pharmaceutical processing. *Eur. J. Pharm. Biopharm.* **1999**, *48*, 199–206. [CrossRef]

37. Gao, Y.; Glennon, B.; He, Y.; Donnellan, P. Dissolution Kinetics of a BCS Class II Active Pharmaceutical Ingredient: Diffusion-Based Model Validation and Prediction. *ACS Omega* **2021**, *6*, 8056–8067. [CrossRef]

38. Rahman, Z.; Khan, M.A. Hunter screening design to understand the product variability of solid dispersion formulation of a peptide antibiotic. *Int. J. Pharm.* **2013**, *456*, 572–582. [CrossRef]

39. Prograf®FDA Label. Available online: https://www.accessdata.fda.gov/drugsatfda_docs/label/2009/050708s027,050709s021lbl.pdf (accessed on 2 February 2023).

40. Hirai, D.; Tsunematsu, H.; Kimura, S.I.; Itai, S.; Fukami, T.; Iwao, Y. Theoretical evaluation of supersaturation of amorphous solid dispersion formulations with different drug/polymer combinations using mathematical modeling. *Int. J. Pharm.* **2022**, *625*, 122110. [CrossRef] [PubMed]

41. Li, N.; Taylor, L.S. Tailoring supersaturation from amorphous solid dispersions. *J. Control. Release* **2018**, *279*, 114–125. [CrossRef]

42. Khames, A. Investigation of the effect of solubility increase at the main absorption site on bioavailability of BCS class II drug (risperidone) using liquisolid technique. *Drug Deliv.* **2017**, *24*, 328–338. [CrossRef] [PubMed]

43. FDA Guidance for Industry—Statistical Approaches to Establishing Bioequivalence. 2001. Available online: https://www.fda.gov/regulatory-information/search-fda-guidance-documents/statistical-approaches-establishing-bioequivalenc (accessed on 2 February 2023).
44. Jiang, W.; Makhlouf, F.; Schuirmann, D.J.; Zhang, X.; Zheng, N.; Conner, D.; Yu, L.X.; Lionberger, R. A Bioequivalence Approach for Generic Narrow Therapeutic Index Drugs: Evaluation of the Reference-Scaled Approach and Variability Comparison Criterion. *AAPS J.* **2015**, *17*, 891–901. [CrossRef] [PubMed]
45. Yu, L.X.; Jiang, W.; Zhang, X.; Lionberger, R.; Makhlouf, F.; Schuirmann, D.J.; Muldowney, L.; Chen, M.L.; Davit, B.; Conner, D.; et al. Novel bioequivalence approach for narrow therapeutic index drugs. *Clin. Pharmacol. Ther.* **2015**, *97*, 286–291. [CrossRef]

MDPI
St. Alban-Anlage 66
4052 Basel
Switzerland
www.mdpi.com

Pharmaceutics Editorial Office
E-mail: pharmaceutics@mdpi.com
www.mdpi.com/journal/pharmaceutics

www.ingramcontent.com/pod-product-compliance
Lightning Source LLC
LaVergne TN
LVHW070159170726
843515LV00007B/1954